Never In Doubt
a history of Delta Drilling Company

OTHER BOOKS BY
JAMES PRESLEY

A Saga of Wealth: The Rise of the Texas Oilmen

Food Power: Nutrition and Your Child's Behavior (with Hugh Powers, M.D.)

Human Life Styling: Keeping Whole in the 20th Century (with John C. McCamy, M.D.)

Public Defender (with Gerald W. Getty)

Vitamin B6: The Doctor's Report (with John M. Ellis, M.D.)

Please, Doctor, Do Something! (with Joe D. Nichols, M.D.)

Center of the Storm: Memoirs of John T. Scopes (with John T. Scopes)

Never In Doubt
a history of Delta Drilling Company

James Presley

Gulf Publishing Company
Book Division
Houston, London, Paris, Tokyo

Never In Doubt

a history of Delta Drilling Company

CONTENTS

To Joseph Zeppa
He may have been in error. . .
but he was never in doubt.

ACKNOWLEDGMENTS

In 1975 while I was researching my earlier book about Texas oilmen, *A Saga of Wealth,* I was fortunate to be able to interview Joseph Zeppa at his office in Tyler. As it turned out, our visit came approximately three months before he died. I did not know it then, but that afternoon session of about two hours, during which he patiently reviewed his life and career for me, was the beginning of this book, and the results of that day's conversation are to be found throughout this volume.

From that moment on, the story of Delta Drilling Company intrigued me. I felt certain it had the makings of an interesting book. Since it was not one I felt free to tackle at the time, I suggested to the staff at the Oral Business History Project at The University of Texas at Austin that Joe Zeppa and Delta Drilling Company merited their attention as a worthy subject of one of their publications. The Project was already inundated with scheduled work, however, and nothing came of my recommendation.

Then in 1979, Keating V. Zeppa, Joe Zeppa's son and successor as president and chairman of the board of Delta Drilling Company, discussed with me the possibility of writing a "warts and all" history of the company, to be published on its fiftieth anniversary in November, 1981. Immediately I was favorably inclined.

I discovered, to my delight, that Keating Zeppa was dedicated to a searching examination of the company's past. He realizes that any account of the past is of little value unless it is truthful, and that if we do not report honestly what really happened, we have not left much of value for succeeding generations. So in this book I have made no effort to sugar-coat events or personalities, and Keating Zeppa, who had promised a nonintervening policy, has stuck to it. I can only hope that I have fulfilled my intentions and that my portrayal of events and persons has been true to life.

Actual work on the book began in February, 1980, and the final chapter was delivered to the publisher in April, 1981. To produce a work of this scope and length in that rather compressed period of time is not possible without the cooperation of a great many persons—in this case, hundreds.

My appreciation begins with Keating Zeppa, whose enthusiasm for the project was responsible for it in the first place. He has consistently cooperated at every phase of the project, not only subjecting himself to my tedious and often personal questioning, but ensuring that all records were open to me and that all employees who were willing to participate—and all were—were encouraged to speak frankly with me, which I believe they have done.

I owe special thanks to Gloria A. Stackpole and her staff in Delta Drilling Company's public relations department. She was in on the project from the beginning, supplying me all requested information, names, and addresses, organizing the transcription of a large portion of the tape-recorded interviews, and facilitating my interview schedule. Her assistants, Martha Wiggins and Tina Barnett DeShazo, did much of this, with most of the burden falling upon the capable shoulders of Tina, who set up interviews as I requested them, photocopied huge stacks of material, and oversaw the transcription of tapes, among other chores.

I must single out Grace Baker for guiding me through Delta's central files. She knows the location of every piece of paper there, and she never failed to deliver every document I requested.

I am deeply grateful to the scores of persons, in and out of Delta, who talked to me at length about Delta and the various personalities involved. They are all listed in the Oral History section of the Bibliography. The reader will find most of them mentioned, to varying degrees, in the narrative. Without these persons I could not have written this book. The next generation also owes them a large debt of gratitude for passing on their memories.

Librarians, as usual, were helpful to me, particularly at the Palmer Memorial Library and the Texarkana Public Library, both in Texarkana, Texas.

Jean Smart transcribed the bulk of the taped interviews relating to the early chapters of the book. The chore was finished by a number of transcribers at Delta Drilling Company working on their own time. The final manuscript was typed by Joan Streit, Kathy Smith, and Louise Lee, with Louise accomplishing most of it, under pressure at the end in a race with the deadline.

Nancy Pierson at Gulf Publishing Company contributed her copy-editing skills.

My agent, Blanche C. Gregory, was, as in all my books, a source of encouragement throughout.

Finally, I owe much to my wife, Fran, my son John, and my daughter Ann for patient understanding and cooperation during approximately fifteen months of research and writing that, for the most part, demanded a seven-day-a-week schedule, with few evenings free, a period when I was away from home more than I had ever been before on a book. Their support was crucial to the work.

James Presley
Texarkana, Texas
April, 1981

FOREWORD

Any Fiftieth Anniversary calls for celebration. It is also cause for reflection and, in some cases, amazement. It could be said that anyone who survives fifty years of life's vicissitudes has a story to tell. It's the same with a company. We have a story to tell which, I believe, is at once interesting and entertaining.

This book was undertaken because Delta Drilling Company has attained a level of experience many companies never reach. We were formed in 1931 by five men who had the vision and tenacity necessary to surviving those years and to building a business which would outlive them all.

But in order to tell this story a special kind of storyteller was needed. Since Jim Presley had met Dad some years before, I felt that he would have insight into this project that would make him an excellent choice to tell our story. Fortunately for us, he agreed. The result is this book. I believe it is an honest, straightforward account of fifty years—fifty years of Delta's life, and nearly fifty years of my father's life and of mine.

But most importantly, this story is not told with statistics and financial data. It is told by the people who lived it. They are Delta Drilling Company. And more than to anyone else, this story belongs to them.

Keating V. Zeppa
Tyler, Texas
April, 1981

PROLOGUE
A DREAM THAT GREW

For brawny, 20-year-old Marcus Jones, struggling to make ends meet in northern Louisiana in 1934, the Great Depression was a grinding reality.

In his hometown, Gibsland, he found it a challenge just to eke out a living. Toiling in a barrel-stave mill, he earned 35 cents an hour—when they needed him. Millions of Americans, though, did not even have a part-time, irregular job. Many had turned to government relief. Others went without food for days at a stretch. Most men clung precariously to whatever means of income they had.

To Jones, young and strong and with his life before him, opportunity lay beyond little Gibsland. His older sister worked as a secretary for an up-and-coming drilling firm headquartered in the brightest spot, economically, on the United States map: the gigantic East Texas oil field, which had been discovered in 1930. He felt sure she could help him.

One day in February, 1934, Annie May Jones telephoned from Longview, Texas, to report that Delta Drilling Company had an opening for a new "hand." If he hurried on over, Marcus could have a job that paid 50 cents an hour. Jones was overjoyed.

"I was on the job, working at a barrel-stave mill," recalled Jones nearly five decades later, "and she called me and I just quit right then, right in the middle of the day, and came home and cleaned up and caught me a bus and came to Longview."

Magically, the Depression seemed to have vanished, relatively speaking, by the time he reached hustling and bustling Longview. Rigs were drilling and men were working. Though there were still more men than jobs, Longview was the liveliest town Jones had yet seen. It had been three years since the explosive opening of the huge East Texas field, and the town was adjusting to the boom. People were no longer sleeping on cots in the street, for one thing.

It was dark when young Jones got off the bus at the station next to the railroad track in the strange East Texas town. Young Jones didn't know where to go. Picking up his luggage, he started toward a filling station up ahead. He spied a young man there who had been washing cars.

1

Jones took out a letter from Annie May, with the Delta address on it, and asked the fellow if he could tell him where Whaley Street was. The man saw the letterhead on the sheet of paper and his face beamed with recognition.

"You looking for Mr. Joe Delta?"

Jones said he was, and the man directed him to the company's office. Jones scurried over to Whaley Street where, around seven o'clock on that winter evening, he found the office on the ground floor of a two-story frame building. His sister, a stout brunette with a no-nonsense aura about her, was still working, winding up the day's business when she glanced up to greet her brother.

The reunion was brief, whereupon she assigned younger brother Marcus his first chore with Delta Drilling Company, that of carrying a sheaf of correspondence and other papers upstairs to two of the firm's owners, Joe Zeppa and Robert Stacy. By then Zeppa lived in Tyler and Stacy in Shreveport, but they kept the upstairs apartment in case one or both of them needed to stay overnight in Longview.

When Jones entered the apartment he saw two men, one short and dark, one tall and stately. They were drinking, and on the kitchen drainboard there was a large fruit jar of moonshine whiskey. Though Prohibition had been repealed when the thirty-sixth state ratified the Twenty-First Amendment in late 1933, Longview was still legally dry.

Jones handed the shorter man—Zeppa, the "Joe Delta" the car washer had mentioned—the paper work, and the two men invited him to have a drink. Young Jones had only tasted whiskey a couple of times; he turned down the offer. The next day he officially started work for Delta Drilling Company, dry-watching (guarding in the absence of the crew) a rig in the East Texas field. A year later, after months of work as a roustabout often unloading trucks, he had a regular job in a Delta warehouse, earning an unheard-of $90 a month.[1]

Time proved the man at the service station more prescient than anyone would have guessed in 1934. "Joe Delta" was, indeed, the most succinct way to sum up the future of the company. Joe Zeppa's name was to become practically synonymous with Delta Drilling Company. Over the years, thousands of persons would be involved with Delta as employees, partners, and associates; and, although Joe Zeppa did not always hold a majority share of stock in the company and never owned it outright, he ran it very much as if he did. For the first 45 years of Delta Drilling Company's existance, as long-time employee Nick Andretta phrased it, "Joe Zeppa was the company."[2] There was no way to separate the man and the legal entity.

Thus the story of Delta Drilling Company is, to a large extent, the story of Joe Zeppa. But, since he was not the sole influence, this story is also more than Joe Zeppa's, no matter how strongly he stamped his personality upon his company.

In those 50 years since that November day in 1931 when Zeppa, Stacy, and three other partners—Sam Dorfman, Sam Gold, and Sam Sklar—established the business, Delta Drilling Company has grown amazingly. This fact is perhaps more impressive when one considers how littered the economic landscape is with the bodies of similar companies that sprang to life in the 1930s, only to fail entirely or lose their identities in mergers with larger corporations somewhere along the way. Contract drilling has had its dramatic ups and downs over those five decades. If roll call were taken, more drilling companies would reside among the fallen than among the survivors. To have prevailed and grown, Delta has swum against the tide to achieve a destiny its founders may never have foreseen in the early days of the thirties.

There is a lot of history between the founding in 1931 and today, but one convenient way to grasp Delta's successes over its life is to compare rig counts and the number of company divisions. In 1931, its two second-hand steam rigs immediately went to work in the prolific East Texas field. That was the extent of Delta Drilling Company.

Today, Delta has 56 drilling rigs operating in the United States, with more than 2,000 employees. A rig today may cost as much as $5 million or more, depending upon its degree of sophistication. But while Delta continues to operate in East Texas, with its corporate headquarters in Tyler, it also boasts four other domestic divisions: West Texas (offices in Odessa and Midland), South Texas (Houston and Victoria), Gulf Coast (Lafayette, Louisiana), and Northwest (Indiana and Pittsburgh, Pennsylvania) divisions. In addition, Delta operates two gas plants, one in East Texas and the other near Ozona in the rugged Pecos River country in Southwest Texas.

And then there are the foreign operations. There probably was no inkling, back in 1931, that the shoestring operation would ever leave the continental limits of the United States. But today Delta subsidiaries have been in Mexico, Italy, Brazil, Venezuela, and Argentina. At one time or another, Delta rigs have drilled also in Canada, Libya, Spain, Australia, the Philippines, Turkey, and South West Africa.

Delta started out drilling for oil and gas for other men and companies; consequently, it had no production of its own. Over the years, beginning soon after its first contracts were fulfilled, the company gradually developed a steady flow of production. In hard times and good this reliable source of income, as much as its contracts for drilling, has proved to be a constituent part of the company's lifeblood. Today, more than in the past, Delta continues to drill for itself and in partnership with others.

The evolution of Delta's ownership brought only gradual changes until its fiftieth year. Sam Gold died in 1936, ultimately leaving Sam Dorfman and Sam Sklar in possession of what originally was a one-third interest shared by the three men. Bob Stacy sold his original one-third interest before the first decade elapsed. In the process, Joe Zeppa gained majority control and, as president and chairman of the board, ran the company until his death in 1975 with the firmest of hands. For 40 years, Zeppa, Sklar, Dorfman, and their heirs owned Delta Drilling Company.

Entering its fiftieth year, Delta could be characterized as the largest privately-owned domestic land-based drilling company in the world. The key words are *privately-owned* and *land-based*. Other companies with bigger, more expensive rigs were working in offshore waters like the North Sea. And most companies that had grown larger than Delta had long before sold stock to the public in order to finance their growth.

In 1981, Delta, for the first time in its history, held a public offering of stock and entered a new and exciting phase of its existence. But the story of Delta as a public company is for the future. This account, of that private company, relates how Joe Zeppa and his fellow entrepreneurs endured some of the most crucial years in this country's history.

How did the company survive the lean years of the oil industry? How did Joe Zeppa imprint his personality upon his corporate vehicle? What were the problems and the pivotal decisions facing the company and its owners? What were the people like who ran the operations and worked for Delta? What were the salient stages in the evolution of the firm?

In other words, how did Delta get from there to here?

It is a story with many threads, the first of which start with the company's founders, whose lives began in Texas and in Europe, in the late nineteenth century.

1
NATIVE AND
EUROPEAN ROOTS

I

Delta Drilling Company was founded on credit with two second-hand rusty rigs, in the depth of the Depression, by a native American and four immigrants: a Texan, an Italian, and three Russian Jews. These men came from different backgrounds, circumstances, and locales—totally different worlds. Yet behind the outward dissimilarities lay common business aspirations that drew them together to build the foundation for a sturdy corporation.

Robert A. Stacy grew up in Houston in the early years of this century, at a time when the city was scarcely more than a hint of its present sprawling self. Joseph Zeppa was born on a farm in northern Italy. Sam Gold, Sam Sklar, and Sam Y. Dorfman all originated in czarist Russia, in a period when pogroms against Jews had become a dispiriting, and regular, feature of life.

To understand the early history of Delta, one must first understand the backgrounds of these men.

How did they, with such discrete heritages, come to meet in East Texas and form a company in a specialized field which traditionally had been dominated by native-born stock? What traits had they brought with them that worked together for their common good? Did their early experiences handicap or benefit them as they strove to make their marks in American business?

Unfortunately, we do not have all the details of these men's early lives. Only Zeppa, who outlived all of his partners, left behind personal journals and correspondence from this period, and only he was interviewed at some length. For Stacy, Gold, Sklar, and Dorfman, we must rely primarily upon the memories of kinsmen and friends to reconstruct the events which occurred 80 or more years ago—necessarily fragmentary evidence, further handicapped by the natural erosion of the decades upon the powers of recollection. And at times we must rely upon hearsay, because in some instances it is the only account left.

II

When Bob Stacy was born in Houston, Texas, on November 17, 1896, there was not the slightest indication that the small town would grow into the oil capital of the world. The great Spindletop field at Beaumont was not to open up until 1901, and the other Gulf Coast boomtowns were little more than sleepy communities, drowsing on the verge of history.

Stacy's mother was a Clayton, a family having some members who appeared prominently in the city's economic and social annals. Businessman Will Clayton was a cousin, and there apparently was some limited contact.[1]

Descended from Anglo-American stock with a strong dash of Welsh and probably Scotch-Irish, Stacy was the youngest of three children; he had an older brother, Jean Clayton Stacy, and a sister Margaret.

The Stacy family was far from being moneyed—and possibly not far above the poverty level at that time for white, native-born Protestant Texans. The elder Stacy held a series of jobs, including one in sales. Bob Stacy, based upon what he later told his son, grew up in Houston and quit school in the eighth grade. This same pattern of a limited formal education crops up in the biographies of Stacy's future partners.

Like his partners-to-be, Stacy went to work at an early age, taking his first job in Houston with the Texas Company doing manual labor, rolling barrels of oil. Slender and handsome, he was more than six feet tall by the time he was 15, which made it easier for him to pass for an older person. He eventually went on to other jobs before joining the Army.

It is not clear whether he went into the Army before the United States entered World War I in April, 1917, or upon the outbreak of hostilities. He once referred to having been in Mexico in his early years. Whether this was as a soldier in Pershing's expeditionary force in pursuit of the revolutionary Pancho Villa following his 1916 raid on the border, or as a sight-seer, is not clear. He said he lied about his age to enter the Army, which gives some slight evidence that he volunteered before the war. We do know that he left home very early and did not return to Houston until years later.

He was in uniform, though, no later than the beginning of the United States' participation in 1917. His height and athletic build probably would have convinced a recruiting officer of his professed age even when he was a tender teenager.

"He was quite, quite young when he left home," said his sister-in-law, Elizabeth Powell Fullilove. "He was six feet three inches and very slender in his youth. He didn't look all that husky, but he was big. He was a little young, but didn't have a bit of trouble."[2]

He trained in Texas camps and subsequently shipped out for Europe from New York, in the infantry. He was in combat, though the details are unknown.

"He had the years in France, in the infantry," said Mrs. Fullilove. "He saw action. I don't know where. I know that he was in France and in Germany, and was in Germany after the war; for some reason he was kept there with the Army, not any great while, but he was kept there some months."

In all probability he returned to the States and was discharged in early 1919. According to his son, Robert A. Stacy, Jr., he was discharged as his company's first sergeant—quite an accomplishment for a man with only eight years of schooling to have achieved in probably less than two years.

After the war he went back to work for the Texas Company as a traveling salesman, calling on service stations with his company's products. To a certain extent he was following in his father's footsteps, who had fed his family with sales work. As a Texas Company salesman, the tall, handsome bachelor gained valuable experience that he would find useful as a "contact man" before and after Delta was formed.[3]

No later than 1921, Stacy moved to Camden, Arkansas, where the Texas Company had assigned him a new territory.

In Camden he met Priscilla Powell, the lovely, tall, and glamorous daughter of Smead Powell, a prominent, well-to-do local lawyer who had once run for governor of Arkansas. They were a storybook couple—a handsome man and his beautiful woman.

The two married in 1921, the year oil was discovered in El Dorado, Arkansas. By this time Stacy was operating a service station in Camden, 30 miles away. This boom in Arkansas was to have an important impact upon Stacy's life, as well as upon most other people's in that region.

Though some believed Stacy had "married money" because Priscilla's family was well-fixed, what substance there was to the belief evaporated by the late 1920s when the stock market crashed.

"My father was well-to-do," said Mrs. Fullilove, Priscilla's much younger sister. "Well, my father died when I was nine, and Mother when I was ten. When Mother and Daddy died there was money, but there wasn't nearly as much as we had thought there would be. A lot of that money was put in the stock market as an investment and, of course, there was the Depression. So that really went by the wayside."

Young Elizabeth, a child of ten, was taken into the Stacy household where she remained until she married at 26. Little Bob, Jr., had been born in 1925, so within about a year the Stacys had added two children to their family.

"No one could have been kinder to me than Bob was," said Mrs. Fullilove, "and this was for 16 years. Bobby was born and they had a family. I was not exactly an easy child either. I am sure this was difficult for them. It is one thing to have a small child and another to have one that is ten."[4]

From the time Stacy was a youth, he had improved himself in many ways. Though he had dropped out of school in the eighth grade, he did what his son, Robert, called "a beautiful job of educating himself." Primarily he achieved this by reading on his own, acquiring an excellent command of his native tongue and a very serviceable vocabulary. Along the way he purchased books. Interestingly, according to his son, Stacy wouldn't "have anything to do with novels or mysteries," but read widely in nonfiction, including books related to the oil industry.[5]

"Though Bob couldn't have had much education," said Elizabeth Fullilove, "he was well-read and widely read, which I think is amazing. My mother said he was much better read and educated than Sister, who had gone to Sweet Briar, and he really was. It was just a natural thing with him. He was not in any way going to try and educate himself. He just happened to like better books."

Wit and merriment were regular features of the Stacy household.

"Sister had a sharp tongue too," said Mrs. Fullilove, "and there was lots of laughter, and they enjoyed a good story. Bob had a keen sense of humor. A really dry, keen sense of humor. He used to sit back and let you talk, but his humor was very sharp. He was real good at it, and he was very subtle."[6]

He also had a flair as an entertainer. He was "a great imitator," remembers his friend P. D. Connolly. Though not an effortless dancer, Stacy would attempt a tap dance; and for members of the family and close friends, he would perform a special "Chinese act," for which he would put on a housecoat backward, to signify Chinese dress, and act out his role in pantomime. He and his older brother, Jean, liked to swap jokes, and they sometimes teamed up to recite Kipling as if they were on stage.

Both Stacy and his wife liked to live well, and he never lost the habit after their marriage went on the shoals. He wore the finest clothes, with custom-made shirts. When he had to go into the oil fields, after Delta was formed, he would wear custom-made khakis. In his affluent years he bought several Cadillacs a year and always hired a uniformed chauffeur.

Around 1950, by which time his marriage with Priscilla had broken up, he bought a lake house near Hot Springs, Arkansas, and spent much of his time there. He liked to cook, specializing in barbequed beef. At times, if he found himself unable to sleep, instead of tossing and turning he'd get up and bake a pie or a cake. On his lake-side property was a homemade bright red baroque trolley which, electrically controlled, conveyed him and his friends from the house to the lake and back.[7] These projects seem to have been the nearest things he had to hobbies. He never displayed any sustained interest in sports, despite having participated in sports as a boy and having a physique which could only have been an asset to an athlete. He never enjoyed hunting or fishing.[8]

He was a big man, holding his weight down to 190-200 pounds, and well-proportioned on his six-three frame. He was proud of his height. One day, seeing for the first time a person whose size he had heard others extol, Stacy exclaimed, "They call that man a giant. Why, he's not any taller than I am!" He liked to wear large hats, which made him seem even larger, and he loved Western music. At a night club in Galveston he frequented, the band, as soon as he walked through the door, would strike up a popular song of the day, "Big Man from the South."[9]

Despite his size and athletic build, Stacy never threw his weight around.

"He drank way more than his share," said his son, Robert, "but he never got out of line. He was a gentleman at all times. I often marvel at the amount of whiskey he could put away and never lose his cool. He'd hold the door open for any lady that came by.

"He looked like he could have been a pretty powerful fellow. If I met him in a strange place I wouldn't have wanted to tangle with him, but he was not a fighter and he was not aggressive or anything like this.

"I only saw him in a fight one time. That was about 1946. He was about 50. He was still in pretty good shape. It didn't last long. It was pretty impressive. The fellow was at our home, and he was quite obstreperous and he was just coming in, and Dad said he wasn't coming in. And he hit the fellow—the fellow deserved hitting—and it lifted him off the ground. I never will forget that. It lifted him off the ground, just one right hand. It was just one of those real bad scenes. The fellow didn't come in.

"Now, of course, when Dad got around 60, he no longer was in good physical shape, and he led a pretty soft, easy life; but I tell you, he was a rounder in his day and he didn't let any grass grow under him."[10]

Stacy was a generous, handsome man with charming manners and a witty mind, so it was inevitable that women would be attracted to him. "The ladies made a fuss over him," said Elizabeth Fullilove. "And he made a fuss over them." But even when his marriage with Priscilla had degenerated into separation, he continued to hold himself financially responsible when illness struck her.

"He was a very honest man," said Mrs. Fullilove. "I think in his dealings with Sister, too, as far as money or anything like that was concerned, I don't think there was ever any doubt that he was going to take care of everything. Once he said what he would do, there was never any doubt that Bob wouldn't take the responsibility."[11]

Stacy's integrity in his business dealings was a salient characteristic. As Stacy, Jr., said: "I suppose his outstanding trait was honesty. I've never seen a more honest individual from a business standpoint. Now he may have had some shortcomings in his personal life, but from a business standpoint, day-to-day, give-and-take situation, one-on-one or group or what have you, the man's scruples were beyond reproach.

"On some deals where he had strangers in it, he would have the money put up before the well was drilled. If then for some reason the well was delayed, he would pay these investors interest on their money for the period beyond the time he should have spent it. If it was a month, or six months, he would pay them the going rate of interest on the money they had put up—which I never saw before.

"He didn't believe in any foolishness in trading or dealing. His word was it. And I always admired him for that, but there were times when people would take advantage of him, and he'd say, 'Well, I got into this thing and that's too bad. This is the way I understood it and this is the way it's going to be'—and he would stick to it.

"He was forthright. He didn't believe in too much flowery speech leading to a subject. He was more inclined to just come to the point, but he wasn't rude about it. He was perceptive.

"He was not truly a good Samaritan. He cared about some people, but most people he didn't care about. He was self-centered to as great a degree as most people. I wouldn't say more or any less. But he still had this knack of bringing people to him that he wanted to. He could really gain their confidence and they thought a good deal of him. Now those that disliked him, disliked him immensely, but it was invariably because of his personal life, not his business life." [12]

III

What of Stacy's European-born cohorts?

Zeppa, Gold, Dorfman, and Sklar led entirely different lives. Their childhoods were spent in either rural or small-town European environments, speaking either Yiddish or Italian, shaped by cultures as different from each other's as Stacy's was from theirs. Gold, Sklar, and Dorfman came from czarist Russia; Zeppa, from rural northern Italy. Though circumstances impelled the Russians and the Italian to emigrate to the United States, those circumstances were so different that separate discussions are required to understand the European conditions.

Of all the Europeans who came to the United States during the first decade or so of the twentieth century, the Russian Jews experienced the least enviable vicissitudes. The waves of immigration to this country from Eastern Europe correlate very closely with the horrendous repression that dominated the refugees' lives in the Russian empire.

The Jew had known nothing but persecution throughout the nineteenth century. The regime of Czar Nicholas I (1825-1855) produced more than 600 anti-Jewish decrees. But the accession of his son, Alexander II (1855-1881), brought hope for amelioration of the Jews' plight, as well as for the conditions of gentile Russians. Political reform seemed possible. Though Alexander II's liberalism was modest, he freed the serfs, relaxed travel bans for some Jewish businessmen, and permitted some Jews to attend the universities.[13] More reforms were expected in the future.

Then came a pivotal event which accelerated the flow of emigrants from Russia. On March 1, 1881, Alexander II was assassinated by revolutionary terrorists. With his murder, all hope crumbled. The dead Czar's son and successor, Alexander III (1881-1894)—"a strong man with a narrow mind," as historian Bernard Pares described him—squelched any expectations of progress in most Jews' minds.[14] The reaction to the assassination, engineered to make Jews a prime scapegoat, made life as unbearable as it was under Nicholas I. Pogroms—the virtual lynching of a community or a group of people—followed, in which entire communities were ravaged, until there was no doubt in most minds that life could not possibly be worse elsewhere.

The Jews of the old Russian Empire were predominantly urban—usually dwelling in towns. The Jewish town, or *shtetl,* was usually a small one, customarily formed of wooden houses clumped together. The town was originally situated in the middle of a backward agricultural society in which Jews, most of the time, weren't allowed to own land. This meant they had to live by their wits, often as traders. As Irving Howe has commented, frequently "the relations between the social strata of the *shtetl* came to little more than a difference between the poor and the hopelessly poor."[15]

The mass exodus began soon after the assassination of Alexander II. The Russian political strong man, Constantine Pobedonostsev, pushed for a police state with a simplistic formula that would solve the "Jewish problem." Guided by the Russian Orthodox Church, the police state would let one-third of the Jews die, one-third convert to Christianity, and one-third emigrate.[16]

Though the fanatic's plan was not systematically set in action, conditions remained miserable until Alexander III died in 1894. Again, under Nicholas II (1894-1917), hope rose cautiously. At first the new Czar relaxed the enforcement of the oppressive anti-Semitic laws. But economic depression and revolutionary attacks soon changed this, as anti-Semitic agitators and Russian loyalists rose to bitterly denounce the Jews as leaders of the revolution and enemies of the Czar and Orthodox Christian religion. This emotional uprising led to the Kishinev massacre, southwest of Kiev just next to Romania and north of the Black Sea, which began in April, 1903, and lasted for three terror-filled days. Another massacre of Jews followed at Gomel, near the River Dnepr and north of Kiev, in White Russia. (Kiev, the largest city of the region, is on the River Dnepr in the Ukraine.) In the midst of this turmoil, the Russo-Japanese war followed in February, 1904, when Japan attacked Port Arthur in Manchuria. As the government dispatched troops and supplies 5,000 miles across country on the uncompleted Trans-Siberian Railway, only to meet defeat at the hands of the Japanese, the Russian people grew dissatisfied with war. The first organized revolution came in January, 1905, when striking workers marched on the Czar's Winter Palace in St. Petersburg to ask for reforms. Soldiers fired on the strikers, killing and wounding hundreds. Another, more serious uprising was crushed late that year. But the Jews suffered an even worse fate, as agitators blamed them for the war and whipped up the peasants' pent-up fury against them: in 1906 the Russian bureaucracy deliberately helped create hundreds of pogroms in Russia, with local authorities carrying them out.[17]

By this time, hundreds of thousands of Russian Jews had fled, most of them to New York, with few or no regrets about the old country. From 1881 through 1910, a 30-year period, 1,110,059 Jews came to the United States directly from Russia.[18] And though immigration peaked in 1906, the émigrés continued to arrive until the time of the Bolshevik Revolution in 1917.

The make-up of the Russian Jews, because of the restricted lives they led, was radically different from that of the gentile Russians. For one thing, the Jews as a group were relatively more literate than the total Russian population.[19] A partial explanation of this is that the Jews, pinched together, turned to one another for education, valued highly by them. Furthermore, Jews were generally urban, in a backward agricultural society, which improved their statistical ranking. Only three percent of the Jews in Russia were employed in agriculture—a dramatic reversal of the gentile trend—and 70 percent were in gainful occupations involving commerce, manufacturing, and mechanical pursuits.[20]

Jewish professionals were rare among the immigrants, while the skilled tradesmen arriving here were likely to be those who had been butchers, carpenters, and shoemakers in Russia. A common Yiddish saying made use of these facts: "*Ver geht kein America? Die shnayders, shusters, und ferdganovim* (Who leaves for America? The tailors, shoemakers, and horse thieves)."[21]

While the Jews tended to be a bit more skilled in their occupations—as well as in dealing with strangers, which could only help them in the New World—there was another characteristic of their Russian experience which gave them a certain advantage over other immigrants of the time. The Jewish merchants took an entirely different approach to business than did the gentile Russian merchants, in effect using the same competitive principles and methods then used in Western Europe and the United States. They sought a quick turnover with a small profit, which could be repeated over and over, and the use of short-term credit. On the other hand, the gentile merchants operated on the basis of customary prices and long-term credit.[22]

These Jews spoke Yiddish, a language which originally had been German, and which had accumulated words from most of the European languages. The holy language for the Jews was Hebrew, which only the educated knew. But though Yiddish was a foreign language to gentile Americans, the new immigrant could find other Jews here, not merely from Russia, who spoke it.

Jews from western or northwestern Russia, making their way out with forged passports or by bribing the guards, would cross the German border and go to Berlin, then to northern ports to board ships bound for the United States. Though it was illegal to cross the border, it could be readily done if one were willing to take large risks. One usually was. One might simply steal over by night, or cross while concealed in a shipment of goods or while in possession of a forged passport. Yet even if one succeeded he still faced the dangers of robbery and theft at the hands of guides and others who might unexpectedly exploit the desperate, fleeing victims.

At the northern European ports, the émigrés usually booked steerage passage across the Atlantic to the Promised Land; in some instances American relatives sent steamship tickets. Steerage referred to the lowest-class travel facilities available on the ships, though it could well be labeled "no-class." And the refugees paid dearly for their passage. In 1903 a steerage ticket from Bremen to New York cost $33.50. Since they had to save and scrape to raise money for tickets, many Russian refugees landed virtually broke in the United States. In many cases the husband would go first, then earn and save enough money to send for the rest of his family later.[23]

Although conditions tended to improve around the turn of the century, steerage travel was exhausting, with hundreds of persons crowded into close, smelly quarters. There was little water and poor—never kosher— food. Some compartments held more than 300 persons; some ships had three-tier bunks on which the passengers slept; and often there were as few as eight toilets and wash basins for more than 200 persons.[24]

Because of the tragic circumstances in Russia, the Eastern European Jews retained no longings for their country when they left. They usually considered themselves fortunate to have escaped with their skins, and multitudes of those who fled carried with them memories of their own narrow escapes or of the tragic ends of their kinsmen and friends. Nostalgia for Mother Russia was not part of the emotional baggage of these new Americans.

Four characteristics of this Jewish immigration were that it was a movement of families, that most were young persons, that they were permanent settlers, and that their ranks included, as opposed to other immigrant groups, more skilled workers, fewer unskilled ones, and craftsmen and artisans.[25] Furthermore, the Jews, unlike many other immigrants, came to stay. They brought their families, and few left the United States thereafter.

Other Europeans, however, frequently set out to make their fortunes in the New World and then returned home. For the years 1908-1925, more than 50 percent of the non-Jewish immigrants from Romania, Hungary, Italy, and Russia returned to their homelands. During that same period, only five percent of Jews returned to Europe, and those Jews returning to Russia and Poland were insignificant in number. One figure sets the Jewish returnees to Poland in 1919-1922 as 0.5 percent.[26]

A further indication of the permanence of the new settlement is that between 1899 and 1910, a total of 267,656 of all Jewish immigrants were children under the age of 14, representing about one-fourth of all the immigrants.[27] This fact relates intimately to two of the founders of Delta Drilling Company who came over from Russia as boys. It might be noted that youths also fled to avoid conscription and the harsh life that followed it. In the early nineteenth century, many boys were conscripted as young as 12 and 13 and for as long as 25 years, the horror of which was not apt to be forgotten. Because of the harsh economic conditions they had endured, the Eastern European Jews—and the Italians, as well—usually had few options when they arrived here. Shortly after leaving the ship, within days if not hours, they would have to go to work if they were to survive. In most cases, they began at the bottom.[28] Generally speaking, however, the Jews, because of a background that had instilled ambition, provided some industrial skills, and given them an awareness of the commercial processes, were as a group able to enter the economy at a higher level than did the Italians.[29]

The "push" which had brought them from czarist Russia's deplorable conditions was strong, indeed, but the fact that such large numbers of emigrating Jews came to the United States suggests that the "pull" was magnetic. Otherwise, why did they not go to a country other than this one? The lure of freedom and economic opportunity was sufficient to direct the Russian natives to these shores. Emigrated relatives and friends sent back word that America was a magic land. Though some discordant notes of a flawed America intermingled with the praise, the oppressed Jews in Russia had reasons for leaving that had never been stronger, at a time when all they needed was the knowledge that another country existed in which life could be secure and safe, perhaps even a pleasure.

By 1905, those leaving were not merely the displaced and declassed; they were also likely to be the energetic, the vigorous, and the ambitious, for these qualities helped them little in the Old World. But with an entire people in flight, there was bound to be a mixture of types.[30] For the Russian Jews, America represented a place to realize their dreams that had been denied. Goaded by the past, they doubled their efforts upon reaching the safety of American shores. As Thomas Kessner has expressed it, "Russian Jews were driven by a demon, seeking the security that had constantly

eluded them in Europe. . . . They placed great emphasis on independence, on being a *balabos far sich* (one's own boss). This ambition translated into an emphasis on professional positions, shopkeeping, and manufacturing. These had been goals in Europe. . . .''[31]

New York City was the primary haven for these displaced persons, but
they came in such numbers that many New Yorkers, including some of the
older, established—which is to say German—Jews began to search for a
way to spread the newcomers around the rest of the country. However,
other locales were not always hospitable to these poor Europeans, and the
incoming aliens were frequently loath to go elsewhere. Nonetheless, in
1900, New York City created the Industrial Removal Office to disperse
Jewish immigrants from New York. In the first five years, only about
40,000 had been relocated. By then another plan seemed necessary. This
led to the Galveston Immigration Plan, based on the contention that
Galveston, Texas, was the ideal entry to the South and the Midwest. By
December 1913, Rabbi Henry Cohen, a leading advocate of the Galveston
Plan, announced that 8,150 immigrants had been dispersed.[32]

According to one source, W. L. Moody, Sr., a wealthy Galvestonian,
funded much of the relocation of Jews to small towns because he was aware
of anti-Semitism in large cities.[33]

IV

The European heritages of Gold, Sklar, and Dorfman were no exceptions to the pattern of persecution and turmoil: the same events which
pushed and pulled hundreds of thousands of other Russian Jews to these
shores also brought these three men.

Sam Gold, a handsome, well-built six-footer, was the oldest of the
three. He was a mature man with a family when he emigrated. His name in
Russia was Simon Goldman, which he retained as his legal name in America, but he adopted Sam Gold as a shorter, more easily remembered name.

Little is known of Gold's early life. He apparently was born around 1871
or 1872, based on the fact that he was about 64 when he died in early 1936.

Gold and his family lived in a town called Iskeritz. "It's very near
Kiev," said Goldie Rappeport, Gold's stepdaughter, "because I remember my mother telling us that whenever there was any real serious illness in
the family we went to Kiev to the doctors."

By the time we have any details of his life, he had had five children by
his first wife, who had died. Four of the children were living. At this point
he married his second wife, Minnie Gardsmane, who had married for the
first time at 14, and had a daughter (Mrs. Rappeport) by that marriage.
Thus there were five children until Sam and Minnie Gold had a child of
their own, which made a total of six. Minnie Gold was much younger than

her husband, perhaps by as much as 20 years, and some of his older children were about as old as his wife.

In Iskeritz, the Gold family was "a little bit" privileged, because Sam Gold operated a factory that made boots for the Russian army. Because he had a skill that was useful to the government, Gold was able to parlay this "little bit" of privilege into favors which helped him escape the brutal pogroms. But despite this slightly privileged status, the family didn't have a great deal.

The Russo-Japanese War, which began in 1904, was a turning point for the Gold family. Sam Gold was conscripted into the Russian army in 1905. During his service he served as a valet to one of the German generals who were in the Russian army to train the Russian officers.

Gold related his experience to his family in later years. Mrs. Rappeport recalled that story.

"This man took a great liking to Mr. Gold and told him that when his tour of duty was over he should do everything in his power to get to America. There would be a terrible war and Germany would rule the world. Anyway, he did that."

His family's circumstances, never really favorable, and the uncertain state of conditions in Russia prompted Gold to take the general's advice.

"When Daddy came home from the war," said Mrs. Rappeport, "he immediately began to make plans to come to America, and he left the country, I guess, in 1909, maybe a little before that, and came to Shreveport, Louisiana, because my mother had a brother living there that we had been in correspondence with. Daddy was in America for two years [before he sent for the rest of the family]. It was just about the time the Oil City/Caddo Lake area became an oil field.* And my uncle persuaded him to open a shoe shop in Vivian, Louisiana, which is 30 miles out of Shreveport."[34]

While Gold labored in his northern Louisiana shop—and may also have sold apparel, according to one source—he saved money with which to bring the rest of his family to join him in the new land.

"Our family was not molested as much as we might have been, because of the fact of what we were doing for the Russian army," said Mrs. Rappeport. "But [trouble] was going on continually. It was not safe for a Jewish person to be out on the street after sunset."

Sam Gold left Russia on a forged passport.

"Anybody that got out of Russia at that time had to have a forged passport and forged papers from the czarist government," said Mrs. Rappe-

---

*The Caddo field in northwestern Louisiana—the first natural-gas production for that state—was discovered in 1905. When Sam Gold arrived in the region, one wild gas well had been blowing out of control since 1906. See James A. Clark, *The Chronological History of the Petroleum and Natural Gas Industries* (Houston: Clark Book Co., 1963), pp. 85, 98.

port. "Some way it could be done. It took a lot of money and a lot of pull, and so for that reason we were all able to leave."

When Sam Gold had saved enough money, he sent for his wife and children, who departed Russia via one of the common means of escape for Jews in those days.

"We left Russia in a wagon under straw," said Mrs. Rappeport, who was six at that time. "My mother and all the children, and we traveled that way. It was still very cold at that time, early April, and we left our home in a sled in 1911. I guess it was late March when [we] left this town, because we were a week getting to Hamburg, Germany, from where we sailed, and we were three weeks on the ocean getting to Galveston.

"That trip took three weeks and almost ended in tragedy, because the two older girls—they were very near my mother's age—they were ahead of the line, and the week before you get to a United States port they begin examination, and unless you are proved to be in *perfect* health, without a *blemish* on you, you cannot get off the boat. People wonder why so many got over here, but those were the conditions at that time.

"And these two girls were at the head of the line and they went right off. Daddy hadn't gotten to Galveston yet because he had been told we would have to be there in quarantine for a week before he could take us home, and he was busy at the store. So these girls got off the ship and were standing at the wharf and Mama was next, and she had a pimple on her chin and they turned us back, all of us. Of course I don't remember what all went on, but finally through the generosity of the doctor—she was a very good woman, young—they fixed it so the pimple did not show and we got off the boat. But this took hours.

"These girls were starving to death down there. They didn't have a dime on them. Evidently they were complaining, didn't know a word of English. Somebody handed them each an orange and that's what they had until we were taken to this quarantine hotel, and there we were very well cared for. We couldn't leave the place. But at the end of a week we were taken to a hotel, and Daddy was at the hotel and he brought us to Shreveport to my uncle's house and then we got some clothes, because the oldest sisters were told that the streets of America were paved with gold, and they took all of our clothing and threw it in the ocean. They weren't going to wear anything that came from Russia.

"So we were dressed and went to Vivian. It was Easter time, April. I remember we came to Vivian and my father said that a family wanted us for dinner. It was a holiday. We didn't know what Easter was, and I'll never forget, if I live to be a million years—I'm an old lady now—but we walked into this dining room and there was a pig with an apple stuck in his mouth. We were still strictly kosher refugees from Russia! I'll never forget that. Well, none of us knew a word of English. So Daddy, in his broken English, he could make himself understood, and he explained that there were certain things that we could not eat, and they didn't force it. There was plenty on the table: fruits and vegetables—you know how people cook for Easter in a country town—and there was plenty of everything."

All of the Gold family spoke Yiddish and Russian when they arrived in Louisiana. However, the new residents took on their Americanization with enthusiasm. The children learned English and in time forgot their Yiddish and Russian. "In fact, by the time I got into high school in Shreveport," recalled Mrs. Rappeport, "I had forgotten my Yiddish. We used it so little."

Though the children learned to read, write, and speak English fluently in school, Sam Gold the businessman learned his through usage. Mrs. Gold had tutors come to the house to teach her to read and write English and to improve her speaking command of it. Sam Gold wouldn't take time, then or later, away from his business to go to school or tutors.

"He learned to write enough to get by," said Mrs. Rappeport, "and I don't think he ever learned to read English; I really don't. But it didn't stop him. He spoke English all the time. You had no trouble understanding him. That's why we forgot Yiddish. It was not even spoken at home. Everybody wanted to speak English well, and after I got into high school in Shreveport I went to our rabbi and had him teach me how to read and write Yiddish because I had completely forgotten it."

The Golds lived in Vivian for three years and then moved to Shreveport.

"The little shoe store did very well," said Mrs. Rappeport, "but my mother was very unhappy because there were no other Jewish families and she wanted us not to forget our Jewish upbringing, and wanted us to have a little more of a social life than we could have in a little tiny country town.

"We moved to Shreveport and opened up a shoe store there. Well, it didn't seem to go so well. So by that time Bossier City, across the river from Shreveport, was kinda blossoming and Daddy opened up a grocery store in Bossier City. He was trying anything to make a living for his family. There was not a really nice grocery store there, and this place was for sale cheap and with living quarters in the back, which for the time being made it ideal.

"But, in the meantime, through getting acquainted with the Jewish people in Shreveport, one of his first friends was Mr. Sam Bender—the Bender Pipe and Supply Company. Mr. Bender was buying junk and pipe at that time, and since Daddy was not even slightly interested in a grocery store he put Mama and the kids in there and he went out to learn to buy pipe and what-have-you. And from that to finally opening his own junk shop. At first it was mostly junk. He didn't go into the pipe. He wanted the pipe, but he couldn't swing it himself, so not too long after that was when he met Mr. Sam Sklar. And Mr. Sklar had some money, so that's when they became partners and when the pipe business really began. Of course, they immediately moved to Shreveport and opened a pipe yard there. Gave up on the grocery store, because somehow this pipe business seemed to blossom right away."[35]

J. R. Parten, the Houston oilman, went to Shreveport in 1919, where he and his partners opened a drilling company. He sees the role of Bender, a dealer in second-hand metal products, as the historical precursor to the firm that was to become Louisiana Iron and Supply Company. "Old S. Bender was really the daddy of them all," said Parten. "He raised them all, was the forerunner of them all. They all grew up there in Shreveport. They were all good operators."[36]

Shortly after World War I—probably during the 1922-1925 period—Gold and Sklar founded the used-pipe business which became Louisiana Iron and Supply Company. Each partner held a half interest in the partnership.[37] It appears Sklar put up the money and Gold contributed his skills as a buyer and salesman and the knowledge of the business he had acquired through Sam Bender. In addition to oil development in northern Louisiana at the time, the boom in South Arkansas, not far off, was on at this time, and the company soon had a pipe yard there too.

"The people that Daddy had gotten acquainted with in the oil field around Vivian, they loved him," said Mrs. Rappeport, "and when he came back to them to buy pipe, all he had to do was ask for it. They didn't even argue prices with him. So it just, you know, took off right away and, thank God, it didn't get worse, it got better."[38]

Sam Gold's personality seems to have had a significant influence upon the success of the new company.

"He was the kind of man who could walk into Humble Oil and shake hands and have a deal," said his daughter, Rose Bishkin, who was married to Sam Dorfman in the 1920s. "He also was the kind of man who could take a purchasing agent into his apparel shop and fit him out in a new suit and tell him, 'I'm doing this because I want you to have it.' He knew how to do it. He was a goodwill ambassador. He did very little work, but knew how to meet the public and deal with people for the company. He was a buyer; he bought for the company. He knew when a well was shut down, and he could buy pipe from them for almost nothing."

Similarly, he stood by his family. When daughter Rose, unhappy in her first marriage, told Gold she couldn't live with Sam Dorfman any more, he said, "All right," and accepted it.

"He seemed to know why," she said. "He was a sweet, kind, gentle man."[39]

Dr. Myron Dorfman, the only child of that union between Rose Gold and Sam Dorfman, perceived his grandfather in a somewhat different, but not contradictory, light. Dorfman remembers him as "a very domineering man," almost Germanic in his Old World, patriarchal attitude.

"He ran things," said Dorfman. "He was very conservative, a great big man, but I think it was more in the middle-European attitude where the father is the master of the household. He was the master of his house, and in business I think he was that way too. He pretty much called the shots, and he was conservative, although he must have been farsighted enough to associate with good people like Sklar and Dorfman and let them have a certain amount of autonomy, because they were entirely different types."[40]

Anti-Semitism existed to varying degrees in those days, but the Golds were insulated from it, particularly by Sam Gold's popularity, and later, his economic status. His stepdaughter, Mrs. Rappeport, remembers him as "the soul of kindness," a quality that no doubt affected his reputation in the business world as well as with his family.

"Everybody knew Sam Gold was honest," she said. "If he said a thing was so, you could make book on it, and the people who knew him knew that. The men he met liked him personally immediately, and I've never known him to say that he had any trouble getting any backing for anything he wanted to do. All he had to do was walk into the bank and say what he wanted, and he got it.

"I couldn't have wanted a kinder father, and when he died we heard from people, and people came to us who we did not know existed, whom he had helped. Nobody was ever turned down by Sam Gold for any reason. It got to the point where he was kidded about it, that he would never carry more than two dollars in his pocket, for whatever else he had, he gave away. They knew if they asked Mr. Gold he would give it.

"Back in the days when people didn't go to restaurants every Sunday, our house was Grand Central Station. If there were 10 or 20 or 50, they knew they could come to the Gold house and be fed, and Mama was prepared and nothing pleased her more than to cook for and feed people. I always say that if Papa had died poor and she could have opened a boarding house, she would have been the happiest woman in the world.

"It wouldn't make any difference whether they were Jewish or gentile. We had *friends*—not Jews or gentiles: our neighbors, the people we worked with, the people we met in school. It did not make any difference. Whoever came by, they knew they were welcome and they knew they wouldn't be turned away."[41]

V

Unlike Sam Gold, the other two Russian Jews who helped found Delta came to the United States as boys.

Sam Sklar came from a small town—*shtetl*—or possibly a village, in Russia called Wolochisker. As far as can be ascertained, he was born August 12, 1890. Few accurate records were maintained in the old country, and many dates were guessed.

We know practically nothing of his life in Russia, for he came over at an early age, when he was about 15 years old. He and three of his four sisters were brought to the United States by a cousin, S. J. Pearlman, who lived in Milwaukee. Dr. Samuel J. Pearlman, the son of that cousin, thinks the year that the Sklars arrived may have been 1905.

"This was common practice in the extended families of European origin," said Dr. Pearlman. "The Sklars lived with us off and on until they found employment or were married. Sam Sklar worked for my father after we moved to Chicago in 1909. It was in a little grocery store."[42]

His son, Albert Sklar, born in Shreveport in 1925, never heard his father speak much about his childhood. "He wasn't that talkative about the old country. He never mentioned it to me very much, and he wasn't very talkative about his past in any way."

Sam Sklar was the only boy in his family, and none of his four sisters is living now. We do know that he must have started to work almost immediately upon arriving in this country, or not long afterward. What education he acquired he gained through self-education as he went through life, working at whatever jobs came his way, and through observation. We can assume, with a fair degree of accuracy, that he had a variety of work experiences by the time he reached Louisiana in his early twenties.

Either in Milwaukee or Chicago he went to work for a manufacturer of carnival items, selling trinkets and souvenirs at county or state fairs. The manufacturer was in Milwaukee. On one occasion Sklar went to a fair in Dallas, Texas, and that turned out to be the connection which eventually influenced him to go to Louisiana. Probably he was around 21 years old when he went to Dallas, which would have made it sometime around 1911. He was traveling with a carnival, selling the manufacturer's wares, and wherever the carnival went he accompanied it. Sklar had a chance to remain with the carnival-trinket manufacturer and become a partner, but decided he wanted to try something other than that business.

"The only thing he ever told me," said his son, "was he got to Dallas and somebody told him in Dallas that he ought to go to Shreveport. There were some things happening there, that maybe he could find something better to do. So from what I've been told, that's what he did. It had something to do with, I think, the early Pine Island fields, and that this might be a good place to go into some kind of business. Now from what they tell me, the only kind of business you could go into in those days was like a peddler. I don't know if the reason was strictly the oil business or any other type of business; he never made that too plain to me. But that's how he got here."[43]

It is uncertain what Sam Sklar did in the Shreveport area or how exactly he came to know Sam Gold, but their common background must have been a factor. Shreveport was not a large city, and the Jewish community was even smaller. It would have been logical for one Russian Jew to have gotten to know another who was new to town. Sam Gold was well-known and popular in Bossier City and Shreveport. Sam Sklar was clearly a bright and successful man in his way. Their making contact with each other, perhaps through the synagogue, perhaps in one of a variety of social situations, would not seem difficult, and once that association was reached it would appear reasonable that they might consider joining forces in some enterprise. After all, Sam Gold already had a business going, and Sam Sklar had a wide range of experience and had saved some money. Sam Gold had learned something of the pipe business through his association with Sam Bender, and Sam Sklar also may have learned from Bender.

Sklar seems to have been in some sort of business when the United States entered World War I. Possibly it was the scrap business, which his son thinks it was, but whatever it was he sold out and was getting ready to leave for the service when the war ended. This is believed to have been with a partner other than Sam Gold, and at a different location.

Sklar married Ida Siegel, whom he had met in Chicago through family connections, on April 11, 1918. By then he had accumulated what in those days would have been considered a substantial amount of money. And, of course, he had the proceeds from the sale of the business. His wife was very young, not long out of high school (she had been working in Chicago when she married), and new to Shreveport. Eventually, all of her brothers came to Shreveport and worked for Sam Sklar as he went into partnership with Sam Gold and subsequently grew prosperous.

The partners Gold and Sklar were, at least physically, a study in contrasts. Gold, a large man, was more than six feet tall, while Sklar was relatively short, around five feet seven inches, bald at an early age, and, later in life, portly. He never entirely lost his accent. A very robust man, he was exceedingly strong.

"Physically, he was the most powerful man I ever met," said Leonard Phillips, his son-in-law. "If you'd ask him to lift this building, he'd move it!"

Probably because of his early experiences, Sklar tended to solve problems in a straightforward, uncomplicated manner. As Phillips put it, Sklar would solve a problem "physically or verbally" in a decisive manner; there would be no doubts or hedging.

"Sam Sklar was the epitome of the expression, 'a diamond in the rough,' "continued Phillips. "He was self-educated. Brilliant! Analytical, but could not explain to you why he was analytical nor how he arrived at the conclusion.

"He was extremely active, and his manner of living was rather unusual. He spent every Sunday, from early morning till late evening, out in the field. He'd go see all the individuals who had worked for him for many years and check out the wells.

"He was liked by every segment of this community, because he was totally honest and knew his business and was receptive to the charitable community actions.

"Sam Sklar worked at every conceivable job until he came to Shreveport. And at that point he dedicated himself to a company, Louisiana Iron and Supply, which was first in the scrap business, and then, because of the technological growth in the area, he found himself in the pipe business, and thereupon found that major companies would dispose of nonproductive wells—in their terms of 'nonproductive.' And he would buy old wells, clean them up, and establish his own crews, his own cleanups, and it was rewarding.

"The shame of it is, that he can't come back today and see what he had accomplished by the acquisition of hundreds of stripper wells that he had purchased or drilled, and watch the price of oil [rise] from $2.65 to $37.00 a barrel."[44]

Albert Sklar confirmed the general impression of his father.

"He was a hard-working businessman. He worked all the time. He didn't have too many pleasures that he indulged in. I would say he worked nearly 12 hours. He was here at this office at six o'clock and made it seven days a week. He might not have stayed all day Sunday, but he was here just about all the time from early in the morning until late in the evening, till suppertime.

"If a well was down and you needed some pipe, then he'd come out and see that it was shipped out correctly. I'm sure he did that many times, but I don't remember. But he was the type of a man that when he wasn't working, he was at home. He didn't drink and he didn't smoke and he got his primary enjoyment out of working. He never really took that much of a big interest in his home or his children *per se*, because he didn't have time. He was always at the office and he was always working, and he came home late at night and he'd go to bed—he was tired.

"So he was gone by six o'clock and he was here loading and seeing that these trucks were rolling, and doing whatever he had to do. So my memory of him was that he was sort of a tough, hard-working man.

"It wasn't the sort of father relationship where we'd have these days on the farm and fish and swim and go out and learn how to hunt or play golf. He wasn't the type that would take his sons around and teach them anything, neither one of us, Fred or myself."[45]

Dr. Myron Dorfman, Gold's grandson, later worked for Sklar and came to know him well.

"Sklar was of the same conservative bent as Gold had been," said Dorfman. "He was the man who worried about the nickels and dimes. Sklar was a very shrewd trader with very few diversions. He got up very early in the morning, about five-thirty or six o'clock to make sure the trucks got out, loaded correctly. He was standing at the door, waiting for his key people to come straggling in at seven-thirty or eight o'clock in the morning. God help you if you got there after that. He would still be working when we all went home at night.

"And of course he'd go home and have dinner and go to bed. So his life style was not exactly the same as those of [his son] Albert Sklar, myself, and some of the others.

"I'll never forget after we had developed Blue Bayou, one of his recreations was to go out in the field on the weekends and watch the oil go in the tanks and look around and make sure that all the equipment was in good running order, and so he would call me up. I may have been a few days and nights working, and he would call me up at six o'clock on Sunday morning.

"'Myron, what do you know?' What could you know at six o'clock in the morning? 'Let's drive down to Blue Bayou and see what's going on. Pick me up in 30 minutes.'

"I'd pick him up in 30 minutes, and we'd go down there and travel around the lease and he'd look around. 'See that valve there, half buried in the ground? Pick that up and throw it in the car.' And we'd do this several times during the course of our visit to the field, and by the time we got back in he'd say, 'You know what this is worth?' and he'd open up the trunk and he'd start naming off what all this was worth and we'd have $1000 worth of equipment that had been just strewn around the lease and everything.

"That was very important to him. He'd look after those little things. Very compulsive doer, hard worker, convinced he was going to be broke half the time, and that may have been his compulsion.

"He did not really understand the oil business very well. Because he knew pipes so well, he knew equipment extremely well, and many of the things we did in the way of oil and gas exploration were done over his protest. Sometimes, being a true judge of character, he was protesting, I think, to make sure that we were convinced that we were right. We had to sell ourselves to him. But he was a very shrewd trader."[46]

Albert Sklar recalled a story his father once told him about a well that Sklar and his associates had drilled in the early days near a little town of Dixon, not far from Shreveport.

"He said they drilled this well, a very shallow well, and the well just came in fantastically, like 2000 or 3000 barrels a day, and it just looked like it was really going to be something—a tremendous field like Pine Island or something of that nature. It didn't cost much money and it looked like he had the world by the tail at that moment, and he had a very substantial offer by, I think, Gulf, and he said, 'Absolutely we would not sell anything like that, because this is how you would become big in the business!' Well, as luck would have it, three days later the well went to water and he didn't take the offer, and because of that damn well he didn't have a bigger interest in East Texas [field] because they decided they were in the pipe business and not the oil business. So they decided with their limited funds they couldn't afford to do it; they had better stick in their business of selling pipe.

"It was a learning episode, but at an unfortunate time, because he said if it had been different they would have had bigger interests in the East Texas field [in the 1930s] than they ended up with. Instead of just selling the pipe to Dad Joiner or whoever he sold it to or all the early operators and taking oil payments for it, they would have had more interest than they would have, just having pipe."[47]

Seymour L. Florsheim recalls that Sam Dorfman would have to argue strenuously to convince Sam Sklar to take advantage of business deals that involved considerable risk. "He was conservative!" said Florsheim. "He was very conservative. Wouldn't take a chance. Didn't like to take a chance."[48]

His early experiences, no doubt, shaped his character in this respect, and the lack of more formal education also played a part.

"Sam Sklar was a very strange person," said Mrs. Goldie Rappeport. "He was a very hard man to know and to get close to. He was a good man, but he just never had an education, and it was difficult for him to be a rich man. He didn't know what to do with money."[49]

Like Sam Gold and, in fact, the other European-born partners, Sam Sklar was patriarchal in the Old World tradition. "Definitely, he was the big boss," his son said. Sklar would seem tough about the pipe yard, shouting at employees, but, the younger Sklar believes, "they knew he wasn't all that tough. He'd act tougher than he was. He used to holler a lot. He and [Sam] Dorfman, particularly, would scream all the time. They would just disagree. They'd yell in the telephone sometime and then when they were together they'd yell at each other."[50]

But there was another, gentler side to Sklar that some of his associates didn't always see. Marguerite Marshall, who was Joe Zeppa's private secretary for many years, used to see him when he went to Tyler for Delta board meetings.

"Sam Sklar was a delightful old man. He was as nice as he could be. I am sure that he was a very astute businessman too, but you would never know it when he came to see us. He just liked to fool around and tease the girls. He was a lot of fun."[51]

At least to his younger son, Sam Sklar had a complex personality. Whether he was quiet or talkative depended upon whom he was with.

"He had a dual personality," said Albert Sklar. "I had a very easy conversation with him: whatever I said, the answer was going to be No. It didn't make any difference. But then I'd see him around some of his business acquaintances or some of the people that he knew, that he was working with, and he'd talk, and he'd talk their arm off. So it was Dr. Jekyll and Mr. Hyde; you never knew whether he was going to be talkative or not.

"As I say, he was not a social man, he was a working man. He just enjoyed working and I think he found more to talk about when he was talking to somebody in his own field than he possibly would to talk generalities or whatever: 'What did you do last night?' Well, he did the same thing. 'Where did you and your wife go?' Well, he went to bed. 'My wife had the cook fix dinner and serve it and I went to bed.'

"Well, I guess if you had to leave a country you were born in and you had no roots and you had no financial security, then you have to have some feeling that would probably make you feel a little different. The important thing in your life is to have financial security for yourself and your children, so that was obviously his goal."[52]

VI

Sam Y. Dorfman, the youngest of the three refugees from czarist Russia, was born in 1899 in the village of Zhitomir, a railway switching center about 30 miles north of Kiev.

The sketchy family recollections have it that he was of rural or near-rural origin, a rarity among Russia's Jews. His surname, however, is suggestive, for *dorf* is the term that Eastern European Jews used for *village*, as distinct from *shtetl* for *town*.

One fragment from his Russian days has it that Sam Dorfman was recognized early in life for his intelligence. Nobles would have him spend the summer at their country estates, teaching their children. This is particularly striking because he could not have been more than 13 at the time, and possibly younger. And while tutoring the privileged children, he used his spare time painting, an artistic bent he was to pursue in later life.

Whatever were his experiences in Old Russia, the fact that he was born in 1899 and that his family lived near Kiev tells us a great deal. He lived through turbulent times. After the Kishinev massacre of 1903, the fate of Jews remained up in the air, with a succession of pogroms providing all the "push" one needed to want to depart. In later years he evaded the subject of his homeland more than he discussed it. He never liked to speak about his early days in any detail, which suggests that they were powerfully traumatic.

According to Abe Rozeman, whose mother was Sam Dorfman's older sister Dora, the Dorfman family came over from Russia in three waves. The father of the family, Labe (later changed to Louis) Dorfman, was a veteran of the Russian navy and had spent time in a naval penitentiary for fighting. This experience had left him determined to help keep his sons from being conscripted into the Russian army. Such resolve led to his sending his older son Yaacov to the United States in 1905 following a pogrom. Yaacov accompanied his sister Fannie, who was married to Barrell (later changed to Barnett) Bronstein. Yaacov went to Montgomery, Alabama, where he died of yellow fever in 1915. The Bronsteins landed at Galveston, from where they went to Shreveport, Louisiana, for reasons unknown today but probably because of Bronstein's or his family's connections.

The second wave came in 1912 when the mother of the family, Rosa Lermann Dorfman, left with her 13-year-old son Shlam (later changed to Sam) Yandell Dorfman, her daughter Dora, and daughter Gussie who was younger than Sam. They made their way from Galveston to Shreveport, probably because Fannie had prepared the way.

A few months later the remainder of the family came, apparently because Labe, or Louis, Dorfman had business to settle, which had delayed his departure. He was accompanied by his daughter Annie, who was older than Sam, and a niece.[53]

"I know that when he [Sam Dorfman] landed in Galveston," recalled his second wife, Elizabeth Fischer, "he always told me he didn't even have a shirt. Somebody had stolen all his clothes in steerage."[54]

In addition to this deficit in personal belongings, he also, as did his other foreign-born partners-to-be, landed with no knowledge of English. This raised no obstacle in his getting started economically. There was, almost always, some link in the New World from the Old, and most frequently there were established businesses where someone spoke Yiddish. He apparently went directly from Galveston to Shreveport, where he went to work soon afterward in a little mercantile store.

Despite the promise of his early years, his schooling virtually ended when he left Russia, though he did attend night classes in English after arriving in Shreveport. From that point on, he continued his education primarily through reading. In later years, he remained very conscious of his accent.[55]

At some point after his arrival in Shreveport, young Sam did manual labor for Sam Bender, breaking up cast iron with a sledge hammer. It was probably this early connection which brought Dorfman into contact with Sam Gold and Sam Sklar. Later Dorfman worked out of Beaumont as a sales representative for Brooks Supply Company, covering oil fields in southeast Texas. In the early 1920s he returned to Shreveport and struck a deal with Gold and Sklar, subsequently running the Louisiana Iron and Supply yard for them at El Dorado. He may have had a working interest in the Arkansas operation.[56]

After El Dorado, Dorfman went to Greggton, near Longview, when the East Texas field opened. In Longview, a separate legal entity was formed—Louisiana Iron and Supply Company, a Texas corporation— with Dorfman having a one-third interest. Thus he shared in the Texas company named after the original partnership, but not in the original Gold-Sklar partnership. Following Gold's death, Sklar and Dorfman were partners in a number of ventures.

The precise moment in which Dorfman became a partner is not known, for he probably held interests in some of the operations before the relationship was formalized in papers of incorporation. He married into the Gold family early, in the 1920s, and had held a responsible position in the business from the time he was a young man. His ability, then, was recognized early, and one may speculate that had not Sam Dorfman been offered an attractive working interest in the business, he probably would have gone elsewhere or even established his own business.

Whatever were the details, he was trusted for his business acumen and reliability. He would make deals in his own name, though on behalf of the company. Technically, the entire deal might be in Dorfman's name. Subsequently, the interests would be distributed according to their oral agreement.

Dorfman, about five feet ten and slender, was handsome, well-built, and determined. "He was very debonair in his dress," said his second wife and widow, Mrs. Fischer. "Very jaunty, very neat."

Seymour L. Florsheim, Mrs. Florsheim's brother and long-time associate of Dorfman, commented, "I don't think he ever wore a pair of khakis in his life, even going out to oil fields. He'd wear a suit or a sport outfit. He'd wear boots, but it was a riding habit. He liked good automobiles. He liked everything good."

"Also the ladies!" laughed Mrs. Fischer. "And they liked him, too. The first time I saw him, it was in a yellow Packard convertible, and I thought, '*Boy.* . . . ' And after I married him, of course, I inherited the yellow Packard convertible. But the way he bought that convertible was, he and one of his big-shot friends were in New York and they one night went to an automobile show and this other guy bought an Olds and he bought this yellow Packard convertible, just like that. You know, I thought that was fun. He knew what he wanted. It didn't take him long!"[57]

Dorfman impressed practically every person he met, in one way or another.

"He was an extremely bright, gregarious, but rather introverted man," Leonard Phillips described him. "He had a great native intelligence. He would walk through a pipe yard, without a piece of paper or a calculator, and he could come within five percent of the inventory, by himself.

"I don't think he ever found his potential. He and Sam Sklar made great partners. Sam Dorfman was the outside man, the ostensibly gregarious one. Sam Sklar was the inside man. It was a hell of a good arrangement! He was perhaps more daring than Mr. Sklar. Mr. Sklar was the balance, the flywheel; he kept everything from hitting the fan. So they made a great team."[58]

Dorfman was frequently willing to take business risks that his partner wasn't. "And he would push Sam Sklar into it," said Seymour Florsheim. "They would fuss and fight and fume, you know—I'm talking about on a friendly basis. He would have to really convince Sam Sklar to do something."

But in certain ways Dorfman was like Sklar. "Business was their hobby," said Florsheim. Neither relished hunting and fishing. (Neither had Bob Stacy, it might be remembered.) Dorfman, more socially inclined than Sklar, did display some slight interest in amusements which Sklar eschewed.

Dorfman did not make life consistently smooth for those around him, though at the same time he was sensitive to other's feelings enough to express what might be described as a guarded empathy after he had given them a hard time.

"We used to call him the Mad Russian," said Mrs. Fischer. "Very quick temper and then really sorry."

Florsheim, who started work for his brother-in-law at the age of 16, agreed. "He'd carry you just as far as he could, then get sorry for you. He was the type of person that had respect for anyone that would stand up to him. In other words, he'd take advantage of you to beat the devil, if you didn't. If you did, then he'd appreciate that. He respected that. I learned that the hard way, 'cause I was a youngster when I came down here. He'd know just when I was ready to break and he'd just make a complete turnabout. Like nothing had ever happened. He'd chew me out to a fare-thee-well, but the minute he saw I had taken all I could take, well, then—he knew."

Dr. Sam Y. Dorfman, Jr., one of two sons born to Dorfman's second marriage, remembers his father as being "very authoritarian," a view redolent of the patriarchal tone of Gold and Sklar. Young Sam had his work cut out for him, trying to do what he thought would please his father. "I really don't think he was hard to please," said Dorfman, Jr., "but he always made out that he was. He never said, 'Well, that's great,' or something when you'd win a debate or tennis match or anything like that, but I always knew he was behind me."

The elder Dorfman's complex personality was sometimes concealed behind a facade of sporty clothes and flashy cars, but facets of it seemed to belie that outward appearance.

"He was a very modest person," said Florsheim. "He did not like to brag or have anybody brag about him. He wanted respect, but very definitely had an inferiority complex. He had no reason to have one, 'cause he was far from being an inferior person. I mean, he could talk about any subject. He was up on everything."

"He was also an artist," said Mrs. Fischer. "We have lots of his work. Watercolors mostly. It just came naturally. He loved flowers and gardens. He didn't work in the garden, but he loved all that sort of thing. When I first married him, I moved into a new house that he already lived in and it was just beautifully done. And it was small, but the garden in the back was just like a little showplace. In Longview."[59]

Unlike Sam Sklar, Dorfman concentrated less on the "nickels and dimes" or immediate aspects of business. "Father was the guy who worried about what was going to happen ten years from now," said his eldest son, Dr. Myron Dorfman.

"He was a very complex, intelligent man. He was one of the most interesting people you would ever want to meet. A very engaging personality, and he had what I call integrated intelligence. He was the kind of man who, given A, B, and C, could see X, Y, and Z. That, to me, is the true test of intelligence—the man who can see a problem and, by logical analysis, solve that problem without seeing what is going to happen next. He had vision, and he was charming and very personable, and one of the best traders you'll ever want to meet in your entire life. He had amazing capabilities. His vision was primarily responsible for much of the growth of the early oil and gas production for our company [Sklar Oil], and probably had some influence on Joe Zeppa and his work, because the two men regarded each other very favorably."[60]

VII

In reviewing the lives of these four founders of Delta Drilling Company, we can see that there were both common characteristics and dissimilarities. The gulf created by the backgrounds of Stacy and the three Russians, however, was not overwhelming, or the men would probably never have gotten together. They held in common their reputations for honesty and doing business with a handshake, and their respect for hard work. They all had known hard times and had grown up with limited formal education. Not one had finished high school. Of these four, Stacy and Dorfman seem to have had the most in common. They both liked fine cars and clothes and, like Gold, were adept in meeting people. All of them were highly intelligent. Stacy and Dorfman had, through reading and their associations with others, educated themselves. All of them seem to have been driven men, determined to succeed. It may be said that the Russian Jews were driven in a way Stacy never was, by virtue of the historical necessity which had brought them to this country.

And within that group, the characteristics of each man seemed to mesh with those of his partners. Though Stacy and the Shreveport group were never as closely associated as the immigrants were with each other, they had more in common than appeared on the surface. It is pertinent that Dorfman, apparently Stacy's first contact with Louisiana Iron and Supply and the first of the Shreveporters to learn of the drilling firm idea, had certain personal characteristics shared by Stacy. "Thinking big" and future prospects appealed to them both.

But an integral key to early Delta history is the fifth member of that founding group, who provided his own special talents at a time when an expanding economy promised golden rewards to those willing to work hard enough to achieve them. Had these men not known him, it is not certain that the company would have been established, for, after all, each partner's contributions were crucial to its existence.

He was the one from Italy.

2

A BOY FROM ITALY

I

Joseph Zeppa's cultural background was altogether different from his partners'. He was Italian, of humble farm background in a part of the world that had never heard of the oil industry. Even after he came to this country, his early business experience was on Wall Street—a rare point of departure for an Italian immigrant boy in those days, and an even more unusual staging area from which to enter the oil business. But his being a native Italian played a strong part in his entering the oil business in the first place.

And, then, there were other, less obvious, elements in his early personal history that would mold him into the type of individual that these partners-to-be would find attractive in a business associate.

All this began in Italy, where he was one of millions who were to swarm to this country in the late nineteenth and early twentieth centuries.

II

The first Italians arriving in the United States in the nineteenth century came from northern Italy's agricultural regions; but in the history of the total Italian migration, these Italians were a distinct minority. Of the 5,058,776 Italians coming to the States during the 1876-1930 period, only 20 percent were from central and northern Italy, with 80 percent from the South.[1] Since Joseph Zeppa came from northern Italy, we will be taking a close look at the difference between these "two Italies."

One thing that immigrants from either part of Italy had in common, however, was the economic and social hardships. This was the "push" that drove them from their homeland: the pressure of the population on the land; heavy indirect taxation; a high cost of living; and an inept, distant national government.

The designation, "the two Italies," was a reality. Friction between northern and southern Italy was long an historical fact. The North, or "Alta Italia," had been influenced by French and German cultures. It had been the heart of Renaissance culture, and in the nineteenth century it had been largely responsible for the nationalistic movement which had resulted in the unification of the country. The South, on the other hand, still was under the sway of the medieval codes of the Byzantines, Normans, and Arabs. The regionalism of Italy even extended to the dialects, for the northerners had difficulty understanding the dialects of the six southern provinces.[2]

Northern Italians—and presumedly this would have an impact on the farmers, or *contadini,* there—were richer in tradition and culture. The northerners were better off in most ways than the southerners. An extremely important statistic is that only 11.5 percent of immigrants from the north were illiterate, while 53.9 percent—more than half—of the southerners were. This would be highly important in adjusting to a new culture, for literacy in Italian would make it easier to become literate in English. On another important point, the northerners had an edge on their compatriots: they had twice as much money upon arriving in the United States than did the southerners.[3] The difference in amount may have been slight, but to one who has little, each lira counts that much more.

But what may be more telling, for our purposes, is the distinction noted by one scholar between the "rural, still semi-feudal" South and the "economically-advanced" North: "The northern Italian is a European; the spirit of capitalism and the acquisition of wealth motivated him as single-mindedly as they do the Swiss or the German."[4] This assessment gains further currency when we add to it the scholar Joseph Lopreato's observation that immigrants from the North "showed the cultural and physical characteristics of the older American stock more frequently than their southern fellows," which left them "less suspect, less foreign, and hence more readily acceptable to the 'old American' group."[5] These statements should be given close attention, for they distinguish, to a large extent, young Joseph Zeppa from the millions of southern Italians who also came to this country during the early years of this century. Zeppa came from near Alessandria, south of the River Po in the Piedmont region of northwest Italy. That also is the general region of Italy that produced the Lombardy bankers.

The "push" that sent the Jews and the Italians away from their native lands, however, was so unalike in each case that different types of migrations occurred. Although the Italians were suffering economically, socially, and politically, their plight, even in southern Italy, was hardly comparable to the systematic persecution of the Russian and Polish pogroms. There were no massacres, no annihilation of entire families and villages in Italy. Whereas the Jews came in families and came to stay, the Italians frequently came alone to earn enough money to return to Italy and make a better life. The large majority of Italian immigrants during the 1880-1910 period were male, with 83 percent of them in the mature 14-44 age group.

There were relatively few children; from 1899 to 1910, only 11 percent of the Italian immigrants were under the age of 14. By the 1907-1911 period, the average annual return to Italy reached 150,000, which represented 73 of every 100 Italian immigrants.

However, if we look closely at the returnees, a significant fact emerges: 80 percent of those returning were farm and common laborers, and 96 percent of the Italy-bound went home via third class. As Thomas Kessner has commented, "Apparently the return movement did not draw its people from the successful classes."[6] Whether those who remained did so because they were successful is not altogether clear, but it does appear that those who did succeed, for the most part, did become permanent American residents.

III

"So far as I know," wrote Joseph Zeppa late in life, "our family was concentrated in Piedmont, particularly the part referred to as Monferrato, which includes Alessandria, Asti, Casale, etc."[7]

The Piedmont, in the upper Po valley, is at the base of the Alps. The Piedmont's capital is Turin. A fertile region, particularly in comparison with the area south of Rome, the Piedmont is strongly agricultural, and was even more so around the end of the nineteenth century. At one time Piedmont was part of the Kingdom of Sardinia.

Fubine, where Zeppa was born, is a little town south of the Po River, between the Po and its large tributary, the Tanaro. Fubine is close to Alessandria; both are in northwestern Italy—far closer to France and Switzerland than to Rome. Fubine is relatively close to Genoa, the port city, in that part of Italy where the hip of the "boot" slims down to a leg.

The Zeppa family was old to this part of Italy, having been there more than 300 years. By some accounts, the first Zeppas came there from Turin in the early 1600s. An ancestor settled at the cross of the roads and started making bricks, and his descendants remained there. To obtain land and hang onto it through the generations, in a part of the world where the nobility and other large landholders held the upper hand, bespeaks the tenacity of the clan.

We can go back to at least the eighteenth century in tracing the Zeppa genealogy. In their culture, lineage is traced through the paternal side of the family; through the years the Zeppas intermarried with families named Rota, Percivalli, and Longo. Joe Zeppa's great-grandparents, Michele and Teresa Rota Zeppa, were born in the late 1700s; we do not know Michele's birth year, but his wife Teresa was born in 1794. One of their sons, Carlo, born in 1823, owned 320 acres of land, a large holding for that time and place. Of their nine children, Carlo and Luigia Percivalli Zeppa had four sons, and when Carlo was about 70—20 years before he died in his nineties—he had his sons draw lots for their shares of the family farm. These sons—Vincenzo, Michele, Giuseppe, and Cristoforo—each obtained 80 acres in this way.

Two of the Zeppa brothers married sisters from the Longo family. Vincenzo married Clara Longo, and his older brother Giuseppe married Delfina Longo.[8]

Vincenzo, born in 1858, and Clara, born in 1863, became the parents of Joseph Zeppa and his siblings.

IV

Giuseppe Zeppa was born to Vincenzo and Clara Zeppa on August 2, 1893, in Fubine, Italy. In all, Vincenzo and Clara produced five children who lived, born during a 17-year period stretching 1883 through 1900. The eldest, Carlo (Charles), was born in 1883; Gilda (Hilda), 1887; Giuseppe (Joseph), 1893; Paola (Pauline), 1895; and the last-born, Cristoforo (Christopher), 1900. All of these children eventually emigrated to the United States; only one, Pauline, was to return and die in Fubine. "They were practically all gone by the time I got grown," said Chris Zeppa, the youngest. "But until they left they all worked on the farm."

Regular, hard work was a normal part of growing up in Italy at that time.

"My father did pretty much like it was done everywhere," said Chris Zeppa. "He did what he could, with the small amount of land available, and they worked with other things to make a living. There was not even a thought in those days of riches.

"When we talk about 'farmer' in those days, in those little towns of northern Italy, we talk in terms of when 40 acres would be just a big farm, because those farms in that part of the country, which were very beautifully undulating and hilly, were very good for certain kinds of farming products. The main one is vineyards, ultimately making wines. It was a kind of cultivation that occupied a lot of labor—hand labor. There was no such thing as machinery in those days.

"My father had his little farm, which was one of four little pieces that Grandfather gave them. And in turn each of them bought another little piece if they could. So did my father succeed pretty well, from a little small beginning to build it up to a reasonable, nice wine farm. Our farm, though it was not a big thing, had wheat, we planted corn, we had vineyards, we had hay meadows. A balanced affair. If you had a stable with animals, you had to have some hay. Our farm there consisted also of a very nice home. There was no such thing as wood homes. It was strictly brick and mortar—and solid."

Vincenzo Zeppa was, by all accounts, a well-liked man in Fubine. Though a disciplinarian, he never raised his voice to his children; young Giuseppe, or Joseph, received only one whipping for neglecting his duties, he told an interviewer many years afterward. Chris Zeppa remembered the elder Zeppa as "an average good father: he expected the children to behave, go to school, be respectful of old people, and keep clean." Both parents wanted their children "to learn all they could in school and on the farm," and they encouraged this "without being tyrants about it." As Chris summed it up, "We thought our Mama and Papa were the best!"

The closest the Zeppas came to learning business while in Fubine was through the transactions that might be a part of daily life, which invariably involved knowing the true worth of a commodity or an object which was being bartered, sold, or purchased. "All farmers had to have certain knowledge to deal with tradesmen," said Chris Zeppa. "What does a sack of wheat cost? What is it worth? All the transactions were on this same level." Although such experiences might be elementary, the emphasis was on mastery of basic economic values.

"We all made our living on the farm because that was the only thing, especially for the young people," said Chris Zeppa. "They just followed a pattern of growing, going to the little school in town, and it happened that we lived in a little neighborhood just almost a mile from the center of the town. So our going to school meant [that in] sunshine, rain, or snow, we walked a mile to the schoolhouse and back.

"It wasn't a question of doing a lot of study, because schools in those days were pretty well restricted to whatever they had. In my time it was still five years for the boys, three for the girls. So if we paid attention to what we were doing in school, in five years we went from the one, two, three, four, five, and we were graduated. That was the end.

"So in those days it was truly what you would call, in English, a rural life, strictly rural, [although] the town at one time reached a population of better than 4,000—*before* the emigrations began to get started."[9]

Some of the conditions under which they lived were never forgotten. Nearly 80 years after he was born, Joseph Zeppa recalled those early days in a letter to his younger sister Paola.

"When I visit Fubine and see the changes that have taken place in the mode of living, I cannot but think of the days when you and I were small and spent several months in the stable where the cattle kept the place warm," he wrote. "We never had enough wood to keep the rest of the house comfortable except for the kitchen, and we had to have a fire there in order to do the cooking. What changes have taken place!"[10]

Giuseppe started to school at six, as did all Italian children in the local tax-supported schools. "He was an exceedingly bright student," said his brother Chris—and that statement has never been denied. Years afterward,

Joe Zeppa said that once he was out of school he worked as an unpaid apprentice in carpentry and cabinet making.

"I do know that he took some lessons in woodcarving, by attending a shop," said Chris Zeppa. "They would let lads come in and just learn the business. At one time he designed, himself, a two-inkwell holder made like a double chair with a back to it. He was excellent in drawing pictures on a piece of wood. Then with a very sensitive little bitty saw he could make all sorts of flowers and things."

By the turn of the century, at which time the bulk of the *émigrés* were beginning to pour forth from the South, emigration was an established fact in northern Italy. In the province of Alessandria, Chris Zeppa recalls, the migration began early, before he was born. "I recall that some of my kinfolks reached the United States before the turn of the century," he said. One aunt—Vincenzo's sister—and her husband settled in New England.

In those days an Italian male could not leave the country until after his twentieth birthday because he had first to satisfy his military obligations. When Carlo Zeppa, the first-born, reached 20, the lottery system excused him from military service and he was permitted to leave. Charlie, as he was to be known in America, needed a sponsor in the beckoning land of opportunity, and his uncle, already working in Connecticut, provided the assistance. He left for the United States in 1903.

With Carlo's departure, Giuseppe was the elder son still in patriarchal Italy. Not long afterward, Vincenzo, while building a house, suffered a severe injury to his leg when the large stones from the structure collapsed and fell on him. While he recuperated, he depended upon his ten-year-old son to carry out for him some of his responsibilities. Although Clara Zeppa, as would be expected of a country woman at the time, also assumed some of her husband's work, to a certain extent Giuseppe, as the elder son, "took over" some of his father's role during the slow recuperation that left Vincenzo with a limp. Though Giuseppe's "taking over" may have been more symbolic than actual, it no doubt encouraged the boy to think in terms of adult responsibilities at a tender age. When afterward, Joe Zeppa said, "I thought, when I was 12, that I was a man," he was referring to this time in his life.[11]

As the father Vincenzo returned to normal health, albeit with a limp, it became increasingly evident that Giuseppe's future would be severely limited in Fubine. The Zeppa family, it appears in retrospect, had gone about as far as they could there. One suspects that Vincenzo Zeppa saw the handwriting on the wall and encouraged the emigration of his children.

By this time Giuseppe was 12 years old.

"So Joe was the one that came to the United States as a little lad because brother Charlie, the older one, decided that Joe could come to America and go to school and make something out of himself besides being a farmer over in Italy," said Chris Zeppa. "The Italian law at that point had already modified the rules to more or less favor the emigration, because our population was just too big. And the economic situation of the 'caller'—in this case brother Charlie—was satisfactory enough that the young lad could be entrusted to him and he could be totally responsible. At the same time, they had to put in a so-called money guarantee, that if anything happened, that lad had to be shipped right back to where he came from."

By now Charlie was married and successfully and steadily employed as a waiter. All of the family, along with other young Italians of the time, had, as Chris Zeppa said it, "an ambition to have a better life than [was] available back home." By this time the United States, rather than Turin or Rome or one of the other large cities of Europe, was the magnet because of the stories that others had sent back in person or by mail.[12]

Unlike so many of the southern Italian *contadini* who sought money in America that they could take back to Italy, the Zeppas seemed inclined, from the beginning, to seek permanent residence in the States.

V

There is a copy of Giuseppe Zeppa's passport, dated May 29, 1906, which permitted the 12-year-old boy (*condizione: contadino* [status: farmer]) to leave Italy for New York. It was not only that brother Charlie had sent for him and that father Vincenzo was willing for him to go; young Giuseppe wanted to go.

Vincenzo accompanied his son from Fubine to Genoa, the same port city that produced Christopher Columbus. Vincenzo, his son related years later, hated to see him go. Giuseppe booked steerage passage on the *City of Milan*. A steerage ticket from Naples to New York in 1904, which should be comparable to that from Genoa in 1906, cost from $27 to $39, depending on the ship and other factors. Though this does not seem like much today for passage halfway around the world, it was very high at the time. In some parts of Italy a grown man might work 100 days, if he could find the work, and earn only $35. When Chris Zeppa followed 16 years later, the voyage took from 20 to 30 days by steamship, and first class fare—which by then he could afford—was $125, and steerage "practically nothing."[13]

Nuova York was, to an Italian of that time, *La mecca del dollaro*.[14] To a people overrun by hardships and overpopulation, the city was a godsend, a fluid society in which, as Thomas Kessner has observed, "opportunity was not so oppressively imprisoned by the past. It was allowed to unfold in the spontaneous atmosphere of an expanding capitalist economy."[15]

Most immigrants, however, idealizing America, set their expectations unreasonably high and, as a result, faced rude shocks when they encountered the slums of Manhattan's East Side where the new immigrants usually went to live. The Fourteenth Ward in lower Manhattan became known as "Little Italy," with communities within it representing different regions back in Italy. This generally comprised the section between Seventh and Twelfth Avenues, from around Thirty-fourth to Thirty-ninth Streets. The part known as Hell's Kitchen, a rough-and-tough polyglot congregation with more than its share of gangsters, was in the Thirties and Forties. It was in this part of Manhattan that Giuseppe went to stay with his brother Charlie and his wife Jenny. By 1900 New York City had 145,000 Italians; by 1920, there were 391,000, with the bulk of them compressed within the Little Italy district. [16]

Though the golden streets many immigrants expected to find in New York had tarnished by the time they arrived, both Italians and the Eastern European Jews steadily advanced themselves as groups. The Jews moved upward quicker and more often. If an Italian remained in New York City a decade, his chance of going from blue-collar to white-collar status was 32 percent according to one study; the Jewish rate was 41 percent. [17] The Italian figures are probably lower partially because of the tendency to seek means of making money relatively fast, usually as blue-collar workers, so they could take their savings back to Italy.

Charlie met his young brother at Ellis Island when the *City of Milan* landed. The boy was bewildered by the strange, teeming metropolis. He knew no English. It was reassuring to see a familiar face and to have a place to go and plans to carry out. But for 12-year-old immigrant boys of that era, there were no leisure-filled vacations. The first summer in this country, Joseph, as he became known almost immediately, went to visit his uncle and aunt in South Glastonbury, Connecticut, near Hartford, and picked berries and chopped wood. Though he had the summer to start learning English while he worked, he hardly knew the language by autumn when he returned to New York City to begin school.

That fall, back in Hell's Kitchen where he lived with Charlie and Jenny, Joseph entered Grade 5-A in Public School 32 in back of Hammerstein's Manhattan Opera House, on West Thirty-fifth Street.

As he put it himself, "I went to school for two years. I started in the fifth grade and I graduated in the eighth, in two years—I skipped a lot of classes. And I couldn't speak English, of course, so I got to learn it all.

"In 1906, Italians were not very welcome in this country, and they used to call 'em wops and guineas and what-not. And, of course, the fact that I couldn't speak English didn't make me particularly welcome.

"But I had a very fine principal, and he became, for some reason, attached to me, and he wanted me to continue going to high school." [18]

The principal of whom he spoke, a Dr. Ayers, had a deep influence on young Zeppa's life, and the grateful immigrant remembered him fondly as long as he lived. Despite Joseph's intelligence and determination, it probably was not easy for a bright Italian boy to be "discovered" in one of America's urban jungles. In New York, ethnic differences tended to ensure that Italians, bright or not, were apt to go unnoticed. Few counselors or administrators were Italian, and one of Irish, Jewish, or Anglo-Saxon ancestry was less likely to perform the function of discovering incipient Italian geniuses. There was, as Lopreato has said, "no one to encourage them to reach for the moon."[19] For this reason, Principal Ayers himself must have been a remarkable man.

Without in any way detracting from Ayers's role, it must be said that it would have been difficult not to have known of Joseph. The boy's achievements would have caught any teacher's eyes. It must be remembered that at first, because he did not speak English, Joseph did not understand the instruction. He soon learned to take his lead, in ambiguous situations, from the boy who sat in front of him in class. But he was a fast learner, and once he had begun to close the language gap he practically skyrocketed up the educational ladder. He moved rapidly from Grade 5-A to 5-B, and, as he mentioned, within two years he had satisfied the requirements of four years of school and was graduated from the eighth grade. Not only this, but he was at the head of every class and graduated with honors. Along the way, his artistic skill won him a five-dollar gold piece as top prize for drawing a picture of a boat.[20]

Among Joe Zeppa's scattered memorabilia at Pinehurst Farm as late as 1980, there was a diploma from the Department of Education in New York City, stating that "Joseph Zeppa has satisfactorily completed the course of study of the elementary schools and has the approbation of the principal and teachers and is entitled to admission to any high school in the City of New York." The date was June 30, 1908.

Although Joseph acquired his high intelligence at birth and did encounter adults who encouraged him, we must not lose sight of the fact that he also was from *northern* Italy, where values were likely to be closer to American standards than those of southern Italy. Despite his *contadino* origins, once he acquired the language he may have been as well prepared for American city life as any native-born peer, for he was endowed with a quality—determination—which, paired with intelligence and a fair amount of good fortune, may overcome a great many barriers.

We do not know exactly how Joseph filled his days and evenings while in school, but we can be certain he kept busy. For one thing, he sold newspapers on the street in an atmosphere in which Italians were made to feel like aliens. Undoubtedly, there were chores to perform at home, as well. But he must have had some time to himself. He liked to read, and he probably read a great deal. In addition, he cultivated friendships from those days that were to last him a lifetime. The Schmiedekamp family was one he was to remember. One of the Schmiedekamp boys was a close friend, and the parents were particularly kind to him. Many decades later Zeppa wrote to the brother of that friend: "In fact, it seems that your flat over the saloon on 9th Avenue and 35th Street was a second home to me, and I spent many hours there when not in school."[21]

Any dreams of a rosy future which Joseph may have entertained, however, were hemmed in by economic necessity.

"After I graduated from grade school," he said several decades later, "I had to go to work because my brother was working for a living, and he couldn't afford to send me through high school. So I went to high school at night and started working. I worked in various jobs, just doing this thing and that. So I scanned the papers and found work, one kind or the other. Sometimes it lasted very long, sometimes it didn't."[22]

Thus he summed up a series of jobs. What proved to be his "big break" was one of those jobs which was temporary, as an office boy at Shearman and Sterling, a law firm and financial counselor on Wall Street, filling in for the other boys as they went on vacation. Obtaining the job was no accident. Charlie's wife Jenny, also born and reared in Fubine, was working as the personal maid of Mrs. George H. Church, whose husband was the financial man of Shearman and Sterling. Through Jenny, Mrs. Church became interested in the bright little boy who was looking for work. "Mrs. Church loved that boy just as if he were her own," said Chris Zeppa. The Churches had no children of their own. That summer Joseph operated letter presses, washed inkwells, and ran errands.

Out of a job in the fall when the other boys all returned from vacations, Joseph read the want ads again. Car fare was scarce, so he walked miles and devoted considerable time to finding work. At one time he worked in a loft factory, making beds for the New Englander Company, his hours from seven in the morning until six in the afternoon, with 25 cents an hour for overtime, always welcomed. With the job went a free lunch accompanied by a dime glass of beer; he gave the beer to his friends.

Later he found a job as an errand boy at the New York Society Library, the oldest library in New York. He would pick up new books, cut pages, and deliver books in all kinds of weather. This, too, became an opportunity because he made it so: he read a great deal, including many old books he might otherwise not have had access to and magazines and newspapers. As a result of his deliveries, he became acquainted with New York City in a way he otherwise would not have. Throughout his youth he read continually, whenever he could find the time.

Sometime during this period, when he was 14, he met another poor boy from Hell's Kitchen who was to become a close, lifelong friend—George T. Keating. Since Joseph was 14, that would have placed this meeting between August 1907 and August 1908. Thus they most likely met in school, or possibly in the summer of 1908 while Joseph was working as an office boy in the financial district. There is one account that Zeppa and Keating met while working in the loft factory, but it may be that both already knew each other by then. At any rate, they remained close friends, with each helping the other and keeping in touch ever after. Soon afterward, if not at that time, young Keating went to work as a messenger boy for Moore & Munger, a company dealing in clay products. Moore & Munger was not on Wall Street, but was close to the financial district of Manhattan. Keating, an extremely hard-working individual, spent the remainder of his business career with Moore & Munger, rising up the ladder into managerial status and eventually, with partners, acquiring the company.[23]

A turning point came in Joseph's life in 1909 when he was 15, when George H. Church, who had been keeping up with the boy's activities through his wife who was regularly briefed by Jenny Zeppa, offered him steady employment at Shearman and Sterling as an office boy. Young Zeppa remained there until 1917 when the United States entered World War I. The job was a continuation of his education in many ways. The firm transferred stocks to brokerage houses, so early in life he came into contact with the financial world that is the heartbeat of America. Shearman and Sterling counted as clients men like Henry H. Rogers of Standard Oil. As he observed the parade of high-placed personages in the business world, he had the opportunity to learn what made them tick and, perhaps, internalize in his own psyche some of the characteristics which may have guided them toward success. He earned $3.00 a week. What he learned and the men he came to know added up to an incalculable plus.

Though he had been forced to drop out of school after the eighth grade, Joseph now enrolled in night school at Stuyvesant High School, to take woodworking and study machinery. He had no specific plans at the time, but as one thing led to another, he continued to attend night school, completing high school and going on to study finance and accounting at New York University. For eight years he went to night school, as clear a proof as one might ask of tenacity, intellectual curiosity, and orientation toward achievement.[24]

VI

Of all the persons who touched the life of Joseph Zeppa, few, if any, did more than Helen C. Adams, the administrative secretary to John W. Sterling, a partner of Shearman and Sterling. Born in St. Louis and a descendant of President John Adams, she had taught school in Denver before moving to New York where she pursued a career as an actress. She appeared with Richard Mansfield in *Julius Caesar*, was in *Mary Magdalene,* and was a member of the original cast of *Ben-Hur*. She also was in the Italian dance group of the *Ziegfield Follies* which featured the Florodora

Sextet. She left the stage in 1904 and became a secretary with Shearman and Sterling. In 1908, Sterling established the Miriam Osborn Memorial Home Association for aging women, and she continued to manage it after his death in 1918, investing the home's endowment funds so judiciously that she became recognized for her financial acumen. After leaving Shearman and Sterling she served variously as trustee, treasurer, executive superintendent, and secretary of admissions at the Miriam Osborn Memorial Home in Rye, New York, until her retirement in 1947.[25]

Miss Adams was 16 years older than Joseph Zeppa, which means she was 31 when he started his regular employment there at age 15 in 1909. She not only encouraged him in various ways but also helped him improve his English. She had been an actress and could provide tips on how to eliminate his accent. As a secretary she gave him guidance on both his grammar and his written language, an impact that was to be long lasting.

He fitted comfortably into the heady Wall Street environment. The intricacies and complexities of finance attracted him from the outset. After studying finance and accounting at New York University, he began to audit accounts. He had found his niche, at last, and he proceeded to sharpen his skills as an accountant. His faculty for numbers was so great—some of his friends later told his stepson—that he could memorize amortization tables on the way to the bank. By 1917, while he was still only 23, he had been promoted to cashier at Shearman and Sterling.

Busy though he was, he always found time for interests other than work and study. He belonged to the West Side YMCA, and he engaged at various times in sailing, camping, golf, and tennis. In the process he developed friendships, many of them to persist for a lifetime. He approached his leisure hours as enthusiastically, and apparently as profitably, as he did his work. He was never idle, he recalled in later years—a condition that never changed up to the day he died.

He seems to have pursued an active religious life; though reared a Roman Catholic, Joseph joined the Baptist Church in New York and was baptised. He, in fact, taught a Baptist Sunday school at a mission there before he was 20 years old. At some point during this period he joined Calvary Baptist Church in New York—John D. Rockefeller's church—and never moved his membership even when he later left the city. Technically, he remained a Baptist to the end of his days.

A number of Protestant churches worked at proselytizing the newly arrived Italians throughout the country. Most of these had facilities or programs aimed at the new residents of New York City. Using English-speaking missionaries, the Baptists organized Italian Sunday schools as early as 1889 in parts of New York state. By 1918 the Northern Convention of the Baptist Church had 82 Italian churches or missions in the United States, with 2,750 members. Four such churches were in New York City, with 680 Italian children registered in Baptist Sunday schools there.[26]

"As a boy in New York City he was quite active in the Baptist Church," said his son Keating V. Zeppa, "because they were the only ones who sort of took him in. The Baptists were apparently strong with their training unions, their Sunday schools, their sports activities. They had summer camps up the Hudson River, this sort of thing, and if I'm not mistaken, somewhere among his personal effects we still have a couple of attendance buttons from Baptist Sunday school, Baptist Training Union."[27]

Undoubtedly Joseph's interaction with other Baptists, lay or clergy, had a strong bearing on his affiliating with that church. At the same time, by joining a Protestant church he took a large step toward assimilation into the larger, native society. Becoming a Baptist made him more "American," and less a foreigner.

Another early decision Zeppa made was in his political preference. He was a lifelong Republican. Although businessmen of the stature he was to attain tend to vote Republican, he may have become attuned to the Grand Old Party not only by virtue of his Wall Street associates but also because many Italians did gravitate toward the Republican party in those years. The predominant reason for this seems to be that the Irish controlled the local Democratic machinery in the major cities, including New York, a fact which tended to repel many Italians.[28] In all likelihood, there was probably more than one influence which led him to make his political preference.

During these years he took positive steps to ensure his permanent residence and citizenship in the United States. In 1913, when he turned 20 and therefore had reached the eligible age for military service in Italy, he went to the Italian consul in New York and secured a release from this obligation. A printed form, directed to those men who had emigrated from Italy before the age of 16, released them from service in time of peace.[29] Less than three years later, on June 30, 1916, when he was 22, and ten years after arriving from Italy, he declared his intention to become a citizen of the United States, renouncing "forever all allegiance and fidelity to any foreign prince, potentate, state, or sovereignity." He further had to certify that he was not an anarchist, not a polygamist, and that he intended to reside permanently in the United States.[30]

By this time Charlie and Jenny Zeppa, and Joseph along with them, had left Hell's Kitchen and the ghetto-like environment of Little Italy for an upgraded life style in an apartment at 18 West 125th Street, just east of Fifth Avenue—a choice place to live at that time. Charlie by now was in charge of a large restaurant downtown, and Joseph was an up-and-coming young man at Shearman and Sterling.

It was during this period, around 1916 or 1917, that Joseph began dating a younger woman, Anna Martin. George Keating had fallen in love with her older sister Harriet, or Hattie, and Zeppa became engaged to marry Anna, an ethereal, very attractive woman. Keating went on to marry Harriet, but Zeppa's love affair was, as it turned out, star-crossed. Anna contracted tuberculosis and died at 19. Her death apparently occurred before the United States entered World War I, which would have made Zeppa 23 at the time.[31] There is nothing left behind in Zeppa's personal effects to indicate how deeply this loss affected him, but one can assume it was a difficult blow.

Reviewing these younger years in which Joseph Zeppa worked on his accent, scurried through grammar school, continued his education in night classes, and joined a Protestant church, the evidence strongly suggests that his assimilation began there in Hell's Kitchen, soon after the ship docked.

"Of course, who can say what's in the mind of a 12-year-old," said Keating Zeppa, "but I have got to believe, from Dad's life and looking back over it, that he didn't come to this country to continue being an Italian. He came to this country to be an American. And I can't think of any aspect of his life that would argue against that. [Later on] Italian was never spoken at home. Of course, Mother didn't speak it, not being Italian. He didn't marry an Italian woman. He didn't stay in the ghetto, if you will; he didn't stay in an Italian community."

It might be pointed out that his first serious romance was with a woman with an Anglo-Saxon name. Furthermore, he went to work in a field foreign to immigrant Italian experience at the time. One might suspect he was the only Italian farm boy on Wall Street at the time, in any role.

"A Jewish immigrant coming at the same time he did could find any number of Jewish firms, for example, to work in," said Keating Zeppa. "Whereas I doubt that there were many Italian firms in finance. That's why he ended up in an Anglo firm, if you will, at Shearman and Sterling. I can't think of two more Anglo names."[32]

It is interesting to speculate where Joseph Zeppa would have ended up, had he continued in his job as cashier at Shearman and Sterling. It seems a strong possibility that he would have spent the rest of his life in New York, probably even for that particular firm. But the United States' entry into World War I interrupted his tenure with Shearman and Sterling and had a significant influence on the direction of his career.

VII

When the United States declared war on Germany in April, 1917, Joseph Zeppa, though not yet a citizen, was like most other young men of his age. He wanted to go overseas, right then. At first he tried to get into the Army's new Air Corps, but received a telegram (sent to "Joseph Zappa") which stated: "NO AUTHORITY TO ACCEPT MEN LACKING COL-

LEGE EDUCATION OR EQUIVALENT ADVISE YOU APPLY ENGI-
NEERS."[33] Zeppa enlisted in the First Reserve Engineers in May.

"In those days everybody wanted to go right now," he said in 1975.
"They wanted to join the Army today and be shipped out tomorrow! Well,
I joined in May. Actually I applied for a place in the Army in April, but
they had to get a waiver for me because my height was only five feet three
and three-quarters inches, and the Army had a rule that wouldn't take anybody who was less than five feet four inches! So the admissions officer
asked for a waiver. It was given to me. I joined in May and [after training at
Fort Totten, New York] we sailed out of this country in July. I remember
that we were on the ocean on French Independence Day—that's the fourteenth of July—and we landed in England, landed in Plymouth. Landed
where the people started from. It was terrible weather. It was raining."

He crossed on the troop ship *Carpathia*, and his contingent was among
the first American troops in England. Shortly afterward, in August, they
sailed for France, crossing the Channel and landing in Calais. There they
boarded a train and were taken to the front.[34]

Zeppa's outfit—Company E, 11th Engineers Regiment of the American
Expeditionary Forces in Europe—was at Ypres, the Somme, the Cambria
offensive, St. Mihiel, the Argonne, and the defense of Paris. After the Armistice in November, 1918, he studied French and international law with
other American soldiers in France. In all he was overseas for two years,
and he returned as a staff sergeant.

These two years are documented perhaps better than any other period of
his life. All during his overseas service he dutifully carried on a correspondence with Helen Adams, his old dear friend and tutor at Shearman
and Sterling. Miss Adams kept all of her younger friend's letters, and upon
her death in 1967, the packet of World War I letters was found and eventually returned to Zeppa, who kept them in his own personal files. Although
the letters undoubtedly give us only a partial view of the young soldier at an
impressionable stage of his life, they do give us a close-up of facets of his
personality which he revealed to Miss Adams. In addition, we have
Zeppa's notebook he kept after the war while still in Europe. These include
notes he took in French and in English while in classes at Toulouse University, and the notes he jotted until his return to New York in the summer of
1919.

His letters to Miss Adams not only provide a view of life near the front,
they also give a sampling of remarkably effective prose. In France he saw
miles of "ruins, desolation and devastation" on a battlefield: "trees
clipped off a few feet above ground or shorn of limbs and leaves; tumbled
walls just showing above the weeds; huts, dug-outs and trenches partly
standing and partly caved in . . . . Troop and supply trains hum over rails
in quick succession, and a constant procession of auto and horse transports

and red cross [sic] trucks rumble over the road to and from the front. Zig-zagging miles of trenches once manned by flesh and blood and now in the full possession of weeds feet high, are as silent lines telling the terrible story of a titanic struggle in which the Germans are surely if slowly being pushed and catapulted (by mines) back over the border which they so heedlessly regarded and dastardly crossed . . . . Here and in England the war is absolutely everything . . . . France is part a battle ground and the rest a military camp."[35] For one just past his twenty-fourth birthday, his powers of description were exceptional. His eye not only caught the significant detail, but his vocabulary and choice of words suggested a literary bent, partially the result of his voracious reading since boyhood, in a language not his native one.

He next wrote from a tent "on a knoll north of a French Cemetary and German graves, South of the Front and a Red Cross Clearance Station." He was in an advance party preparing for the rest of the regiment, close enough to hear the German artillery. He had witnessed two aerial battles. There had been some gas. He moseyed around a French cemetery and gave a vivid description of what he had seen.[36]

Apparently Zeppa was not the only member of his regiment who devoted much of his spare time to correspondence. He wrote Miss Adams: "After reading the newspaper reports about the 'Fighting Engineers,' the officer who does the company censoring remarked that we ought to be called the 'Writing Engineers.'" In the same letter we gain some insight into how important Miss Adams had become to him. "Don't you realize that I haven't talked to you in six months and that I was in the habit of seeing and talking to you most every day in the week for about seven years?" By then he had had a white Christmas in France, with a fine dinner to mark the occasion.[37]

Spring came, but behind the combat lines he seemed to fret. "This place is about as exciting as Sleepy Hollow on a Sunday afternoon during the days of Rip V.W."[38] He seems to have spent even more of his energies on his letters than appeared on paper. "I think out long letters before I fall asleep at night, but thus far they have remained ethereal," he wrote. And his thoughts turned to philosophy in that letter. "Is religion intensely real to the fighting man? Something is very real but I doubt if we call it religion. A great many of the wandering boys to whom 'home' was but a house; 'love' an expression much used on Valentines; 'mother' a busy woman who darned your socks, scolded you for running around with the giddy Jones girl and hanging around the street corner in front of Flanigan's; 'marriage' something one does after he is tired of being happy, do come to realize the true meaning of these things, and after the squabble is settled will make good use of their new knowledge. However, sad to say, there will undoubtedly be others whose experiences will not be as profitable."[39]

Zeppa never romanticized war, and saw things clearly—a good observer, if his letters are any indication. "Modern warfare is a monotonous humdrum thing," he wrote, but he had learned the region in which he was stationed. He knew the Somme district "like a book." He spent his leisure time, it appears, studying the environment about him, reading, writing letters, and in discourse. Miss Adams sent him books by O. Henry which he both read and shared with his comrades. ("The crown of the short story writers belongs to him. He can paint a picture with less strokes than any writer I know."[40] Actually, John W. Shearman had sent them to Zeppa, but Miss Adams had made the selection.[41]

On the Fourth of July, 1918, he visited Paris for the first time. It was a gala holiday. ("The atmosphere was surcharged with a spirit of comraderie undescribable.") A sports event attracted "Plenty of pretty girls, flashy colors, and fragrant perfumes." While there he met two old friends from New York, one an officer in the Red Cross, the other with the YMCA. Upon returning to his assignment, he found it necessary to emphasize his relative safety. "You can be within five miles of the line (closer than hospitals usually are), and after you are accostomed [sic] to the noise you can undress at night and sleep without knowing that there's a war going on. As you have noticed, we are enjoying undisturbed peace way behind the scene of strife. About as interesting as milking cows and raising chickens."[42]

He seems to have studiously taken every opportunity to see and experience everything he could in France, a country he came to love. He described the streets, the buildings—whatever he ran into. He sampled both the scenes and the victuals of the country. "Four and a half francs bought me a good meal at the smaller of the two hotels: soup, ham, beefsteak, *plenty* of salad, cheese, prunes, and a pint bottle of wine. Don't you think I should be ashamed to take credit for being in the Army? This is the life Reilly lead [sic] before the war."[43] Or he'd spend a few francs for biscuits and ice cream, or visit with a girl from Iowa at the canteen, where the entertainment involved a French soldier with two stomachs who swallowed a half-gallon of water, four bullfrogs, and three goldfish, then spat the contents back out, with the fish and frogs still alive.

Zeppa seems always to have managed to find ways to fill his spare time, despite his frequent lowered financial status. Sometimes he borrowed the little he might need. "You see, in my days of flowing milk and honey and francs I played the good Samaritan to the extent of about 200 francs, but now that I need it it's a case of 'try and get it.' I do—but not from my debtors."[44]

In a sense, his Army hitch was a training period for his business career to come. Because he served in Europe during the stresses of wartime, he gained a wide range of experience in judging character that might have been denied him at 55 Wall Street. "Since I entered the employ of our esteemed Uncle I had the opportunity to associate [with] and ponder over men of all ages, dispositions, inclinations, and qualifications," he wrote. "When we get together, I'll match them against those you have recently catalogued and will draw conclusions."[45] While it is doubtful that he saw this practice in assessing character likely to be helpful in a business way later on, it was one of the intangible assets that he gained from the war. Another plus, it should be said, was the spurring of a philosophic turn which seems to have been the logical development of a curious, enquiring mind: "Helen," he wrote, "what is war? Why is it? Who makes it? Who pays the cost and who suffers the heartaches?"[46] If he provided no specific answers in his letter, his mind at least was in ferment, probably in a way it would not have been had he spent the war in Manhattan.

But, judging from his letters, he seems to have had no firm postwar plans as late as six weeks before the Armistice. The Army was paying him 250 francs per month. He often was broke, but he took advantage of passes to travel and see what he could of the country. He noted that "my plans would be post bellum if I had any—meaning that I haven't . . . . just now I feel like being a hobo or just a little above it. In my different moods and tenses I have considered 55 Wall Street (if they would have me); a try at a college education; roaming around the world just to see things, etc. What I shall do will depend entirely on conditions at the time peace is declared, for I am quite confident that industrial and financial as well as social conditions will be much different than they were when I entered Uncle Sam's boarding house.

"You have probably heard me say that I have little use for money except for the pleasure it may give others in the spending of it. Consequently I hardly think that I shall go back to the canyon of the money grubbers just for the sake of making a mere living. I am of the opinion that I can do that in doing things which will be a pleasure as well as a duty. Of course, I may not be able to own a Ford and play golf, but that is not on God's schedule of a man's necessities."[47] At the time, he had been reading *Beloved Vagabond*, Kipling's *Kim*, and a history of France, and his thoughts often turned to the canoeing he had done on the Adirondack lakes in peacetime.

Already he had turned down a promotion to sergeant major at the regimental headquarters, because he preferred to stay in the field with his company. His commanding officer, Captain H. W. Holland, had already left the outfit, headed for the States, as the result of a promotion to major. Zeppa, due for promotion, was then recommended for artillery school.[48]

The last several weeks of the war were spent in movement and hard work "hot on the heels of the Yank advances." Shortly before the Armistice, he noted to another correspondent that "The place we are in at present was 'no-man's-land' for about four years, during which time the only things sowed appeared to be dug-outs, barbed wire, trenches, shells, etc." He was at Varennes, though he could not say this for the censor's sake, at the time. The nearby town was a mass of ruins, and "one walks thru weeds without even realizing that perhaps he is trampling over the tiled floor of the room wherein Louis Sixteenth waited to be returned to Paris as prisoner . . . . I often wonder what France's last king would think if he entered the town now and saw the Yankee 'M.P.' standing in the village square directing the never-ending procession of auto-trucks, guns, tanks, etc., with the nonchalance of a N.Y. 'cop.' I wonder!"[49]

In retrospect, it seems clear that by this time, when Zeppa was 25, he had already formed essential aspects of his personal philosophy. In one letter he stated his belief that membership in the YMCA could be a factor in future success, for there one would "meet, play, and fight with all sorts of boys from the best to the worst—the boys whom he will have to meet as men when he goes out in the world to make his way—and his experiences will make a stronger man of him. Swift current makes strong fish. The days and evenings spent at the YM are some of my most pleasant memories."[50] There in five words, he summed up a basic tenet: *Swift current makes strong fish*. By then he had already shown signs of being a "strong fish;" displaying independence, imagination, perceiving the world and people around him with realistic eyes, and working hard. One might argue, in fact, that he was offering as philosophy the reality of his own experiences.

Slightly more than two weeks after he wrote these lines, the Armistice was signed—"it is all over, including the shouting." He summed up his role in the Great War: "I am not surprised to find that I have little to cause an undue chest expansion. All I can say is that I did my best in the service I enlisted and—that's all!" He and his fellow soldiers were building huts for the winter. He intended to try to linger in France and perhaps Italy for some time after peace was settled "to get a good idea of their ways in business."[51]

Weeks after the Armistice he realized that his view of the war had been necessarily one from an unprivileged vantage point. "Events have transpired so rapidly that we have been somewhat like a boy watching a five-ring circus—bewildered and unable to fully realize the importance . . . each act bears to the entire show." He was still in Varennes and expected to remain there till the spring thaw. "We are mighty comfortable, as fat as ground hogs, and soon there'll be enough huts for all of us—though most of us prefer to live in tents," he wrote Miss Adams. They were encamped on a former battlefield which the Germans had dug in and fortified and the French had mined. The trenches of the opposing armies were close enough to each other that one could stand at a lookout box and throw a grenade into the enemy's trenches. While the French had constructed their entrenchments of dirt and sand bags, the Germans used concrete and even had electric lights and running water.[52]

A few weeks after the Armistice, Joseph's mind turned tentatively toward an assessment of his wartime service and the postwar world. He believed that "freedom is only a relative term; I have found it in the army . . ." When he turned to his prewar employer, Shearman and Sterling, he commented, "In the first place, I have not yet made up my mind; and in the second place, if I come there it will not be as a slave to Debit and Credit and Balance. Though I worked hard I never lost the use of the pronoun 'I,' and do not intend to do so in the future." He tentatively considered working his way toward the top there, but noted that older men seemed to be standing in line. Even then, so far from 55 Wall Street, he seemed to harbor doubts that he could commit himself permanently to Shearman and Sterling. And while a decision was possibly building, dependent upon the conditions he found upon his return, he still hoped to travel more in Europe. He was waiting for approval of his application for a leave to visit Italy.[53]

In early 1919, by which time most American soldiers were home, Zeppa entered Toulouse University under an educational program for servicemen. He studied French and international law. What we have of his correspondence indicates that his letter-writing flagged as he entered serious study at the university. Instead, we have an assortment of notebooks, mainly filled during his classes in international, constitutional, and commercial law; and some notes in French, most in English. There are notes on lectures on torts, banking, commercial associations, contracts, maritime law. In little four by seven-inch notebooks with ledger-like pages, he took well-organized, thorough notes on topics ranging from prohibition to marriage laws.[54]

But in the midst of the school notes, he also left a record of some of his personal thoughts. In what apparently was a first draft of a letter he wrote to George H. Church back at 55 Wall Street in late April, 1919, he reexamined the question of his return to Shearman and Sterling. He apparently had been offered a job upon his return, and he worded his response carefully.

"I enlisted in the Army voluntarily and without legal obligation, nor without other consultation than personal introspection," wrote Zeppa in his notebook draft. "Consequently when I left your employ it was with the clear impression and understanding on my part that I was leaving for good and that there was absolutely no obligation toward me on your part. Therefore when your letter reached me shortly after the armistice it gave me much pleasure and a feeling of appreciation for the kindness which prompted the offer.

"However, since no mention was made of it I could not help to wonder in what capacity I would return to your office. My common sense told me that having been from office matters for two years and being the youngest in years of age and time of employ I would have to take my place at the foot of the list. This is only fair, yet the prospect did not appeal to me over strongly—at least not on the basis of permanent employment. But since I must work as soon as I get my discharge I shall naturally come to you first since you are the one who knows me best, and if you have any opening I'll put my hand to it—if not I shall ask you for a recommendation."[55]

All was not work, though. He visited Italy that year, returning to the old hometown of Fubine for the first time since he left in 1906 as a boy of 12. After 13 years, he saw his mother, his father, and his younger brother Chris, home on leave from the Italian army. It was a glorious homecoming.

VIII

Chris Zeppa, five years old when Joseph had sailed to America, remembered little of him. One day brother Giuseppe was there with the other children, the next day he was gone. And as the years passed, Cristoforo, or Stofu as he was sometimes called in Italy, had grown up much as Joseph might have if he had remained. He had completed the limited schooling available in Fubine. Soon afterward the wars had come, halting all movement from the motherland. First, Italy's war in Libya disrupted the plans of any would-be emigrants. Then in 1914, World War I erupted. That August, Chris was 14 years old, not yet old enought to be called up but old enough to know what to expect if hostilities persisted long enough.

"Military service used to be at 20 years old, but the war changed all of that," said Chris Zeppa. "They could call you at 15. So I couldn't get out of Italy anyway on account of all that situation. I went into the Army in 1917 in April. I was called way in advance. So I entered the Army at about 16 years old, a little over, inducted the same year, and served in it, training and so forth. By the time I got through the training and got ready to move into the trenches, why, the Armistice was registered.

"So I'd never got beyond the lower trenches in northern Italy. I was inducted and I spent my training time in the town of Bologna, Italy. Not too far from the East Coast. Not too far from Venice. In fact, we were so close to the trenches there that we could count every shot fired. At one time some planes came in on our encampment—some German planes and Austrian planes—but luckily they missed the whole bunch and there was no killing. In effective service in the trenches I saw but very little.

"So the war ended pretty quick there, and [Joe] came home—he didn't stay there long—and our class, by the way, being still young, they let us go home and stay home six months and then come back and finish our military service. So I had the six-month leave, and I went back into the Army and I was released from the Army. And that was the end of the big wars. So I went back to the farm home and helped Father doing the farm business.''

A few days before Chris left the Army an unusual episode occurred. By then he was in the Italian Air Corps, with the *Terzo Gruppo da Bombardamento*, or Third Bombardment Group. He worked part-time in the group office, so the other soldiers around the *caserma*, or quarters, knew him well. One day while he was out, his sister Pauline, just back from the States, dropped in to see him without writing him in advance. When he returned to the camp, the guard at the gate told him of his visitor.

"You know how soldiers are, young—they make a big to-do about any girl, and Pauline was a real pretty woman. The guard told me, '*Zeppa, una signorina bellissima!* Man, that was a dame if I ever saw one!' You know, they used to kid one another. He told me the hotel she was staying [in].

"So I gave her a call and then later we had dinner together. But two days later after that I was already scheduled to leave the Army for good. So she stayed there till I was ready and we went home together.

"That was quite a deal. And Pauline stayed until—I forgot what year it was—she returned to the United States. I myself, after I got over the Army and one thing and another, began to look around. I said, 'Now that we've got the war finished, what is it to do, besides farm? Farming, that's all.' Naturally we were all enthusiastic about the brothers and sisters here in the United States! So I about decided then that since everything was going good in the United States, the brothers were successful, so forth and on [I would go]. And Father reluctantly said, 'Well, if you must go, we'll make out.' That was about the most critical thing that I've ever done. Naturally, I must say it was a sorrow parting between me and my daddy, my mama. Course, Pauline was there at that time. And that made it a little easier for them to turn loose.

"Incidentally, at that time two of the children of my sister Gilda were still there. Mama raised them. And I made plans then, and in 1922, in September if I'm not mistaken, I left Fubine with the boys of my sister Gilda and took the *Giuseppe Verdi* ship to the United States. And that ended the episode of a youngster in Italy."[56]

IX

Near the end of his Army stint, Joseph Zeppa, now a staff sergeant, made the most of his waning days in Europe. Besides going to Italy, he took at least one other trip. At the end of May, 1919, he used four days off from the University of Toulouse to visit Arras and Cambrai in northern France and Brussels.[57] Considering the location of Toulouse, on the Garonne River in southern France near the Spanish border, his was a quite active itinerary, for he had to travel north of Paris and into Belgium.

By early July he was headed back to the States, and in his personal notebook he documented the return passage on the *S. S. Dakotan*. It is a notebook such as one might find among a professional writer's effects. Zeppa recorded his last hours in France and jotted down items of local color on the "tramp boat," as he called it.

In a section he labeled "Day in and Day Out," under the date July 1, he noted that "Yesterday we kissed half of Toulouse and waved the other half good-bye . . . . One began to realize just how many attachments the past four months had produced. Young girls & old mothers stood by the trolly, tears in their eyes, chatting & wondering when their departing American friends would return. Kids of all ages stood along the way, shouting a warming good-bye, and which their most superficial thoughts were expressed by some of the gayer hearts as they waved bye-bye and shouted *'fini zen zen gum* [the last of chewing gum]!' I have no doubts the kids were also thinking of the many hours spent in the parks playing with 'Ze Americaines' who were wont to bounce them up and down on their knees and feed them chocolates.

"The girls felt the parting most. Never before had they been so joyfully treated, catered to and allowed to have so much freedom in their promenades, dances, etc. They had had a good time without feeling that they were supposed to be toys for the men.

"Mme Laffitte filled my canteen with some choice wine, and after making many promises to write and to be sure to go to see them again when I returned to France, I kissed her and Jeannette good-bye, picked up my pack and made for a car. While waiting for the car, the Barrouse family came out and I had to go thru the good-bye scene for the second time . . . ."[58]

At 2:00 p.m. on June 30, the train pulled out from Toulouse. It was the morning of July 3 when they woke up in St. Nazaire. On July 9 their ship left for the States.

As the ship steamed out into the Atlantic, his thoughts turned toward summing up his Army experience. On July 14, 1919—French Bastille Day which he had marked two years before in the Atlantic, going *to* France—he wrote:

> Two years ago today I set sail, as the historians would put it, for the western shore of Europe in search of fame and adventure and to fight for the cause of liberty. After two years of military servitude, I am returning to my adopted country with four gold service chevrons to show for my labor. Tis the only souvenir I could afford to buy and will serve as the reminders of the gold I lost while searching for an ideal. Was it worth it? No doubt!
>
> But can I help wondering why so many others should enjoy the benefit of our efforts and likewise the abundant gold accrued

to them by the very fact of our presence on the field of battle? Will the time ever come when the men who are willing to sacrifice their lives will not also have to sacrifice their future and that of their dependents? Will the day ever dawn when the able-bodied stay-at-homes will not be given the financial advantage of those who go forth to stop the enemy's bullets?

Even the Glory of Victory has been stolen from the soldiers. One reads more of the deeds of those of who 'did their bit' at fabulous wages for eight hours work, far from the noise of guns, than of the achievements of those who struggled 24 hours each and every day in the face of death and disease.

Oui! they did their bit, these sons of the steam hammer and drill, and now they can sit back and laugh while those who did their duty at $1.10 look for jobs they themselves would not take.

I am not angry; I am just thinking in plain words and stripping the thin sugar coating of sweet sentimentalism from the bitter facts.

I joined the army with an idea of sincere duty and not of gain.

I did my best and expected nothing—I am not disappointed.

And then, as an accountant might, he made two columns, for debits and credits, which he labeled "Profit and Loss a/c June 1/17 to July 14/19."[59]

Profits	Losses
Learn to read French RR time tables.	A job. $? 0 Respect for the officers of the USA.
Learned taste, color, & solidity of soil of France & felt its climate.	A wife (if I should have married). 26 mos of a useless life.
Learned that though all woman may have been angels, a hell of a lot of them have fallen from grace, & that if a lot of the men were devils, Satan need never worry about race suicide.	The certainty of what is evil & what is good for the other fellow. Some of my rare & beloved hair. Some of my eye sight. Guess I didn't lose much after all—outside of gold, which I never could keep.
Lots of sins are the creation of own mind—thinking makes them so.	I might be a CPA by now, but then I

continued on next page

Profits	Losses
A pair of hob-nailed shoes and an OD uniform.	shouldn't have anything to study about.
Some French friends.	
An appreciation of what it is to have friends.	
A love for a hat that is a hat.	
The knowledge that we men don't want military training.	
A hate for the man who inflicted the overseas cap on us.	
Knowledge that one can find a better friend in the house of the red lamp than on 5th Ave.	
That there's infinitely more good in the world than one believes, and likewise more evil.	
That France hasn't a monopoly on immorality, but that she does not try to deceive the world or herself as to what her condition is.	
France believes that if you want to do a thing, do it without being ashamed of it.	
A love for France, the land 90% of the soldiers will calumnize.	

Profits	Losses
All immaterial assets—the kind the bank will not accept as collateral.	

His account of his conversations aboard ship reveals a man with strong opinions who doesn't hesitate to argue his views. In fact, a penchant for disputation may be detected. The double sexual standard was one such controversial issue; he opposed the hyprocrisy it implied. His insight into the woman's lot was coupled with a talent for lucid observation. In America, he pointed out, one hides "all that is sinful and deceives ourselves that it does not exist." He took up for the French woman whom American soldiers blamed for lax morals: "women who have been living for four years under the hardest of conditions socially and economically—who have lost families, earthly possessions, prospects of marriage, etc. and who have had to slave for a livelihood—[in] a country where she is a pawn in the hands of man. People of America, how about our moral young men who went to France preaching the sanctity and the inviolability of womanhood and who practice just the opposite and who lost not only their sense of morality but also their sense of decency, fairness, and good breathing [sic]?

"Can persons violate the Christian rule of sex morality and still remain gentlemen and ladies in the full sense and upright citizens? The French believe 'Yes,' and they show it. The Americans believe 'No' and they show it.

"I am constantly drawn into arguments on this subject against multitudes of the short-sighted apostles of the man-propounded doctrines of the double standard. Man, the powerful ruler of the world is, they plead, the victim, the poor, helpless victim, of woman—the delicate little creature who god placed under the benevolent protection of man."[60]

Nor was this all he chose to defend.

"Yesterday I was almost mobbed for daring to argue in favor of the system of compulsory military training, and at that . . . under a plan which presupposes the reformation of the present army system into a democratic institution and the period of training to be very short. Primarily for the purpose of picking out and classifying the fit and assigning them to military units to which they would report in case of need. The plans call for the establishment of a small standing army, well paid but in which a man cannot serve longer than, say, five years, and schools for training officers who would serve a short period in the army and then would be relieved unless chosen for staff duty.

"The present officers seem to be the only ones in favor of the military training plan. The enlisted man will hear nothing with the word military in it."[61]

Arriving back in the States a few days later, he was mustered out of the Army and received a $60 bonus, with which he bought new civilian clothes. One of his first official acts as a civilian was to become a naturalized citizen, before the Supreme Court of New York County in New York City, on September 4, 1919. By then he had made his way back to 55 Wall Street where, true to his expectations, he found an older man ensconced in his old position as cashier. He did return to Shearman and Sterling, but the same dissatisfaction which had struck him earlier in France now manifested itself again. He had practically signaled himself, through his letters from France, that he didn't expect to remain long at 55 Wall Street if, indeed, he did return. It was a matter of time before he surveyed the conditions and made a more permanent postwar decision. As it turned out, a curious set of circumstances soon drew him from Wall Street and, eventually, into a new life and role entirely, one he hardly could have predicted from his tent near the French battlefields, but one for which his thoughts and his assessment of the world had fittingly prepared him.

3

THE ARKANSAS CONNECTION

I

Exactly 20 years to the day after the great Spindletop discovery near Beaumont, Texas, a wild oil boom exploded in El Dorado, Arkansas: Busey No. 1 blew in, late in the afternoon of January 10, 1921.

Enjoying the luxury of historical retrospection, it is easy to see, now, that the El Dorado boom provided a staging area for the founding and development of Delta Drilling Company in East Texas a decade later. Joe Zeppa and his later associates came to know each other during the 1920s in these oil fields. Men for whose companies Delta would drill were in El Dorado in abundance. Experienced oil-field men moved almost directly from El Dorado to East Texas, echoing the tendency of workers to follow the major booms. Some rushed in close on the heels of the East Texas discovery; others drifted down later. In the first decade or so of Delta's existence, the influence—both qualitative and numerical—of men who had earlier been in Arkansas was a dominant one.

In almost every way, El Dorado was a typical *roaring* boom town. The oil gushed from the top of the wooden derrick and announced itself forcibly to the town of 2,000. Capped two days later, the discovery well flowed 5,000 barrels daily. People of all descriptions and from almost every origin soon packed the little southern Arkansas town until 32,000 could be counted. By the end of the year more than 500 wells had been sunk, few of them dry. Workers, speculators, and tough guys slept where they could, some of them in chairs, some on the grass. Hijackers poured in along with entrepreneurs.[1]

It was in this setting that relationships were built which would prove crucial to Delta's early history.

II

Joe Zeppa arrived in El Dorado through an indirect route. When he returned from World War I, he went back to work for Shearman and Sterling. But his absence in Europe had made a difference.

"Well, I went back," he said. "I realized then I had been gone for 26 or 27 months, and they had rearranged their staff and I didn't feel quite at home. I knew all the people working there, of course, but they all had their occupation. The job I had as a cashier was occupied by that time—a fellow by the name of Teasdale who had been put in my place. Well, he was a man rather advanced in age—at that time, about 50, you know, was rather an advanced age. If you were 50 years old, you were an old man then.''

At this point, when Zeppa was ill at ease in his old surroundings and, no doubt, casting about in his mind as to what he should do next, his old commanding officer, Harry W. Holland, called him. Holland had been a captain in Zeppa's Army company in France, and Zeppa had thought highly of him. Apparently the esteem was mutual. Holland suggested Zeppa meet him for lunch, as he had something to discuss with him.

At lunch, Holland said, "Joe, how would you like to go to Italy?"

A bit surprised, Zeppa replied, "What for?"

"I've got a contract to go to Italy to work on a deal that has been started and is working fine, but it's got to be completed. It's for the purpose of converting the Italian railroads from Naples north to burn Mexican oil."[2]

Mexican oil was then selling for 10 cents a barrel, and this obviously presented a large opportunity to those involved in the transaction. The English entrepreneur, Lord Cowdray, who had begun developing Mexican oil production years before, headed the group which Holland had connected himself with.[3]

Holland continued, "I need a man who speaks Italian, because I don't, and you do."

The fact that Zeppa spoke both Italian and French no doubt brought him quickly to Holland's mind, but Holland's relationship with his former sergeant must also have left a memory of Zeppa's general efficiency, as well. If the offer of a two-year contract seemed like a godsend to Zeppa, at the same time Holland must have realized Zeppa's background was exactly what he needed. Not only did Zeppa have command of two helpful foreign languages, but he had studied international and commercial law at the University of Toulouse and had accounting experience on Wall Street.

"Let me go back and talk to the man I work for," said Zeppa, "and tell him I'd like to leave, because I can't see where I fit in there anyway."

The deal was made. Zeppa announced his decision to George H. Church at Shearman and Sterling, who expressed regrets at losing the young man, and in September 1919, the month after his twenty-sixth birthday, he was ready to sail for Italy.[4]

"We worked in Italy for about six months," said Zeppa. "We sent the equipment to convert two locomotives to burn fuel oil. They didn't know if it would burn fuel oil in locomotives. See, this was in 1919. They burned coal, and they were paying about $40.00 a ton. So that proposition was very good, you know, to sell them ten-cent fuel oil in place of $40.00 coal."

The group Zeppa represented as interpreter demonstrated the usefulness of fuel oil on a rail line between Naples and Rome, but other difficulties arose.

As Zeppa put it, "Well, we got her going pretty well, and we were working with the biggest bank in Italy at that time, the Banca Commerciale Italiana, with its principal office in Milano. And they wanted in on this deal. Shortly after that, the rate of exchange kept declining and the lira kept going down. On top of that, they had a election, and the minister with whom we'd been dealing was not reelected. So we had to start all over again. At the same time they had a strike of the post and telegraphs. So we couldn't get messages back and forth.

"It got to a stopping point where we couldn't go any place. The Italian company said, 'We can't go in and keep putting up more lira for the same interest we agreed to take.'

"Anyhow, to make a long story short, I came back to the United States to tell 'em the facts of life. And in order to get back to the United States, I had to catch a strike-breaking train and go from there to Milano and wait in the station for about a day and catch another night train that went from Milano to Switzerland. And then caught another train that went to Paris. I had no reservations. I got to Paris where I managed to give one of those ticket sellers a little money and I got a place on a French boat, the *Lorraine*."

One positive gain Zeppa received from the assignment was that he got to visit his old home and see his younger brother Chris. But the Italian business venture came a cropper, and he returned to the States in 1920. Mussolini was on the rise, and nothing would come of the attempt to sell Mexican oil in Italy.

"Well," recalled Zeppa, "I came back and I told the story, and they said, 'Well, there's nothing we can do now except wait a while and see what happens.'

"So I said, 'Well, what do you want me to do?'

"He said, 'Go on out and play golf or do anything you want to do.'

"I was really a golf player in those days. I had taken it up when I was about 17 or 18 years old. I had an old set of clubs, had about four or five clubs. You know, we didn't make much money, so we bought balls from those kids that hung around the golf courses.

"So I hung around there several months. Now this was in the latter part of 1920, and in early part of '21 they discovered oil in El Dorado, Arkansas. The man I worked for in the United States—his name was Payne—was from Virginia, and he officed with Henry L. Doherty and Company at 60 Wall Street. By this time Holland had come back and he sent Holland down there [to Arkansas]. He went there himself and bought a lot of leases. You know, everybody got excited. Big flowing wells, 2,000 feet deep—and it was a deep well then.

"So he then called for Holland to go down there and take charge of the operation and get some wells—get ready to start drilling.

"Meantime, I guess he spent most of his money buying leases. Apparently he was running short of cash. Anyhow, Holland then asked for me to go down, because he had to have somebody to handle the office, and I went down in June, 1921."[5]

Not realizing how foreign the Arkansas boomtown environment was to the orderly business atmosphere he had known in Manhattan, Zeppa arrived in El Dorado in June 1921, wearing a derby hat and a wool suit. The New York Central Railroad ticket agent hadn't even known the location. Not only was his attire rather warm for the climate, it also, as he soon learned, was hardly suitable for the clouds of dust that permeated the town. It was his first contact with oil exploration and a boomtown, and it was, at that time, his farthest excursion west.[6]

The Arkansas town was as wild as any, with all the attributes of the classic boom.

"In those days El Dorado was a boomtown," said Zeppa. "I mean, a typical boomtown. I've never seen one [like it] since. Gambling joints all over the place, in pine thickets and everyplace. El Dorado was just chuck full of promoters. They had these 'chicken houses,' you know, scattered here and there, where they stayed awake all night and slept all day, that sort of thing. You didn't dare go out without a gun. You didn't have any roads. You just drove through a field, and when you got stuck, why, you got pulled out by somebody [for a fee]. In fact, there was somebody waiting at every mud hole to pull you out! None of the streets paved, except right around the square. Whenever it rained, you just sank into mud, and if it didn't rain you got smothered in dust. And no water, except certain times of day. You'd try to get water around dinner time and you couldn't get it because everybody either was washing, taking a bath, or washing dishes."[7]

Probably the most graphic boomtown symbol, however, was the teams of mules, following heavy rains, regularly getting stuck in the mud of Washington Avenue in downtown El Dorado.[8]

Zeppa's employer, El Dorado Petroleum Company, soon found itself with gushers—and serious problems.

"We fought it out," Zeppa recalled. "Oil was selling for 35 cents a barrel. We had fifty 5,000-barrel tanks and didn't have money to pay for 'em. We filled them up and sold the oil for 35 cents a barrel. Well, that didn't even pay for the tanks."

The venture faltered because it became difficult to sell the oil.

"The company finally went into bankruptcy," said Zeppa. "Holland was appointed receiver."[9]

With his company in receivership, Zeppa hardly had an enviable position. He had $50.00 of his own, which he scarcely viewed as constituting the foundation of his future fortune. He soon restricted his lunch to two hamburgers, later reducing it to only one, as he cut corners any way he could.[10]

"After the receivership closed, Holland went back to New York. I saw him several times after that, but then I lost track of him."

As the work was finishing and Zeppa prepared to return to New York, Colonel T. H. Barton, an oilman from Texas whom he had met earlier, approached him with a new offer. Barton was associated with two other oilmen, Dick Sowell and W. M. "Morris" Coats, in a small gas company, Natural Gas and Petroleum Corporation.

"Well, one day [Barton] wanted to talk to me about some work, and so I went to see him. He said, 'Joe, how would you like to go to work for us?'

"So I said, 'I don't have any objections to it. I don't have a job, actually.'

"I don't know how much money I had. I don't think I had very much, because, you know, I worked for a receiver and the chances are that I was always broke. Because people would come in that needed some money and just take it out of your pocket and so forth.

"He said, 'How much you want per month?'

"I said, 'Well, Colonel, I don't know what I'm worth to you. Let me work for a few weeks and see. If I do any good, why, you can decide on how much you are gonna pay me.'

"Well, actually, I went to work for them [as secretary/treasurer] and they were in bad shape too. They had started without any money. They had one or two gas wells and pipelines to the field and they were selling gas to the drilling rigs."

But Barton and his associates were expanding too fast for their small gas company.

"So then I got up against the same situation I was in before: scrambling for money. And Colonel Barton went to New York and finally made a deal to sell control of the company. He sold the stock that Coats and Sowell [had given] him an option on—I think it was only $50,000 each—and he sold it to Cities Service. He got Cities Service in. Then they started putting money in. Well, they did all right then, but about that time they started a deeper zone, in Smackover, which was also prolific, so we were getting along fine."[11]

After Cities Service gained control of Natural Gas and Petroleum Corporation, the latter entity merged with Arkansas Natural Gas Company in 1928, with Zeppa the acting secretary/treasurer. Along with Dave W. Harris he directed the merger and set up Arkansas Natural Gas Company; this move sent Zeppa to Shreveport, its new headquarters, in 1928.

Zeppa was to maintain friendships with these men. In later years he was associated with the energetic wildcatter Morris Coats in several drilling ventures and Barton, who became a prominent El Dorado oilman who subsequently had a number of business associations with Zeppa. In 1929 Barton took over the management of Lion Oil Refining Company, which had its start in a small refinery near El Dorado shortly after the discovery well came in, but which had been headquartered in Kansas City, Missouri, since 1923. When Lion Oil merged into Monsanto Chemical Company in 1955, it boasted $150 million in assets. [12]

III

Bob Stacy, one of Zeppa's co-founders of Delta Drilling Company, also had, as indicated earlier, his Arkansas antecedents.

Stacy had arrived in Camden, Arkansas, before or about the time of the oil discovery in southern Arkansas, and in 1921, while he was operating a service station there, he married pretty Priscilla Powell, whose parents were prominent there.

Because the oil development became an economic magnet, Priscilla's father, Smead Powell, soon moved his law practice to El Dorado, 30 miles away.

"Camden was a dignified little town," said Elizabeth Powell Fullilove, Smead Powell's daughter and Bob Stacy's sister-in-law. "Smaller but somehow not touched by the oil boom. Daddy had practiced law there, but then with the oil boom he went to El Dorado, and then he would come in on the weekends. He commuted between the two towns because El Dorado was such a rough town at this time. You know, they didn't allow women out after dark, and it was an oil boomtown."

Stacy soon joined the procession to El Dorado, where he managed the El Dorado Gas Company, a local utility. He was in El Dorado by 1925—and probably before—for his son, Robert, Jr., was born that year. Priscilla Stacy went to a hospital in Little Rock in order to have the best medical facilities in the state. It was also around 1925 that Stacy's elder brother, Jean Stacy, managed the competing gas utility in El Dorado, and the brotherly rivalry was played up for all it was worth in the newspaper advertisements of the day.

One can see how easy it would be for Stacy, in this atmosphere and in his position, to meet almost anyone in El Dorado at that time. As Smead Powell's son-in-law, he undoubtedly came into contact with a great many of Powell's associates. Not only was Powell a prominent attorney, he also ran for govenor of Arkansas at one time and must have had a great many friends there. And with the advertising in conjunction with the "utilities war" in El Dorado, few people probably escaped knowing who Stacy was.

During this period Bob Stacy met two of his future partners: Joe Zeppa and Sam Dorfman.

And had it not been for a series of events, it is possible that Bob Stacy, with his newly-acquired ties to Arkansas, might have remained there. But both his father-in-law and mother-in-law died, and the Stacys took Priscilla's younger sister Elizabeth into their home. The stocks that Smead Powell had bought took a nose dive with the crash in 1929. Within a few years, Stacy's fortunes had plummeted.

Then, in 1929, Stacy and his family left Arkansas.

"At that time my father was pretty well busted," said his son Robert Stacy, Jr. "I think he had an illness and, for reasons that I don't recall, was not with the gas company any longer, and I believe leaving El Dorado was a necessity, because he acquired, through means I don't recall, a job in Los Angeles, California."

In spite of the vicissitudes of life, Bob and Priscilla Stacy knew how to live well—and they usually did, whether it was easy to do so or not. They struck out from El Dorado for the West Coast in a chauffeur-driven Cadillac. Stacy had a friend with the Texas Company who he thought might be able to help him land the job he desperately needed. In California, Stacy's Texas Company connection failed him, but he did renew his friendship with Joe Zeppa.[13]

IV

The connection between Joe Zeppa and his three Shreveport partners also began in El Dorado. After the boom came, Louisiana Iron and Supply Company opened a branch in El Dorado, which Sam Dorfman apparently managed. Dorfman seems to have had the full confidence of Sam Gold and Sam Sklar.

"He sort of ran it himself," said Goldie Rappeport. "Nobody knew what he did or why he did it. He was greatly trusted. He made a lot of money for the company and for himself, and nobody asked him any questions that I knew anything about."

Dorfman was one of the early ones to sell second-hand pipe to H. L. Hunt on credit when, as Sam Sklar remembered later, nobody else would do it. It was also in this period that Dorfman married Rose Gold, the daughter of Sam Gold.

It would appear to be almost inevitable that men like Zeppa, Stacy, and Dorfman would meet, once they had arrived in a place like El Dorado, for their businesses would tend to bring them together. And these men's personalities would also increase the likelihood of their getting to know each other.

"Joe and Sam were very, very social," Sam Dorfman's second wife said. "They both had the knack of making friends, and the right kind of friends that, you know, could advance them."

It is probable that Gold and Sklar, during business visits to El Dorado, also got to know Zeppa and Stacy in these days. In fact, it probably would have been difficult for them *not* to have met at that time.

Louisiana Iron and Supply maintained a yard in El Dorado long after the El Dorado boom had faded. Some of the pipe and other supplies were moved from El Dorado when the East Texas field opened in the early 1930s. The El Dorado yard was finally closed in 1947, by Albert Sklar and Seymour Florsheim.[14]

V

As it turned out, El Dorado was a good place to be in the 1920s if one were to move up in the oil business. Many latter-day executives and independent operators passed through the boomtown at one time or another. H. L. Hunt first entered the oil business in El Dorado; Zeppa met him in 1922. Back then Hunt was operating a gambling joint in El Dorado. Later he started in the oil business himself with one drilling rig.

Hunt also liked to gamble, particularly at poker. "He and Colonel Barton and Dick Sowell and a group in El Dorado used to play once a week," said Zeppa. "I never did [play with them]—I wasn't in their class. I couldn't afford to!" Sometimes the pots went as high as $1,500 or $2,000, a huge pile of money for that time.[15]

A crucial connection which Zeppa developed in El Dorado was with Lee Felsenthal, head of the Exchange Bank and Trust Company there. Zeppa personally borrowed money from the bank to purchase what appeared to be lucrative stocks. He also made transactions for the companies he represented, all the while building a reputation for integrity—an essential quality when one seeks financing in the oil or any other business.

Furthermore, Zeppa made friends with men in businesses integral to the oil business. He knew, for instance, Earl Wood, who was with Continental Supply which had its regional office in Shreveport.

Zeppa, Stacy, and Dorfman were each in a position to know scores of independent oilmen. One of these was Dr. Jerry Falvey, who practiced medicine there and, after buying some leases and finding them profitable, became an oil operator, as well.

In the informality of the oil boom, friendships blossomed from seemingly trivial incidents. One morning Joe Zeppa walked into the Randolph Hotel Coffee Shop for breakfast. Just as he arrived he overheard a man near him order a "three-minute egg."

"That's what I've been looking for in El Dorado—a three-minute egg!" said Zeppa to the waitress.

This inaugurated a friendship with Sylvester Dayson, who was in El Dorado to set up a refinery. Dayson, a French native whose first knowledge of refineries had been in Romania before World War I, and Zeppa were to remain argumentative but close friends until Zeppa died.[16]

Interestingly, J. R. Parten, who was to become a business associate of both men, met Sylvester Dayson under circumstances similar to the way Zeppa met Dayson.

"Sylvester Dayson was a new man in charge of the Lion Oil and Refining Company, brought to El Dorado by E. C. Winters of Kansas City and recommended by the Texas Company," said Parten. "They had trouble refining the Smackover crude. Smackover came in in 1922. Smackover crude had a good deal of sulphur in it and they couldn't handle it. Dayson was brought in for that purpose. I was living in the Garrett Hotel, which was the old hotel.* I usually had breakfast every morning in the Randolph, the newer hotel, across the street. They served better breakfast over there. I walked in one morning and the young lady said, 'Mr. Parten, there isn't a vacant table this morning. Would you mind having a seat by Mr. Dayson over there?'

"I said, 'Certainly I'll sit by Mr. Dayson.' That's where I met Sylvester Dayson. I remember he was eating eggs like my mother used to. My mother was from a French family. Soft boiled eggs—they'd just cut the top off and eat out of the shell with a spoon.

"Well, Sylvester and I got to be good friends. He went on and whipped the problems in the Lion Oil Refining Company, made a go of it, and made Smackover crude. El Dorado crude was a light crude and very sweet— very much like East Texas crude. Smackover crude, instead of being 38 of 40 gravity, was about 20 gravity."

Parten met Zeppa about the same time, in a more conventional manner. Parten was a partner in the Woodley Drilling Company operating out of Shreveport.

"They were both very bright young men and I was very favorably impressed with both of them," said Parten. "I know I was toolpushing on three or four drilling rigs running in Smackover at the time I met Joe. Right about 1923 or '24.

"Joe and I had gotten to be good friends in El Dorado. As I recall, he later commenced buying crude from us. And I got very well acquainted with Joe in El Dorado, liked him very much, considered him a very good friend of mine."[17]

---

*The Garrett Hotel had long knotted ropes in the rooms. In case of fire, one threw his rope out the window and climbed down to safety. That was the fire escape. Both hotels were always full. (Joe Massie Interview.)

VI

There were so many persons who became Delta employees or officers who had an "Arkansas connection" that it is impossible to list them all. Many of them Joe Zeppa or Bob Stacy knew from the El Dorado days, while others came from other parts of Arkansas or worked there before going to East Texas in the 1930s.

A review of the early lives of a few of them will give us a sampling of the personalities who later linked up with Delta.

Annie May Jones, who rose to become assistant corporate secretary, was born in Monroe, in North Louisiana, in 1898, the first of a family of five. Her father sent her to college in Tennessee and then to Northwest Louisiana State College where she finished. She taught school at Mansfield, Louisiana, until a meningitis epidemic closed the school. Leaving teaching, she went to Homer, Louisiana, and from there to Camden, Arkansas, with the J. M. Meeks Oil Company doing clerical work. She worked there for a few years until she was hired by H. L. Hunt as his secretary. It was then, undoubtedly, that Joe Zeppa came to know her and to learn that she was a competent person with a wide range of experience, particularly in the oil business. She moved to Longview, Texas, shortly after the boom opened there.[18]

Guy B. White, who retired as a vice president, was born in Pike County, Arkansas, in 1902. His father was a millwright who followed the sawmills when he was not farming; his mother taught school before she married. Young Guy left the farm in Prescott, Arkansas, where he had finished high school, and headed for the newly opened oil field in El Dorado in the summer of 1921. This was in the midst of a depression, and he worked at odd jobs, whatever he could find. Soon he went to work for the ice company in El Dorado, becoming general superintendent at the age of 21. It was a small company at first, and his vast experience with horses and mules, which were essential to delivering ice, made him a natural. The company expanded into little surrounding towns such as Junction City and Norphlet. White remained with the company until 1929, when the owners sold out to Arkansas General Utility Company; he continued as general superintendent over the El Dorado area. From there he became manager of the ice company and water systems at Smackover. In 1929, White decided to go into the transfer and storage business in the oil field and resigned, despite the promise of a promotion to assistant division manager of the company and a salary of $325 a month, a munificent sum.

"And when we had just about got started, the crash came that fall," said White, "and we didn't only go broke, we lost everything and owed everybody, like a lot of other people."

Starting over, White headed for Hobbs, New Mexico, another oil town, where he worked for a teaming company. When he heard of the boom in

East Texas, he left the West and returned to a part of the country where he felt more comfortable.[19]

One of Delta's pioneer drillers, Louis V. "Frenchy" Portier, came from South Louisiana. The El Dorado boom drew him up too. He was working for an independent contractor, Edward M. Jones, when he met his future wife, Ada Viola, or "Tinsie" or "Shorty" as she was variously called. Portier already had his nickname, along with a French accent, that would follow him to the end of his days. He'd been living in a tent when Ada Viola met him, and he was coming out of it when she first saw him.

"He'd been sleeping," she recalled. "He worked at night and his hair was long. I thought he was the awfulest looking guy I ever looked at in my life. Then wound up marrying him."

Later they met formally at a picnic near Norphlet, at which time Frenchy Portier appeared more presentable to the young lady. They were married in 1923, when she was 17 and he nine years older, and she accompanied him as he followed his drilling rigs over Arkansas, Louisiana, and Texas. Often they lived in tents, sometimes they lived with relatives. It frequently was difficult to rent apartments or houses, for many landlords refused to rent to "oil-field trash." But starting out during the El Dorado boom as the wife of an oil-field worker, Shorty Portier came to know at an early age many of the men and their families who later would be working for Delta.

When the Depression hit and the East Texas field opened up, Frenchy Portier headed there. Despite his drilling experience he hired out first as a roughneck, at a time when a man, whatever his experience, was happy to have a regular job.[20]

Nick Andretta, one of Delta's earliest employees, was born in Hartford, Connecticut, the son of a banker-engineer who came over from Italy in the nineteenth century. The elder Andretta immigrated as a full-fledged engineer and taught English in night school only two weeks after he arrived in the United States. As an engineer he was in charge of Indian and Polish immigrant laborers during the building of a bridge over the Connecticut River. His son, Nick, attended Choate School and later Dartmouth College, where he graduated in 1922. Nick Andretta met his future wife, Angie, while she was an opera singer, and pursued her to Europe, where she finally capitulated to his matrimonial proposal. They were married in Florence, Italy. Angie Andretta was of Indian and Irish extraction, from Pawhuska, Oklahoma. Nick turned his hand to dealing in oil royalties, then banking, but neither satisfied him. By 1926 Angie's family was living in El Dorado, where her stepfather, A. T. Woodward, was president of the Exchange Bank and Trust Company. "So I went to visit there and they told me of an up-and-coming Italian chap that was really a guy I would like," recalled Andretta. "His name was Joe Zeppa." Zeppa was working for Morris Coats and his associates at the time.

Then the crash of 1929 came. "It seems that nothing went right for anybody in 1929," said Andretta. "I had invested in the market and lost my shirt, as everybody did during the Depression, and my wife's mother and stepfather moved to Tyler, Texas—he was in the oil business—so we came to Tyler, Texas, in 1931."

Andretta started out for Delta as a roustabout and later roughnecked. "But," as he explained, "Joe Zeppa said my hands were too tender for that." For many years Andretta was to be the only scout that Delta had, as he scouted oil possibilities, leased land, did the title work, and checked abstracts. An excellent contact man, he came to know the men who headed the drilling departments of the major oil companies operating in the East Texas and other regions. These relationships frequently led to lucrative drilling agreements for Delta. During his career, he had a hand in a large number of the deals Delta made in seeking production of its own. Andretta retired as a landman in January 1967.[21]

L. E. "Red" Merritt, who became Delta's first drilling superintendent, was also in Arkansas for the El Dorado boom. In his career he worked as roughneck to field manager, and, as a hard-working man who insisted on doing things his way, he became known as the fastest driller around those parts.[22]

Luther Milton "Ark" Carter, one of the first three drillers hired by Delta, was born at Louann, Arkansas, in 1898. His father had been a sawmill man, but was killed in a boiler explosion when his son was 11. A few years later, at 14, young Carter went to work at a sawmill. In 1920 he went to the oil field for the first time, in Homer, Louisiana, as a roughneck. When the boom came to El Dorado, he followed as a driller, despite his scant experience. Carter gained his nickname when he first went to the oil fields in Louisiana. At first the other men called him "Arkansas," which was later shortened to Ark. When his younger brother Lloyd came along, he was also called Ark. Since this meant there were two Ark Carters, their co-workers soon arrived at a clarification: Luther Milton became "Big Ark," and Lloyd, "Little Ark."[23]

Other employees made the trip from Arkansas later, after Delta was on its feet. James Milton "Joe" Bevill, born in the "little old bitty town" of Austin, Arkansas, about 30 miles north of Little Rock, had no contact with the oil industry until he hired on with Delta years after its establishment. The youngest of four boys, Bevill lost his father when he was ten months old, and his mother moved her family to Kensett, Arkansas, where she raised the boys. They had a small piece of land where the boys raised hogs to supplement their mother's income and to put some meat on the table. Young Bevill graduated from high school in Searcy, Arkansas, in 1930, and managed to study accounting for two years at Ouachita College before he had to drop out during the Depression. He took extension work from the

University of Arkansas and a correspondence course in accounting from the Alexander Hamilton Institute. He worked for the Works Project Administration for a while. He headed for the larger cities to seek work—to Little Rock, then as far north as St. Louis and Memphis. Work was hard to find. Nothing lasted very long. Money was scarce, and transients seemed to cluster everywhere, making the competition fierce for what little there was. In May 1933, as the Depression ground on, Bevill headed for Tyler, Texas, where he had heard things were "a little better economically than they were in any place I had been." He had $20 when he hit town. Five years later, in 1938, he joined Delta.[24] When he retired, he was corporate secretary and remains a member of the board of directors.

Two other "latecomers" from Arkansas were Gus Jones and Robert W. "Bob" Waddell, who both came to Delta from Lion Oil in El Dorado in the 1940s. Jones, later to become field production superintendent for the East Texas division of Delta, was born in Donaldson, Arkansas, one of 11 children. In 1924 he went to work in the production department of Lion Oil Company. While there he met Joe Zeppa and, later, Chris Zeppa. After 11 years with Lion, Jones headed for California, but only remained three months before returning to El Dorado and a job back with Lion. Then he left Lion Oil for East Texas, where he worked for two oil companies before joining Delta as a pulling-machine operator in 1940.[25]

Waddell, the youngest of a Methodist minister's 11 children, was born in a little town between Pine Bluff and Little Rock in 1909. Because of his father's occupation, young Bob lived in a variety of towns over southern Arkansas, finally graduating from high school in Norphlet and attending Arkansas State Teachers College. He returned to Norphlet and taught school there for five years, until the Depression turned his teacher's pay into scrip which had to be discounted before he could get his money. Learning he could make more money elsewhere, he went to work for a wholesale grocery company, with his own territory.

Meanwhile, because his sister and brother-in-law lived across the street from Chris and Gustine Zeppa in El Dorado, Waddell came to know Chris. Chris Zeppa was in charge of building a gas-sweetening plant near Magnolia, Arkansas, for Lion Oil, and he got Waddell a job in the construction. Subsequently, Zeppa took Waddell with him to a gasoline and recycling plant in the Schuler field out of El Dorado. Zeppa resigned from Lion Oil in 1944 to join Delta. A year later, Zeppa called Waddell one evening, to see if he'd like to work for Delta.

Waddell said he would. "Knowing him," said Waddell, "I didn't even ask what it was I'd be doing, because he knew I didn't know anything about the drilling business, but, you know, he knew me."[26] Waddell eventually retired as a vice president.

Though there may have been more "stragglers" through the years, the arrival of Chris Zeppa and Bob Waddell in Tyler, more than a dozen years after Delta's founding, marked the end of the major migration from Arkansas.

VII

Though his brother Joe's presence in Arkansas became a factor in Chris Zeppa also going there, his route was circuitous. When he came to New York from Italy in 1922, he first stayed with his brother Charlie in a comfortable apartment on 125th Street. Charlie was managing a swank Wall Street club, a position he had secured through the recommendation of a rich member. "At that time," recalled Chris, "125th Street was still the Fifth Avenue of upper New York City."

With no specific long-term goals, Chris worked at a variety of jobs in New York City. He had arrived in the States knowing only a few words of English. But this was soon remedied because of his employment. "I was forced to learn English whether I liked it or not," he chuckled. At the same time he was making a concentrated effort at gaining an American education. Attending night school, he—a 22-year-old man at the time he left Italy—completed elementary school at P. S. 119 in New York and went on to night classes at DeWitt Clinton High School, where he took courses until 1925.

"When I finally decided that school was over and that I had to get seriously into some good employment, which at that time was hard to get, I said, 'Well, I'm going to have some fun.' I took a job on a ship that made regular trips from New York to San Francisco. So I made a nice trip, went through the Panama Canal, just looking the country over, worked my way. Pay wasn't big, just working in a place where they kept the food and everything in a warehouse.

"Meanwhile, Joe had already come *out* of New York City, into Arkansas, due to his employment. And about the time that I came back from my voyage—and I was looking for another position—why, he comes to New York City on something about what he was doing in El Dorado. And we met; naturally, he knew where I was. Called up and said, 'Let's go over here and have lunch and talk a little bit about it.' So the first thing he said [was]: 'What are you doing?' He always said, 'What are you doing?'

"I said, 'Well, right now I'm not doing anything. I just got back from California and I'm looking for a job.'

"We were close to the Fifth Avenue library there, when we met, on Forty-Second Street.

"He said, 'Hell, why don't you come to El Dorado?'

"I said, 'Me? Come to El Dorado? What's over there?' So we got to talking, you know, gabbing, one thing or another.

"'Well,' he said, 'there's a big oil field there and there's lots of jobs. They're hard work, but you'd like it. You're used to it. You're a farmer.'

"Things were not exactly right in New York about that time. So I bought a ticket at the Pennsylvania Station in New York City all the way to El Dorado, Arkansas! And I got to El Dorado the Fourth of July in the evening. At that time, you know, we used to get off at Gurdon, Arkansas, and take a little shuttle and go through down to Monroe. And I left New York City about six o'clock in the evening and got to [the] El Dorado station the next day about four o'clock, four-thirty. It was warm, I know. That began my life in this part of the world. That was in 1925 when I came to Arkansas.

"Naturally, due to my relationship with Joe, when I got to Arkansas I got a job immediately. It wasn't one to be so proud of—all it was was one where you'd get a pick and a shovel and from then on, that's your job. But at that time in the oil field of Arkansas, it was right at the stage of the high development. They were laying pipeline, they were building plants, they were doing this and that, and the company that I got the employment was the company that Joe was employed in, too. Arkansas Natural Gas, at that time, was a subsidiary to Cities Service Company.

"So they send me to the town of Louann, Arkansas. You'd never know where it is. It's right north of Smackover on the highway. It wasn't much of a town. Just a country place. And about that time the company was building a gasoline plant, and I got right into it, head and feet, and began my oil-field career. Right there, 1925, and things moved along from one job to another, to the office, you know, just keep on going up the ladder; and then along about 1928 is when I married my wife, Gustine, from Gurdon, Arkansas, of all places; and the only reason she was in that town was that *she* had a married sister there.

"I was fortunate, for a little farm boy, to come to the United States—and by my own bootstraps, without having benefit of ten years of university as most kids have nowadays, I was able to [go that far]—in fact, it even surprised me.

"I landed in El Dorado, and two days after that somebody in an old beat-up pickup took me to Louann, Arkansas, and says—well, there was a boomtown at that time—'Here you are. Go find yourself a place to sleep. Tomorrow you go to work over there,' and that was about a mile and a half, on foot. That didn't bother me.

"It was hard times, and only one time I recall that I ever doubted whether I chose right or wrong. Started out building a gasoline plant north of Louann. We got it built and then it started into operation and there was something that didn't exactly suit me just right, but at that time was the beginning of the oil fields in Venezuela, beginning to drill, and I happened to pick up the newspaper. In those days, the Standard Oil of New Jersey had an office in Shreveport, Louisiana, pretty close to the river in about a four-story building, and they put out an ad in the paper that they needed some oil-field workers in Venezuela. So I was, like I said, a little bit not totally convinced that where I was was where I wanted.

"So I borrowed Joe's car—he had a Dodge coupe, a gray Dodge coupe. He was in the office in El Dorado. I said, 'Loan me your car. I'd like to drive to Shreveport.' At that time there was no such thing as pavements, nothing but good gravel. So I went to Shreveport. Got there all right. Went up to this address. It was Standard Oil of Louisiana, but it was talking for Standard Oil of New Jersey, in New York. I went in there and I say, 'I saw your ad in the paper that you need folks in—' I didn't even know where Venezuela was.

"He said, 'Heck, yeah, we need lots of them. We can't tell you the salary or nothing else. But if you care to apply and go into Venezuela, all you've got to do is go to Baton Rouge, get into a Standard Oil tanker, and you'll be on your way. Or you can go to New York and get the first ship there.'

"Well, offhand, it didn't sound too encouraging, this getting on a ship and no money. So I went back, gave Joe his old Dodge, and while I was studying just a little bit there came some changes where I was; in about a week or ten days there came some changes that sort of quieted my ruffles a little bit and I decided [to stay]. I went into Operations, firing boilers with gas, no controls, just by hand, you know, those things. In those days, controls were unknown in the oil field.

"And I sort of liked that. I said, 'Well, that'll do for a while and maybe I'll go from this into something else. And I made a good decision. In fact, that's just the way it happened. I went from firing the boilers to an operator; from there I went to loading gasoline in the cars in the railroad track—and I loved that! Not long after that, [after] I began to learn my way around, why, they put me in the offices, to do what little local office work there was. And so from then on it went on, always a notch better, a notch better. That took me out of the idea of going to Venezuela!"

While going up notches, Chris Zeppa also worked on his English as a part of his personal self-improvement program, paying out a correspondence course by the month, at a time when he was earning 50 cents an hour.

"So I stayed with Cities Service then. I graduated on up to plant superintendent, to regional superintendent, and then I began to take over all the duty of constructions, engineering and all, not exactly doing all the engineering work, but just following the engineering work and attending to the construction. It was a really good improvement [from the pick and shovel]. So I learned the oil fields—gas-wise and gasoline-wise, production-wise—the right way.

"And finally went on to 1937, and I changed, went to work for Lion Oil Company, and I was in charge of all the maintenance in that big refinery in El Dorado for several years. Gas facilities, plants, and one thing and another. And then came a time—course, we kept pretty good contact—that Delta Drilling Company was formed here in East Texas and I began to hear from them and Joe; naturally, he'd come over to Arkansas pretty often. And I can say that he talked me into quitting Lion Oil and to come to work for him [in 1944]. So that's what happened."[27]

Chris Zeppa knew first-hand what he was talking about when he said, years later, to Gus Jones: "When we hire these Arkansas boys we know what they can do. They can work."[28]

VIII

In 1928, Joe Zeppa went to Shreveport as secretary/treasurer of Arkansas Natural Gas Company, controlled by Cities Service. At first he continued to live in El Dorado, driving to Shreveport each week. Later he found an apartment in Shreveport. But he was not happy with the way the company was being run, and he was in the market for a better position.

From time to time he had received inquiries from the McMillan Petroleum Company as to his availability to go to work for them. McMillan, headquartered in California, was "in pretty damned good shape," as Zeppa put it. The company had a refinery in El Dorado, another in Borger in the Texas Panhandle, and one in Long Beach, California.

Zeppa put off McMillan for several months before he finally decided he needed a change of jobs and geography. Then he left Arkansas Natural Gas on good terms. They gave him a farewell party.

"Of course," he said, "I hated to leave, but I just felt I had a good opportunity down there in California, because of the fact I felt I could go ahead a little bit faster than I was doing over there in Shreveport. So, anyhow, in 1928 I went to California as the vice president of McMillan Petroleum, in charge of sales."

The Depression hadn't hit yet, but the country was moving closer to it.

"You know what conditions were then," said Zeppa. "They were just getting on the edge of the worst break we ever had in history."[29]

4

EAST TEXAS—
THE BEGINNING

I

The Great Depression, hardly celebrated as the Cradle of Success, nonetheless provided the setting that brought together the men, equipment, labor, and money it took to form the corporate entity which became Delta Drilling Company.

Although the men had come to know each other in El Dorado, they did not move directly from Arkansas to East Texas. The road between those two points, at least for Joe Zeppa and Bob Stacy, was to be long and indirect. While Sam Gold, Sam Sklar, and Sam Dorfman continued their lucrative supply business in Shreveport and El Dorado, both Zeppa and Stacy wound up in California before they went to Texas.

Zeppa moved to California in 1928 as the vice president of McMillan Petroleum Company, in charge of sales. To a certain extent, this period of his life, from 1928 until 1931 when he was dispatched to Texas by the company, is relatively undocumented. We do know that, following up on a long-held interest in flying (he had, after all, first applied for the Air Corps rather than Engineers when World War I broke out), he acquired his student pilot's permit in February 1930 from the U.S. Department of Commerce while in Los Angeles.[1] Among his effects in later years he kept an old leather flying helmet and goggles from these days.

As an executive of McMillan, he was also in a position to exercise his sociability, and as a result developed a number of friendships on the West Coast, as he had in New York and Arkansas. Among his associates from this California era were oilmen Edwin W. Pauley and J. Paul Getty; though Zeppa was never to join the later billionaire Getty in a venture, he did subsequently share a Delta deal with Pauley.[2]

More important, however, he married Gertrude Ruppel, nee Wells, while in California. Divorced from her first husband, a contractor-inventor named Frederick Ruppel, she had a young son, Frederick Howes Ruppel, the third child of that marriage. Her other children had died as infants. Her son, incidentally, was born August 2, 1926, thereby sharing his birthday with his stepfather, Joe Zeppa, born August 2, 1893.

Gertrude Wells had been an independent person from her earliest years. Born in Cincinnati, she moved with her family to Chicago while still very young; from there the family moved to Berkeley, California, where she grew up as a teenager and went to the University of California there. After taking a bachelor of science degree in biology, she took a job as administrator of a railroad hospital at Albuquerque, New Mexico, before studying nursing at Johns Hopkins University in Baltimore, where her aunt, Helen Barnard, had been a widely-known superintendent of nurses. Awarded her Registered Nurse pin after a few years of training, Gertrude Wells was appointed director of nursing at the YWCA in Chicago. After marrying Ruppel and the birth of her third and only surviving child, she and her family went to Europe for few years. Upon returning to the States, the couple divorced, and some time later, in 1931, she married Joe Zeppa.

Gertrude Zeppa was an asset to her husband from the beginning, particularly in helping him entertain his associates in the McMillan company. Her son from the previous marriage, then around five, has recalled a few of those scenes in California.

"All I remember, I would be put to bed in Mr. Zeppa's apartment in a back room, and they would all be meeting in the front room. My mother would fix dinners for them all. She would drive over from Pasadena to Los Angeles with me and would fix dinners for them. And they would have their business meetings in the apartment."[3]

Not only did Zeppa acquire a socially-alert mate who was already noted for her entertaining in this country as well as in Europe, he also had accumulated a wealth of valuable experience, along with a great many business friendships, by this time. And soon he was to obtain some unexpected experience.

"Well," Zeppa summed it up later, "business went very well for a year or so. Then it seemed we were getting into difficulties financially."

Because of his experience and reputation, Zeppa was called in to take over the financing end of the business, for, as he put it, "Things got awfully bad" and the man who had been handling that job "was getting nervous and he couldn't run the company."

"So I went in and I found out that he hadn't handled the paper like he should have, and then [I] made a trip to see all the people we owed money to and made arrangements for deferred payments on the basis that we had estimated we could pay."[4]

A bank in California to which the company owed money saw things somewhat differently, however, and its officers wished to place one of their men in charge of McMillan's finances. The leverage of the bank proved to be decisive. The other man was put in charge of financing, and

Zeppa was dispatched to Dallas, where he was to handle the products of the Borger and El Dorado plants.

At this point in 1931, Zeppa had been married a short time. He went to Dallas to manage the company's affairs from an office in the Magnolia Building, with his wife and stepson to follow later, by automobile. Gertrude Zeppa's Willis proved to be a serviceable car for the long trip from California to Texas over roads that were often unpaved; but her small son was car-sick a great deal, spending much of the trip prone in the back seat. Once settled in Dallas, she enrolled her son in kindergarten at Highland Park where they lived.[5]

Zeppa opened the Dallas office at the time McMillan was developing and advertising its Ring-Free Motor Oil.

"Well," he recalled later, "I hadn't been in Dallas but two or three weeks when I got a message to go to Los Angeles, that things were not in good shape. So I went over there, and I found that the deals that I had made with the banks—when [the other fellow] got in he said, 'No, we can't pay 'em. Can't pay them anything.' He wouldn't pay 'em anything. He simply said, 'Well, to hell with Joe!' So he managed finances to suit himself, and the first thing that happened, why, these banks just grabbed all the money they had on deposit, and all the checks that were written on the account— we never kept very much money in it—were all turned back. So the question, What the hell to do about it?

"'Well,' I said, 'I don't know what you can do about it, except to get in touch with them and tell them that you have made a mistake.' Oh, no, they wouldn't do that. So, they were going to go into receivership when I got to California. I talked them out of it and I showed them where they could go ahead and weather that little flurry that they had .

"I went on back to Texas and hadn't been there but four or five days when I got a telegram that they were going into receivership. No thank-you or nothing.

"I stayed there [in Dallas] for about a month. I was waiting for something to develop. I didn't get paid for my expense accounts; I didn't get paid for my salary; and I'd just got married. And we had rented this house furnished. But the question was, what to do next?

"Well, I found out soon enough that they had no idea of using me in the company and that I'd better look for something else. Of course, this was in 1931—during the worst, very worst part of the Depression. You couldn't get any money any place, [though] the banks were still open."[6]

II

When the crash of 1929 left times hard in El Dorado, Stacy took his family to California, but unlike the migration of farm families from Oklahoma and Texas to the Golden West, the Stacys went in style—in a chauffeur-driven Cadillac. George, a black man whose surname has been lost to memory, was one of a series of uniformed chauffeurs that Stacy hired even at times he could ill afford any luxury.

Stacy remained in California for about a year.

Elizabeth Powell Fullilove, Priscilla Stacy's sister who was taken into the Stacy household when her parents died, recalled:

"Bob had a connection with the Texas Company that he thought might work out in Los Angeles."

When the original expectations failed to materialize, Stacy did what he could to support his family. His son, who was nearly five at the time, remembers only fragments of the era.

"He acquired, through means I don't recall, a job in Los Angeles, California, as a manager of a driving range in Hollywood. I recall the range, a fenced-in affair, and evidently it was popular and he was manager of it."

"It was next to the Ambassador Hotel on Wilshire Boulevard," said Elizabeth Fullilove, "and George [his black chauffeur] worked for him."

In California, Stacy renewed his friendship with Joe Zeppa. Zeppa's status in the world in that Depression year impressed the Stacys and lingered in young Bobby Stacy's memory ever after.

"I recall that Joe was one of the few people we knew who had some money," Stacy, Jr., said. "He didn't have a whole lot, but he had some. I remember he always drove a big car."

Joe Zeppa was frequently in the Stacy home in Los Angeles.

"I am sure I must have seen him in El Dorado," said Mrs. Fullilove, "but the first memory I have of Joe is in California. Joe was a friend and there was much communication between them. He was a bachelor. He was going with Gertrude and she lived in Pasadena, but Joe took me to my first night club and I must have been 12, maybe 13. He was short and attractive and positive. He was considered intelligent and smart and, I think, was respected by the men he was associated with, even at that time."

During his visits to the Stacy home, Zeppa apparently felt himself to be in familiar surroundings.

"I remember Joe going to sleep," said Mrs. Fullilove. "He would go to sleep after supper. Right at the house. He would eat supper and then he would stretch out on the couch. It was just funny and nobody paid any attention to it. It didn't matter what company was there, he would just stretch out on the couch and take a nap."

The California adventure did not work out for Stacy.

" Bob and Sister were there maybe about a year," said Mrs. Fullilove. "He did one thing and then another that didn't suit him at all, and we came back to El Dorado.

"[Colonel T. H. Barton] was quite helpful to Bob, because Sister called on him, I think, when Bob went to work for United Gas in Dallas, and I think Colonel Barton really got Bob that job. That is when we came back from California and came back to El Dorado, and I think it was Colonel [who] made that connection for Bob. They were good social friends.

"Times were still terribly rough and Bob lost money on the stock market and had a hard time. In Dallas he worked as some sort of sales representative for the gas company, a kind of position where he belonged to all of the clubs."

As Robert Stacy, Jr., recalled, "Then we came back to Dallas—this must have been 1930—and Dad got a job with the spin-off of a little production company from The Electric Bond and Share, called United Gas, that wasn't much back in that particular time. As a matter of fact, I think he worked directly under [founder N. C.] McGowen. It was a small company in those days.

"My dad did have a job and that was something, I suppose. I did recall, though, that we had this rent house in Dallas at the time he had a job with United, and it was a three-bedroom house down on Fairfax, and we rented out two bedrooms and we had kitchen privileges. Now I remember our kitchen privileges were between five and six in the morning. We had to get in there and clear out. That didn't mean anything to me, because at that age children don't give a damn about what's going on.

"And the car we had was blocked. We'd go play with it. The wheels were gone and so things must have been tight. It wasn't much of a car, but it was a car and I think we finally got the wheels for it and put it back in use. It seems to me that he did have some sort of company car at that time."

Mrs. Fullilove had a similar memory of the move to Dallas, where they lived in a cottage on Fairfax in Highland Park.

"This was about two blocks from the Dallas Country Club," said Mrs. Fullilove, "and it was a one-story gable cottage. Then as we got a little harder up we moved to University Park near S.M.U. and there was another small cottage there."

The Stacy family's only means of transportation was the company car Bob Stacy drove in his work for United Gas.

"We looked much more affluent that we were because he had a company car that was a Buick."

The Stacys had arrived in Dallas from Arkansas in the same chauffeur-driven Cadillac that had taken them to and from California.

"It was old, but it still looked good," said Mrs. Fullilove. "It was one of those cars that kept looking good, but it cost us too much to drive it. It used as much oil as it did gasoline. So it was up on blocks in the garage. It was just terrible. We couldn't quite use it and couldn't get rid of it, I guess. And when we sold it, the radiator fell out on the way downtown. The man who was driving it had a sense of humor and he called back. He said, 'It is just setting in the street,' and we just laughed."

Despite these episodic flirtations with economic peril, some observers gained the impression that the Stacys were better off, financially, than they were.

"Both of them were just as extravagant as they could be," said Elizabeth Fullilove. "Both Sister and Bob, from the beginning of time. It didn't matter what they made. Bob owed a lot of money over a period of time, and he had to pay that money back when he started making money and he was determined to pay it back, was determined not to take bankruptcy, and did not do it. I am sure he felt strapped, knowing that the money was coming in and that he had to do all of this, but they were just the kind of people that let money get away from them.

"It wasn't done for the effect. They were just naturally that way. I can't say they lived rich at all. They lived well. More than they should have. There were cars and there were chauffeurs and this kind of thing, but they didn't have that kind of money. Even though they went in debt they weren't spending that kind of money. They would have enjoyed it if they had."

But by the fall of 1931, the days of uniformed chauffeurs and fine-looking cars were a memory for Stacy. "Gosh," said Mrs. Fullilove, "we were lucky to eat about that time." Stacy was a driven man by the time he returned from California, but he never, then or at any other time, made any attempt to use his connections with the well-off branch of his family in Houston—the Claytons—to borrow money or for any other reason.

"I don't think it ever occurred to Bob," said Mrs. Fullilove. "What he would do, he would do himself."

In Dallas, while working for United Gas, Bob Stacy saw Sam Dorfman frequently. They would talk about their friend Joe Zeppa, who was expected to be coming to Dallas from California. Whether they talked specifically of organizing a company is not known, but it is possible; and if they did discuss it, it is probable that Stacy and Dorfman both were interested in bringing Joe Zeppa into it.

"When Bob was back [from California] and in Dallas and working and not satisfied at all with what he was doing, wanted to do better and more, this was when it really began to take shape," said Mrs. Fullilove, "and it is my feeling that Sam Dorfman got in on it [at the beginning] and, you see, when it became clear some money could come—they went to Sam Sklar."[7]

III

While Joe Zeppa was fighting McMillan Petroleum's financial battles in California, the biggest oil field known to man at that time had been discovered in East Texas near Henderson when C. M. "Dad" Joiner's well—the Daisy Bradford No. 3—spewed forth a fountain of oil on September 5, 1930, from the Woodbine sand at 3,536 feet. The Joiner well was completed in October, and other rotary bits were busily turning in the vicinity. A few months later, on December 28, 1930, a second important well, drilled by E. W. Bateman ten miles north of the Joiner well, also brought forth oil from the Woodbine. Less than a month after that, on January 26, 1931, a third extension well, operated by J. E. Farrell, W. A. Moncrief, E. A. Showers, and Arkansas Fuel Oil Company found oil five miles northwest of Longview. It was not at first apparent to but a few that these three wells, so far from each other, were all in one field; once that was recognized, the boom that already was large increased in intensity. By the time Joe Zeppa arrived in Dallas in the latter part of 1931, East Texas was throbbing with activity.

By January 1, 1932, about six weeks after Delta Drilling was formed, the East Texas field boasted 3,612 producing wells, and its peak of activity came in May of that year when 783 oil wells were completed in that month alone. Seven years later, in November 1939, there were 25,987 producing wells within the confines of the field which stretched 42 miles long and from four to eight miles wide.[8]

When, later, independent oilmen drilled on the Kilgore townsites, with 20 or so wells to a half-acre, the frenzy was redolent of some of the wildest boom scenes in American history. In fact, some businesses, including a bank, were vacated to accommodate drilling rigs. W. S. "Buck" Morris, who in 1936 came to East Texas from Oklahoma City where townsite drilling had made millions for operators in that city's boom, particularly remembered the bank scene.

"The bank had a terrazzo floor," said Morris. "And they drilled a well right in the lobby, where the lobby of the bank used to be, and I used to take visitors by there and tell them, 'This is the only well in the world that had a terrazzo derrick floor.' They drilled right across the street there where J. C. Penney is now; there was a restaurant and they shortened that restaurant and drilled two wells where the kitchen used to be, and the wells were so close together that they used just one beam—they had a well on each end of the beam. And the idea of putting a well on each end of a pumping beam caught on, and right there in the Kilgore townsite within a stone's throw of this location there must have been a dozen wells [with] two wells to the beam."[9]

Thus, when Zeppa arrived in Dallas in 1931, the greatest oil boom in history was feverishly under way, almost under his nose, and with most of its opportunities yet to be exploited.

The phrase, "being in the right place at the right time," is one whose accuracy is invariably measured in retrospect. In Joe Zeppa's case, if Mc-Millan Oil hadn't sent him to Dallas and then dispensed with his services, one thing is highly probable and almost certain: he wouldn't have been in on the founding of Delta Drilling. For while he was going through the vicissitudes of his McMillan experience, a new stage was building about 100 miles southeast of his Dallas office, one that would bring him business success such as he had never before known it. The East Texas field was that golden opportunity of which millions of immigrants have richly dreamed.

IV

In Dallas in the fall of 1931, at this low point in his life and at an equally low ebb in the nation's economic history, Zeppa ran into his old friend Bob Stacy. At this juncture, Stacy—who probably thought Joe Zeppa had money as well as ability, if we are to judge from the impressions Stacy, Jr., gleaned from his father's views—approached Zeppa with the idea of starting a drilling firm.

Stacy had been impressed years before by Joe Zeppa, and it is hardly surprising that he thought of him then. "Bob felt that Joe was real smart," said Elizabeth Fullilove. "He liked Joe and he thought he would make money in the oil business and he would like to be associated with him. I suspect then it was more important to Bob than it was to anybody else, though Joe may have wanted it. But to Bob this was vitally important at that time. This was his vehicle to move on from where he was." [10]

Of the original partners, only Zeppa has left behind his own straightforward, though condensed, memory of the events which led to the fulfillment of the idea.

"Well," he said years later, "I had an old friend of mine by name of Stacy who had lived in El Dorado and was working for United Gas Company, and another chap by name of Dorfman. Dorfman was a second-hand dealer; you know, he bought old pipe and sold it and so forth, and Bob [Stacy] got the idea of getting in the drilling business.

"Well, we didn't have any money. I had no money at all. In fact, I guess I had just barely enough to pay the rent and grocery bills. And Bob didn't have anything, apparently, though he did have a job. Now Sam Dorfman was well-fixed, because he was in a partnership with Sam Sklar: Sam Sklar, Sam Dorfman, Sam Gold—three Sams!

"Bob Stacy said, 'We can get Dr. Jerry Falvey.' Jerry Falvey was an old friend of ours from El Dorado who had moved to Longview. He had a lease that he wanted developed. He had two or three wells on it. He had been selling oil, getting paid. They were up in the north part of the field." (Falvey operated as Sabinas Oil Company.)

"So we arranged it. Sam Dorfman had a second-hand yard in Longview, just outside of Longview. He had a lot of junk—just piles of junk. Drawworks and pumps and everything. Course, you couldn't tell what condition they were in.

"We bought two drilling rigs from him for $20,000. Complete. We were to go and pick 'em." [11]

The rigs which Dorfman put into the company, for his (and his partners') one-third interest, represented probably the most crucial step in the formation of the venture, though Zeppa and Stacy still had to scrape up money for their own investments. The Louisiana Iron and Supply Company yard in Longview was apparently an extension of the Shreveport headquarters, for the Texas corporation by that name was not chartered until April 1933. However, from the 1933 charter which left the shares divided equally between Sklar, Gold, and Dorfman, we can infer that this same working arrangement was already in effect in Longview in 1933, though it had not yet been formalized in an official document. Dorfman managed the company's yard in Longview, where it had been moved after being briefly based in Gladewater. Though Louisiana Iron and Supply's Shreveport yard was the largest one the company had, the Longview yard boasted ten acres of used pipe and oil-field equipment—a treasure trove and a logical starting point for men empty-handed at the beginning of such a venture.

Without the drilling rigs, there obviously would be no company. Though both Zeppa and Stacy, no doubt, already had discussed the matter with Dorfman, Zeppa left nothing to chance as he enlisted his wife Gertrude's entertaining skills in his well-orchestrated performance to persuade Dorfman to contribute the vital equipment.

As Mrs. Zeppa's older son recalled, "Her famous fried-chicken dinners are how he got the money out of Sam Dorfman to get the rigs. She gave an excellent fried-chicken dinner one time."[12] One may assume it took more than a fine dinner to gain Dorfman's participation, but at the same time one may be certain a splendid meal with pleasant company in comfortable surroundings left no negative memories with Dorfman.

Delta's first old rigs had been used in Arkansas, and some of the parts had come from Louisiana. "It hadn't come out of East Texas 'cause there wasn't any [rigs] in it," said Marvin Williams, whose company later built rigs for Delta. "It was coming from where they had surplus stuff. And that's where Delta got theirs."[13]

Zeppa and Stacy at this point needed a man capable of knowing equipment as well as drilling procedures. He was near at hand, one who had earlier been in Arkansas too.

"Well, we knew a chap by the name of [L. E. "Red"] Merritt, who was a good toolpusher. We hired him to pick out the rigs," said Zeppa.

The company was called Delta Drilling Company. "Delta is a D, you know," said Zeppa, "and so we tried to figure out a name for three people. Well, we just picked the triangular [Greek letter] D."[14]

There also is a good chance that his association with Cities Service may have partially contributed to the name of the company. As Chris Zeppa—who was still in Arkansas at this time—has pointed out, the Cities Service symbol was a three-section expanding circle. Joe Zeppa had worked for a

Cities Service subsidiary in Arkansas. "I think that's where the word Delta came from," said Chris, and it seems reasonable to speculate that at least a partial basis for the later logo design was born in Joe Zeppa's mind as a result of his experience with Cities Service. We can almost hear the clicks in Joe Zeppa's mind as he tied in the triangular symbol with the fact the ownership of the new drilling firm would be divided three ways.

"I remember the day [Bob Stacy] came home and said they had chosen Delta as the name for it," said Elizabeth Fullilove. "Evidently he was try-ing to raise the money, but he knew it was going through. He was saying that each part of the triangle meant something. He thought Delta was a good name for it. There wasn't anything wrong with it. This would have been when they were starting to write everything up and, I guess, got the lawyer in. I do remember talk about that and I do remember much, much excitement about the whole thing. There was much excitement in our home.

"He immediately left his job with United Gas. He may have left even before then. He knew it was a going thing by the time we came to Shreve-port."[15]

Officially, the charter for Delta Drilling Company was approved as a Texas corporation and filed on November 17, 1931. A telegram to that ef-fect was signed by Watt L. Saunders, assistant Texas secretary of state. The date was certainly an auspicious one for all the new partners, but espe-cially so for Bob Stacy, the one who had first come up with the idea. No-vember 17, 1931, was Bob Stacy's thirty-fifth birthday, and he had double cause to celebrate. It is probably no coincidence that the birth date of Delta was the same as that of Stacy. It seems likely that he chose the day and did what he could to ensure the chartering on that day.

On November 17 when Delta Drilling received its charter from the Texas secretary of state, its official headquarters was at 1310 Magnolia Building in Dallas. That afternoon at two o'clock, the first directors' meet-ing was held as the company's paperwork got under way. Only three per-sons were present: Zeppa, who ran the meeting, and J. W. Hassell and Aline Kraemer. Apparently Aline Kraemer worked in Zeppa's office; the role of Hassell is not clear. In effect, Kraemer and Hassell served as prox-ies for the other partners. The three directors formally adopted by-laws so that the company could be organized and operated as a corporate entity.[16]

In that first meeting, Joe Zeppa was elected president and treasurer. He was to hold both positions for decades, and he was to maintain the presi-dency the remainder of his life. At the second board meeting on December 9, Bob Stacy replaced Hassell on the board and Sam Dorfman took the place of Kraemer. Stacy subsequently became secretary of the company and the office of vice president continued to remain open.

The first order of business in the December meeting was the matter which had brought Delta into existence—the contracts with Dr. Falvey's Sabinas Oil Corporation for drilling five wells on the T. M. Collins C lease. The company agreed to open an account with the Republic National Bank in Dallas, but, in the depth of the Depression, it is interesting to note that officers were restricted in spending or committing the company to an expenditure of more than $500 "without the unanimous approval of the board of directors," although once a matter had been approved, the officers had the authority to carry out the mandate.[17]

"We formed this company without a dime," recalled Joe Zeppa. "Of course, Bob had a job. Sam Dorfman was well-fixed, so he didn't need any money. I was broke, but I borrowed money from a friend. We gave him a note for $20,000 and then we borrowed from Dorfman about $2,000 or $3,000 to put in the bank. He was the only one who had any money.

"We started out. We signed a contract with Falvey. Well, that involved a lot of ifs, ands, and buts. We made arrangements to have certain equipment. We were supposed to give him a turnkey contract, you know, furnish tubing and everything. Made arrangements in Dallas to get the stuff, and got the first well down after changing the pump and this and that. You were constantly, always changing things because we couldn't get anything that would run very long. Now you can imagine what you could get in second-hand equipment."[18]

The first well was drilled in the north end of the gigantic East Texas field, in what was known locally as the Big Woods, north of Seven Pines. Problems appeared immediately.

"You know," recalled Nathon Bacle, an early employee, "when they first started in the oil business, you drilled a well before you got your pay— drilled three or four wells and get enough money to pay their salaries. And the first well they drilled, I think, they hauled a rig out of that Louisiana Iron and Supply and something went wrong with it. They couldn't get enough parts for it and they hauled it back in and carried another one out there, you know, to finish the first well."[19]

That was scarcely the beginning of difficulties.

"So we had to get *pipe* for these *wells*," Joe Zeppa recalled. "You know, in those days it took 20 days to get a well down. Now you get 'em down in four or five days. We didn't have any money. We not only didn't have any money, but we owed money to the bank at El Dorado, where we had borrowed against stocks that we had purchased and which were worth a good deal of money—but the demand market went down until they weren't worth anything."

At this point, when he had no money but needed it desperately in order to generate a living, he turned to some basic capital frequently overlooked in computing one's assets: friends.

"The people that I knew," said Zeppa, "and that I met in the oil business, I can say unqualifiedly, with few exceptions, they were all the finest people that I ever met! Good friends, and trusted each other and that sort of thing. But, of course, when you've got a man that is broke for a friend, you can't borrow anything from him, you know!

"But I did have a good friend that I'd known since I went to school in New York, who was very prosperous, making lots of money in New York. And so I managed to borrow what I needed from him. I didn't know how I was gonna pay it back! Because, you know, when you start with a couple of junk rigs . . . . "[20]

Though Zeppa did not specify the name in this 1975 interview, George T. Keating apparently was the "good friend that I'd known since I went to school in New York, who was very prosperous, making lots of money in New York." Their closeness had persisted since their impoverished boyhoods in Hell's Kitchen. As boys, each had sworn to help the other. By 1931 when Zeppa was starting Delta, George Keating was well-established at Moore & Munger, a New York wax- and clay-products company.

Keating, a very hard worker, had started as messenger boy with Moore & Munger, and by dint of unremitting application had worked his way up from "the bottom of the barrel" and, with a friend securing the money necessary, had acquired Moore & Munger. Moore & Munger sold clay mined in Georgia that was used as filler in paper and china. With their products in demand, Keating and his associates made money even during the Depression. At least modestly affluent in those dire times, Keating had "made it" first, and was a logical person to call upon, and Zeppa did. Since childhood, the two had been dedicated to helping each other and, as Leo T. Sides, who was married to Keating's daughter, explained, "There was never any accounting of who did what for whom."[21]

George Keating and Fred Snyder, also from New York, "funneled money down here to Mr. Zeppa at the time of the East Texas oil field," according to Zeppa's stepson. "They were investing down here at his behest." According to this source, Keating subsequently would advance money to drill wells, and he and Snyder and other New Yorkers would participate in deals. "They were among the few people in the country who weren't so ruined by the Depression."[22]

Keating came through at the critical moment and Zeppa never forgot it. Nor was it a one-sided arrangement by any means, for Zeppa also made a great deal of money subsequently for Keating through astute oil deals.

Zeppa had, again, drawn upon his richest basic capital—his friends. Like practically all of his other friendships, he maintained his ties with Keating to his dying day.

But this was not the end of Zeppa's financial needs. He needed credit and more money for the company's operations.

"Of course," he said, "the thing that helped me more than anything else was that the people I had done business with, in El Dorado and with Arkansas Natural in Shreveport, thought so much of me, because when I went to them and said, 'Look, I've got to have this pipe,' the representative of Continental Supply in Shreveport said, 'Joe, I'm not supposed to do it. I have got instructions not to do it. But I'm going to do it if I have to lose my job. I'm going to let you have two strings of pipe. You say you can pay for it, say, 30 to 60 days?'

"I said, 'Yes, I think I can.'

" 'Now I'll let you have those two strings of pipe, send them out, so you can finish those wells, but I don't know if I can get by with it or not.'

"He did. Earl Wood, his name was. Those are the things you remember for a long time. We are still doing business with Continental [Emsco]. We've been doing business with those people 42 years. They've been, in fact, our main supply house. Now we can buy from *anybody*; I mean, we're not restricted in our purchases to anyone at all. I don't know how much money that those two strings of pipe have brought Continental Emsco."[23]

This statement gives us a clear insight into why Zeppa's treasury of human capital was so well-stocked. He never forgot one who befriended him. Earl N. Wood, the Shreveport office manager of Continental Supply company, was one of a number of men who had a deep faith in Zeppa's ability to make good his promises. More than 40 years after the event, Zeppa vividly recalled the uncertainty of the circumstances in a letter to Wood, by then retired.

"I could not help but look back on the first string of pipe we purchased from you for the Jerry Falvey well which we were drilling and had no casing or tubing for," Zeppa wrote Wood. "I told you then that I would see you got paid, although I didn't know how, and the fact that we have been doing business with your concern ever since is indicative that you have given us good service and we have managed to pay most of our bills. This is another opportunity to say 'Thank-You' for the faith you showed in me when I worried you for that string of pipe. I do not know of anyone else who would have done what you did."[24]

It is instructive, when later we come to analyze Zeppa's personality and style of operation, that Zeppa personally wrote Earl Wood shortly after Delta's fortieth anniversary. After all, Wood was retired by then and it hardly mattered, objectively, whether Zeppa acknowledged the good turn again. But he operated through a personal network of friendships and relationships, and that was his strong point when it came to achieving his personal and business goals. Very early in life he seems to have had the insight to realize business, like everything else in life, is a matter of interpersonal relationships. He expressed his appreciation in gentlemanly fashion, and his friends never forgot him.

"In any event," recalled Zeppa, "I went to El Dorado, and I talked [Lee]Felsenthal, who was the head of the Exchange Bank and Trust Company, into loaning me enough money to carry out our contract. I assigned him the contract, and he was gonna advance us enough to help us carry it out, although he himself was in bad—I mean the bank was in poor shape; he'd just sold some preferred stock, sold it to the government to get the cash necessary to stay in business. And then, of course, I got Colonel Barton, who was a friend of mine till he died. I mean, he and I got to be awfully good friends. So I went to talk to the Colonel and asked if he would endorse the company's paper. The Colonel still had money. Lion Oil was still in business, and they were making some money. And, yes, he said, he would. Anyhow, I worked it out so that we could get the pipe that we needed."[25]

When he went to El Dorado, however, it was not at first apparent that Zeppa would receive the loan. It even looked for a time that he would not. He had approached Felsenthal because the Exchange Bank and Trust held his note for money he had borrowed to purchase stocks which the crash of 1929 rendered practically worthless. Despite this dismal reality, it appeared possible that a banker who had thought enough of Joe Zeppa to lend him money once might be inclined to do so again. Ordinarily this would be excellent logic, but the circumstances were hardly ordinary. After all, Zeppa still owed the bank $30,000, with which he had bought stocks, and the bottom had fallen out for everyone. Zeppa's request for another loan hardly kindled the banker's enthusiasm. Why send good money after bad? Felsenthal was adamant. He refused to risk another $1,000.

But as stubborn as the banker was, Joe Zeppa was at least his equal. Zeppa now acted upon a sudden inspiration. He told Felsenthal that, conditions being what they were, the bank stood little chance of collecting the note. However, if he, Zeppa, had the $1,000 loan he could make his business a success and eventually pay off all the debt. It was the banker's best hope of getting his money back! The banker did not budge. Still, Zeppa did not give up and head empty-handed back to Texas. As Felsenthal ushered him out the door, still intending to maintain the friendship, Zeppa sat down on the bank's doorstep and announced:

"I won't leave until you let me have it!"[26]

Documenting that at least one capitalist could use the sit-down strike as effectively as labor leader John L. Lewis and his coal miners who were to make it famous, Zeppa's tactic succeeded where all else had failed. Felsenthal relented. Zeppa obtained the $1,000, a hefty sum in the land of little cash, and headed back to Longview with a new lease on his business life. Elemental determination had carried the day.

"I think he was very determined about what he wanted," said Elizabeth Fullilove. "Also he was full of words. He was very full of words, and so I guess this would be very persuasive."[27]

The story of Joe Zeppa camping on the bank's steps until he received the loan, told over and over within the company, became a part of the Delta folklore. Without that successful trip to Arkansas, Delta might never have been. But, being the type of man he was, one suspects Zeppa would have found the money somewhere else had he been forced to leave empty-handed. Essential as the loan was, the episode symbolized all the crucial negotiations upon which the company's survival depended in the beginning, for few approached the El Dorado scene for dramatic intensity.

V

Delta began drilling its first wells in December. The first contract, according to Nick Andretta who joined Delta soon after it was formed, involved oil payments for the drilling. Oil payments were lucrative, but there was a delay in receiving the cash.

"It gave the Delta Drilling Company a big start," said Andretta, "for the simple reason they turnkeyed those wells and they got most of the supplies from Louisiana Iron and Supply. Second-hand equipment was damn good in those days, you know, and you could turnkey one of those wells, I would say, probably about $17,000 a well for them in the tanks—turnkeyed, you know, pumping equipment or anything else—and they were all good wells."

Delta discounted the oil payments it was receiving for the drilling work in order to secure cash with which to operate.[28]

Though prospects looked promising, nothing was certain. The venture could have gone broke at anytime.

"It was the well-known shoestring," said Elizabeth Fullilove. "I think $25,000 was what they had to start with, and I remember Bob walking the floor on those first wells, because if there was a big fishing job or if something like that happened you really were in the soup. He would get home and just walk the floor over that.

"I guess the reason we left Dallas was because he couldn't get the banking connections in Dallas that he wanted, so why be in Dallas? And at that time the banks were pretty hard-nosed about the oil business, particularly when you were out on a limb a little bit this way. Anyway, I do remember the banking connections didn't work out, and that was one of the reasons we went to Shreveport, where he could make a loan."

In addition to taking on the new economic responsibilities, there was the matter of Stacy's personal debts which had built up during the hard days of the Depression.

"It seemed an astronomical debt at that time for him to owe personally," said Elizabeth Fullilove. "These were just personal debts, and these were things that occurred during the Depression. You are very young and, suddenly, you have all of this responsibility, plus your wife's sister, plus everything, and you are indebted—and this is a lot to put on a man, particularly [in] those years, and you have to give Bob full credit for that. When things were not working out right and he kept working at them and working at them, that took a lot of determination.

"I think Bob was absolutely determined. Joe Zeppa certainly was, and apparently the three Jewish partners were too, although they were not really as active in it. All the others could say No, because they had the purse strings. The ultimate purse strings."[29]

Most of the early Delta minutes reflect these trying times and detail how some of these financial obligations were met. In December 1931, barely a month after the company had been organized, the stockholders of record (Zeppa, Stacy, and Dorfman) met to consider the solution of some urgent problems. For the next several years, money was to continue to be the major item on the agenda. In that December session, for instance, the first item was the assignment of $27,500 worth of oil to be paid Delta by the Sabinas Oil Corporation upon completion of its contract. This session was also to authorize Zeppa, as president, both to execute assignment on the remaining oil properties the company would receive from the Sabinas contract, with which it would repay loans from the Exchange Bank and Trust Company in El Dorado, and to borrow $15,000 from George W. James and T. H. Barton, old associates from El Dorado days, at eight-percent interest. This, Zeppa explained to his partners, was "absolutely necessary" in order for Delta to meet its obligations under the Sabinas contract. As in the overwhelming majority of instances thereafter, the stockholder-partners acquiesced.[30]

But as uncertain as the company's first days may have been, it also was making money, enough so that at the stockholders' meeting in February, 1932—three months after the founding—the partners declared dividends, which in that money-scarce time must have seemed like all the money in the world.

Dorfman, Zeppa, and Stacy each held 100 of the 300 shares of outstanding stock, and two dividends were declared—for $75.00 per share and $16.67 per share, providing each of the trio with payments of $7,500.00 and $1,666.66, to come from oil assignments. In addition, Zeppa and Stacy were each paid $5,000.00 to reimburse them for expenses and payments made out of their own pockets during the drilling of the first five wells for Sabinas. In addition, Stacy was advanced an additional $2,500.00 as a loan.[31]

The Delta founders were yet a long way from being millionaires, but they were on their way, provided they could keep the momentum going in a time, place, and business that guaranteed as much opportunity for bankruptcy as for riches.

As Joe Zeppa summed it up many years later, "Well, anyhow, that was the beginning of the company, and I assure you that it was not an easy path to travel for quite a while."[32]

VI

The next year an opportunity came which also can be seen as pivotal in Delta's eventual success.

"I made a deal to buy a lease, 10-acre lease, for an oil payment," said Joe Zeppa. "It was on the east edge of the field. Drilled a beautiful flowing well on it. Brought it in, I remember, on my birthday [in August 1932]. Now this was on ten acres. Then there was another lease adjoining it of 20 acres. I made a deal with the same people. These people are very fine people. Their name is Hughey. He was working for the highway department, and he had a big farm right in the middle of the field! This was a little remnant, leftover, and that piece was owned by him and Sam Ross. Sam Ross was a Kilgore boy that did very well in his day, because he had 1200 acres right in the middle of the field. He and Hughey.

"We got to be very good friends with everybody we did business with, and we could always go back and trade with them."[33]

Throughout his life Zeppa was cautious about taking an interest in a well he was drilling. Typically, he would take one-sixteenth in the early days; one-eighth was large at that time. "Usually tried to cover his bets, you know," said Joe Bevill. "And always did."[34] But the Hughey-Ross transaction, the company's first such acquisition, was atypical. First, Delta leased surface rights in order to put equipment there, then secured an oil and gas lease entitling the company to seven-eighths of these minerals. Subsequently, Delta purchased the surface rights to the land, north of Kilgore, and set up a campsite. The maintenance yard was built upon the property and remains there to this day.

The Hughey-Ross acquisition, which put the contract drilling firm into the oil business for itself as well, was readily recognized as "the *diamond* of this whole organization," as Chris Zeppa expressed it. Although the strike didn't make the company rich and secure overnight, it did represent "money in the bank," so to speak, that held the company together in these difficult times.

"It was an outpost [lease]," said Chris Zeppa. "It was, you might say, what they call a dangerous proposition. Well, they drilled that well. It was the biggest producer you ever saw. That was east of the oil field, as many geologists used to think about it. So the Hughey-Ross is really the one that put Delta in business.

"They had production before they started! And there was another lease with two wells, another lease with four wells. They had three leases in town, right in the gut of town, in the middle of Kilgore. So that was the start. Up to that time they were working for other people."[35]

Opportunity ripened before one's eyes, but the means with which to exploit it were usually scarce. As Joe Zeppa remembered it, "If a man had not much money—$100,000—he could have bought stuff that was proven! All he had to do was to punch a hole in the ground. Knew he was going to get wells. And then all he had to do was hope to sell the oil. But nobody had any money. Well, you know what happened shortly after we started in business? Roosevelt was elected [in 1932]. Closed the banks. Couldn't get a dime from anybody.

"The leases in the East Texas field were cheap, because nobody expected them to cover that big territory. And then finally when they brought in the third well, north, people said 'Well, it's bound to be one field, because it's all the same zone.' Then it was just a question of where you wanted to buy. Because there were a lot of people that held leases way down the road without selling them. And many of them sold those leases after the wells came in, around $25 or $50 an acre.

"First of all, you couldn't sell the oil, to speak of, very well, and the fact was, nobody had any money to drill the wells. So you could get any number of wells to be drilled where you would get three-fourths of the oil and you would give the man that owned the lease a fourth. As a royalty, an overriding royalty. Course, those were not *big* tracts, but there were a number of small tracts that you could get; and even after the field was three, four, or five years old, you could still make deals of that kind. Quite a while."[36]

Meanwhile, Delta's contract drilling—which was to be its "bread and butter"—continued energetically as the East Texas field was developed. From the Falvey contract, the company spread to other associations through various personal connections made by Zeppa, Stacy, or Nick Andretta, who joined Delta in the early 1930s. Delta drilled a number of wells for large companies like Shell Oil, Sun Oil, Sinclair Oil, British-American Oil Company, and Lone Star Gas.[37]

VII

Although the first few months must have been heady ones to Zeppa and Stacy as they completed their contracts and moved on to other locations, enjoying dividends along the way, in some ways those events, despite the lack of cash, were the most promising. Financial juggling was required to ensure Delta's solvency from the beginning; it was to be a way of life for some time.

"You look at the early minutes of Delta Drilling," said S. L. "Bubba" Florsheim, Jr., a latter-day board member. "It makes some interesting reading—borrowing from Peter to pay Paul."[38]

The minutes during the 1930s refer repeatedly to new loans to be made as the company operated, met its commitments, paid its employees and executives, and sought growth. Besides the El Dorado bank, loans were made with the First National Bank in Shreveport, its namesake in Dallas, the Commercial National Bank in Shreveport, the Continental-American Bank and Trust Company of Shreveport, and several others. One thing that necessitated such financial maneuvers was that, in most instances, Delta was drilling not for cash but for payments out of the oil. By that time a well sunk in the East Texas field was a sure producer, for the field's limits had been defined. Cash was still scarce; oil was plentiful. Because of the wells that could be drilled fast and relatively inexpensively in that huge field, the "black giant" of East Texas had been instrumental in bringing the price of oil down to a dismal low. A man might pay handsomely out of the oil his drilling crew found for him, but could he always sell the oil—and for how much? These were the uncertainties that plagued oil operators and drilling contractors alike during the Depression years when oil was not always a substitute for money.

In the troubled thirties, even the best-laid plans sometimes went awry, but the Delta management, perhaps with occasional good luck, seemed always to meet the challenge. A case in point is a 1933 problem. Delta had drilled three wells for the Mazda Oil Corporation in the East Texas field, but the customer was unable to pay an $18,000 balance owed Delta. To avoid "a costly and long drawn out litigation," Zeppa negotiated with Mazda for a one-half undivided interest in the 13½-acre leasehold and the three wells. Then he turned around and sold this interest to Sam Sklar for the $18,000, half of which was paid in cash and the other half applied to Delta's indebtedness with the Louisiana Iron and Supply Company.[39]

As Florsheim suggested, Delta was scrambling for money periodically during the 1930s, particularly in the early years. At one point late in 1934, Zeppa called a special board meeting to gain authorization to borrow $137,500, apparently already negotiated by Zeppa, and an additional $57,500 if needed. For this purpose, the company mortgaged various oil and gas leases in Gregg and Upshur Counties, in the north end of the East Texas field, including an interest in the Hughey-Ross lease.[40] As can be seen from this transaction, the Hughey-Ross lease not only flowed cash dollars into Delta's coffers, it was also near at hand, like Old Faithful, when a banker required solid collateral. Since the loan presumedly would be repaid in time, the collateral could be used over and over. As Chris Zeppa indicated, the Hughey-Ross lease was a jewel in Delta's crown.

A glance at some of Delta's drilling jobs will provide some insight into the nature of a small company's "keeping its head above water" around 1932. In the spring of that year, for example, Delta drilled several one-well contracts on small tracts in return for oil from the wells: one well for Faustine Dotson on a 5½-acre tract; one for W. G. Humphreys on a 3¾-acre tract; and another for J. W. Sewell, as trustee, on a 3¾-acre tract. For each well, Delta was to be paid $31,500 from oil the well would produce. In other words, they were paid nothing till the oil was produced, and $31,500 worth of oil in the earth did not necessarily translate into $31,500 in the bank. As a matter of fact, Delta sold to a C. E. Florence of Gilmer, Texas, $15,500 of the Faustine Dotson oil payment for $8,250 in cash. While this represented a slashing discount by Delta of its property—about half of its presumed value—it was a fair deal for both parties. Florence would almost double his investment in due time, while Delta had the dearly needed cash from the transaction with which to operate. [41]

VIII

"Sam Dorfman had his residence in Longview, had his store over there. We had our office in a two-story apartment house, small apartment house, and we had rooms over the office where we slept." [42]

Thus Joe Zeppa described Delta's modest headquarters, after the initial focal point soon shifted from Dallas. The center of the company's activity was in the East Texas field, which made Longview a logical office site. Another, even more pertinent factor was the closeness of supply houses, especially the Louisiana Iron and Supply yard. Zeppa and Stacy first operated at 107 East Whaley Street, but in 1932 moved into the next block, at 209 East Whaley. Except for one small apartment on the first floor, Delta leased the rest of the frame building, maintaining its offices in the remainder of the first floor, and also using the entire second floor with its three bedrooms, a dining room, and two baths. A large garage housed the company's two trucks, purchased subsequently.

The central staff was few in number. Annie May Jones was in the office, taking care of the payroll and certain correspondence, along with a clerical assistant, Evelyn Taylor. B.W. Freeman, the company's only mechanic, worked in the shop at the back of the building. Johnny Kelly, a black youth, worked upstairs as a cook and waiter.

Zeppa and Stacy were usually in the office. Occasionally Dorfman would join them there. Stacy spent a lot of time in the field keeping tabs on drilling operations and maintaining contact with customers and prospective clients. Stacy frequently stayed in the apartment overnight so he wouldn't have to drive nightly to Shreveport, where, a little later, he stationed his family in a splendid home on Fairfield Avenue. When the Zeppas moved from Longview to Tyler in 1933, Zeppa also would occasionally stay overnight in the apartment.[43]

From this command post, the handful of individuals performed the essential duties that later, once the company had grown, scores or hundreds would be required to do. It was a work-filled atmosphere, with Zeppa the sometimes impatient center of activity. On one of those busy days, as early employee Herman F. Murray recalled, Zeppa yelled out to B.W. Freeman in the shop, "Freeman, I have to go to Austin. Service the car." A few minutes later, while Freeman was still complying with the assignment, Zeppa energetically strode out again and enquired, "Freeman, the car ready yet?"[44] Time, like money, was a commodity one did not spend lightly.

In 1933 the Zeppa family moved to Tyler, a more established, sedate city outside the rough-and-tumble environment of the oil field itself. Many oilmen tended to move to Tyler. In many ways Tyler also made a better "center" than did Longview, Kilgore, Gladewater, or Henderson. After all, Tyler had more experience at this sort of thing. In the nineteenth century, the city had been known as the "capital" of East Texas.

Automobile travel was particularly uncertain on the winding, narrow roads of the early 1930s, and Joe Zeppa, like many other oilmen or company executives, spent much of his time on the road. After the Zeppa family moved to Tyler, it was frequently late at night, after a long, tiring day's work, before he got home.

"The Delta office was still on Whaley Street [in Longview] there, and Mr. Zeppa was traveling back and forth," said his stepson. "But he would get tired and he would roll his car over several times. Never was injured. He used to routinely roll his car over the highway—demolish the thing. Traveling late at night, and tired, [he would] go to sleep. And in those days there was always a lot of wagon traffic on the road, and they would keep a coal-oil lantern burning on the tailgate of their wagon, and you'd come on them unexpectedly and have to suddenly turn off, and the shoulders were always narrow, and the roads were winding and you'd just roll over. But he was never injured."[45]

This personal move paved the way for a corporate move a few years later. In 1937 the stockholders officially moved the principal office to Tyler, at 402 Citizens National Bank Building.[46] In addition to Tyler being a larger city and more of a business center, Delta by then was looking outward, beyond East Texas. At the same time, the company could continue its activities in East Texas, where it owned oil properties as well as enjoyed drilling contracts. But another, perhaps dominant, factor was that, operating from Tyler, Joe Zeppa would not have to risk highway accidents on his way home from Longview, tired, late at night.

IX

What was it like in East Texas? Joe Zeppa, comparing the boom there to the earlier one at El Dorado, remembered East Texas as "orderly" in comparison. The East Texas field, by virtue of its size, was more spread out, which possibly tamed it some. But it was more wild than mild.

"There was, of course, a lot of rabble-rousing and chicken-houses and night clubs and things of that kind," Zeppa said. "And of course there were no roads in the oil fields in those days and consequently you used mules. Mules and horses. Mules, principally. And you couldn't go hardly any place without having some mules to haul you out. That is, in the winter time. The first winter we were in business—1932—it rained, it seemed to me, forever. We couldn't go *anyplace* hardly without carrying boots and raincoats."[47]

A traveler crisscrossing the country in the early 1930s would have observed a remarkable pattern. All over the United States, with few exceptions, misery and depression sapped city after city, region after region. But East Texas—a region historically habituated to hard times—was full of hustle and bustle and, for many, prosperity. Oil made the difference.

A traveling salesman who regularly drove his 1930 Model A Ford coupe through towns as he called on wholesale grocers has given us an indelible impression of the times. The contrast Floyd Sexton found between the boom region and the rest of the countryside was striking.

"You see," said Sexton, "during the thirties in East Texas, the field was booming while the country as a whole was having a depression. A bad depression, and you could go from, say, Tyler or Ardmore [an oil boomtown in Oklahoma at the same time] to a town 75 or a 100 miles away where the Depression was going on and you'd pay 25 or 35 cents for a meal, and in the oil-field areas, why, it would be at least a dollar, maybe more. All these boomtowns were really something."

The mud didn't affect Sexton much, for the main highways then were graveled, which took him from one town to the next, and the highway generally ran through the center of town. But if he turned off either way from the main road, he likely would be in trouble. In the oil-field towns, it was a common sight to see four-mule teams pulling boilers and other kinds of equipment, and it was not unusual to see a six-mule team.

His first impressions remained with him nearly a half-century later.

"The first thing that impressed me so very much was in Longview— that's where I first saw people sleeping on the streets on cots. They were still doing that in 1931. I'd never seen people living in those conditions and so many people in so small an area. They were just off the streets, and as far as I remember there were no sidewalks and very few paved streets. It was just mud.

"And then the next trip I came around was six or eight months later. There were more wells and there were some facilities that had been built in the meantime. Tin buildings were put up quickly. And rooming houses. Siding was like these little metal buildings. Seems like the thing you saw quite often was an oil-field supply. They were tin buildings."

Later he went to Kilgore, where he saw oil wells closer than he was ever to see them anywhere else. At a residence, there might be one in the front yard, another in the back yard.

"You were very much aware of the fact that everything was being done in a hurry. Seemed like the people who were putting down these wells over here, they wanted to beat those who were putting those down over there. And that's why they had the tin buildings and that's why they didn't have paved roads or streets. They just plowed through the mud. Big, big hurry, seemed like, and I don't know why they were, because the field lasted and is still going!"

Different towns within the East Texas boom region had different characteristics.

"In Longview," said Sexton, "there were a lot of supply houses built. In Kilgore there wasn't very much. As I recall, there was a wholesale grocery—a tin building—and then there was an oil-field supply house there in Kilgore. But Gladewater and Kilgore were pretty much the same. They just took care of the actual workers—rooming houses, eating places—and Longview had the supply houses. Of course, Gladewater was on the railroad. Longview, you see, was right on the railroad, and they got a lot of their supplies in by rail. Truck traffic was not so strong there.

"I didn't stay at Longview or Kilgore. I always drove on to Tyler because Tyler had good hotels, about three there, but the big one was fairly new—the Blackstone. The Blackstone was the last word in hotels. It compared very well with hotels in large cities. It had more rooms with baths, for example. And they had, as I recall, ceiling fans at that time. It was a very comfortable hotel. Good restaurant. It wasn't always easy to get a room. You'd have to call ahead. They had an influx of people, but, see, most of the people who stayed there were just traveling men or oil-field executives. None of the oil workers stayed there; they were out in the field.

"You'd see men making deals in the lobby of the Blackstone. You'd see that all the time. Well-dressed, of a particular-type dress—it's kinda like the man who runs the big cattle ranch. He has a distinctive-type dress. Most of them wore boots and big hats, but a different kind of suit. More or less just a business suit, you know, no fancy frills like the cowboy suits.

"Well, there was no Depression in that area. There was no Depression at all."[48]

Despite the prosperous surroundings, the Depression still assailed many. Others remembered that "everybody was broke," but friendly. Many were looking for a small loan, five dollars, even less, just to survive on, to buy a bowl of chili and find a place to sleep, while they were trying to close a deal—after which, they promised, the money would be repaid. Very frequently the money was never repaid even when the deal was closed, and usually it wasn't.

"But everybody was more or less in that fix," said Nick Andretta, "and if there was anything that was hot they'd lean over your shoulder just like [in the TV commercial] 'when Hutton speaks, everybody listens,' you know. And the Blackstone Hotel was the focal place. That was it, in Tyler. Everybody hung out at the Blackstone Hotel—brokers and people coming to town. You couldn't hardly get a room there and everybody trying to make a deal, listening, and if you had any money everybody was after you, trying to promote you. There were a lot of deadbeats, and some of these deadbeats got rich."[49]

When the East Texas field came in, "it flashed all over the United States," as one man put it, because there wasn't anything anywhere else. People thronged in on hope—"on hope of finding something."[50] Most of them headed for Texas any way they could. Marvin Williams saw the boom as a ticket to a better life and struck out for it after finishing high school in Arkansas.

"I read about it in the paper, and I decided that I didn't want to be a-walking behind a mule. I grew up on a farm, and I was tired of the cows and the mules. Then I left, a-walking, on May 7, 1931, coming to East Texas. In fact, I walked 16 miles before I caught me a ride and I got it right out of Magnolia, Arkansas, and the man, he was coming to East Texas in that Model A Ford. He said, 'If you don't give no problem, we'll go right on.' And he pulled an automatic pistol, one of the U.S. Army issue automatics, out of the side pocket of the door."[51]

The unemployed and the desperate seemed to throng in from everywhere. "They came in riding freight trains in there," said W. L. "Buster" Medlin, who witnessed it at Longview. "I have seen a family that had five kids, and I'm pretty sure she had a baby in her arms, and they didn't have a dime in the world. At night the people would sleep in the streets if they could, and in old buildings. Anywhere they could find a comfortable place to sleep without standing up. They had them old rooms and rented a cot for about 10 or 15 cents a night. There would be one man renting them and about four or five men would come in and lay on the floor beside them, as many as could stack in one room."[52] Others slept on the grass in the park.

Some of them, like Dewey Strength, dug piplines with sharpshooter shovels for 20 cents a foot. "Somedays I would go out and make $3.00," he said. 'Come back in a get a bowl of beans over across the railroad in

Kilgore for 15 cents, all of the cornbread and beans you could eat—if you could get the 15 cents. People slept on pipe tops, and cardboard houses all over that part. They finally put a few flop houses in there for a quarter a night. I slept wherever there was a rig going. They flared every well they could bring in. They had flares all around them and you could get under them. By the hundreds, people would sleep out there to keep warm."[53]

Ted T. Ferguson remembers the fierce competition for even the hardest jobs at common labor. "They'd pay 20 to 50 cents a joint to dig a ditch, a foot and a half deep, to lay their line in, and according to what you dug, that's how much you got. And there would be anywhere from 10 to 100 men out there every morning in case you didn't show up or fell out or got run off—why, they was ready to step in your place. It was rough. People think they've got it rough now. They have never seen rough times until they go back in that country, back in that era.

"I've never even read of it being like that, myself. They don't know about those 'good old days.' You could go down and buy a sack of groceries that would last you all week for three or four people for $5—if you could get the $5. But you'd work all week for $10 or $12, too. Didn't have no taxes to pay and wouldn't have nothing to pay it with if you had it. So it was rough, I tell you. The younger generation don't know what hard times are—put it that way.

"I raised a girl and boy during that. Both of them came along during that, one in 1931 and one in 1933, and boy, it was rough to make a living. I mean just a living; wasn't no what-you-needed, it was just what you had to have to exist."[54]

Thousands sought work on drilling rigs. A roughneck could earn $6.00 for a 12-hour day. A driller could make $10 a shift, and was glad to get it. A family might even save a bit on a driller's pay. Nobody complained about the hard work. Most of the early oil-field workers had lived on farms as boys or young men, and consequently they were used to long hours and hard work. The roughnecks of that day frequently had to be ready for any emergency. If a truck rolled in with a load of pipe and got stuck 75 yards or so from the rig, why, the roughnecks went over and carried the pipe by hand to the rig. "You reached and got it," as one said.[55]

The East Texas boom was "kind of wild and woolly" for the laboring man in more than a few ways. Danger lurked not only in honky-tonks and gambling dens, but also on the job. "They had them old steam engines down there," said one veteran of those days. "First of all, they had no safety devices. If a man got killed, they'd drag him off and bury him and hire another man. It was just about that bad."[56]

The roads were among the most memorable characteristics of the boom region. Eula Carter, an oil-field wife who lived with her husband "Ark"

Carter on a lease being drilled during the first part of the boom, remembered they couldn't even get into Kilgore because of the mud—a distance of only three and a half miles. Heavy rains and heavy loads of equipment kept the roads practically impassable. It might take nearly all day to go to Longview and back, a distance of 13 miles, over the oil-fields roads. Sometimes the men would have to walk the three and a half miles into Kilgore to buy food.

But as hard as the conditions were for men, they seemed worse for the mules that kept the oil fields going. "I've seen mules with mud up to their ears where they'd fall down." said Mrs. Carter; and Dewey Strength, who had been in El Dorado before moving to Kilgore, recalled parallel sights in both towns. "They moved in boilers right through the streets [of Kilgore] with mud up to the belly of the mules," he said. Marvin Williams observed one harsh scene that could be duplicated in a dozen or more other oil booms that preceded East Texas. "I've seen in the streets of Kilgore where mules would drown trying to pull a boiler through. In that mud."[57]

Exploiting the situation, many small-time entrepreneurs would construct toll "bridges" over the worst mudholes or provide alternative routes over their property—all for a fee. Even if the toll over such a route was only 10 or 15 cents, that was a great deal of money, and before one had finished his journey he might have to pay five or more dollars, an exorbitant sum.

"You didn't see many cars because the roads was so bad" said Buster Medlin. "The roads would have real big mudholes in them, and if a little car'd drive through it, them little cars would just disappear, the hole was so big. If there was a big mudhole, the farmers would just cut a road on his place around the mudhole and charge people to use it, and they'd have to use it—that was the only way around the mudhole. One old boy would haul water all night long to fill up a mudhole down in the summer so he could keep his toll road open where they could keep them coming around it. It was sad. He'd pour all that water in that mudhole."[58]

But even toll roads were not always a certain route. Marvin Williams recalled the time he and others, disgruntled at the toll bridge of a few boards one entrepreneur had erected over one mudhole, decided to take an alternative "toll" road.

"I went there and it pouring down rain, and here was a man that had an ax on his shoulder and said follow him, he had a short route and we wouldn't have to go across the bridge," said Williams. "And we—ten of us—followed him back of a house through a field, and he was going to show us around. He carried his double-bit ax, had it on his shoulder, and he took two bits from each of one of us, took up his collection and we followed him, and today I've never seen the man again. He never did stop. He just kept going. He kept walking with the ax, and it pouring down rain and we was all stuck. He made it fast, and he was gone. He had a slicker on, and a double-bit ax. And finally L. T. Campbell, a trucking company, came out from Greggton and pulled us out. They did a lot of work for Delta Drilling Company later on."[59]

At the height of the boom, the mass of humanity that crowded into Kilgore, Longview, and Gladewater frequently precipitated the kind of friction that booms have generated from the beginning.

"There was a lot of honky-tonks and lots of home-brew joints and some few fast women and stuff like that," said Nathon Bacle. "It was just typical oil field. You throw a bunch of people together like that, and there's no telling what you're liable to come up with. They's lots of people in there had their cars locked to the railroad tracks—you know, for not having the license tags on the cars—by the law enforcement agencies during the boom. So they chained the cars until they could get enough money to register 'em, put some tags on them. If they'd arrested him, they couldn't get no money out of him. They'd just impound the car."[60]

Some of the newcomers were "tougher than a boot."

"I was over at Kilgore one night they had a chain over there and [were] a-chaining people on it," said Bill Dailey, an early Delta employee. "They didn't have no place to lock'em up. They'd arrested 'em and they didn't have no place to put 'em. Well, they was fighting and drunk. It wasn't nothing to walk into a place and see two or three fights."

Two badge-carrying tough guys from those days were Texas Ranger Manuel T. "Lone Wolf" Gonzaullas and a railroad detective known widely as Texas Slim or Texas Bull.

Dailey recalled: "I don't know whether it was a policeman or deputy, whoever it was, locked a pregnant woman to that chain, and, boy, [Gonzaullas] liked to got him. That old Gonzaullas told him, 'Hell, take her home with you if you want to, but you're not going to leave her tied out there.'

"And he liked to got old Texas Slim, that 'bull' on the railroad, down there at the junction one night. All them hobos coming in on them trains and I wouldn't say whether it was a woman or kid or what it was, Texas Slim slapped him off a train down there one night and Gonzaullas went down there and jumped him up. Texas Slim tried to tell him he was doing his job, and Gonzaullas said, 'Well, you'd better do it right, then, and do it careful, because if you ever do another stunt like that you're going to kill me or I'm going to kill you.' He said, 'Remember that.' That junction was a mean place and they was hobos of every kind.

"Well, you could go on up in there between Longview and where I lived and you could hear rigs running—sounded like these wood sawyers [larva which burrow under the bark of dead pine trees,] you know, in a pine tree, sucking that sap; they call them sap suckers, just sawyers going zap, zap-zap-zap, rigs running in every direction. Everywhere, and you'd see people laying out there asleep. They didn't have nowhere to go and they'd just sleep where they could, expecting to go to work."

As with previous booms in history, there also was the expected influx of gamblers, bootleggers, prostitutes, and criminals.

"Plenty of them," said Daily, "and, you see, beer wasn't legal at first there and when it did come in they had what they called 3.2 [percent alcohol] beer. You'd get sick on it 'fore you'd ever get drunk on it. Wasn't nothing but a belly wash. But now there was plenty drinking. They had all the unlawful liquor you wanted—moonshine, home brew."[61]

Some people were so desperate they broke into grocery stores at night for food. "They would get anything they could," said Elmer S. "Whitey" Young. "Get them something to eat. Canned goods . . ." His wife added, "We lived through that. It was horrible."[62]

Occasionally robbers would hijack a rig for the men's wallets, but times were so hard that crime paid poorly even for those who got away with it. Buster Medlin remembered: "Two old guys came out one time and robbed the whole crew of four men out there. Robbed them of all they had and got about three dollars off all of them, and [took] some of their best clothes off the rig."

Hard times wrought hard men, and sometimes honest men faced risks as great as crooks did. Medlin was on his way home from Gladewater one dark, cold winter's night when he unwittingly almost became a victim in an ambush.

"I was coming out of Gladewater and I had this old car and the windshield wiper wasn't working," he said. "Boy, it was cold, about eight-thirty that night, good and dark. It started freezing and my windshield was froze up. I saw a light and I pulled up by this little old liquor store. It had a upstairs living quarters. I pulled around up in that little old lot and stopped to take my knife out and scrape the ice off my windshield so I could see.

"And what had happened was an old boy had escaped from prison and he come up there and talked to his buddy and what he wanted his buddy to do was help him and another friend rob that liquor store that night. His buddy told him, no, he didn't think he wanted none of the action. He tipped the cops off.

"Ol' Ray Watson, the sheriff of Wood County, and the Sheriff of Upshur County and [Texas Ranger M. T.] Gonzaullas were all upstairs in that little old place. Ol' Ray Watson was the sheriff up there at Gilmer. Man, they were ready.

"I didn't know when I stopped that they were up there, but later they told me. They had their eyes dead on me, but ol' Ray Watson hollered and told them, 'That's Buster Medlin!'

"About ten minutes later I was driving down the highway and I hear a siren blowing and guns started shooting and they shot both them old boys. You ought to seen that car. Them old boys were loaded for bears. They had sawed-off shotguns, three or four rifles, pistols, and everything in that car. They were going to knock that liquor store off and get away. They were prisoners; they had broke out of the penitentiary. They'd killed three people. So [the officers] weren't gonna give them a chance to kill more.

"Old Gonzaullas was the first man to come down the stairs. They fired at him and he shot that car full of holes. This old boy jumped from the front seat to the back and laid down; he had already been hit. Old Gonzaullas walked up to the car after the shooting and this guy raised up and shot the strap on Gonzaullas' shoulder with a .38 caliber. It was that close to him. He was already shot, but he just waited for someone to walk out. That old Gonzaullas unloaded on him. Bullets went clean through that car, seem like it busted the motor on one side. They had them .30-.30 high-powered rifles."[63]

"It was rougher than hell then," said Marvin Williams. Honky-tonks and beer joints attracted oil-field workers, particularly around the Longview area. The Bluebird Dance Hall featured juke boxes, beer, and lots of girls. Then there were the cheap beer joints, and the brothels where the top price was $1.00. In Kilgore, as one observer put it, "They had one whole street down there—well, there was no other way to say it: they was just whorehouses." Most of the virile young laborers didn't bother with commercial sex, though, for there were usually enough unattached young women in the beer joints and dance halls.

The night clubs in the region then were Mattie's Ballroom and the Palm Isle. Cover charge for a couple came to $5.00 a princely sum at the time.

"Mattie's didn't have the big celebrity bands," said Marcus O. Jones. "She had mostly just local bands, but occasionally she would have a big band out there. The Palm Isle is the one that had the big bands, like Bob Crosby, Cab Calloway, Jan Garber. Mattie's was out there on the old Henderson-Longview highway about halfway between Kilgore and Longview. That place was crowded every night."[64]

Seymour L. "Bubba" Florsheim, Jr., later to become a member of the Delta board, entered the pipe business with Louisiana Iron and Supply several years after the East Texas field opened up. The atmosphere still hadn't calmed down a great deal in Longview.

"That was a rough-and-ready town then," he said. "I went there in '37. Everybody was drilling. We knew no hours. We worked . . . till we finished. It was *still wild.* Longview in those days was 10,000 people, dirt streets, and the only place to go for entertainment when I first went down there [from Dallas] was a place called Mattie's. It was infamous. On the Longview-Kilgore highway. The original Mattie's down there. And you'd go out there and the richest, most influential person in town would be sitting next to the roughest roughneck. It didn't make no difference. That was the only place there was to come. Except for the *real* honky-tonks, where you wouldn't dare to walk into. And we had pretty much open gambling. Longview was wide open, bootleg, gambling."[65]

East Texas, the last of the mule-drawn booms, came in the twilight of steam rigs and wooden derricks. The petroleum glut also brought cheap oil and gasoline. Oil frequently sold for ten cents a barrel, and a lot went for as little as two and a half cents a barrel.

Buster Medlin spoke for everyone when he said, "We will never see those times again, never."[66]

X

With producing oil wells scattered over the East Texas landscape like grazing cattle over a huge ranch, it was common for an oilman to spend his leisure time looking over his petroleum domain as would a rancher over his cattle spread. Joe Zeppa was no exception, regularly taking his young stepson, who has left us a memorable picture of those days.

"Every Sunday I would go with Mr. Zeppa in the field for years, beginning when I was seven, until I was nine or ten. I used to go with him around the drilling rigs of the whole East Texas oil field. I'd be in my little boots and standing there by some drill site while all these busters were standing around, telling about what they were gonna do. So I met all sorts of people.

"I would go to the rigs frequently with Dr. [Jerry] Falvey's bodyguard, George Glover, who everyone was scared of because he had been accused of manslaughter in several other states. And when a well would come in he would take me and hold me over the blow pipe and paint my face with drilling mud and oil—said it was to give them good luck. Well, it would scare me to death, but I wasn't scared of George. He would hold me out over the thing. He was huge. So I would travel around the field a good bit of the time.

"The biggest thing for me was the change in scenery and variety and different people. I'd never seen unpaved streets, coming from Pasadena, California, and Longview's streets weren't paved except for Main Street, and downtown was paved with wooden blocks next to the railroad depot. And pigs and cows and mules wandered all over the place. And the big casing wagons being hauled with various spans of mules. The mud. . . ."

It was around this time he met Myron Dorfman, the son of Sam Dorfman, "who was the first civilized Texan that I had ever met." Later he would go over to the company's headquarters on Whaley Street and play with Bobby Stacy, just slightly older than he, in the warehouse and shops behind the office.

"I spoke with a Boston accent in those days because my father was from Boston, and that was a problem with me in East Texas. It wasn't very popular.

"In Longview I was known by my original name [Frederick Howes Ruppel]. When we moved to Tyler two years later, why, then my mother and I picked my present names before I went to register in the public school. She didn't want to move to a new town with a child from a previous marriage with a different name. I wasn't adopted till much later. We just started calling me Trent Harvey Zeppa on the day I went to register for public school in Tyler. We just had a list of names, and decided those would be the best ones. The most euphonious."

Joe Zeppa legally adopted Trent in 1940 when the boy was fourteen and had just returned from boarding school in California.[67]

On January 28, 1933, Keating Vincenzo Zeppa—the only natural child of Joe and Gertrude Zeppa, named for his father's close friend George T. Keating and for Zeppa's father, Vincenzo—was born in a Dallas hospital. George Keating became his godfather. That summer Gertrude Zeppa took her two boys to New Jersey to escape the heat and mud of Longview. Upon her return, the family moved to Tyler.

That first year in Tyler, the Zeppas lived in a stylish apartment on South Bonner; the following year they purchased a white house on Chilton Street, a block away, and Nick and Angie Andretta moved into their old apartment. Later, the Zeppas bought an old Victorian house at 714 South Broadway and remodeled it. They were to own it for decades. Situated in a fashionable section of town, the house was close to H. L. Hunt's home on Charnwood, and Mrs. Hunt became Trent Zeppa's Sunday school teacher.[68]

Keating Zeppa was introduced to the oil patch even younger than had been his brother Trent. His mother took him to the oil fields in a basket and, even if he didn't recall those early events, most of the old Delta hands knew him almost from birth. In fact, to these veteran employees he soon became known by a nickname, Pinny or Pin, that they were to remember him by for decades. But, oddly enough, only Keating Zeppa seems to recall how he obtained the sobriquet.

"When my brother Trent was a little baby," he said, "he was very roly-poly, and at that time there was a popular comic strip called 'Puddin' Head.' Puddin' Head was a little roly-poly boy. And so in his early years he was known as Puddin' Head. But this was just when he was a baby and it disappeared thereafter. When I was born, believe it or not, I was shaped like a toothpick. And so they called me Pinhead, Puddin' Head's little brother! And I went from Pinhead to Pinny to Pin. I have no idea how. It evolved."

The young scion's formative years were spent growing up in Tyler in the 1930s and afterwards. His memories were of a busy household as he went to elementary school and, later, Hogg Junior High School within a few blocks of the house.

"Probably one of my earliest recollections was the winter of 1938-39 when we had a very bad ice storm in Tyler," he said, "and the reason I remember it is not because of the ice storm but because my mother's sister, my aunt Audley Lyon, came to stay with us for a few months. My aunt Audley was a rather remarkable woman. She was quite a teacher, not by profession, but by inclination. And as a result of her visit I spent about, oh, three months with Aunt Audley walking every morning. Didn't mean much if it was raining or shining or hot or cold, we went for a walk! We generally went for about an hour's walk, and I can remember the route that we followed—north on Broadway to Charnwood over, round the block on Fannin, back. And she taught me the alphabet, the numbers, and just taught me a lot of things during those walks. She was truly a pedagogue. And, in fact, because of that, I started school in the second grade. And really in many ways that was one aspect of my childhood that has shaped my life ever since, because not only did I start in the second grade when I was six years old, but I also was caught in the change in the Tyler public school system of going from an 11- to a 12-grade system. And so I skipped the sixth grade—it happened when I normally would have been in the sixth grade, so I went from the fifth into seventh.

"The reason I recount these two things is because I feel that because of those—you see, I was barely 16 years old when I got out of high school, and, therefore, I entered Rice Institute when I was 16. And I was no more ready for that than the man in the moon! Emotionally, psychologically, I just didn't have the maturity then.

"And I've reflected on it many times, and how different—I don't know how different things might have turned out. But very simply I feel that my aunt Audley, along with the Tyler school system changing grades, inevitably influenced my development, for better or for worse.

His childhood was different from that of most children.

"I grew up around adults, for one thing. There were not a lot of children in our neighborhood. Trent was six years older, and he was off at school. Because he went off to boarding schools, the years when a big brother would have been more influence upon me, he wasn't there.

"And I was always treated as an adult, and expected to behave as one. In one of the daily newspapers in Buenos Aires [when I lived there later] there was a little comic strip. I forget the name of the principal character, but he was rather globular shaped and also like one of the characters in *Smilin' Jack* if you remember, but not like *that*. The subtitle of the strip was *El hombre que nunca tuvo enfancia* (The man who never had a childhood). And when I'm called upon to reflect back, I really don't feel that I had a childhood as it is commonly defined, simply because I only had adults around."

Of those adults, the most important were, of course, his parents. Though Joe Zeppa was both the real and symbolic head of the family, Gertrude Zeppa exerted her own influence, particularly in the rearing of her sons. Keating Zeppa has referred to his mother, brought up in a strict family, as "the last of the Victorians." He said, "And I mean that in the true sense of the word, not just identifying a period in time but rather an attitude of life and family."

She seems to have been a highly intelligent person, well-read, at times hard-headed, and quite outspoken by some standards.

"She didn't go out of her way to offend people," said Keating Zeppa, "but asked for her opinion, would give it with no punches pulled. If it suited the listener, fine; if it didn't, that was all right too."

A second-generation American, her earlier forebears had immigrated from England, some through Canada, others directly to the States. A brother once traced their family through various pathways as far back as 750 A.D., to a Scottish clan.

The stiff-upper-lip British attitude toward life was reflected in the rearing of children. "It was sort of along the line that children are seen and not heard," said Keating Zeppa. "Not in an unkind sense, but that's just the way it was. You didn't cry if you were physically hurt. You were not encouraged to be, as she would put it, weepy, or as the British might put it, sort of wet and weedy. She was quite proud of her British heritage and was very active during World War II in the Bundles-for-Britain campaign, [even] before America entered the war."[69]

Even before then, at a time her younger son would not have remembered, she had taken a personal part in helping victims of catastrophe. When the New London school building exploded in East Texas on March 18, 1937, taking the lives of 296 persons, Gertrude Zeppa was one of the many volunteers who did what they could. Since she had been trained as a registered nurse, her assistance was highly valued.

"Mrs. Zeppa worked very hard in this New London school disaster," said Milton Winston. "She was up night and day during all that time over there. She was at the school, actually on the scene, doing what was necessary there. And then she went from there to the hospital."[70]

Both parents were "strong people," remembered Keating Zeppa.

"I might change my opinion in the next ten or twenty years," he said, "but my present feeling is that, of the two, Mother was probably the more clever. Mother was more the rapier; Dad was more the battle-ax. Both very effective. Dad more plodding, methodical—sort of like the inevitable push of a Caterpillar tractor or a big bulldozer; Mother, much more agile, much more witty. But both in their own ways intelligent, well-read, curious people—tremendous curiosity."[71]

XI

At the beginning of the East Texas boom, Annie May Jones left her secretarial job with H. L. Hunt in El Dorado and headed for Longview. She found a job at the First National Bank in Longview, and this was where she was working when Joe Zeppa, on the verge of founding Delta, "borrowed" her temporarily—a temporary position that lasted until her death at 68 in 1964. Plump, brown-eyed, cigarette-smoking, and sharped-tongued, "Miss May" virtually was "the office" for a long time, though another woman was soon hired to help out with the clerical work. A person of independent mind, she could hold her own with any of the roughnecks or drillers when it came to an exchange of words.

"She was pretty hot-headed and she didn't take anything off of Mr. Zeppa and them either," said her brother Marcus O. Jones, who also worked for Delta in the 1930s. "They were jawing each other all the time. She'd go to work whatever time she'd get up in the morning and got down there. Sometime it'd be 10:00 or 11:00, and then she'd work until way in the night, you know, in order to get through. I've seen her a lot of times, when she was balancing books, if she was a penny short she would work and work and work until she found the mistake, just for one penny. She would stay there sometimes way up in the night before she would ever quit."

In the office she was tough enough to hold her ground with her boss or a rowdy employee. At home on her own time, she had a taste for fine things. She owned crystal and had a full set of strawberry-diamond cut glass. She played popular songs of the day on the piano.

Playing the piano was one of her ways of relaxing, but like most people who worked around the oil fields, she also liked a drink of scotch or bourbon after work, before her nearly always late dinner.[72]

Few roughnecks or drillers who invaded her domain at the Delta office escaped her wrath. Strangely, there seems to be no record that she stirred their anger. Ada Viola Portier remembered one time May Jones told off her husband, Louis V. "Frenchy" Portier.

"Oh, brother, she got on Louis one time and I laughed till I hurt. He come home just dying laughing. He didn't get mad at her. I don't know what it was, he wouldn't tell me, but she got on Louis and he come home just dying laughing. He said, 'That May Jones eat me out today.' She'd eat out Uncle Joe, too."[73]

Delta's owners weren't exempt from the consequences of her fire either.

"One morning Sam Dorfman came into the office, raising Cain about something," said Herman F. Murray, "and Miss May told him, 'I'm running this in here,' and he popped off something or other at her, and she grabbed a book and run him out of that office! She hit him over the head with a book. It was just a little pad, didn't hurt him, but he got out of there. Some of 'em said, 'That woman ain't got no sense. Why, she'll kill you!'

"She'd attend to her business, and you'd *better* let her business alone, too! That office in there was *her* business. And she'd chew Mr. Zeppa and Mr. Stacy out if they'd come in there messing with her.

"You didn't mess with Miss May. She'd tell off *any* of them. She was friendly, she was likeable, but she was just that way."[74]

One of the few men who seems never to have had a run-in with her was Joe Beasley, later to become corporate secretary of Delta, who found her "very pleasant." As he recalled, "We got along famously." Beasley, who hired on after making Delta's first tax return in 1932 as an employee of an accounting firm, worked with her for about eight years. His formula for maintaining the pleasant relationship: he minded his own business, primarily accounting and tax work, and left hers, primarily the payroll, alone.[75]

Eula Carter, whose husband Ark Carter would also argue with May Jones and then go home laughing after she had chewed him out, acknowledged that May was "fractious" but that "she knew what she was doing and they knew she knew what she was doing."[76]

XII

Delta's early crews were hard-working veterans of the oil patch. They knew their business, which invariably was the main reason they were hired. William B. "Jack" Kennedy, who came to work for Delta in the middle 1930s, quit, then returned in 1937 to stay until he retired, had helped drill the discovery well for the East Texas field in 1930—the Daisy Bradford No. 3 at Joinerville—on an old steam rig. All of the early roughnecks and drillers had been seasoned at fields in Arkansas, Louisiana, and West Texas.

Zeppa and Stacy had chosen wisely in hiring L. E. "Red" Merritt to put together the two original rigs, see that they were run well, and hire the crews. Sandy-haired and hot-headed, as Nathon Bacle described him, he was "a pretty good man—he was 'bull of the woods,' you know." Though an "independent guy," Merritt was highly experienced, having started as a roughneck about ten years before and worked himself up to a field manager's position. As his widow assessed him, "He was known by friends as the fastest driller around that part of the area, and was in demand by many drilling contractors. Being young, energetic, and a man that was not afraid of hard work, he was successful in everything." Merritt was so successful, as a matter of fact, that he subsequently left Delta to form his own company, the Merritt Drilling Company, with one rig.[77]

The first driller Red Merritt hired for Delta was Louis V. "Frenchy" Portier, from South Louisiana. Portier was the day driller on the first rig; Blackie Watson was the night driller. The first driller hired on the second rig was E.W. "Mutt" Hays; the second on that rig, Frank Garland. Garland and Watson didn't stay with Delta long, but Portier and Hays did, along with a great many other roughnecks and drillers hired in the 1930s, men like Luther Milton "Ark" Carter, Jess Petty, and Homer Keel.

Keel, one of the company's three day drillers a few years later, then became both a toolpusher, supervising several rigs, and superintendent of drilling.[78]

Men like Carter, Portier, Hays, and Petty all had in common their affinity for hard work and dedication to their employer. Yet each was distinctive in his own way.

Portier, a rather long-faced man who walked and talked fast, was already in East Texas when Delta formed, roughnecking when he couldn't get a drilling job. Red Merritt, who had known him from El Dorado days, sought him out for the first driller's job at the new company. Portier, set in his ways and characterized by an odd habit of licking his right thumb when he talked, had a Cajun accent he never got rid of.[79]

"A lot of people couldn't understand him," said his widow. He was quick in his movements and, as she said, "He never asked his roughnecks to do nothing that he couldn't do—never would. Now he'd hire you and he'd tell you what your job was. If you didn't do your job, you got that little pink slip. He worked hard all his life. I've seen him lay on his back to put his shoes and socks on and his britches, and get up and go to work. His back was hurting so bad. He got hurt pretty early [in his career in the fields]. And he went to work. There wasn't no lazy bone in Louis."[80]

His liberties with the English language became the subject of many anecdotes. One day Portier went hunting, and upon his return he reported, "I kill this gray squirrel three times and he get in the hole and get away." But it didn't matter how colorfully he handled his words, he managed to have the last say. One of his rigs during the 1930s had a "knee-busting clutch" that had to be kicked in with his knee instead of his foot because it was up on the drawworks shaft. Sometimes the action would send Portier sprawling across the rotary floor. One day when this happened, the derrick man hollered down to Portier, "Hey, Frenchy, why don't you put some sand on that clutch?"

Portier licked his thumb before he said anything, then he shouted back up at him, "By golly, you know so much, how come you no drill?"[81]

But while the language barrier in South Louisiana was a problem for some of Delta's transplanted crewmen, it wasn't for Frenchy Portier. As an apparent "outsider" arriving with a Delta rig, Portier at times had his fun "passing" as an Anglo-Saxon at Ville Platte.

"Let me tell you what my husband did," said his widow. "He could talk French and understand French, and the nasty stink liked to play poker. Oh, he loved poker. And he got down with them Frenchmen and they didn't know he could talk French, and he was playing poker with them.

"He come home one night laughing, and he said, 'Shorty,' and I said 'What?' He says, 'Them damn fools up there ain't got no sense.' I said, 'Why?' He says, 'They're setting up plotting against me to cheat me in poker and they're talking French.' Says, 'They don't know that I understand what they say.' And I said, 'Daddy, that's not right. You ought to tell them you could talk French.' And he went back up there one time and he played poker with them, and they got to doing him the same way, and he started talking French to them. He cleaned them out good. They broke the party up. But he told that everywhere he went. He told that, I want to tell you right now! Oh, Lord. After we went to bed, I'd lay there and laugh."[82]

Among a group of men dedicated to labor, Ark Carter distinguished himself as being the "toughest" of the lot. In some ways, Carter, more than six feet tall and around 200 pounds or so, was almost larger than life.

"Ark was a hard-working Arkie from Arkansas," said H. J. "Red" Magner. "Mutt Hays and Frenchy Portier, they came off the farm and knew how to do things and were strong. Ark was particularly strong. He never wore gloves and he'd grab hold of one of those steam lines with his bare hands and you could hear that old meat a-frying in there and he'd keep right on going. He was tough.

"They were all big strong men in those days. Labor didn't have that good a deal and there were guys on street corners that had a raggedly pair of gloves and a lunch all packed, waiting for you to quit or whatever. And so drillers could be pretty mean and pretty nasty. All he'd have to say was, 'Now I want this done,' and you'd jump or else you got replaced.

"Ark was of that caliber. He wasn't, obviously, an educated man, and he always had somebody that could figure his books for him, and Guy White used to do a whole lot of his report things for him, but they knew how deep they were [drilling at anytime] and knew what was happening down in the hole, and they knew mechanics and how to fix it. They were just really intelligent men."[83]

A hard-working man who wore a size 16 1/2 shirt and whose broad shoulders tapered down to a 34-inch waist, drawling Carter never met a stranger. Once he met a person, he never forgot him and would always remember where they had first met. Guy White, who roughnecked for him in the early 1930s, called him "the hardest-working driller I ever saw in my life, and he expected everybody else to work hard." The one thing about him that everybody remembers is that he always worked without gloves;

he said they got in his way. Consequently, he would nick his hands now and then, and once, in later years, he lost a thumb to a piece of machinery in Wyoming. After it healed, he went back to work.

Carter was the day driller on the discovery well in the Schuler field in southern Arkansas in the late 1930s. But, strangely, he never wanted to be a toolpusher, promotion though it was, because he didn't want to be running around from one rig to the other, and up nights.

Joe White, who started roughnecking for Delta in the late 1930s, remembered, "He went bare-handed all the time. I've seen him walk up to that steam line and rub his hand on it. As far as toughness, he would get right in there and work with you just like one of the crew. He didn't stand back and watch you when you were rigging up and tearing down. Them days there was a lot of work on them old steam rigs. We all had to work *hard,* and he'd get right in there and get the biggest load there was."

"He was a good driller," said Herman Murray. "He was a good man to work for. He just told you what he wanted done, and if you done it he was happy. He knew what he was doing, and how to do it. He was a company man—he wanted to do what was right for the company.

"You didn't burn him out. No way. And he didn't say, 'Go do it.' He'd say 'Let's go do it' and he'd go with you. If you were toting that pipe, he'd get out there and tote with the roughnecks. Never put on a pair of gloves— no way would he put on a pair of gloves. Said he couldn't work with gloves on.

"And going in the hole, you could might' near see the steam coming from his hands, he'd hold his hands on that *brake* drum. He was tough! He was just a big ol' tough, raw-boned guy, tough as nails. Never seen him with a pair of gloves on in my life."[84]

Jess Petty, another old-time driller, was a friendly, jolly type, something on the order of Ark Carter. Hi Cole remembered once, when he was a salesman for Continental Supply, taking his five-year-old daughter when he called on Petty's rig. As soon as Petty spied Cole's daughter in the car, he summoned one of the roughnecks over to take his place at the brake. Petty started talking to little Beverly Cole, and it was apparent he loved children.

"Do you like watermelon?" he asked.

Certainly she did.

The big driller said, "Well, let's go out here. I am going to pick one off the tree for you."

He went out by some trees and returned with a melon he had brought with him that morning, and called some of the roughnecks so they could cut the watermelon. Beverly ate some of the melon, and the event stayed in her memory forever after because of the wonderful man who had told her watermelons grew in trees. Big, hard-working Petty, it appeared, had an imagination equal to a child's—an uncommon trait in his rough, sweat-filled world.[85]

Most of the early drillers had limited educations, but, like E.W. "Mutt" Hays, used their native intelligence to get the job done.

"Most of them didn't have any education," said Red Magner, "Mutt Hays [one day] said, 'School Boy, what does that sign say up there, that they put on that?' and I said, 'That's "Delta Drilling Company, Rig No. 6, Tyler, Texas," ' and he studied a minute and said, 'Hmmm, it looks like a cafe sign down there across from the Emory Hotel.' All he saw was a mental picture. But they had fantastic memories and they could remember to the nail or the size of the footage or the depth or whatever you needed for 15 or 20 years, exactly."[86]

Hays, a "great big" raw-boned 210-pounder, identified so strongly with Delta that he seemed to equate the company's well-being with his own, if we are to judge from one anecdote which Hays later told Joe White about an incident during the Depression.

"He was using six-inch drill pipe in East Texas to drill with," said White, "and they twisted it off down in the ground and they didn't have what they call an 'overshot' to go in there and get it. Now Mr. Zeppa wasn't there at the time; I believe they said he was in New York. And Mutt Hays went to a machine shop in Kilgore and put up $150 of his own money to get an overshot to go in there and fish that drill pipe out of the hole. Well, Delta, their credit wasn't good enough them days—these rental companies during the Depression, they wouldn't do it unless you put the money. That's what Mutt did. He went to the bank and got $150 and put it up to get his overshot to fish this drill pipe out of the hole. Homer Keel, one of the old drilling superintendents, he verified that, too.

"Course, Mr. Zeppa refunded his money. But at that time he needed it quick. See, if he had waited a day or two, his pipe would have been stuck down there, but when they got this overshot on it, they could pump through it and come on out of the hole. That must have been along about 1935."[87]

This incident doesn't surprise men who knew those old-timers.

"These guys were dedicated to Delta," said Hi Cole. "They were thoroughly dedicated men. Delta was them. That was their company, because they knew Joe Zeppa personally. They believed in him."[88]

XIII

When Herman Murray went to work for Delta in the 1930s, as he recalled, "We were just about one big happy family then. Even the owners— they were just like the rest of us."[89] Some remembered that when a crew was drilling in the East field at El Dorado, on occasion Joe Zeppa would drive up to the rig at midnight and visit with them.

Dorfman

These five gentlemen in 1931 formed a drilling company in Longview, Texas, with two junk rigs and a lot of ambition. Fifty years later, their company has grown into a thriving business and a leader in the contract-drilling segment of the energy industry. Although none of them are still living, their spirit continues to pervade Delta Drilling Company. These men were Sam Dorfman, Sam Gold, Sam Sklar, Robert Stacy, and Joseph Zeppa.

Gold

Sklar

Stacy

Zeppa

Sam Y. Dorfman was a persistent partner, always voicing his opinions and reviewing the company's operations. Sometimes his opinions differed from Joe Zeppa's, but Dorfman remained active in company affairs until his death in 1957.

*Mr. and Mrs. Sam Sklar at their home in Shreveport, Louisiana. Sklar
was a partner in Louisiana Iron & Supply, a company which provided
Delta with much of its equipment and drill pipe during the company's
early operations.*

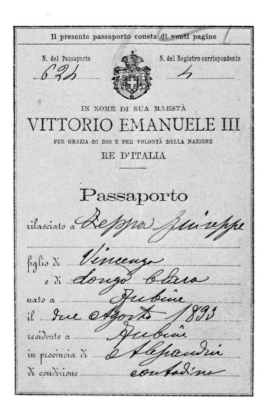

This passport was issued to Joseph Zeppa when he emigrated to the United States from Italy at the age of 12.

This letter is from the correspondence which Zeppa maintained with Miss Helen Adams, with whom he worked at Shearman and Sterling on Wall Street.

This excerpt from one of Joe Zeppa's diaries during World War I illustrates his facility for observation and invention, a quality he exhibited throughout his business career.

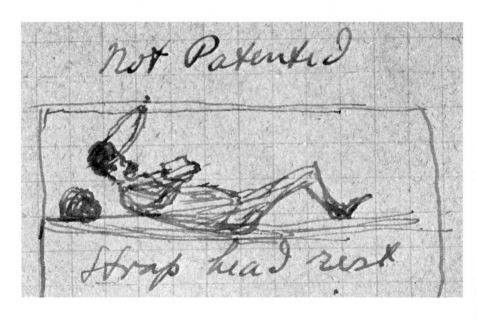

-5-

After Reveille.

A day of rest.

The

these might

Sunday.

Why is it

Pays the

heart ache

France,
Evening of Aug 31/18

Dear Helen:

Your letter of the 25th came

daily expecting

Joseph Zeppa
Co. E. 11th Engrs
Amer E.F.

Miss Helen C. Adams,
55 Wall Street,
New York City,
U.S.A.

PASSED AS CENSORED ★ A.E.F. ★ A 593

given us much the same opin
of him — and ours is honest because
disinterested. I am not surprised
that his passing should have given

Vincenzo Zeppa, born July 21, 1858, and Clara Longo Zeppa, born July 17, 1863, were the parents of five children: Carlo, Gilda, Giuseppe, Paola, and Cristoforo.

This gathering of the Zeppa clan took place during a visit to Fubine by the Joseph Zeppa family. In the back row (second from left) is Mrs. Gertrude Zeppa. Next to her are Vincenzo and Clara Zeppa, Joe's parents.

Chris Zeppa was a member of the Italian army during World War I, but never saw military action.

Chris Zeppa joined Delta Drilling Company during the 1940s, and retired some 20 years later as executive vice president.

Gertrude Wells Ruppel married Joseph Zeppa in 1931 while he was working for McMillan Petroleum in California.

Trent Harvey Zeppa, left, and Keating Vincenzo Zeppa, sons of Joseph and Gertrude Zeppa.

During a visit to his home in Italy, Joe once again worked the oxen through fields he had plowed as a boy.

The Zeppa home on South Broadway in Tyler, Texas, was a familiar gathering place for friends and associates.

Though Joe Zeppa took a close, personal hand in almost every facet of the company in its early days, Marcus O. Jones remembered at least one occasion when Zeppa, unknowingly, failed to have his way.

"We didn't have any bathroom facilities out there [at the production warehouse] at all, and Mr. Zeppa told me to put in one and sent me to Sears Roebuck in Tyler to buy the commode and the hot-water heater. We got an old hot-water heater that didn't have any insulation on it. It wasn't automatic. You had to light it and put it out after you got the water heated, and he had me put in a septic tank. I dug the septic tank where he showed me he wanted it and we had a joint of pipe on the rack. It was bent in a C-shape and we wanted to get rid of it. So the production superintendent, Bill Baggett, told me to put the crooked joint down first and dig my ditch to fit it.

"Course, I did that, and Mr. Zeppa came along before I had a chance to cover it up. He said, 'Using a thousand dollars worth of pipe on a two-dollar crapper!' And he made me take it out. The production man came back and I told him what happened and he said, 'Well, put it back in there and we'll cover it up so he won't see it.' And we did that and it's still in there, as far as I know. He said he might be able to use that piece of pipe, and we didn't have any way in the world to straighten it. After we got it covered up, he didn't know the difference."[90]

From the first days, Zeppa's tightest grip was on the company's finances. He insisted on passing judgment on any major expense and many minor ones. C. L. Vickers, who was in charge of credits and financing for Continental Supply's Shreveport district at the time, recalled one crisis this policy precipitated in the late 1930s.

"Joe and Mrs. Zeppa were in Rome for a month. It was when the Rodessa [Louisiana] oil field was discovered. Bob Stacy, who was vice president of Delta and handled the field operations, procured a contract covering several wells, and he and Sam Sklar called me for a conference. I sold them a $125,000 rig, and when Joe returned, the rig was running and the invoice covering the rig was on his desk.

"Sam Sklar called me and said, 'Mr. Zeppa is blazing mad.' I went to [Zeppa's] office and he said, 'Okay. You sold us a rig without my knowledge or consent. I'm going to have my say right now. I will pay for this rig when I'm damn good and ready.'

"That actually took place. And then he carried through with his threat."[91]

When Joe White joined Delta in 1937, the company still had two of the rigs—"two little bitty steam rigs"—it had started out with six years before. But when the Rodessa, Louisiana, field came in and Delta gained drilling contracts there, the company bought two rigs from Roberts Brothers, a drilling firm, which brought the rig count to four. A fifth rig was leased from the F. H. Brown Drilling Company in Rodessa, for by that time there was enough work to keep the extra rigs and crews busy.

Joe White had a special personal reason to remember his roughnecking days in the Rodessa field.

"Mr. Joe Zeppa was one of the nicest persons that you could ever have any dealings with," said White. "I never will forget this. He gave us a dinner one time in the Continental Supply store over in Rodessa, and he told us—see, we wasn't getting but $6.00 a day; $6.00 a day was a lot of money in them days—he said, 'I'm gonna try to up the wages.' He said he was going to *try* to. And he says, 'All the rest of these contractors are paying $6.00 a day, but [he] said, 'I'm going to try to pay $7.00.' And he did, and that was a dollar more a day than other outfits were making. And he didn't have no trouble getting good hands. We had the best. We were known for the best bunch of hands in the oil fields. And he was known for the best drilling contractor in the oil fields.

"Firstly, and just every way you could take him, he was a wonderful man, Joe Zeppa was. When he told you he'd do something, he would do it. If he didn't, he'd tell you why. That's what everybody liked about him. And I don't think he had a man working for him that wouldn't jump off [in] that slush pit for him if he needed to. I don't think there was a man working for him, them days, that wouldn't done it."[92]

What was it like on a drilling rig in those days?

"Well," said Nathon Bacle, "the early years it was give and take. You had a lot of time off, lot of working time too, between the locations. When they first started in the East Texas field, you could get somewhere around 20 days on a well. But it wasn't long after that till you could drill one a week. They moved every Sunday. Drill a well and set pipe and move on to another one in a week. And that went along until the field was drilled up. Then the Rodessa [Louisiana] field opened up."

The wintertime brought the worst working conditions. Along with the mud, there often was rain—and ice.

"Well, it was pretty rough," he said. "I've worked with a corduroy cap on and icicles were hanging down and no slicker on. You know, we wasn't making enough money to buy a slicker suit or rain suit or all the good things they have now.

"I can recall one time we was down here right out of Gladewater, and it was cold wintertime and Joe Zeppa come out there and we're all about to freeze to death and he went up there and got a half-gallon of corn whiskey—we wasn't doing too much—and brought it back down there and warmed us all up. He was a sensible man, though. I'm quite sure he didn't know every employee. He knew them when he seen them, but he didn't know what position they played or nothing like that, 'cause he was too busy doing other things. But he never forgot his employees."[93]

XIV

Joe Zeppa looked back on those beginnings from the vantage point of four decades and assessed the managerial hardships that the company encountered.

"Course, when you don't have but two rigs and are trying to save every cent you can so you can meet the payroll, in spite of the fact that the men get 50 cents an hour and the drillers get 75 cents an hour—that's what they were getting paid—it was still hard to meet. But I was thinking the other day that in all the years we've been in business and the lack of cash that we've seen—the fact that we never had much—we've never missed a payday. And I think that's a pretty damned good record all these years. And we have gone through some tough times.

"We were drilling as contractors because that's what we had—we had two old rigs. But we got 'em fixed up after a while so that they would run."[94]

Oil payments were a frequent vehicle for financing Delta's operations. In the fall of 1932, Delta sold a $30,000 oil payment, from 7/32 of oil produced from the company's Hughey and Ross lease in Gregg County, to El Dorado banker Lee Felsenthal, as trustee, for $12,000 cash.[95] While Delta was forced by circumstances to sell oil payments at great discounts, the happy side of the situation was that Zeppa and his partners by now were on excellent terms with a number of banks, so there was a market for their oil payments as well as a variety of sources for loans. In addition to the relationships Zeppa had built in El Dorado and since, the company benefitted greatly from the Shreveport partners, who were well-known to the Shreveport financial community. Sam Sklar was a stockholder in the First National Bank in Shreveport. By linking up with a man who not only had money but had access to more, Zeppa and Stacy had established a solid foundation for the company from the beginning.

Had Delta not acquired the leaseholds and production early in the East Texas field, the company might have faced more perilous times. But between the cash income the production provided, either through sale of oil or oil payments, and its value for purposes of mortgage, the oil-producing leaseholds provided Delta the financial "cushion" it needed during the economically bumpy rides. In 1937, for instance, Delta borrowed $113,750 from the Motor Finance Company in Shreveport on a note payable in 13 monthly installments at eight percent interest—amounting to $8,750 each month. On this transaction the company not only mortgaged properties but also planned to partially liquidate the debt through sale of oil and gas runs.[96]

In 1938, Joe Bevill joined Delta in the midst of a hectic growth period, when it was trying to expand under adverse circumstances. The company had five rigs, as he recalls.

"Joe Zeppa certainly met every obligation that he ever encountered," said Bevill. "And I remember distinctly that there were many, many times when the payroll was coming—in those days, paid twice a month, the fifteenth and first—and Joe Zeppa had to borrow money to meet those payrolls. He was constantly juggling finances and borrowing and stretching the point to get funds with which to carry on the company's activities. Not just a matter of payroll; he had to have money for exploration and money for new rig equipment, most of which was purchased in those days on credit.

"Well, the company never gave a chattel mortgage on any of its equipment. That's the way a lot of people financed their business. But he'd always go to the bank and borrow the money. Most of it was borrowed on some type of collateral: some of these little old oil interests that we had scattered around, or he'd sell an oil payment against one of those leases. Done that a number of times. But they were small sums of money compared to modern day. You know, I remember distinctly oil payments in those days were $25,000-$50,000."[97]

Sometimes dismal events had bright sides. "I burned the [Delta production] warehouse down," recalled Marcus O. Jones, who lived there at the time. "We were getting our gas from a well here, just raw gas, and it was cold weather, in December, and I went and left my heater on. It was just barely burning, and I didn't think the heater was on, and I went down and opened the well to get more gas, and when I did that warm gas coming out of the ground hitting the cold pipelines condensed into gasoline. And, boy, it shot a charge of gasoline through that heater into my room. By the time I got back up there the whole thing was just engulfed in flames. We didn't have any way to put it out, and it just burned to the ground. I was living in the warehouse. I lost everything I had except the clothes I had on, and I had a batch at the wash woman.

"Oh, man, Mr. Zeppa had a fit. He asked me, he said, 'What's the matter, Jones, did you get cold?' Then I told him what happened. He didn't fuss too much, but I thought he was going to run me off. And about two months after that, Mr. Bob Stacy came out and he had a long Delta stationary envelope with him and I thought sure that was my time he was bringing. And he surveyed the whole yard over and looked over the remains and he said, 'Well, Jones, I think it's going to be a pretty successful fire.' The insurance paid off, and he told me later, he said, 'Man, that money we got out of that fire really came in handy!' "[98]

XV

Around the middle 1930s, Red Merritt went into the drilling business for himself. Friction had been building between him and Bob Stacy. As Guy White recalled, "Bob would try to suggest or tell him kinda what to do, and Red didn't like that because he knew he was the only one who knew how to drill a well." Claude Johnston followed Merritt as field drilling superintendent.

In 1935, Delta bought a somewhat larger rig from Plains Drilling Company and moved it to Beeville, Texas, to drill a well for Lion Oil and Refining company. Lion's people thought they had a large field in South Texas. Colonel T. H. Barton told Zeppa he thought there would be about two years' work there, which was extremely promising at that time.

Delta's crew rigged up about 40 miles west of Beeville on a ranch. Delta's first automatic firing control was on the rig. The project also was Delta's first drilling assignment outside the East Texas region. The well was drilled to 6,012 feet, which was another first for the company—the deepest well Delta had yet drilled. In the East Texas field, it only took 3700-3800 feet before the rotary bit hit the oil-bearing Woodbine sand.

Oil-field people then worked hard, played hard, and manufactured their amusements on the spot. While Bob Stacy and Nick Andretta were driving to Beeville to check on the work there, they stopped to pick up a hitchhiker who turned out to be a "tattooed lady" who could have landed a job with a carnival. She checked in at the hotel in downtown Beeville. A sudden inspiration seized Stacy and Andretta. A Lion Oil executive was also in town, seeing after the drilling. Stacy and Andretta told him of "this lovely debutante" they had him fixed up with for a date. They piled on the adjectives, and the visitor's enthusiasm steadily grew. He spruced up his clothes, slicked down his hair, and was ready to go out on the town. During the process of their selling job, Stacy and Andretta engaged him in a drinking session that soon left him exceedingly inebriated. By the time he met his lovely debutante, it was questionable whether he knew she was tattooed until he had her in his room. The denouement of the romantic episode was viewed with contagious hilarity through the transom by the gleeful pranksters.

The drilling contract was short-lived.

After three attempts to find oil, the prospect in South Texas dried up and the rig was shipped by truck all the way to Rodessa, Louisiana, where activity was humming.

The rig arrived in Rodessa in late October, 1935. To the crew coming from balmy South Texas, North Louisiana was chilly. Homer Keel was the toolpusher in Rodessa; he later became drilling superintendent and, before he died, general superintendent. They drilled a well in the Rodessa field to around 6,000 feet.

"And on the first well we drilled in Rodessa, Delta Drilling Company bought a new drawworks and put on the rig. First new drawworks Delta Drilling Company ever owned. In fact, I suggest it was the first piece of new equipment they had ever owned," said Guy White, who was working derricks at the time.

"And then after they bought the drawworks they decided they needed a better engine to pull it, so they bought a new engine."[99]

In the mid-1930s Delta added a large, long-wheelbase White truck said to be the largest one in Texas at the time. As might be expected, the owners of the company took a close interest in such equipment.

"The old boy that drove that truck didn't know Bob Stacy and he broke down between Shreveport and Longview and he was down under that truck sleeping," said Marcus Jones. "That's where he slept. Stacy came along and he—Bob Stacy—got out and looked the truck over and wanted to know what the trouble was, and this guy wouldn't answer his questions. Bob Stacy said, 'You don't know who I am, do you?' And [the driver] said, 'I don't give a damn who you are,' and [Stacy] finally told him, he said, 'Well, I'm the owner of the Delta Drilling Company. You come into the office when you get back to Longview and get your time.' But he didn't fire him. He [the driver] was pretty rude to Stacy. He wasn't in any mood to answer questions to a stranger."[100]

Drilling in Rodessa was followed by the Shreveport field, which was then followed by the Schuler field at El Dorado which brought Delta back to that city in 1937. Nick Andretta was sent to open an office in the Exchange Bank Building in El Dorado, as two rigs were moved into the field. That same year, Andretta, as the company's agent there, hired H. J. "Red" Magner there, among others, to roughneck out in the field.[101]

The Schuler field in southern Arkansas in the late 1930s proved to be a profitable undertaking for Delta. With Lion Oil Company gaining a lion's share of the field, and with Joe Zeppa a long-time close friend of Colonel T. H. Barton, the Lion's chief, Delta obtained choice drilling contracts, moving in some rigs from Rodessa and leasing additional ones. "If they had a location and wanted to drill right quick and think they might be losing a lease," said Joe White, "Mr. Zeppa'd lease a rig off F. H. Brown or Bob Allison Drilling Company, somebody that'd have one down. Mr. Joe, he'd get the contracts and then he'd just lease a rig off them to drill it and use their men. Their men would come with it."[102]

Joe Bevill, who began working in the Delta office in the late 1930s, remembered, "Delta drilled, I know, at least 58 wells in there, because we had one contract for drilling in the Schuler field and we listed those well numbers down the side of that manuscript cover on the contract and the date of completion. And I remember very well it got up to 58 and we had to add another sheet on there. But during that period of time we bought additional rigs to put up there, I think, three rigs going up there. And they were much better and bigger rigs than the ones we used in East Texas. That

Schuler operation, I think, was about 6500 feet drilling or somewhere in that neighborhood—7,000. This was in '38, '39.

"That was a big step and we had to borrow the money to do those things, and seemed to me it was the Commercial National Bank in Shreveport, Louisiana. And at one time we did business with a bank in St. Louis, Missouri. [Zeppa] had become acquainted with someone while he was with Shearman and Sterling that later was senior vice president in a bank in St. Louis, and at the time he was borrowing funds to expand his business.

"It was a big step. It was really a big step. We got larger and better drilling equipment at that point and more of it. I think we went up to eight rigs there. I remember we had three good rigs at Schuler. We were up there through '41 or '42."[103]

XVI

When Delta initiated the development program in North Louisiana, Bob Stacy began devoting more time there, supervising the Rodessa and Shreveport fields operations.

To the crewmen there, as well as to others, Stacy was an impressive figure. "Oh, he was a great big, tall, stately-looking man," said one. His neatness stood out in another observer's memory. "Just as clean as a pin," said Nathon Bacle. "You'd think he'd taken a bath every 30 minutes, just to look at him. Nice dresser. He wasn't talkative or nothing like that, but he just was real polite."[104]

"He was in the field a good bit," said his sister-in-law, Elizabeth Fullilove, "because I think they all were at that time, but I don't think the field was 'his job.' I would say, primarily, Bob's was meeting people, making deals, and I am sure Joe's was the same thing. Bob was awful good at this. This is what they both did, and this is what they had to do in order to keep it going, and then they both were in the field whenever they felt they were needed."

After leaving Dallas, the Stacys had moved into a small apartment in Shreveport and soon owned a new Graham-Paige and had a stately home on Fairfield Avenue. Stacy still had debts, but had whittled them down substantially. And though he had established a reputation for living high, Mrs. Fullilove suspects it was exaggerated.

"It really wasn't all that much money involved. I guess the size town Shreveport was, and the time they lived in it, it may have seemed more. I remember when Bob was making $18,000 a year. That was just more money. He was just delighted, I can tell you. He had never had any, so it was all just great. But when you are like Bob and Sister, it just gets away from you."[105]

XVII

The one-third interest in Delta belonging to the three Shreveport partners—Gold, Sklar, and Dorfman—was held, as we have seen, in the name of Sam Dorfman. No mention was made of either Sklar or Gold in the Delta minutes until August, 1932, when Sam Sklar joined the board for the first time—the stock previously held by Dorfman having been trasferred to Sklar.[106] No explanation for this change was given, but based on other evidence it appears that this was done for the convenience of the partners, for some unstated reason. It is clear from other evidence and from subsequent events that Dorfman continued to maintain his interest in Delta affairs, and later his status as a stockholder was reassserted.

Despite the fact that Sam Dorfman was not formally either a director or stockholder, he actively helped Delta obtain drilling contracts as well as secure oil properties. At the end of 1933, the directors voted to reimburse Dorfman for his expenses over the year in behalf of Delta. The sum came to $1,200.00.[107] The next month, in the new year, Dorfman was named assistant treasurer and placed on the payroll at $8,000.00 per year, or $666.66 a month, at a time when Zeppa and Stacy were each being paid $1,000.00 per month in salary.[108]

Square-faced, well-built Sam Dorfman was the highly visible representative of Louisiana Iron and Supply in East Texas; he managed its yard in Longview and lived in the same city. As at least one observer saw him, Dorfman had two quite different sides to his personality—one in the marketplace, another in his leisure time. Bill Dailey, who worked as both roughneck and pumper in the 1930s for Delta, saw him in both roles.

"Sam Dorfman was one of the friendliest guys you ever saw," said Dailey. "When he was tending to his business, talking about business, you'd think he was turning into a mad dog, he'd get so rough. Get so mad, tear all to pieces. Well, when you'd meet him off anywhere, he had time to talk to you, go buy you a drink—just as peaceable as he could be. You wouldn't think it was the same feller.

"He was something else. Sam come down in the pipe yard [one day] and I was over there to get some connections. His welder, Gordon Stansell, was a rooster fighter like me, and I was down there where he was working on some pipe, talking to him, and Sam drove up. He said, 'Gordon, what are you doing?' and he said 'Hell, I'm setting down, can't you see me?' [Dorfman] says, 'I do, and when you go by up there, get your check.' Says, 'You don't work here setting down, you work working.' [Stansell] says, 'Okay, I'll get it as I go out.' So he went back up there, I went with him, he was laughing about it. Got up there and Sam Siegel [Sam Sklar's brother-in-law who worked at the yard] says, 'Gordon, you got a little raise.' [Stansell] said, 'Well, Sam told me to come by and get my check. He's laying me off.' Siegel said, 'Naw, he came on up here and said you boys were a little lazy.' He was a sight. Gordon didn't ever take him seriously."[109]

XVIII

Sam Gold—Simon Goldman—died on February 12, 1936, thereby dissolving his partnership with Sam Sklar formed in Shreveport back in the early 1920s. How, though, did it affect the three partners' Delta interest?

Sam Gold's role in the early Delta history poses a certain degree of mystery, because his name is never mentioned in the Delta dividend record. The fact that Sam Gold does not appear in the Delta minutes cannot be considered conclusive, however, as Sam Sklar is not listed in the beginning, and then there is a gap during which Sam Dorfman was not a stockholder, officially. In fact, for 16 years following August 18, 1932, when Sam Sklar replaced Dorfman in the Delta dividend column, Dorfman received none of the profits. Then in 1948 the minority interest was divided evenly between Sklar and Dorfman, on a 50-50 basis. This provides evidence that Gold was never an owner of Delta, but it is ambiguous evidence because the one-third interest shifted, first from Dorfman, the man on the scene in Longview, to Sklar. Though Dorfman ostensibly had no formally stated interest in the company for a number of years after 1932, we know, though, that he was, in fact, an owner. Might not the same situation have applied to Sam Gold?

The evidence that Sam Gold was an equal partner with Sklar and Dorfman in a one-third interest in Delta includes the statements of the second generation of Gold's family, the ownership arrangement of the Longview operation of Louisiana Iron and Supply, and Joe Zeppa's testimony.

Zeppa, interviewed in 1975, said: "Sam Gold, Sam Sklar, and Sam Dorfman—they were a partnership, you see, and their third was in Sam Dorfman's name."[110] Though it may be argued that Zeppa was looking at the Gold-Sklar-Dorfman partnership from the outside and was not privy to their private understandings and therefore may have misunderstood the facts in the case, this contention seems doubtful. Zeppa was not the sort of person *not* to know precisely what the financial arrangements would have been. He knew the three men personally, and had since the early 1920s. If he understood all three to have been partners, it appears highly likely that they were.

As we have noted, the Shreveport partnership of Sklar and Gold, in which Dorfman did not share, was dissolved by Gold's death under Louisiana law. At the time, each held a half-interest in Louisiana Iron and Supply Company, a Louisiana partnership.

But more relevant to the Delta interest is that on March 15, 1933, Louisiana Iron and Supply Company was issued a Texas charter, doing business in Longview. This, a separate legal entity from the Shreveport partnership, encompassed the pipe yard at Longview. The papers were signed by the three partners—Sklar, Gold, and Dorfman—with each holding one-third of the stock. Since Delta first came to life, one might say, in that Longview yard, this ownership scheme of the Texas corporation—Louisiana Iron and Supply—infers that the same arrangement carried over into the Delta transaction, even though only one partner held Delta stock at a time until 1948. This ownership pattern in Louisiana Iron and Supply (the Texas corporation) continued after Gold's death in 1936, with the Sklar family holding 1,000 shares, Dorfman 1,000, and Gold's heirs 1,000 (divided between Minnie Goldman, with 500, and each of Gold's five children with 100 each). Although any Gold ownership of Delta was not a matter of record, as far as can be ascertained it seems to have been an understood part of the Louisiana Iron and Supply holdings in the Longview venture. That, at least, is the memory of the Gold family.

Mrs. Rose Bishkin, Sam Gold's daughter, for example, corroborates Zeppa's view of Gold's having one-third of the minority interest in Delta through his ownership of stock in Louisiana Iron and Supply, the Texas corporation.

Whatever were the details of the arrangement, Gold's death apparently triggered a series of transactions, spread over a period of years. Some members of the Gold family sold out their interest immediately and the others later. The result was that the Gold family gradually got out of all the different business efforts. The transactions did not go smoothly.

"They [the Gold heirs] wanted out, as I remember," said Albert Sklar, Sam Sklar's son. "I was a child then, but I remember there was a lot of yelling and screaming and all kinds of stuff when they got out."

According to Mrs. Bishkin, following Gold's death his heirs were "railroaded into selling their shares" in Louisiana Iron and Supply because Sam Sklar argued that " the company is at a standstill and couldn't move" unless they sold. Mrs. Goldie Rappeport, Gold's stepdaughter, also recalls a drawn-out, bickering settlement that involved lawyers on both sides.

At any rate, the negotiations resulted ultimately in the Gold heirs' interests being bought out, leaving Sklar and Dorfman in sole command of holdings that once had been divided three ways.[111]

XIX

In the early 1930s, a while after Delta's debut, Zeppa helped organize Premier Oil Refining Company in Longview, a company in which he was one of the three major partners and in which Delta, as a corporate entity, subsequently held stock. His role in the refinery came about because of his friendship with two men he met in El Dorado—Sylvester Dayson and J. R.

Parten. "At that period," Continental Supply executive Fred Mayer said of Dayson, "he was, in my opinion, one of the best refining engineers there was in the country."[112]

The venture grew organically out of a deal Dayson and Parten were in together at Baird, Texas, in the western portion of the state in the late 1920s.

"About 1929," said Parten, "Sylvester came to my office in Shreveport and he said, 'I just wanted you to know I have quit Lion Oil and I'm going out and build a refinery somewhere if I can find a crew. I've got financing and I'm gonna build a refinery of my own.'

" 'Well, Sylvester,' I said, 'We've got about 50 shallow wells out in Callahan County, Texas, and no market. They're making about 2,000 barrels a day of sweet oil. How would you like to build a refinery out there?'

"So he went out and looked at it, brought a sample back to El Dorado, got his old laboratory at Lion to test it, and was told that it was sweet and would make good gasoline fast. So he came and told me he wanted to go into it.

"It was right after the Depression, in '29, and I said, 'Sylvester, how much money do you have to put into this thing?'

"He said, 'I don't have but $4,000 in musty $100 bills that I took out of my safety deposit box, because I've got such a loss on my Lion stock.' The Lion stock he had, broke him.

"And he put that in and we financed a refinery the rest of the way, virtually carried him for a quarter interest in it. My company, Woodley Petroleum, did because we wanted a market for that crude. Built that refinery at Baird on the Texas Pacific Railroad just west of Cisco—between Cisco and the little town of Baird, the county seat of Callahan County.

"And then the East Texas field came in. We [in Woodley] saw immediately that it was a big oil field. I had been watching very closely and got ahold of some substantial leases in it and I went out of the drilling business right there. I sold my six drilling rigs—decided I was going to contract wells from then on—and I met Joe [Zeppa] up there in the Blackstone Hotel at Tyler, and Joe was just going in the drilling business. He organized the Delta Drilling Company.

"Sylvester came over there later, and other wells came in, and it was clear that it was gonna be a giant field. He wasn't very busy with a little ol' 2000 barrel plant out there at Baird, so he got ants in his pants to build a refinery at Longview, and chose Greggton, this little switch on the TP Railroad west of Longview. So my company didn't want to go in another refining business. I got permission from my company to go it alone with Sylvester and build that refinery because I knew he didn't have enough to occupy him with this one little refinery in central West Texas. We invited Joe in and the three of us built that refinery. We were the original stockholders—Sylvester, Joe and myself. My company didn't want any part of it, and I wanted a third, so we just divided three ways.

"It was called Premier Oil Refining Company. Sylvester came up with the name, so far as I can recollect. And then we invited Myron Blalock in. Myron Blalock, my friend and prominent lawyer in Marshall. He was a great deal of help to us, and we decided that we would invite some local people in. We invited some local people to join us on the ground floor, and three or four other prominent citizens there joined us.

"Sylvester, Joe, and I did all the work, and Sylvester did most of it, because he was the expert in refining. Joe had just gotten started in the drilling business and he started off doing well. We started out refining. It was early in the days of the field. It was about a 3,000 barrel plant, and it was an old thermal-cracking plant—did not have any of the new stuff in it that came along later, the catalytic cracker. It was just a duplication of the refinery that he and I had already built at Baird, but a little larger."

Zeppa's interest was as an individual—an investment on the side—as Parten recalled.

"I became chairman of the board, Sylvester became president of it, and we built it and operated it for several years, and Sylvester, Joe, and I were on the executive committee all the way through it. It made us some money."

Gradually the refineries involved grew in number. In addition to the one at Baird, one was built at El Dorado; another one at Cotton Valley, Louisiana; one was purchased at Fort Worth; and another one bought at Arp, Texas, in the East Texas field. The Cotton Valley toluene plant produced the basic product for 100-octane aviation gasoline during World War II; before the war, a Premier plant there had produced naphtha.

The purchase of the refinery at Fort Worth gives some insight into how Dayson operated.

"Lo and behold," said Parten, "he came into my office one day at Houston and said he wanted to buy this Marathon plant at Fort Worth and he said he could buy it on credit. Well, I was just swamped with refineries, with two, and one being talked about. We weren't making any money. And I felt like I could throw him out of the fourteenth story window of my office. But he says, 'You've just got to go up there with me and look at it, because it's a buy. It's a bird nest on the ground.' So I went up there to look at it and I saw immediately the way he had it lining up. He was going to buy it for $75,000 down, and the balance was on credit. The balance was damn little and they had a pipeline that took in the Ranger field, the source of the crude—10,000 barrels a day—and they were making a good grade of lubricating oil, as well as gasoline. So we bought it. Then we built a plant at Cotton Valley.

"Another refinery came on the market down at Arp, Texas, and, lo and behold, while I was away in Washington during the war [as director of transportation for the Petroleum Administration for the War, building the Big Inch pipeline], he bought that for a song. Now Joe helped him on buying that Arp plant, I'm sure. I wasn't in on that because I was in Washington building the Big Inch pipeline. It was a small plant down at Arp—had been running hot oil.

"We had a reputation as the only refinery in the Longview-Kilgore area that never ran any hot oil, and we were very proud of it. I think it's true today that we were the only refinery that went through that field and did not run any hot oil. So many of them, that's all they did. People made fortunes out of it. We decided that our plan was to play the game straight and run only legal oil. We felt like we could do it. It was rough. I'm sure we were the only refinery in the East Texas oil fields—80 there—that never ran any hot oil. It was against our policy. That's hard to do. We're very proud of that reputation. That was Joe's idea, it was my idea, and it wasn't Sylvester's idea all the time. Between Joe and myself we kept that refinery on legitimate oil.

"We sold out to a group from Minnesota and Illinois who were farming people. They were afraid we were gonna run out of oil, and they wanted a direct source of refined products. In the spring of '48, my wife and I and the Daysons were at Naples, Florida, and Sylvester came in and said to me, 'You know, we've got a chance to sell these refineries.' The Longview refinery, the Fort Worth refinery, the Baird refinery, and the Arp refinery. The war had ended, and we had abandoned the one over at Cotton Valley, Louisiana. He said, 'We can get about $9 million for them.' I said, 'Are you sure about that?' He said, 'I want you to go with me to meet these people at Kansas City and I'll have our plane in here day after tomorrow morning to fly us to Kansas City.'

"Well, I was having a good time and the last thing I wanted to do was to go to Kansas City. But that Frenchman was very persuasive and so nothing would stop us from going to Kansas City and meeting these people who wanted to buy these refineries. They were agricultural co-ops in Illinois and Minnesota. So we flew to Kansas City from Naples, Florida, and I think Joe joined us up there.

"We met these people, and I discovered they were serious about buying this property. I was pleasantly surprised. I got interested immediately. We drove that deal through. They didn't give us the $9 million, but we got $8.5 million, spot cash. That was a lot for little old, largely obsolete thermal-cracking plants. The most complete plant was in Fort Worth, but it was an obsolete plant. It was a 10,000 barrel a day plant.

"So that put a lot of cash money in our hands, and Joe took a good slice out of it, Sylvester did, and I did too. All the stockholders took a good slice, but the three of us were the largest stockholders."[113]

XX

As an active contractor and producer, Zeppa and Delta Drilling Company were intimately involved in most facets of the East Texas field, including solving its more pressing problems. One worrisome puzzle was the disposal of salt water produced from the Woodbine sand along with the oil. As the oil was taken from the wells, salt water from below moved into the oil sand to replace the extracted oil. By 1935, the field was producing 15,000 barrels of salt water daily from 1,000 wells. Two years later, the figure reached 100,000 barrels, and by the end of 1940 it had doubled that number. In April, 1941, the daily production of salt water soared to 300,000 barrels a day.

This huge excess of salt water created a cluster of crises. First, it deleteriously affected the reservoir pressure and the rate of oil production. Something needed to be done to prolong the life of the field, which risked premature aging and exhaustion. In addition, most of the solutions employed by many of the operators of various leases were unsatisfactory, as surface disposal of the salt water in pits and other means only led to severe pollution of the local terrain and streams.

By the latter half of the 1930s the salt water had entered surface drainage, to the extreme irritation of the affected land-owners. Then in 1939, when it noticeably invaded the rivers of the area, people who lived along the Neches and Angelina watershed filed an injunction suit in Dallas against oil operators in the southern portion of the East Texas field to prevent the salt water pollution.

The suit dragged, but it precipitated action from all operators in the field, not merely its announced targets.

The most practical method of water disposal was to return it to the Woodbine sand, where it not only would become a positive factor in maintaining reservoir pressure and an aid in the production of oil, but it also would eliminate water pollution and injury to vegetation. With the injunction as a catalyst, social pressure now was upon the field's operators to find a way to do what needed to be done.

Joe Zeppa was one of the forerunners in working out a solution. On July 29, 1941, he applied to the Texas Railroad Commission for a permit to drill a salt-water injection well, into which would be put salt water from his and others' wells. He also suggested that the regulatory agency award an extra barrel a day allowable to the owner of each well who returned all of its salt water to the field reservoir.

As Zeppa recalled it, some of his colleagues in the industry didn't accept his proposal for what it was, and even looked upon it with suspicion.

"I applied to the Commission for a special order to permit us to gather [water] in the neighborhood and put it in the ground, and keep on getting the allowable from the wells," said Zeppa. "Well, the companies didn't like it. They said there was some kind of a racket attached to it. Well, I didn't see any racket to it, because if a man has got a small lease—if he's got four wells, he can't afford to turn one and put water in it. On the other hand, if he had a big lease, he could afford to go ahead and take one, put the water in the ground, and keep on getting his allowable from it. And I tried to show 'em that by doing what I'm trying to do, just one well would do the work of quite a few wells if you were running your own business in a large lease.

"Well, they opposed it, so I said, 'Fine.' I dropped it. In fact, we had a lease, a piece of ground we had bought, and [were] ready to drill the well. Well, all right. Then the war broke out. And Bill Murray*—he was then in Washington—came to see me and said, 'Joe, I think we ought to do something about the idea you had about the salt-water business. Because, you know, I've been thinking about it a lot, and I think it's the one way we can keep the federal government from sticking its nose into the East Texas field.'

"Of course, I said, 'Well, Bill, I'm ready for it. All I have to do is get the old papers and draw up a new set of rules, work them over so that there will be no gimmicks in it, form a company, and just take the whole damned field in! But [with the proviso that] you don't have to go in if you don't want to. If you have your own outfit, why, you can go ahead and do the same thing you've been doing.'

"Well, that's the way it started. We got the group together, the major companies together, and all backed it! The ones that were fighting it—it was just me! They all signed; they started getting back of it, and we put it over! And it's been a *marvelous* thing. If we hadn't had that company, we couldn't have operated."

W. S. "Buck" Morris, a pioneer petroleum engineer who, at the time, was with the East Texas Engineering Association, which was funded by all the major companies to pool knowledge of the East Texas field, remembers that Sun Oil Company was one of the first to realize something had to be done. Sun drilled the first salt-water injection well in East Texas in 1938. Although the Engineering Association's staff members were virtually the only people who had any hard data on the salt-water production, it was soon recognized that another separate organization would be needed. By

---

*W. J. Murray, Jr. a member of the Texas Railroad Commission.

this time, eight water-gathering systems already existed, but altogether served only a fraction of the field.

"So we got about ten companies to come in to this deal, and the chairman was a Sun Oil Company man," said Morris. "And the second in control was a Shell Oil Company man. They said, 'What we need to do is to go out there west of the producing limits of the oil field and drill us two wells—drill one well to produce a whole batch of salt water and put it in a pond and study the salt water, not mixed with oil, just virgin salt water. Then we ought to have another well drilled, in which we would attempt to inject this water.' "

The group of ten companies would finance the operation separately from the Engineering Association. The water was pumped into huge pits, one of which held 12,000 barrels, where it was studied for chemical changes after it was brought to the surface under atmospheric conditions.

While some of the majors, through this means, were looking into the situation and setting up their own water-disposal systems, the independents, without any such facilities, were left in a bind. The court gave the parties involved in the injunction a short period of time in which to comply with its requirements.

"So, man, that was a real rush!" said Morris. "A lot of people said, 'Well, we'll just dig a bunch of pits.' The first thing you know, we had the whole damn countryside full of pits. Well, the rate of rainfall in East Texas is greater than the rate of evaporation, so the pits would never get emptied, they'd always be full, so you'd have to keep making more pits. We had all kinds of studies emanating out of the Engineering Association. 'Let's lay a pipeline to the [Gulf] coast and pump this water down to the coast.' We figured that'd cost $47 million. 'Well, let's dig a ditch.' So digging a ditch and buying the land and so forth ran something about almost $47 million. They said, 'Well, we can't do that.'

" 'Why don't we put in a system like this salt-water disposal system that the Sun, Tide Water, Magnolia, and Sinclair are using and take care of everybody?' So, we began to talk that. Now, the major companies had the edge on the thing because we'd been talking all this in our Engineering Association meetings and the independents had been left out. But Joe Zeppa was a friend of mine, and I kept Joe informed about what was going on. So as things built up over the days, pretty rapidly we decided we had to go into a field-wide salt-water disposal system, in the nature of a public utility. So the group started meeting with the Engineering Association as a nucleus.

"We started meeting, and John Suman of the Humble Company was the chairman. And we invited these independents to come to the meetings, and Joe was one of the first. Bryan Payne got in there and we decided that the field-wide utility system was the way we had to go."

The group engaged R. E. "Bob"Hardwicke, an oil lawyer from Fort Worth, to handle the legal work connected with the chartering of the corporate vehicle.

"So we got the [East Texas Salt Water Disposal Company] organized with $25,000, token subscription, and Joe [Zeppa] was one of those, and Bryan Payne was another independent, and Judge C. R. Starnes, up at Gladewater—we just called him Judge; he was just an entrepreneur of the first water.

"So we had these meetings. John Suman was chairman, and we got to the point where we had to decide how much money we needed to start with, and the minimum amount we could need to get started was $2 million. So we decided to expand this $25,000 corporation into a $2 million corporation, and at that time there were 25,000 wells in the field. We figured that we could persuade the operators of two-thirds of that number to buy stock. Now if we needed $2 million, then each one of the owners of those wells would have to buy $120 worth of stock. The stock was to be issued at $100 a share, but to get $2 million we had to buy more than $100 worth.

"So the Sun Oil Company was a big pusher. They assigned a fellow named Buck Warren to work 24 hours a day on this project. Bryan Payne was an independent operator, and he was president of the Texas Independent Producers and Royalty Owners Association, and he was a loud-mouth and he was needling the major companies for their handling of the salt-water problem and so forth.

"Joe Zeppa was among the forerunners in this thing. Although Joe wasn't a needler and he wasn't a loud-mouth or what not, he was a very shrewd operator. We decided that we would—instead of trying to go to everybody's office and sell them the stock, we'd have a meeting in Dallas and ask all the Dallas operators to meet at two o'clock at the Adolphus Hotel, at which time this project would be explained and to please send somebody that would be qualified to say Yes or No, as far as their company was concerned."

In order to please the independents, Bryan Payne, as an established leader, was assigned a major role in presenting the matter to the various companies and individuals.

"We had these meetings in Dallas, Shreveport, Tulsa, Oklahoma City, Houston, San Antonio—various places over a period of a month," said Morris. "The modus operandi was, Payne would open the meeting by telling them what it was all about and the need for a field-wide system, and then Buck Warren [from Sun Oil] would come in with his public-relations thing that this was a project where the major companies needed the independents and the independents sure as hell needed the major companies to make this thing work. And if we'd work hand in glove together, we could do it. Then I would take a chalkboard and draw a picture of the reservoir and tell what had happened—how much water was being produced and something about the necessity of doing something about it before the field was shut-in completely and everybody would lose.

"And we sold that $2 million worth of stock in less than a month."

The charter was approved on January 20, 1942, to put the East Texas Salt Water Disposal Company in business. By August the capital was increased to $2 million, which was approved by the Texas secretary of state.

To smooth over any ill feelings from the independents, Bryan Payne was elected president of the salt-water disposal company. Morris was hired as vice president and general manager.

Almost from the beginning, friction appeared between Payne and Morris. Morris wanted the offices to be in Kilgore, a better location for the work that was to be done, since he would be virtually running the show. Payne, who lived in Tyler, insisted that the corporate offices be in Tyler. A committee meeting in Dallas was to settle the matter.

Morris approached his friend Zeppa, who was on the committee.

"Joe," he said, "I need your advice. You know just as well as I know that this organization needs to be together in the beginning, anyway. And part of it in Tyler and part of it in Kilgore isn't going to work. So are we going to let Payne bulldoze us, or are we going to do what's right from the beginning?"

Zeppa, an oiler of troubled waters, replied, "Now, Buck, I don't want you to go over there to Dallas to this first meeting of the executive committee and project a big rift between you and Payne. Now, I don't think that's right. I'll agree in premise [that] you are right in your conception of how it ought to be done, but putting it in reality, well, let's go along with Payne's deal. If it comes to a vote, now, I'll vote with you, but you asked my advice and I'm telling you that I think it would be the thing to do to go along with him."

So, Morris reported years later, that's what he did.

"Joe was a director of the company from the beginning until the day he died," said Morris. "He was a good one and he attended every meeting that he could and he had a better attendance record than anybody [else]."

On October 1, 1942, the company injected its first barrel of salt water in the field.

During the period 1942-1957, more than 1,700,000,000 barrels of salt water were injected by the company into the Woodbine sand in the East Texas Field. For its time, the experience was without parallel in the oil industry.[114]

XXI

There are several ways to measure the success of a company. Probably as good a way as any, during the Depression years, is to chart the changes in executive salaries. If such salaries are any indicator, then Delta can be seen as "getting over the hump" and perhaps stabilizing around the mid-1930s. For in 1934 Zeppa and Stacy, the principal officers, earned $12,000 salaries, something of a high point; yet in 1935, they set their salaries at $18,000 each, a significant climb, and Dorfman's expenses were set at $2,000.[115] However, the direction of their salaries was not uniformly upward. The following year—1936—Zeppa's and Stacy's salaries were cut back to $15,000 each, suggesting retrenchment, while Dorfman's rose to $6,000. In lopping off $3,000 each from Zeppa and Stacy, an additional $4,000 could be handed to Dorfman, with the net decrease only $2,000. Whether this revision was dictated by economic conditions, or whether Dorfman had demanded additional compensation, is not known.[116] But if the 1936 revision represented a setback of any kind, it was a temporary one. Two years later the Zeppa and Stacy salaries soared to a munificent $20,000 each, with Sam Dorfman as assistant treasurer, and Sam Sklar as assistant secretary, each taking $5,000 per year from the payroll.[117] By this time Delta had stabilized, grown, and was looking eagerly outside the East Texas region that had spawned it.

Another means of measuring growth is through capitalization as a result of increased stock. While in some instances this may reflect frantic efforts to mend a dire economic situation, over a period of time it is likely to be a result of expansion. At Delta's inception there were 300 shares of stock of no par value held in three hands—Zeppa, Stacy, Dorfman. Five years later, in late 1936, the stock was increased by 750 shares to a total of 1050 shares, the additional shares being sold at $100 each to, as it turned out, the stockholders of record at that time—Sklar, Stacy, and Zeppa.[118] This process raised Delta's capitalization by $75,000. A year later, in December 1937, the capital stock rose once again, this time by another 750-share increment at $100 per share, to a total of 1800 shares, with each of the partners holding onto 600 shares, or one-third.[119]

Another gauge of success is that of stock dividends. Of course, the declaration of dividends may not necessarily equal continuing success, and it may not represent growth, but at least it indicates the company has earned a profit for the given year. The company distributed dividends very soon after the company was formed—three months after receiving the charter, as a matter of fact. Zeppa, Stacy, and Dorfman each received a $75 dividend on each of their 100 shares, each benefitting by $7,500, paid by monthly oil assignments. Though this must have looked like all the money in the world during the Depression, it must be remembered that this dividend-taking can also be seen as a form of salary, and the further dividend, on the same date, of $1,666.66 to each of the three partners made the total $9,166.66. This year—1932—was an excellent one for the stockholders, for in August another dividend came in the amount of $51.70 per share, which left each partner with $5,170.00 from the transaction. It was to be nearly four years before another dividend would be declared. But when it came in 1936 it was a healthy one—the highest dividend rate ever paid in Delta's history: $250 per share. At that time there were still only 300 shares of stock, held equally between Zeppa, Stacy, and Sklar, so that each one received $25,000. Though no dividend *rate* would reach this high again, the following year—1937—was even more profitable for Delta's stockholders, for two dividends, of $10 and $75, left each stockholder $3,500 and $26,250, respectively, for the transactions—a total of $29,750 in dividends for the year. Evidently this was what Zeppa was referring to that year when he reported that the company's operations "had resulted in very substantial profits" and that he expected "substantial net profits" for the last half of the year, as well. The 1937 dividend was to be the last one for ten years, as the company entered a period of growth and expansion as Delta reached out into other regions. [120]

And, meanwhile, the rig count was steadily growing. When Herman Murray left Delta in 1942, the company had "twelve to fifteen nice rigs." [121]

XXII

By the late 1930s, Delta was already firmly on its feet and looking outward. "It was a small—very small—company and it could go broke as well as it could succeed," Joe Bevill reminds us, but it also was prepared for opportunities to grow and for other fields in which to drill. [122] Delta—with seven people in the office and not more than 75 in all at that point—was to remain in East Texas, with its corporate headquarters in Tyler and drilling operations covering the region; but from that point the company, through its officers, commenced looking outward. The logic was simple. If one were to continue in the contract drilling business, one had to go where there was drilling. Though drilling opportunities continued to exist in East Texas, the peak had been passed as the gigantic East Texas field had been tapped and brought under control. New booms beckoned elsewhere. To keep its rigs busy and to grow, Delta had to follow other booms. In the late 1930s and early 1940s these led Delta into a variety of locations, particularly back to southern Arkansas in the Schuler field and to the Illinois Basin.

And a development in 1940 was to have a lasting impact upon the company, with the consequences to stretch across the decades.

5

THE STACY INTEREST

I

As Delta entered the new year 1940, on the way to rounding out its ninth year in business, virtually the same partners remained stockholders. Sam Gold had died, of course, but Joe Zeppa, Bob Stacy, Sam Sklar, and Sam Dorfman still held onto their original interests in the company—divided in thirds between Zeppa, Stacy, and Sklar/Dorfman (with the stock in Sklar's name). Though Zeppa, in effect, ran the company, he did not control its stock; his position as president and treasurer was secured by his persuasive personality and financial skills, both of which had played hearty roles in Delta's surviving the economically fickle thirties.

But all was less than serene within the Delta "family," and this single fact was to be translated, before the end of the year, into a major modification of the company's ownership that would shape it down to the present days.

Stacy and Zeppa had virtually run the company since its inception. Zeppa tended to supervise the office and handle the financial arrangements so essential to Delta. Stacy usually kept busy making contacts with customers and potential customers, as well as overseeing field operations. As a "contact" man, he helped acquire new drilling contracts, which meant he spent a great deal of his time visiting oil operators and oil-company executives, and perhaps entertaining them. Likeable Stacy was the perfect man for this role.

Though stories of Stacy as *bon vivant* began to appear in these years, any financial difficulties he may have had do not seem to have impaired his standing at Delta, though some of his debts may have had an influence on his subsequent decisions. He, like Zeppa, conducted transactions with the company when it was to his and the firm's mutual advantage. For example, in 1937, a fruitful year for Delta, Stacy offered to sell 111 shares of stock in Premier Oil Refining Company of Longview, which he valued at $125 per share. Premier, of course, was the company in which Zeppa, as a partner of Sylvester Dayson and J. R. Parten, was a major stockholder. Delta's directors agreed to the deal and paid Stacy the sum of $13,875.[1]

Although Stacy also incurred debts during this period, he was not the only one owing money. The company itself, Delta Drilling Company, contracted growing debts as its business grew, and, in fact, it was through credit that it continued to operate.

As the years stretched on, it appears that Stacy's drinking created the greatest problem among the partners. From all accounts he was never obnoxious or overbearing, nor did he exhibit any ungentlemanly or similar overtly negative traits while drinking, but he sometimes overindulged to the extent of impairing his effectiveness for a day or two.

Nick Andretta, who was with Delta from the early years in various functions, at times helped Stacy recuperate from the effects of alcohol. One incident proved to be pivotal.

"I made this contact with Lone Star [Gas Company]," said Andretta, "and since Joe Zeppa was going to California [that week], I got Bob Stacy to go to Dallas with me. We had an appointment on a Thursday morning at nine o'clock with the Lone Star people about talking contracts, about drilling over at the Opelika field [in East Texas].

"Well, [we] went Tuesday; wanted to get a head start. He got on one of those binges and I had to call the hotel doctor and everything else to get him out of it. Had a long distance call to Joe Zeppa out in California, and I said, 'We've got an appointment with Lone Star Gas at nine o'clock in the morning, and I don't think Bob's going to make it.'

"Joe said, 'I'll be there.' He came in that night, Wednesday night, and we made that appointment the next morning and got a contract with two rigs to start with, and drilled all the rest of the wells for Lone Star in the Opelika field. That was where Stacy tore his pants. I couldn't speak for the company because I wasn't in an official capacity to make a deal for the company of that kind. They wanted one of the top dogs, and Stacy and Zeppa were equal partners.

"That was a case where Joe Zeppa really got stirred, and it was an ultimatum and it couldn't continue the way it was."[2]

Stacy died without leaving his version of the events of 1940, but we do have Joe Zeppa's condensed account.

"Stacy got to drinking quite a bit and spent a lot of time in Dallas, got to running around with the women," said Zeppa, "and so I got tired of it and told him, I said, 'Bob, I think you had better make arrangements to buy me out, or I'll buy you out, one or the other. If you'll just tell me how much you want, how much you feel you should have for your interest, I'll see if I can't get Sam Sklar to join me and I'll get the money somehow to buy you out, because I can't go along and have you carousing in Dallas, when we owe so goddamned much money.'

"Well, he thought it over, came back a week later, and told me what he would take."[3]

Stacy reported to his family that his partners at Delta wanted him to sell out. As his sister-in-law Elizabeth Fullilove remembers, "Bob said they wanted him to get out, and I think the reason they did was: 'If you're not on the spot when we expect you to be on the spot and you continue to be that way and you are out of touch for three or four days and we don't know where you are. . . '. I really think that was it, because there was no [other] reason in the world that Bob would have had to get out.

"Of course, there was no picture of what Delta would do, but he knew it was terribly valuable."[4]

We do not know all of the details of Stacy's response to Zeppa's ultimatum, but we do know that it came in the latter part of 1940, in November when Delta was nearing its ninth anniversary, and Stacy, his forty-fourth birthday.

II

In recounting the transactions which led to Stacy's selling his interest in Delta, Zeppa compressed his version into a few words. Actually, the negotiations were spread over a period of several months, as evinced by the documents in the Delta minute books for that period.

On November 21—four days after Delta's anniversary—Stacy committed to writing the terms he proposed for 450 shares of Delta common stock. This, it might be noted, amounted to only 75 percent of Stacy's 600 shares and, if nothing else, suggests he was reluctant to bid farewell to the company altogether. It also might be noted that if Zeppa had acquired all 450 shares for himself, he would have held 1050 shares compared with Sklar/Dorfman's 600 and Stacy's remaining 150. That single transaction would have left Zeppa the commanding stockholder, whether Stacy stayed or subsequently left.

The letter of proposal which Stacy wrote, however, did not stipulate that Zeppa would so gain the stock, but that it would be offered to the corporation. Among the stipulations on the 11-point proposal, Stacy included cancellation of all his indebtedness to the company, including his open account and notes, while returning to him collateral held by the company; payments of $30,000 cash and $20,000 in notes at the rate of $1,000 per month, the notes to draw five percent interest; the company's interest in a Union County, Arkansas, lease and the three completed wells on it; Delta's interest in four tracts in Nick Springs, Arkansas; one complete 8 1/2-inch Emsco drilling rig, with 8,000 feet of 4 1/2-inch drill pipe, tools and equipment, a complete blowout preventer, and a derrick with timbers and substructures; payment of his salary and expenses to December 31 of that year; title to a Lincoln Zephyr which he then drove, along with the office furniture and fixtures in the Shreveport office; all life insurance in his name with Delta as beneficiary to be transferred to him, and the beneficiary changed as he wished; three nonproducing leases, two of them in northern Louisiana and an 80-acre lease in Rusk County, Texas; and release from any liability for any notes or bonds of surety companies which he had endorsed for the company. Zeppa and his partners apparently had asked Stacy to maintain an association in an advisory capacity for the first nine months of 1941 at a $500 monthly salary. The letter gave the company ten days in which to exercise the option offered.[5]

Nine days later the directors—Zeppa, Sklar, and Stacy—met to consider the letter. The directors agreed to accept Stacy's offer on the terms and conditions he had set out, with his 450 shares to be held as treasury stock.

Zeppa then suggested that 285 shares of this stock, upon being received from Stacy, be offered to the other stockholders (Zeppa and Sklar) at $175 per share, in proportion to the stock each already owned. From this sale, Zeppa noted, Delta would receive $49,875.

At the same meeting, Sam Sklar was elected second vice president (Stacy still being a vice president and board member for the time being) to oversee the company's business in Louisiana.

One last item of business at the November 30 meeting gives us some indication of how Zeppa intended to pay for the stock subscription he surely already had in mind. He owned an undivided one-twelfth interest in a 50-acre lease (C. E. Christian) in Rusk County, Texas. Ten wells were situated upon the lease, as the average spacing by that time in the East Texas field was a well to every five acres. The company already owned a one-fourth working interest in the lease, and Zeppa offered—and the company bought—his interest for $12,500 cash.[6]

Thus in one session, a decision had been made to buy out the stock Stacy was willing to sell at the time, and Zeppa had been provided with part of the means by which he would purchase more shares.

The next formal move on the part of Sklar and Zeppa was to subscribe to the Stacy stock coming into the treasury. Sklar subscribed to 135 shares, paying $14,175 in cash, and gave notes for the remaining $9,450, to be paid in three installments during 1941 in amounts of $3,000, $3,000, and $3,450.[7] At the same time, Zeppa subscribed to 150 shares, paying with a $15,750 check and notes for $10,500, payable in ten monthly installments which would not begin until October 31, 1941.[8]

At this point, Zeppa became—no doubt, with Sam Sklar's approval—the majority stockholder in Delta, a position he was to hold for the rest of his life.

It might also be noted that Sam Sklar was in a better position to buy the stock than was Zeppa, for he gave notes maturing much sooner, and he paid in $3,000 installments rather than $1,000 ones. This further substantiates the testimony that second-generation members of the Sklar and Dorfman families have given: that Zeppa gained stock control of Delta with the full approval of his remaining partners. The rationale seems obvious: Zeppa, after all, was actively running the company and already making the major decisions. He had a strong logical case for being allowed to gain stock control. At the same time, it must be remembered that Zeppa was a very persuasive man, and if there had been any reservations on the part of Sam Sklar, one suspects Zeppa would have persisted indefinitely, as he had with the banker Felsenthal in El Dorado back in 1931.

In the midst of this first round of transactions, Stacy further proposed on December 1 to sell his remaining 150 shares of Delta's capital stock, which would take him completely out of the company's ownership. In return he asked for three oil wells on the Della Crim lease in Gregg County. There were liens upon the wells until September 30, 1941, and Stacy's ownership of the oil would not begin until Delta had satisfied its obligations to the First National Bank of Tulsa on that date.[9]

Stacy's proposal gave the company 60 days in which to exercise the option. On the basis of these documents alone, one might wonder why Stacy chose to sell out serially, as it were, instead of all at once. He may have been slightly reluctant to let go altogether of his interest in a company he helped bring into existence; but, also, there is some indication that this timing may have been for the benefit of Zeppa, who, more than Sklar, probably had some difficulty in putting together the cash for the investment. Thus Stacy may have held off on the last 150 shares, doing Zeppa a favor by delaying, until Zeppa could see his way clear to handling the matter more conveniently.

III

Within the required deadline, Zeppa, as president of Delta, responded formally to Stacy's first proposal regarding the 450 shares, agreeing to the proposal and stipulating precisely how the company and individuals would meet the terms of Stacy's offer. In that letter, Zeppa specified the collateral which the company had held as security for Stacy's December 31, 1936 note for $3,335.72. This included 55 shares of stock in the Exchange Bank and Trust Company, El Dorado; two diamond pins acquired from the First National Bank of Camden, Arkansas; and one diamond dinner ring. In addition, the company acknowledged as settled Stacy's indebtedness to the company of $26,233.56 on his open account. On the same day, Stacy acknowledged receipt of the items and delivered his 450 shares of stock to the company.[10]

The remaining portion of the Stacy interest was liquidated in late January, 1941, when the board members accepted Stacy's proposal to sell his remaining 150 shares of capital stock for three wells on the Della Crim lease in Gregg County.

An interesting sequel to the Stacy transaction came in the form of two letters in mid-December, 1940, in which Sam Sklar and Sam Dorfman each cancelled Delta's indebtedness to them for unpaid salaries. Sklar had $6,000 coming to him; Dorfman, $7,000. Each letter read identically:

> Inasmuch as the Corporation cannot afford to pay this sum of money without handicapping its operations, and in view of the fact that I am a substantial stockholder in the Corporation, I hereby cancel this indebtedness due me and hereby authorize you to write this off the books of the Corporation as though the salary had not been credited to my account. [11]

Did this indicate the company would be hard-pressed to pay Sklar and Dorfman the $13,000 due them? It does suggest it, since a considerable amount of assets, for that time, had been paid out to meet Stacy's proposal: the drilling rig, cancellation of his indebtedness to the company, and oil-producing properties, particularly. But more interesting is the allusion in Dorfman's letter, as in Sklar's, that "I am a substantial stockholder in the Corporation." There is no record in the Delta minute book of that period to indicate that Dorfman was formally a stockholder, though he apparently had an understanding with Sklar, which Zeppa, at least, was aware of. In fact, a month later, in January 1941, at the annual stockholders' meeting, the complete list of stockholders came to three: Zeppa (750 shares), Stacy (150), and Sklar (675), with the total shares coming to 1,575. [12]

When the directors met that same day, little change was apparent in the election of officers. Zeppa remained president and treasurer; Stacy, vice president and secretary; Dorfman, assistant treasurer; Sklar, vice president and assistant secretary; Annie May Jones, assistant secretary; and J. W. "Joe" Beasley, assistant secretary and assistant treasurer.

Stacy's salary, until October 31 of that year, was set at $500 per month, as agreed in the purchase of his 450 shares of stock. Zeppa's salary was fixed at $15,000 for the year, perhaps suggesting that times were not as affluent as they had been those few years in the 1930s when his salary had soared beyond that figure. [13]

Summing up the deal, Zeppa said, years later, "At that time we had some 20 acres north of Kilgore; we had five or six wells on it. We gave him three, I think, three of those wells. Took the best wells on the lease. Paid him I don't know how much money, to tell you the truth. It wasn't very much, for the company wasn't worth much." [14]

Thus these 150 shares went into the company, rather than being purchased by either of the stockholders of record, who at that time were only Zeppa and Sklar. Since these 150 shares had gone into the company treasury, this reduced the total outstanding stock to 1,425 shares, of which Zeppa held 750 and Sklar 675.

The final page closed on Stacy's role in Delta when Zeppa then presented Stacy's letter of resignation, dated December 10, 1940, to be effective whenever the directors decided. Zeppa stated that "the resignation was tendered due to the fact that R. A. Stacy was engaging in the oil and drilling business individually, and for that reason did not deem it advisable to continue serving in these official capacities with the Corporation." The resignation was accepted "with regret," effective that date, and Zeppa offered Stacy the directors' good wishes and their "desire to extend to him every cooperation of the Corporation which may in any manner help to promote the success of his future endeavors." Certainly on the surface and in the official record of the event, the former partners split up amicably. [15]

One stipulation which Stacy insisted upon did not make its way into the official minutes. When he sold out, it was agreed that Flora M. Davis, who had served as the very efficient personal secretary to both Zeppa and Stacy, would be permitted to enter Stacy's employment as he went into the oil business on his own. [16]

J. W. Beasley, an accountant for the company who had been handling its tax reports, was elected director and secretary in Stacy's place. Zeppa had actual voting control of the company with 52.6 percent of the outstanding shares of the stock, and Stacy's years with the company had ended.

IV

Stacy's settlement with Delta, from all evidence, had been a fair one for him and had left him with substantial income. Ironically, however, his production in all but the three East Texas wells soon turned to water, and he was faced with the need to start out again. At this point, during World War II, he was fortunate enough to hire a highly capable man in Shreveport, Douglas Whittaker, who knew the oil business thoroughly. The connection was a fortuitous one, for by the end of the war they had found, in the words of Bob Stacy, Jr., "some damn good production," leaving Stacy well off. Whittaker went on to a job with Delta, working out of Shreveport and reserving the right to make his own deals, and becoming wealthy himself. [17]

The surface reasons for Zeppa's buy-or-sell challenge to Stacy seem to be sufficient explanation for the events in the winter of 1940-1941, and they may have been all that was in the minds of the principals at the time. But beneath the surface there probably were other issues: both Zeppa and Stacy liked to be Number One, plus the fact that Stacy and his partners were unalike in so many ways.

As Bob Stacy, Jr. said of his father, "He wasn't like them at all. It was night and day. My gosh, there was no similarity of personality, no similarity in anything except a common goal of business and profit."[18] Thus the common qualities that had drawn together the native-born Stacy and his immigrant partners were not as strong, in the final analysis, as the conflicting factors that ultimately separated them.

From that point on, if there had ever been any doubt previously, Delta Drilling Company was, in fact and in spirit, Joe Zeppa's outfit.

6

THE HAWKINS DEAL AND DELTA GULF

I

A decisive event in the economic growth of Delta Drilling Company came in the late 1940s with the acquisition of the holdings of J. C. "Jerry" Hawkins, an independent drilling contractor and oil operator. This single transaction constituted Delta's "big jump," for in the process the company doubled its rig count, gained additional production and leases, and extended its operations permanently into South Louisiana and West Texas.

The Hawkins deal was hardly consummated in a few swift sessions. It was an exceedingly complicated matter festooned with tedious details demanding close attention. It took approximately six years for the new property to be fully absorbed by Delta. The complexities repelled many otherwise interested parties. Joe Zeppa was attracted to it, however, perhaps seeing at the beginning the camouflaged opportunity Hawkins's troubles represented to someone willing to unravel and sort out the multitude of problems inherent in the deal.

II

Alabama-born Jerry Hawkins was a large, heavyset man, slightly more than six feet tall, with a mustache turning gray, and about 55 years old at the time of this transaction.[1] He had a colorful past and, as indicated by his accomplishments and apparent goals, was a highly ambitious man.

"Jerry was a good man," said G. I. "Red" Nixon, who worked for Hawkins and knew him well, "but he was never still. He had to be doing something. He could be in the car with you and just tap his fingers off on the dash.

"At one time he had a plaque in his office that showed that he was the high-volume salesman for 'Kohler Light Plants' in the U.S. before they had rural electricity. It was a battery-operated generator."

Early in his career in the oil business, Hawkins "went through some very rough times." At one point he had a partner, H. H. Howell, whose one drilling rig in the East Texas field was manned by members of his family serving as drillers and roughnecks—four brothers, two sisters. Howell's wife virtually ran the outfit. They were dubbed the Egg Shell Drilling Company because in those days when almost everyone was poor, eggs—selling for five cents a dozen—constituted the main food for the family crew. Hawkins reported there were egg shells everywhere around their camp.

"But Jerry made a deal with Howell: every well that Howell drilled— and he drilled a lot of them—Jerry paid him a dollar a foot for a quarter interest in the well," said Nixon. "And they drilled a lot of wells here in East Texas, and a lot of good ones. Well, they hit pretty good here [in East Texas] and then H. H. moved to South Texas, between Robstown and Corpus [Christi] and he was very lucky. He discovered about three good fields around Alice, Corpus, Robstown, and that's the way Jerry got in the oil business down there, and he later bought Howell out—he bought everything Howell had."

By the late 1930s, Hawkins had about 40 good wells in the East Texas field which he still owned when he found himself in crisis in 1949.

Successful promoters of drilling ventures in the money-scarce thirties frequently had to parlay brashness and bluff to ensure that the rotary bit kept moving toward pay dirt. Jerry Hawkins was one who could produce the right ingredients at the right moment to replace the money and confidence that might be missing. Red Nixon remembers the story told him of a crisis reached by Hawkins's partner H. H. Howell, whose "Egg Shell Drilling Company" was then in South Texas.

"Howell had rustled around and got enough money to buy one of the first power rigs that came in. They had all had steam rigs. H. H. had got enough money, or promoted and got ahold of a little rig, and drilled this well. Well, he was busted, didn't have enough money. [Creditors] were about to take his rig away from him. His sister was a driller, and I think another sister pushed tools and everything. He was a character. He had a very smart wife. She was the brains. She ran it. Four brothers, two sisters, and they all lived together and shared and shared alike. H. H. was the head man; he was a funny fellow. He had a fine voice—talked like a woman. But he made himself a million two different times. He did that. And I suppose when he died he was a millionaire.

"They went down [to] the [oil] sand, had no pipe, no tubing, no credit. [Creditors] were threatening to take the rig away, and he called Jerry in Tyler. Jerry and his attorney, Ben Clower, got in a Model T coupe Jerry had and drove down there. And, of course, Jerry didn't have any more money than H. H. did, but he took along a check. And they got down there and there were a lot of people out there, and they had this well ready to complete, and it was a big thing. It sure was a big thing, one of the biggest fields in South Texas.

"And H. H. got up on one of the big trucks there and he introduced Jerry and the man from Tyler, Texas. Said he had a check from Jerry for

$150,000, and he waved it around all of them and told them he was going to get some money, not to worry, and Jerry and Ben both said the check wasn't worth ten cents.

"Jerry gave him the check, and Jerry and Ben, his lawyer, came on back to Tyler that night, and the next morning Jerry got on the phone and talked to a fellow that he knew in Chicago, talked him into it over the telephone, gave him a good enough story on the well they had down there, and they passed that check, and that's where H. H. and Jerry really got started. He drilled seven or eight wells there, and he drilled them for nothing.

"He didn't use mud, and other people came in there and tried to drill with clear water and the wells would blow out on them and they would lose their rig and everything. H. H., said he went from there to Alice.

"They called him 'Clear Water Howell.' He never bought a sack of mud in his life. He was just lucky. I know how lucky he was, 'cause I completed a well just across, offset wells to him, and I knew how much easier he could do it than I could. He just had something that most people didn't have. And then he went from there and discovered the Alice field, which is a big field. And then he sold out to Jerry. Jerry had got on his feet and bought him out.

"I know one time H. H. was selling out to Jerry. He had got ahold of some production in Duval County [Texas] and had a bunch of worn-out rigs—bunch of junk—but had some pretty good leases, and he wanted to sell them. Everytime he got something he didn't want he would sell it to Jerry.

"And so Jerry and Ben [Clower, his attorney] came down to Alice, and Jerry had me look all this stuff over. It didn't amount to too much [in value], but Jerry had some money and he wanted to buy it. And he made a deal with H. H. He paid him $500,000 for the whole, had some city property and some leases and drilling rigs.

"I was talking to [Howell] back in my office—Jerry was out in front, they had closed the deal—and the lawyer had the papers and all. [The lawyer] said, 'H. H., I don't understand how you make these deals like this. You said you never had an education.' [Howell] said, 'No, I said I never went to school.' "[2]

Ben D. Clower, Hawkins's counsel, was himself one of a vanishing breed of colorful old-time lawyers. He lived as many anecdotes as he told. "Ben was always getting into something," Red Nixon said. "Clower was one from the old school. He looked just like W. C. Fields. He had his briefcase and had a fifth of scotch with him all day, and when five o'clock came—I guess [even] if he'd been in church—he would stop and buy a Coke, pour the Coke out, fill the bottle half full of water and the rest with scotch, and he was drunk by nine o'clock. He was something. He would bet on anything. Just worry you to death. He and Jerry were something together.

"[He and his wife Bess] had this big colonial home on White Rock Lake [in Dallas], out there near H. L. Hunt. They were old buddies. Ben ran into some liquor salesman, and the man also sold this White Rock mix. Him and Ben got to talking. He wanted to take a picture [for an endorsement], and Ben was for it. Bess wasn't for it. I asked Ben what he was going to get out of it, and he said, "Hell, I was going to get White Rock for the rest of my life. Never run out of White Rock!'

"Ben was over here in Canton [in East Texas], and they went to this filling station. They had a case of whiskey in the back. He got his Coke and fixed himself this drink, and this sheriff asked him how much whiskey he had, and he said, 'I've got a case. Scotch.' Sheriff said, 'Let's go around to the courthouse.' They pulled out of the drive; Ben thought he was going to arrest him. [The sheriff] said, 'I had to do something because of the teenagers. You are sure you're not selling this whiskey?'

"Ben said, 'Call anybody you want to and they will tell you Ben Clower never got enough whiskey in all his life to drink, let alone sell!' "[3]

Jerry Hawkins seems to have been one of those men who stirred different emotions in different men. Many of his creditors viewed him darkly, but his employees frequently saw him favorably. Ted T. Ferguson, who started for Hawkins in the production department in 1948, about a year before the deep trouble came to light, called him "one of the finest men I ever knew in my life" and "the friendliest guy you ever saw."

Ferguson added, "And he could meet you right here today and he can see you a week from now and as far as he can holler, he'd holler and call you by your first name.

"And I went to work for him, I'd say, in April of '48, and July that year he asked me when I was going to take my vacation. I said, 'I haven't got a vacation coming, I don't guess.' He said, 'You sure have. You've got two weeks. You've got that.' I was driving a '47-model company station wagon. 'And you've got that station wagon and if you haven't got enough credit cards, I've got some more in my pocket. And you go in there and go where you want to.' And that's the kind of a guy he was to work with, and I guess that's what broke him, more or less. He was just too free for his own good. And there's a lot of people that took advantage of that. He was taking everybody to be as honest as he was, I think. And everybody's not. They were eating him up."[4]

III

Jerry Hawkins had the raw materials for a success story that, in time, might have rivaled that of, say, Sid Richardson, the Fort Worth centimillionaire. The vital distinction between the early Hawkins and Richardson was less opportunity than how each one managed those opportunities—and assets. At least, at one point events seemed to be rolling Hawkins's way. In the Slaughter field in West Texas, Hawkins had drilled around 100 wells without a dry hole. But, instead of launching him into the rarified air of oil-rich Texans, the holding instead marked the start of his troubles. He sold out his Slaughter field wells to Stanolind, for around $4 million, a good sum then. But instead of settling immediately on his income tax, he bought drilling rigs with his fresh profits, and ended up with 17 drilling rigs. He had five rigs before he made the transaction.

G. I. "Red" Nixon, who was in charge of Hawkins's overall production, explained, "Instead of paying his bills, he had bought drilling equipment and drilled wildcat wells, and he had as much wildcat acreage under lease as many major companies, and a big part of it was no good. Jerry had it everywhere. He bought leases. He had a person to do the work for him that felt the same way that Jerry felt about it. They bought them up.

"And we had a fleet of trucks for quite a while, for moving rigs. He would have been much better off if he had contracted them, [rather] than trying to be in the trucking business *and* the drilling business. But that is the way he wanted to operate.

"Jerry was—don't get me wrong—he was a gambler. He didn't care what it was, he wanted to gamble. He would rather drill a wildcat well whether he made a well or a dry hole; he just liked to drill a wildcat well."

He never seemed to play cards or other games of chance. His stakes were higher.

"He gambled on his production and his business. I know one time I went with Jerry to a bank in Dallas. We went in and talked to Mr. Dan Rogers at the Mercantile Bank. Mr. Rogers was head of the petroleum department. Jerry's offices were there. Jerry wanted to borrow $150 million, and they wouldn't let him have it.

"And he told Dan Rogers, 'Well, you've lost confidence in drilling contractors.'

"Mr. Rogers said, 'No, Jerry, I haven't. I never did have any confidence in the drilling business.'

"I've seen Jerry, he would be riding with me somewhere over the country, and he'd be using a cancelled million-dollar check for a scratch pad. That's the way he was."[5]

Hawkins had been involved in a number of companies in the 1940s. One was co-owned by him and a man named Ben Roberts. Roberts was killed in an automobile wreck, and Hawkins took over the operation. He bought new drilling rigs from National Supply Company, and this was the beginning of his obligations toward that firm.

Around the same time, Hawkins had gone partners with a lawyer named Ed Graham in West Texas. When debts threatened that business, Hawkins bailed it out and took over the company.

"And," as Red Nixon said, "it loaded him up to where he couldn't [go any further]."[6]

One of Hawkins's most lucrative trades was with the Mallett Land and Cattle Company in West Texas, a ranch of many thousands of acres that might be thought of as the local version of the King Ranch. The man who had started it, old man Mallett, was a former sea captain who had gained the property in a grant. The land had passed down to his two daughters and a granddaughter. They had leased much of the land for oil and gas exploration, but had clung stubbornly to 3,200 acres, despite all offers. But where major companies had failed, Hawkins succeeded.

"Jerry was a trader," said Red Nixon, "and he made a deal with them. Without any money. Gave them so much cash for every well that we drilled on the 3,200 acres, and that's the way he got into the oil business out there in West Texas."

He gave them a one-eighth production royalty, and when oil went to $1.50 a barrel the interest was raised to three-sixteenths. At this time, according to Nixon, Hawkins was drilling a well for $11,500, and the Republic National Bank in Dallas would lend him $25,000 on the basis of it, with which Hawkins would buy more new equipment and leases to go with what he had.

And he just got himself into shape where he couldn't go any farther," said Nixon. "But there were two of the major supply companies that would have bailed him out if he had done what they wanted. Continental and National. They wanted to take control of the company, which eventually Joe [Zeppa] did, and it was the best thing ever happened to Jerry.

"In 1949 he had gone as far as he could go in production. He still had money coming in from production, but it was going out faster than it was coming in. He just overextended his credit, and Jerry was a man who ran his business from his hip pocket. He didn't want any organization. And, didn't have one. He trusted everybody. That was one of his big failures. He had people who were not a great help to him."[7]

Delta's deal with Jerry Hawkins came about when Hawkins overstepped, as Nick Andretta put it. "He tried to follow every boom. He overstepped and overspent and got in trouble. And through an attorney, his attorney, I understood that he needed some help. So I put Joe Zeppa wise. Ben Clower was the lawyer, and I told Joe Zeppa. We got together and we drilled some wells out in the Slaughter field and we made a deal. Hawkins went in there and got a lease from the Mallet Land and Cattle Company and drilled a bunch of wells. When we took Hawkins over, we took those wells in."[8]

Hawkins overextended himself through a repeated pattern of borrowing and buying. "He'd buy a brand new rig, you know," said Joe Bevill, "give them a 90-day note. He'd give them a 90-day note and then he'd fly. Just one right after the other. The people knew he had all that production. They just assumed he knew what he was doing."[9]

Except for a couple of old steam rigs which Hawkins had owned for some time, his equipment, including his drilling rigs, was either new or as good as new.

"Jerry's trouble with his drilling company was his personnel," said Red Nixon. "You know, when you get that many people involved, a lot of them spending money, you need organization. I've been on location where they would move the rig off in South Louisiana and I'd go by the well and [they'd] moved the rig off and maybe left seven or eight drill collars laying in the weeds out there. Just forgot about them. Joints of drill pipe—not do it on purpose, but just mismanagement. That's all it was. It all added up. If it hadn't been for production, he would have went way before then."

Hawkins's production department was, in contrast to his drilling operation, relatively well organized, which apparently pleased, and surprised, Joe Zeppa when he took over the entire Hawkins properties.[10]

The late John D. Hall, who started as Hawkins's office manager in Dallas in the spring of 1949 and later made the transition to Delta Gulf, gained some insight into Hawkins's deficiencies as a businessman before the crisis developed, or, as he put it, when "the roof caved in." Hall saw Hawkins as a "most unsuccessful purchaser of wildcat leases. He'd buy them and never would drill them."

As Hall viewed it, Hawkins had a large amount of profitable production scattered in the East Texas field, near Alice in South Texas, at Quitman in the East Texas region, and some in West Texas. "In addition to that," said Hall, "he had gotten into the drilling business, and that's what killed him. If he had stayed in production he could have just lived on Easy Street the rest of his life and never had to worry about anything. But he didn't know anything about the drilling business. And it got away from him. He wasn't taking any interest in wells. He was just losing too damn much money on the drilling business, bidding too low or costing too much money. They just took some bad drilling contracts. You had to have people in management making the bids who were familiar with the region. You've got to know how fast you're going to be able to drill."

Hall illustrated his analysis with an example from Hawkins's Louisiana operation.

"I remember one well they bid. The footage bid was all right, but they didn't think they'd be on daywork very long and they bid it low and, my gosh, the company they were drilling for stayed on that thing forever and ever on daywork. We were losing—I don't know how much—every day. It was just a combination of things like that."[11]

Certain facets of Hawkins's personality seemed to have been at work helping to create his problems. For one thing, he was an incurable optimist who never wanted to let go of something he owned.

"Jerry and I would get into an argument about a dry hole," said Red Nixon, his production superintendent. "I know one time we got into a deal that I tried to turn down and couldn't. We went in there and drilled a well with a lot of trouble. Tested six or seven times, and Jerry found out what they were doing and he said, 'Let's set pipe.' He was getting in bad shape then, and I said, 'We can't afford to set pipe.' He said, 'I tell you what: Let's set pipe, and if it makes a well, I'll tell them it was your idea, and if it's a dry hole, I'll tell them it was my idea!' They went ahead and set pipe on it, and by the time it was over, Jerry walked off and left it with about $300,000 in it.

"Joe Zeppa was just about as hard-headed as Jerry was. After Joe took over Delta Gulf Drilling Company, these people in San Antonio sold Joe on the idea that there was still an oil field there, and darn if Joe didn't go with them when they drilled, I think, three dry holes. They would make a little oil, but they wouldn't last long. It was a bad deal.'[12]

Chris Zeppa seems to have summarized the weakness of the operation when he said: "There was a little bit of irrational administration in it—by Jerry and by the other people involved. They tried to run faster than they could go. Hawkins had leases everywhere. The fact is that they were too fast and they committed themselves too heavily. Hawkins was a tremendously good oilman and so was Winn, his brother."[13]

IV

Jerry Hawkins's careless attitude toward records had a long history. Red Nixon remembered a story Hawkins told him of one outside examination of his records in the 1930s which substantiates this.

"Jerry had an office in Tyler. Lived somewhere in Tyler and had a basement. He said a couple of fellows came in one day and told him they wanted to borrow his books. They looked around and wanted to know where his files were.

"He said, 'They are at home. I will take you out there.' I suppose it was the truth, knowing Jerry Hawkins. He said he had his files in two wooden apple barrels. He said, 'There are the files.' He said they dug around in there a couple of days and told him they guessed everything was all right. That was one of Jerry's stories, so if you knew him, you would have believed it."

Did Hawkins keep better records later on?

"God, no," replied Nixon. "They would mail out checks before the time sheets came in. One time in West Texas, they were supposed to keep daily gauges on oil run—what had happened on this oil. I had the daily production reports, and the oil hadn't been sold, but it had been burned for fuel oil. Steam rigs, we used a lot of oil. Anyway, they wanted those daily reports. Federal rules, monthly reports. I had mine since I'd been there. And they asked me where the rest of them were. I said, 'Probably mailed to Dallas,' 'cause we mailed in each month. And they asked Jerry about it, what happened to these daily gauge sheets. There was a road ticket made for so many gallons of oil. They asked Jerry about this. He said, 'Do you realize how many sacks full of tickets that would be? That would have been a hell of a lot of sacks full of tickets.'

"That was in the Slaughter field. They investigated it for two or three years. They questioned everybody, and got statements from everybody. They went to Magnolia Pipeline, and Magnolia had a ticket for every run. Two men worked for months making a copy of these tickets. So when they came to indict us, they didn't call me. Dan Rogers was head of the federal grand jury. Finally Rogers said, 'I think we have spent all the taxpayers' money we should spend.' They no-billed Jerry Hawkins and me."[14]

V

The Hawkins deal not only forged a key chapter in Delta's history, it also provides us a window into Zeppa's personality and manner of operating, which an ordinary transaction might not offer.

For an overview of the transaction, we have, first, the compressed account from Joe Zeppa himself.

"J. C. Hawkins was an individual and had gotten deep in debt—couldn't pay his bills," Zeppa recalled in 1975. "And there was a question of whether he was going to go into receivership or whether somebody would put enough money in the thing to get him out.

"So I fooled around with it and, you know, you couldn't tell much about the company, because he didn't keep a set of books like anybody else would. He had a jillion leases and royalties scattered all over God's kingdom, for which he paid money. And he got in such bad shape that nobody wanted to touch him. Except me, of course. You know, I'm one of these fellows that always thinks there's a way of saving something.

"So I went in and fooled around with it. I went to Dallas and spent *weeks* up there. And finally I decided on the basis of what I saw, and what I could do with the property, that I could afford to work it out. We agreed to put a certain amount of money in it, and I think we let him keep 40 percent of the stock. We took 60 [percent]. But it's not really worth anything. And I went in and borrowed money on the properties themselves, and it was kept as

Delta Gulf Drilling Company. We called the new company Delta Gulf. Up until that time it was just Jerry Hawkins, Individual.

"And, of course, it turned out to be a good deal, you see, because it worked out very well. So in two or three years, why, we had the company worked out of debt. Then we had to pay ourselves back.

"Then Jerry Hawkins died. But Jerry was a great trader. He made deals in West Texas after he made the deal with us. One was offered to us, but I told him it was just too steep. We couldn't have done it. And he made a deal out there that I wouldn't touch. But he made a lot of money out of it. I didn't realize that stuff was as good as it turned out to be.

"Of course, we've been in West Texas now for quite a while. We took over all his rigs. He must have had about fifteen, twenty rigs. All of them were in bad shape, but we got it straightened out then. That formed the [division] that we have in Lafayette, Louisiana, and it formed the [division] that we have in West Texas."[15]

Zeppa was speaking without records before him and had indicated he was not certain about the division of stock in Delta Gulf. As a matter of fact, he left Hawkins with 45 percent of the company rather than 40 percent, which suggests Zeppa was more generous than he remembered.

VI

Essentially, Zeppa's narrative outlines the salient points of the transaction and perhaps the emotional features, as well. But the details, provided by others privy to the deal, spell out some of the things he was referring to.

None was closer to the deal than C. L. Vickers, a lawyer who was then general credit manager for The Continental Supply Company. Continental, as it turned out, was Hawkins' major supply creditor, and Vickers's role in the matter was crucial to the solution finally reached.

Vickers had met Zeppa in 1934 when he was transferred from Houston to Shreveport and placed in charge of credits and finances for Louisiana, Mississippi, Arkansas, and the East Texas field. Vickers, who was to become vice president of Continental and its subsequent corporate form, Continental-Emsco Supply Company, came to know Zeppa well and appreciated his ability, and they became friends. He was in a position, from his privileged vantage point, to chart Delta's progress over many years.

"The company continued to grow each year," said Vickers, "and at the end of each year you would find that they had added several drilling rigs and a lot of production, and the production income began to creep up to where it was around $200,000 a month. This continued, actually, from 1934 until 1949, when a great change took place in the company.

"This Jerry C. Hawkins deal put Delta in the big time and prepared the company for the world operations which followed. The two companies were relatively of equal value, insofar as gross assets were concerned. However, as far as net worth is concerned, it was questionable, as Hawkins was loaded with debts. Hawkins had 18 rigs and producing properties in the East Texas field and all over Texas and Louisiana; Delta had 18 rigs and producing properties all over East Texas and Louisiana. The deal doubled the size of Delta.

"This was in 1949. Jerry Hawkins had been in the oil business for many years and was quite a plunger. I hate to put that in there, but it's actually the truth. He made a fortune out of being a plunger, so God bless him. Mr. Hawkins had built his complement of drilling rigs from 2 to 18 in a period of about three years, and this expansion resulted in the accumulation of very heavy indebtedness, owing to several hundred creditors. Debts became delinquent and one small creditor filed a lien on one of his leases. The word spread and others filed liens. The major oil companies refused to pay on the completed contracts, for fear their properties would be subjected to a lien by the creditors. The bankers became concerned and refused to make further loans or release any of the oil runs, so a very wealthy man was frozen and threatened with bankruptcy.

"General American Oil Company, the Hunts, and several other large companies took a look at the Hawkins situation and were not interested in the deal. Hawkins owed Continental Supply Company about $200,000, so I was very familiar with his entire situation.

"Ben D. Clower, attorney for Hawkins, gave me complete information on the situation and asked me to talk to Mr. Zeppa. I arrived in [Zeppa's] office at 2:00 p.m. and told him I had to testify in court at three o'clock. He suggested that I go to the court first and come back to his office. I did this, and I walked out of his office to drive to Dallas at 12:30 a.m.

"When I left, Mr. Zeppa knew all that I did about the Hawkins situation, and we had already arrived at a plan to rescue the company. I tell this to show that he had an excellent and brilliant mind and was quite quick in his grasp of situations.

"On my return to Dallas the next morning, I made an appointment for Thursday of that week for Mr. Zeppa to interview Jerry Hawkins. When he called on Jerry and Mr. Ben D. Clower, who was Hawkins's attorney, they told him that apparently they would make a deal with Trinity Drilling Company, so any further discussion with respect to Mr. Zeppa would be postponed. That was on a Thursday.

"On Saturday morning our office was closed, but a number of us were around the office doing special jobs. And Mr. Zeppa came in. And while he was sitting at my desk talking to me, my phone rang and Mr. Clower, the attorney for Hawkins, says, 'Vick, the deal with Trinity is off and I'd like to find Joe Zeppa. Do you know where he is?' And I said, 'Just a minute.'

"And I told Joe, 'This is Ben Clower and he says the Trinity deal is off. Do you want to talk to him now or do you want me to tell him I'll get hold of you?'

"And he said, 'No, I'll talk to him.'

"But just then is the only time I ever saw any deal or fact affect Joe's emotions. Just changed the expression on his face. During this time when I turned and said, 'This is Ben Clower and the Trinity deal is off and he wanted to talk to you,' for a moment Mr. Zeppa showed some extreme emotion. It just went all over him. It was a big deal, you know. I mention that because of all the things that happened and the deals we've been in, you never saw him change. He was just steadfast the whole time.

"But by the time I handed him the telephone he was a bold, hard trader. He first said, 'Well, Ben, I've been waiting here since Thursday and I told Gertrude I was coming home, but I'll stay over and I'll meet you this afternoon.' He stayed over; he didn't go to Tyler.

"They met on Saturday afternoon—Joe Zeppa, Ben Clower, and Hawkins. And then they met again on Sunday, and Joe had his lawyer [Jack Blalock] come up from Houston. They made a deal and prepared a contract and signed it on Sunday. They got with it. I do not recall all the exact terms of the contract, but it provided for the creation of a corporation to be owned by Delta and Hawkins—which was the Delta Gulf Drilling Company— with a voting contract whereby Delta would be able to control the company for a period of years. The substance of the deal between them was that Delta Drilling Company purchased from Hawkins properties that were free and clear to the extent of $500,000, with the provision that this money would be used under the control of Delta in settling some small accounts owed by Hawkins to keep the situation from going into bankruptcy.

"[Zeppa] sized up that deal. He had a suite of rooms at the Baker Hotel, and every night the lawyers, the accountants, or some of the bigger creditors would be there, and every night when I left my office before going home I would go by the suite, and in many cases I stayed until after dinner, and I would check with the other creditors and confirm to them that we were going along on the deal. I believe that this continued for quite a long time. He didn't move out of that suite, although when he went home to Tyler he still left everything in the suite. And this went on for the entire month of July. He had absolutely terrific concentration. He didn't have a divided attention on anything.

"There was provision that if the proper settlement could be made of all the debts, then Delta would reassign the properties to [either] Delta Gulf Drilling Company or the corporation to be organized, and would be reimbursed out of the corporation for the money they had advanced. So Mr. Ben Clower was instructed to try to appease his creditors as best he could, but if any of the small creditors filed a suit, to file an answer to delay judgment as long as possible. And this brought us down to the question of what we were going to do with some 500 creditors that were owed several million dollars.

"There never was a suit filed. The first thing that Mr. Zeppa did was send in his lawyers to examine titles. He also sent in [representatives of the accounting firm] Peat, Marwick, Mitchell & Company to examine the books, and after about three weeks of hard work, day and night, they were able to arrive at a reasonably correct detail of existing debts. And after this was arrived at, we went over the list and marked certain creditors that had accounts of less than $500, to be paid, and the rest of the creditors were to enter into a contract whereby they accepted notes payable over 30 months."

An agreement was drawn and signed by the creditors which provided for execution of notes by Delta Gulf over a 30-month period. Then at creditors' meetings held in Dallas, Houston, Corpus Christi, and Oklahoma City, Zeppa, Vickers, and others outlined exactly what was going to happen, how the debts would be met.

"And within a period of about 20 days," said Vickers, "we had an agreement with the acceptance of these notes. And that cleared the entire situation, because the rest of it was with banks. The banks were the biggest creditors."

At that time current assets exceeded liabilities by about $500,000, but the creditors' confidence in Hawkins had collapsed, leading to his untenable situation. Once Zeppa entered the picture, his standing immediately brightened the prospects for the Hawkins operation.

"Actually," continued Vickers, "what had happened was that after the agreement was signed by all the creditors, Mr. Zeppa and his attorneys were able to go to the major oil companies and present the facts and ask that they pay their drilling debts which they had been withholding for fear of liens being filed. And when this money came in, actually, Delta Gulf was in very excellent condition. And another thing: the basis of this whole situation was the banks, and the creditors were a little bit tough on Jerry Hawkins—and he always had my sympathy—but the cooperation of the banks and all the creditors shows the confidence that the public had in Mr. Zeppa.

"Now I don't know exactly what this settlement between Mrs. Hawkins and Delta was, but Jerry Hawkins died with a heart attack about 90 days or so after the contract was signed. And Freeman Burford, Hawkins's brother-in-law, went on the board of directors [of Delta Gulf] in Jerry's place. Freeman Burford's wife and Mrs. Hawkins were sisters. And after a year or more of operations, an agreement was reached whereby Mrs. Hawkins was given free and clear a certain number of oil wells in the East Texas field. They got the best ones. And it was sufficient to make her very wealthy. Still left a good deal for Delta Drilling Company. [Eventually] Delta Gulf just dissolved and conveyed the properties to Delta; and then with the same division of ownership, Delta owned all the properties."

How did Zeppa's co-owners Sam Sklar and Sam Dorfman react to the Hawkins deal at the outset?

"The Delta Drilling Company always had an annual meeting of stockholders," said Vickers, "but it was very seldom that they ever had a deal whereby a particular deal was ratified and approved by the stockholders. But in this particular case [Zeppa] did call a meeting of the stockholders and submitted the facts to them, and they approved it."

It appears everyone got what he wanted out of the transaction, once it was finally worked out. There was even a bonus for C. L. Vickers, beyond collecting the money due his company, Continental Supply. The Delta Gulf deal made Vickers the dean of the finance family in the supply industry. Because he had served as a kind of midwife to the transaction, he gained as much reputation out of it as Zeppa did.

"Now that deal started on June 30, 1949, and it was closed on July 31— final signing—of 1949, and the first financial statement and operating statement was put out August 31," noted Vickers. "The thing could have gone on into bankruptcy."[16]

VII

Among the entrepreneurs viewing Hawkins's assets and situation was Clint Murchison, the Dallas multimillionaire.

"Clint was one of them [who was interested]," said Red Nixon. "They couldn't get Jerry still long enough to deal with them. And Continental Supply or National Supply offered to bail him out. But he had to let them straighten it out and he could never be still long enough to do it."

Hawkins had known Joe Zeppa back in the early Arkansas days, and this was an entrée which caused Hawkins to look more favorably upon Delta's participation, for he knew Zeppa's reputation for integrity. And by then the pressure on Hawkins had increased greatly.

"It had come down to the point where I was paying mud bills out of my pocket," said Red Nixon. "Jerry had no credit at all. Everybody had cut him off.

"Actually, there wasn't anything wrong with our production. Our production records were straight, and I think that is what surprised Joe Zeppa. When I came up to Dallas the first time to meet him, to get the deal worked out, they were not too good about keeping records, but in my department we had complete records of what we had done and I think it surprised them.

"It didn't take Joe long to go ahead and pay off everything and get it operating like it should be. Joe could go ahead and borrow the money, and they could get their money. Joe Zeppa was a wonderful man. So was Jerry, but he and Joe could never hit it off together. It was a shame they couldn't."[17]

The Hawkins deal had frightened off other men who had a chance to tackle it.

"Well, it was a mess," said Edward Kliewer, who handled some of the legal work for Zeppa. "Other outfits that made a run at Hawkins either just couldn't come to a meeting of the minds with him or just threw up their hands at the complexity of it. It took a real competent numbers mind to sort that thing out. And that, I think, was the reason Mr. Zeppa was able to do it. It was an unadulterated mess.

"Oh, it was a challenge to him. Once he got into it, he was determined to make it fly. That was a great trait that he had and it was sometimes as maddening as it was heartening."

Hawkins himself was a frustrating ingredient in the complexity, according to Kliewer. Kliewer, for one, never perceived Hawkins as a personable fellow. "I never thought so. He was a big fat fellow and, if not nervous, he was quite disconnected in his thought processes. I could never put a finger on him, but as part of the transaction he would agree to execute to Delta Gulf Drilling Company routinely assignments of all these properties. It was a fight to get every piece of paper signed. He aborted any way he could. And it was just a constant struggle.

"As far as he was concerned, it was his property. He made *this* deal but King's X on *that*. He'd been bailed out, a saviour had come along— 'Now give it all back to me. I'll get you your money back. Let's call a deal on it. You've been unfair to me.' That's not the way it worked.

"Well, the transaction as such was concluded. [Hawkins] was a shareholder in Delta Gulf and I thought Joe gave up too much on that. His agreement was that Hawkins would have 45 percent of the stock of that company. He could have made a better deal than that. They were over the barrel, but he did what he felt was fair.

"Joe's instinct was to believe what people told him. Now he could tell a two-bit broker a mile away and he was selective in how he handled those people, but in dealing with a businessman. . . "[18]

While most potential buyers were frightened off by the economic disarray in Hawkins's company, other investors had easier-fathomed designs, which in the final analysis probably helped Zeppa gain the inside track, not only with Hawkins's creditors but with Hawkins himself. "A lot of the big buyers, when they go looking, they want a 100-percent buy," explained Chris Zeppa. "Not too many are willing to take *half* and then maybe get the other half, you know."[19] Joe Zeppa, offering to leave Hawkins more at the outset, putting his solid reputation behind the deal and striving to resolve the economic chaos, must have seemed as silver a lining as the embattled Hawkins had ever expected to see.

Mark Gardner, well-versed in the personalities of the oil industry at the time, said, "I will say this for Joe: There were vultures circling over Dallas by the dozens, going to pounce on the J. C. Hawkins Drilling Company, that wouldn't have left any of that family a dime. They [the family] were scared to death that they might not even end up with their homes. It was terrible the way some outfits were, but the problem that these other outfits had was that they couldn't see through the horrible mess that J.C. Hawkins was in. And [talk to] people like Fred Mayer [of Continental Supply Company] and people at Republic National Bank and others in the industry, and they will tell you there was only one man ever able to see all the way through that mess, and had the guts to go through with it. But in spite of the fact that he had the field to himself, although he might not have known it, he still left them 40 percent [45 percent] interest in the thing.

"It was typical of Joe Zeppa. Joe was a master of that. You see, the thing was on the verge of complete bankruptcy. But Joe put up some money, he moved over to the Baker Hotel and stayed five months. He was faced with some of Hawkins's personnel that were resentful, and he was faced with completely chaotic records. He put up money enough to keep it going; in fact, he paid off hundreds of small creditors. You know, the small creditor could foreclose just as easy as the big one."[20]

VIII

Joe Zeppa and Jerry Hawkins were a study in contrasts: Hawkins, the tall flamboyant six-footer, heavyset, with an impatience to hit the big time, growing by leaps and bounds through purchases of leases and drilling rigs but careless in his bookkeeping and record keeping; and Zeppa, a low-keyed accountant who exhibited meticulous attention to detail, conservative and cautious in expanding his business. Ironically, it was due to this meeting of Zeppa and Hawkins that Zeppa's company doubled its growth almost overnight—its most rapid spurt forward—yet it was conducted in the most careful, conservative fashion.

John D. Hall, like many others in a position to know what he was talking about, called Zeppa "a financial genius." Working day and night with Zeppa in the Baker Hotel after the Hawkins transaction, Hall observed Zeppa's exhaustive, scholarly pursuit of the smallest details of assets, liabilities, and creditors connected with the Hawkins operation. The new owner's penetrating examination, recalled Hall, was "almost like an audit."[21]

If the oil industry throughout Texas was abuzz over Hawkins's economic condition, employees within Delta knew little at first. Joe Zeppa was closed-mouthed about the Hawkins deal, even to his close aides in Tyler. Joe Bevill got the first wind of it through a Continental Supply officer rather than from his boss. This was a pattern Zeppa maintained when it came to matters involving other persons or other companies. Bevill knew Zeppa was driving himself very hard and spending a lot of time in Dallas, but no details were given him until Delta Drilling Company checks were issued to the Jerry Hawkins payroll fund to the tune of $20,000, then repeated. By this time the Hawkins rigs were stacked, and only essential services were being performed.

"It didn't involve a huge sum of money," said Bevill, "but we never did have $40,000, you know, to give somebody else, but this built up to maybe a couple hundred thousand."

Eventually, Bevill became concerned about the apparent lack of collateral, and one day quietly asked Zeppa if there was some special arrangement he should know of. At that point Zeppa said, "Well, sit down and let's talk about it." It was the first detailed information Bevill, the company's secretary, had of it.[22]

IX

The agreement of Hawkins, his wife Mary F. Hawkins, and her sister Jacqueline F. Burford with Delta Drilling Company and Joe Zeppa was made on July 1, 1949, and executed on July 29. On August 9, the new company, Delta Gulf Drilling Company, ratified the agreement with the signature of Zeppa as president.[23]

In the agreement, Zeppa as an individual was pledged to do what Delta Drilling Company might be unable to do as a corporate entity. The First Parties—Hawkins and his family—received 450 shares, or 45 percent, of the stock in Delta Gulf in return for conveying or assigning certain specified properties. Refinancing of notes with several Dallas and Houston banks was agreed, with an additional $1 million to be borrowed, and Delta agreed to lend Delta Gulf $425,000. Hawkins conveyed petroleum properties in Acadia Parish, Louisiana, to Delta for $75,000, to meet payroll and other pressing needs.

Delta could vote the Hawkins *et al.* stock, and the First Parties would not be able to sell their stock to outsiders except to Delta. Delta, as trustee, was not to receive any compensation for its services. The Second Party (Delta and/or Zeppa) had four years in which to purchase 550 shares (55 percent) of the stock in Delta Gulf, as well as preferential right to buy any stock which the First Parties wished to sell to third parties, over a period of eight years.

In the schedule detailing the Hawkins properties which were to be transferred to Delta Gulf, the exceptions were specified, and these give us some idea of the magnitude of the negotiation, as far as Delta was concerned. The stipulations were that Hawkins would transfer all properties except personal effects and other specifically exempted properties. The schedule did not name the properties which were included, but it would encompass all those not specifically exempted—all gas, oil, and mineral production and interests, all payments coming in, nonproducing interests, royalties, real estate, drilling equipment, warehouses, other equipment, stocks and bonds, accounts and notes receivable, and any other property not exempted.

The exemptions make up a rather long list, adding up to several single-spaced pages. Among the assets which the Hawkins parties could retain, for instance, were the oil-payment obligation of Saratoga Production Company; royalties in certain tracts in nine counties in Texas, four parishes in Louisiana, one county in Arkansas, a county in Mississippi, and one in Oklahoma; Hawkins's residence in Dallas, two farms in Alabama, a lot in Los Angeles, California, and Mrs. Burford's residence in Dallas; Hawkins's Oldsmobile, Chevrolet, and two Cadillacs, and Mrs. Burford's automobile; various holdings which included some stock in an Alabama bank, membership stocks such as in the Dallas Country Club and Los Angeles Country Club, as well as 200 shares of Liquid Carbonic Corporation held by Mrs. Hawkins; and a number of personal notes and accounts receivable. From these exemptions, one can see that the other properties would have been extensive, indeed, as they were.

Although assets were ascertained by merely eliminating what was to be left out of the deal, Hawkins's liabilities were compiled in considerable detail. In all, they came to $7,346,318.93, which was to be assumed by Delta Gulf. Since Hawkins's assets were commonly acknowledged to have exceeded his liabilities, his problem was liquidity. Therefore, while assuming liabilities of more than $7 million, Delta was acquiring considerably more than this amount. The solution lay in reorganization of the operation and the orderly liquidation of the liabiltiies. Hawkins had $3,549,050.79 in notes payable secured by oil properties, of which the largest sum was to Republic National Bank for $2,474,326.31, which made that financial institution his largest single creditor. However, other large notes secured by oil properties, in the range of one-fourth to one-third of a million dollars, included those to Second National Bank in Houston, National Bank of Commerce in Houston, and H. H. Howell, the old driller-operator of "Egg Shell Drilling Company" fame. Other notes were secured by drilling contracts (Bank of Manhattan, $359,337.90); stocks and bonds (Dallas National Bank, $231,620, and others); drilling rigs (National Bank of Commerce, Houston, $311,967.07; Bethlehem Supply Company, $280,366.88, etc.); unsecured notes (Hawkins Graham creditors, $357,420.06, plus smaller creditors); accounts payable, $1,609,147.88; various accrued liabilities such as payrolls, withholding, and other taxes, and interest, $153,622.49; and federal income taxes and interests, $153,000. Not spelled out in this schedule were such matters as the $200,000 owed to Continental Supply, which had become the prime mover in solving the Hawkins liquidity problem. It was a massive inventory of liabilities, and a burden no company would assume lightly.

The corporate life of Delta Gulf can be traced in the contents of about a dozen selections from the minute books of Delta and Delta Gulf. Near the time of the original negotiation, a series of special meetings of the Delta board provided Zeppa with the authorization he needed to carry out the agreed-upon decisions. Although Sam Sklar was absent on many of these occasions, Joe Zeppa and Chris Zeppa constituted a quorum, as two-thirds of the directors, but Sklar obviously gave his approval, since he did sign the minutes subsequently. In some of the minutes, it was noted that Zeppa had conferred with Sklar by telephone and secured his approval.

The first order of business came on July 7, 1949, less than a week after the agreement was made but before it was executed, when the Delta board met to authorize a $900,000 loan from the First National Bank in Dallas for additional working capital that would be needed in connection with the Hawkins deal.[24] A month later at the Delta directors' and stockholders' meetings which Sklar did attend, the Hawkins deal was approved, including the support of Delta Gulf. A particularly important agreement was Delta's obligation to purchase and pay for 13 drilling rigs which Hawkins had used as security for a $625,000 note with the National Bank of Commerce in Houston.[25]

The J.C. Hawkins Drilling Company had its offices on the twenty-second floor of the Mercantile National Bank Building in Dallas. When the organization transmuted into Delta Gulf, the new company merely assumed the Hawkins space. Interestingly, when Delta Gulf moved to Tyler in January, 1950, Ed Kliewer, who had been doing much of the legal work during the transaction, moved his law office into the same space and occupies it today.

Zeppa kept his partners, Sklar and Dorfman, informed about the Hawkins deal as it first appeared and as it progressed. This is illustrative of the way he operated, for he could have handled the matter somewhat differently, perhaps even, because of his standing among the people who were in a position to tie up Hawkins's assets, taking a portion of the Hawkins interest, up to 100 percent, for himself. After all, once Zeppa moved into the picture, the pressure from the creditors receded. Conceivably he could have kept the Hawkins deal for himself as an individual, though undoubtedly it meant a great deal to have the financial power of Delta behind him. But he apparently did not consider this and shared it immediately with his partners, in effect making a gift to them which almost doubled the value of their stock in Delta.

"He wasn't the kind of man who would do that to his partners," said Bob Waddell, retired vice president of Delta. "Some people might have.

They would have confiscated the whole thing and left their partners out, but not Joe Zeppa. He was a man of principle."[26]

By October, Delta had made available $500,000 cash to Delta Gulf for the payment of debts assumed by Delta Gulf as a result of the agreement with Hawkins. This was in addition to other obligations assumed by Delta Gulf, including the bank notes.[27]

Although the initial sessions and acts must have been the most hectic in clearing up the Hawkins deal, the financial transactions continued, on a large scale, for some time. In May 1950, almost a year after the first papers were drawn up, Delta Gulf stockholders authorized two very large loans, $2,375,000 each from Mutual Life Insurance Company of New York and the Republic National Bank of Dallas, for a total of $4,750,000.[28] By that time, Zeppa had brought about enough economic regularity to impress the bankers.

Near the end of 1952, several months of negotiations between Zeppa and the owners of the Hawkins stock in Delta Gulf culminated in a final agreement. Jerry Hawkins had been dead for some time. Of the 450 shares held by the Hawkins estate and family, 439 shares were exchanged for lucrative oil wells in the East Texas field, and the remaining 11 shares were sold for $48,000—nine shares going to Delta Drilling Company, one to Joe Zeppa, and one to Jack Blalock, Zeppa's attorney and associate who had been involved in the transaction from the beginning. Zeppa judged the development to be "most advantageous to the Company, as it would eventually make possible the ownership by the Company of all of the outstanding stock of Delta Gulf Drilling Company."[29] Such ownership was now a foregone conclusion, even though it took another four years.

X

The different personalities came to have their own deep impact upon the negotiations. William D. "Bill" Craig, who was with Continental Supply at the time, remembers the Hawkins deal as being "rough to handle," primarily because of Hawkins's elusiveness.

"You couldn't catch everybody [when you needed to]," Craig said. "Unlike Joe Zeppa. He was there. You could see him anytime. [But Jerry Hawkins] wouldn't be in. In fact, I caught him not being in." Craig laughed. "Looked him in the face not being in."

Many of the personality traits belonging to the two men were translated into meaningful symbols by the financial community.

"Actually, [C.L.] Vickers and I were talking about it," continued Craig, "and I said, 'Well, Vick, looks to me like on the one hand we've got a man here that's struggling'—that's Joe— 'and doing well with it, and we've got another man here that's going bankrupt with more potential, actually, than Joe has.'

"He said, 'You've put your finger on it.'

"That's exactly what happened in the case of the assets. Had they been in the hands of Joe Zeppa, it would never have happened. Because the difference is, Joe could go to the people. He could go to the banks for whatever he needed and do business as a man would like to do. The other man lost the credit of his own character. He couldn't do it."[30]

The creditors grew less restive once Zeppa supplanted Hawkins as the central figure. "They looked to Delta Gulf Drilling Company and Joe Zeppa for the payment of those obligations," said Joe Bevill. "We refinanced some of it. We borrowed $3 million from the Republic National Bank in Dallas and the big insurance company in New York, Mutual of New York. We borrowed it through the bank, but the bank participated part of it out to the MONY people.

"That was the cash they needed to pay off those obligations that could not be refinanced. Continental refinanced their whole package. The National Supply Company refinanced theirs, and a bank in California had to be paid off—it was a bunch of small people. Halliburton refinanced their obligations and all those big people. With Joe Zeppa's backing and Joe Zeppa's guaranty—Delta Drilling Company's guaranty, but mainly Joe Zeppa's word. They didn't have written guaranty from Delta Drilling Company. They trusted Joe Zeppa.

"That was just the cash we needed to get those little ones out of the way, and out of that $3 million we paid Delta back those payroll advances that had gotten to be $200,000.

"The acquisition involved 22 rigs and God knows how much production. They were good rigs, better than we had. Yes, sir, they were fine rigs. *Lots* of production already in existence, and numerous, thousands of acres under nonproductive oil and gas leases. Some of it wildcat, wild as hell. All over."[31]

As Joe Zeppa had noted, Hawkins had "jillions" of acres under lease not only in Texas, but elsewhere too. To these, Delta fell heir. As Nick Andretta recalled, "We took I don't know how many thousands of acres of minerals in Mississippi, Alabama, and Georgia. Joe Zeppa wanted me to try to peddle the whole darn bundle some years ago, right after we got it, for $100,000. I couldn't find anybody to buy it. So he said we'd keep it.

"First thing, off the bat, well, they started getting minerals that Hawkins had bought. Well, those things have made them a ton of money, for the simple reason they came along—some of these companies did—in these different areas and paid a good lease bonus to develop production. And they got that money back and then some over in Mississippi. Same thing in Alabama and other places. In other words, these minerals that we took over from Hawkins and tried to sell at the time, they have been justified in retaining them because they made a lot of money leasing those minerals, and then the income from production from those minerals more than made up for [it]."[32]

Many of the leases, of course, profited Delta not at all. According to Joe Bevill, the leases had a variety of valuations. "Many of them, you know, without a title. It was a lease in name only, just a piece of paper. Several other people might have a lease on the same piece of property. Some of those things were not recorded. Some of them were taken from people who did not have the mental capacity to grant a lease. And titles hadn't been 'cured.' Things were in a state of raw confusion throughout the Hawkins operation."

But Zeppa seemed to enjoy bringing order to this chaotic situation. How did he do it?

"With that mind of his," said Bevill, "that tremendous intellect. Determination, patience. It was tedious, but that was the kind of thing he loved. He thrived on that sort of thing. [It would] drive other people to the wall, but not Joe Zeppa."[33]

Although Delta Gulf let many of the leases go or lapse, others were drilled and proved to be lucrative holdings. In October 1950, for instance, slightly more than a year after the deal with Hawkins had been signed, Delta Gulf kicked off the Hulldale field in West Texas—described as the "best oil field in the county"—with the drilling of Delta Gulf Drilling Company's No. 1 H. F. Thompson well. Over the next 25 years the field sent, for its various owners, 23,912,452 barrels of oil into the pipelines.[34]

From beginning to end, the Hawkins deal with a tedious series of transactions in which Zeppa was the central figure holding the key to a successful resolution of all the problems and loose ends. It was, as Leonard Phillips, Sam Sklar's son-in-law and today a member of Delta's board, describes it, "strictly Joe Zeppa."

"I know of *no* one [else] who could've worked it out," said Phillips. "He just radiated confidence, and his integrity was such that people would go along with him. He liked to work. He was a machine that worked 20 hours a day. Fantastic human being! I don't mean it in the sense that he was nonpersonal, but he was a machine in the sense that he could work longer hours than any of his employees, which is a long-forgotten art."[35]

XI

The Hawkins records were an accountant's nightmare, because of the nature of the operation and the way Jerry Hawkins handled his affairs. As John Justice, a young 25-year-old CPA who came to work for Delta and Delta Gulf in August, 1949, explained, "Well, the accounting records probably were just not kept as carefully as we would have kept them. One account that I don't guess I'll ever forget any time [was] 'Suspense.' When we didn't know what to do with something, 'Just throw it in the "Suspense" file.' And so it had built up into a rather sizeable account, you know, that was completely unidentified [as to] whether it [was] an asset or our own expense. But you are limited to what you can do in terms of supporting data and how much of an explanation you can give to it.

"By and large, it was the kind of thing where maybe they didn't have all the facts at the time of the accounting, and this may have been one of the problems, perhaps, with somebody like Jerry Hawkins who is not inclined toward record keeping or placing that much emphasis on it, being more a man who was apparently very interested in making deals. So the people in the office really didn't know what the transaction was, and maybe he wasn't available to answer the questions, but I just didn't have the facts. You've got maybe a transaction where some money has been disbursed, so that has to be accounted for. So you have the disbursement of cash, and double-entry account bookkeeping requires an entry somewhere else. You don't know where else to make it—well, you made it to Suspense and there were a lot of those. And it might be anything. It could have been an asset or it could have been expense. We just did the best we could.

"Mr. Zeppa was entirely different than J. C. Hawkins. Mr. Zeppa came up through the finance, accounting route, in terms of his own career, and he was very inclined toward doing the things that had to be done in this respect."[36]

At the time of the change of ownership, the Hawkins properties were inventoried and an attempt was made to identify all of the items.

"But there never was a matching of the investment records with the inventories that I'm aware of," said Alton Blow, who went to work for Delta in the accounting department in January 1950. "I never did really correlate the two exactly; there was matching to a certain extent, and they were trying to identify the existence of property. In other words, on the accounting records they had a description of a certain piece of property and the amount of money paid for it. And then just take an inventory of the rig and they identified this same piece as being on a certain rig somewhere. Now we never did really take the inventory to the accounting record and match up item by item.

"Now the land department work was pretty much involved, I think, in getting all these titles changed over from J. C. Hawkins, and as I understand it they pretty regularly would discover a property that they didn't have any record of. In other words, somebody would write 'em about it and we didn't know we had it, you know. As I understand it, J. C. Hawkins didn't have very good records."[37]

XII

Hawkins's income-tax problem was finally cleared up after he had made his deal with Zeppa and Delta Drilling Company.

Then, while still in his early sixties, Jerry Hawkins died of a heart attack. He had just been released from the hospital after a checkup and apparently was in good health, except for the weight problem he had most of his life. Otherwise, he seemed to take good care of himself, and according to one close employee, never drank alcohol.

Red Nixon remembered Hawkins's last moments.

"He died in Dallas. He got up to put his clothes on, dressed, and his secretary brought some mail in to him about seven o'clock and he sat down on the bed and died."

Certain characteristics of Hawkins were even noted in the eulogy, reported Nixon.

"The fellow that preached Jerry's funeral said that Jerry Hawkins was a man who never let his left hand know what his right hand was doing. He said that, preaching his funeral. Just as dry as he could be. Jerry was a good man, but he was never still. He had to be doing something."[38]

Despite his earlier economic vicissitudes, the transactions with Zeppa and Delta had definitely left Hawkins in a healthy financial condition.

"He wasn't hurting when he died," Ted Ferguson said.[39] And though Hawkins left his widow in rather good financial shape, Joe Zeppa contributed further to her welfare.

"Mrs. Hawkins wasn't suffering," said Red Nixon, "because Jerry, after he made up his mind that he was out of Delta [Gulf] Drilling Company, why, he went to wheeling and dealing again, and he was getting right on back up. [After Hawkins died] Joe Zeppa really done a wonderful thing for [Mrs. Hawkins]. He swapped her some of the best wells in East Texas for her stock. He let her pick the wells. She wound up with 14 wells here, and there are 13 of them still producing."[40]

XIII

Upon the unexpected death of Hawkins, his brother-in-law, Freeman Burford, entered the picture.

"So when Jerry died [Burford] took charge of the properties, running them and so forth," said Joe Zeppa. "And he was very unhappy about the fact that he wasn't getting any money. And we were paying bills, paying debts, and I told him, 'Look, Freeman, we're running this company just as we would if we'd owned it all the time, but I think we're running it honestly. No charges made against it. And we're not taking any salary out of it. Now, if you're not happy with it, you tell me what you want. Maybe we can arrange to buy you out. Or tell me what you'll pay us. We'll turn it over to you.'

" 'Well,' he said, 'I don't want to get into the business. . . . '

"I said, 'I can see where you don't want to get into the business.' Because the funny thing about the fellow is that when I made the deal, he said he was going to get 40 percent of the stock.

" 'Oh,' he said, 'hell, I'm ready to get out of it. I don't give a damn what we get. I'm just gonna get out from under. Because the whole thing is tied up, and Jerry . . .'

"And he didn't think the stock was worth a dime.

"Well, anyhow, I made a deal with him and gave him—I think we gave him 15 wells in East Texas for 40 percent [45 percent] of the stock. And that's a pretty goddamned good deal, don't you think? Now I knew I wasn't making a good deal, of course, because I didn't want him to go around saying, 'The son-of-a-bitch just cheated me out of it!' There was no way he could say he was cheated out of it when he told anybody what he got for it. No, there was no way.

"Actually, I offered him that. I said, 'Here is what I'm giving you.' And I explained, 'I'm giving you everything that I think it's worth.'

"Well, he came back and he said, 'We'll take it.'

"Well! After we made the deal, he was very happy.

"So we got rid of him. Then we put the two companies together. It was quite an increase."[41]

XIV

The new company—Delta Gulf—was operated separately from Delta for the next seven years. Although some personnel were officers or otherwise involved in both companies, their working time was prorated, and all equipment was maintained separately.

Red Nixon remembered several occasions when this policy brought him into confrontations with Chris Zeppa, executive vice president and as staunch a watchdog of the Delta treasury as any executive could ever be.

"I like Chris," said Nixon, chuckling. "He was a good company man, but he had a bad habit of saying No. Didn't make him any difference. I told some of them one time that Joe and Chris were the best combination. You could run and tell Joe you needed a steamboat and he would say go ahead and buy it, 'cause he knew damn well Chris wouldn't let you launch it.

"Joe Zeppa would promise you the moon because he knew that Chris wouldn't let you have it. That's the way it was.

"Chris and I were good friends. He and I have had some times. We had a lot of equipment when Delta Gulf was formed. We had abandoned some wells and Jerry would never sell anything. Whatever he had, he kept it. We had second-hand tanks and wellhead equipment and had a lot of pipe, and after Delta Gulf was formed, why, Joe, he made it very plain that it was two different companies. If we used one another's material it would be a sale—it would be transferred, but it would involve cash. You couldn't borrow this, that, and the other. And I'd have equipment that I wanted to sell and [that] Chris needed, and, boy, you would just fight it right down to the last penny. Chris really watched those pennies for Joe.

"Joe had been in more or less the same shape Jerry Hawkins had been in, as far as field personnel taking care of his business, until he got Chris to do it. When he brought Chris in, it went to running right."[42]

Some indication of the work generated by the formation of Delta Gulf can be had by examining the relative workload assigned to a number of key employees at the Tyler offices where both were quartered. Interestingly, there was no single pattern that characterized all departments, indicating Delta still demanded more time from some employees, Delta Gulf from others. Marguerite Marshall, secretary to both Joe Zeppa and Chris Zeppa, and H. S. "Hi" Cole of the purchasing department, for instance, divided their time on a 50-50 basis between the two companies, but most other employees spent more time with one or the other entity. In the land department, N. L. Webster and Tommy Blackwell each divided his time—according to the source of the funds for their salaries—75 percent with Delta, 25 percent with Delta Gulf; Robert W. Waddell, with an official title of safety engineer, followed the same pattern.

In the geological department, however, H. C. Matheny, his secretary Martha Jones, and H. H. Yoakum gained 60 percent of their salary from Delta Gulf and 40 percent from Delta, which suggests that Delta Gulf's drilling prospects were more promising than those Delta already had or was considering—or at least that more time was being devoted to sizing up the leases which Hawkins had transferred.[43] In fact, it is almost certain that additional time was being allotted to the former Hawkins leases to decide which ones to drop and which ones to retain, as Zeppa doubtlessly did not wish to continue paying rental money on all the leases.

Delta paid 60 percent of the office-space rent; Delta Gulf 40 percent.

In 1951, J. M. "Mark" Gardner was hired to run Delta Gulf, with 75 percent of his salary coming from that company and 25 percent from Delta. Gardner, a personable petroleum engineer who had been drilling superintendent for Phillips Petroleum during World War II before entering the Army, had known Zeppa through industry meetings over the years, as Zeppa was a founding member of the American Association of Oilwell Drilling Contractors. After the war, Gardner ran Loffland Brothers, the world's largest contract-drilling company, for two years as executive vice president. He was to remain with Delta for about a dozen years.[44]

XV

Though the mode of operation necessarily changed, many of Hawkins's personnel continued under the Delta Gulf regime.

Chris Zeppa explained the personnel policies followed during this period of transition.

"Well, the guidelines and the way that we went about, we tried to choose the best man capable for a certain job. Men that we *knew* were capable. Fact, a lot of them were out of our own organization.

"We didn't go like some would: when they acquire some property they just sweep out all the old ones and put in new ones. That's the biggest mistake anybody can make. We used all the Hawkins people we could possibly use, and for a long time we made no change at all. In fact, a good deal of their top men worked for us for a long time before we made any changes. The manager in the Gulf Coast—all we did was change his office from one place to another in Houston. The manager in West Texas—we didn't make a change. In West Texas, for instance, they already had a department. We just took over, changed the name, and didn't change a person. They had their office and a yard out there. It was Jerry's yard, you know. They kept the equipment and all. We made no change there. In South Louisiana we did the same thing for a while, and then we made some changes, and it just went on and on. And the production man [Red Nixon] still lives in Kilgore. We used him as the production superintendent. It's a mistake to go out and say, 'Well, I want to sweep it clean.' That's the worst thing they can do."[45]

XVI

Though the process of taking over Delta Gulf involved substantial financial commitment from Delta, the parent company appears to have suffered no setbacks other than perhaps a temporary one. In the summer of 1951, Delta declared a dividend, amounting to 80 cents per share on 128,500 outstanding shares of stock.[46]

It is less certain, however, that the transactions had not strained the finances of Joe Zeppa himself. The addition of Delta Gulf clearly increased, perhaps doubled, his responsibilities, and may have increased his own personal expenses without a concomitant raise in salary. Although his personal worth undoubtedly rose as a result of the enhanced value of his Delta and Delta Gulf stock, this was not expressed in the form of liquid funds with which he could pay additional expenses. Consequently, he drew on a personal account with Delta.

In late 1952, in an effort to repay the company for his overdrawn personal account, he sold to Delta certain stock which he held in Lone Star Steel, Pan American Sulphur, and the Tyler Hotel Company, at a total cash value of $81,395. His salary and dividends, he noted, were "inadequate to meet his personal obligations, which are necessarily large because of the position he occupies and because of the necessity of carrying large amounts of life insurance for purposes of inheritance taxes."[47] Periodically over his career, he made efforts to reduce his personal indebtedness to the company.

XVII

During this Delta Gulf period, two small oil companies—Algord and Creek—were added to the Delta corporate family. One of them, Creek Oil Company, came about indirectly because of the Hawkins/Delta Gulf transactions, as a result of the attorney, Ben D. Clower.

As Red Nixon recalled it:

"In 1954, [Clower] got into a deal with a couple of fellows in Dallas who owned a little oil company in Oklahoma. And they got into trouble, and Joe [Zeppa] took over their obligations. They had partners and everything, and the partners had kept their bills current to the operator, but the operator hadn't paid the bills, so they were stuck to pay their half of the production—they couldn't do it. And Joe bailed them out. So we had Creek Oil Company and drilled some good wells.

"It was the biggest mess. [Clower] had gone to New York and sold that quarter at 14 different interests. One of them was a Thomas Lee, mayor of Chinatown."

At one time, recalled Nixon, Creek Oil had nine wells. Roy Bentley operated the company.[48]

XVIII

It took, in all, about six and a half years for Delta to acquire and completely absorb the Hawkins properties. By the end of 1956, Delta was ready to liquidate Delta Gulf and the two other small companies—Algord Oil Company and Creek Oil Company—involving relatively minor holdings which had come under Delta's control during the Delta Gulf period. On December 31, 1956, the assets and liabilities of these companies were transferred to the parent organization. Among the liabilities were four mortgages totaling $3,675,000, one for $400,000 to the Republic National Bank of Dallas, the other three to The Mutual Life Insurance Company of New York.

Despite the assumption of Delta Gulf's debts, there was still enough left for a dividend of $1.25 per share on 128,500 shares of Delta stock and bonuses for Joe Zeppa ($25,000), Chris Zeppa ($17,500), five officers (total $29,000), and all employees (total $82,000).[49] It was a fitting climax for a camouflaged opportunity that Joe Zeppa had recognized at the outset and which had finally paid off.

In the papers conveying the properties to Delta, some idea is gained of the widespread nature of the Delta Gulf operation, which had properties in

12 states—Texas, Arkansas, Alabama, Colorado, Florida, Georgia, Kansas, Mississippi, Oklahoma, Utah, Louisiana, and New Mexico.[50]

For legal purposes, Delta Gulf continued to exist as a corporate entity until a few lawsuits had been settled, but for all practical purposes it had dissolved and faded into history.[51]

When Delta Gulf dissolved into Delta at the end of 1956, the original company had doubled in size. The move also planted Delta permanently in West Texas and South Louisiana, where Hawkins had been operating. A number of key personnel who had been with Hawkins also joined Delta and remained to play important roles. The Hawkins acquisition was a historic step in Delta's growth that might be likened to a prize cow's troubled pregnancy which had climaxed in the birth of lively, healthy twin calves, at least doubling the value of the herd.

7
DOMESTIC OUTPOSTS

I

While the Hawkins acquisition virtually established Delta's divisions in West Texas and South Louisiana, the company had already spread its operations beyond East Texas years before that. In the late 1930s Delta had been busy in the Schuler field in southern Arkansas and had moved rigs into the Illinois Basin to follow that boom. While Delta was never to set up a formal division in Arkansas (those rigs were under the Tyler office), it did establish one in the tri-state region of Illinois, Indiana, and Kentucky, headquartered in Evansville, Indiana. From this first "domestic outpost," Delta was to add divisions in West Texas and Gulf Coast/South Louisiana as a direct result of the Hawkins deal, and also in the Northeast, the Rocky Mountains, for a short time in the Southeast, and, in the late 1970s, South Texas. Another corporate entity, Delta Marine, operated in both domestic and foreign waters.

In the '30s and into the '40s, there was still so much oil and gas undiscovered that fields were always being found. Delta continued to grow and kept its rigs busy going from one field to the next. It also began to build up its production. Before the Hawkins deal, the significant fields, in order of Delta's involvement, were the Schuler field in southern Arkansas; Chapel Hill, east of Tyler; the Quitman field, north of Tyler; and the Slaughter field in West Texas. During this time Delta was also drilling in the Illinois Basin.

No new divisions were necessary to drill the Schuler, Slaughter, and northern Louisiana fields, despite the geographical distances involved; the effort all came from the Tyler office.

In 1938, the Delta rigs went to the Schuler field in Union County, Arkansas, near El Dorado, to drill for Lion Oil Refining Company. Colonel T. H. Barton ran Lion Oil, and Joe Zeppa and he had been friends since the 1920s. According to Joe Bevill, a top Zeppa aide in those years, Delta drilled at least 58 wells in the Schuler field. Additional rigs were sent there. The Schuler drilling necessitated borrowing more money with which to carry out the contracts. As a result of the Schuler experience,

Delta's rig count went to eight, with three in good condition, and that field was worked until the early 1940s.[1]

The pattern from the early 1930s has been one of continuing growth and expansion, which helps explain what Joe Zeppa meant when he said in 1975: "Well, of course, we never have been stabilized, financially, because we are always buying something, and always drilling."[2]

II

Chris Zeppa officially joined Delta as vice president and director on August 10, 1944, when he was about to have his forty-fourth birthday. He replaced R. C. Johnston in those positions.[3]

Trent Zeppa has described his uncle Chris as "the first administrative regularity that came into the company, other than Mr. [Joe] Zeppa, who could administer a project to its conclusion."[4] While there may have been other "administrative regularities" at Delta before Chris's arrival which Trent had been unaware of, there is no doubt that Chris Zeppa soon proved himself to be a stellar performer as he took command of operations. He was to become his brother's right hand, and as executive vice president, which he was to become, he offered unswerving loyalty, complete integrity, and a high degree of capability which had been finely tuned in the Arkansas oil fields. He was very hard working, analytical, and he could be trusted implicitly, for he was "family."

The two brothers complemented each other in ways that were becoming increasingly important by the 1940s. Joe Zeppa ran the company's business in the board room and represented the company to the outside world— customers, other companies with which Delta did business, bankers, and partners. His financial acumen kept the business going and growing. On the other hand, Chris ran that part of the business at which Joe was less adept—the nuts and bolts of a contract drilling operation. He saw to it that the drilling bids were advantageous to the company, that equipment was maintained properly, that personnel were used expeditiously, that nothing was wasted, and that the company gained good measure for its expenditures. He was a self-educated engineer especially well-versed in gas production, which proved to be very helpful in Delta's two gas plants. Joe Zeppa kept Delta afloat and solvent. Chris made it work.

Through the years Joe Zeppa had kept in touch with Chris back in Arkansas, and after World War II broke out he urged him to join Delta. By that time Chris was a man of considerable experience and administrative skill. He was more than a brother; he was a highly trained and capable brother, with just what Delta needed at the moment.

"Mainly," said Chris Zeppa, "I came into his organization just to look and see if anything needed doing. Well, about that time, he was about to build the gasoline plant at Chapel Hill [in East Texas], and a few other things were boiling up."[5]

"Joe was a big man in any field you took him in, but he was not one for detailed field operations at all," said William D. Craig, who viewed the scene from his vantage point with Continental-Emsco. "Great man of finance and association in business and otherwise. Joe recognized, in the first place, Chris's qualifications. Second place, he probably realized some of his own shortcomings. Chris absolutely made just the perfect fit with Joe in the operation. Because Joe, frankly, was having personnel troubles—the company was—at the time he brought Chris in. I don't mean they were great troubles, but it wasn't running as Joe would like to have it run. He knew that, but he didn't have time to do it. But when he brought Chris in, that thing leveled off just beautifully."[6]

Chris was like an extension of his brother Joe—"an alter ego," Keating Zeppa put it. "A man in whom he had implicit trust. And if Chris made mistakes or bad decisions, it was obviously unintentional. It was just one of those things; no matter how trusted a colleague may be, it's just not the same as being family. Chris, of course, didn't know anything about the drilling business; he did know about machinery and he knew about people, and he could learn about anything else he needed to know. And, above all, he had that *absolute* loyalty and the confidence in Dad.

"It was very fortunate that Chris was available and could come down here to be with us. It has definitely shaped the course of the company. Dad was always the entrepreneur: looking ahead, looking for opportunity. Dad was *much* better at dealing with the outside community, such as bankers. Dad was the businessman; Chris was the operations man. Chris knew machinery, he knew people, was very cost-conscious. Chris may not know much about how to make a loan, because he never borrowed any money in his life, but he certainly knew that a nickel spent had to be earned somewhere, and Chris was a very genial person. He was tough as a boot in many respects and very demanding, but he was genuinely interested in people and their families and, to me, he was a classical working supervisor—and that's not putting him down a bit: the sort of person who can identify in life style and this sort of thing with the people he supervises. He can really enjoy knowing them, the families, and still remember the kids's names.

"And yet they had no trouble taking orders from Chris. There was no question who the boss was. As frustrating as it can be at times, there was never any question who was running things, and they loved him."[7]

Because he had a limited knowledge of drilling operations, Chris systematically set out to master both equipment and procedures, down to minute details.

"I never will forget," said Ed Gandy, "I was running out here on our Rig 14 shortly after Chris came to work for the company, and he came out one afternoon and he'd say, 'Gandy, can you tell me how many teeth such and such sprocket's got on it in that drawworks?' And I'd say, 'No, sir, Mr. Chris, I can't tell you how many teeth's on all the sprockets in this drawworks.' I said, 'There's several sprockets in that drawworks, and actually I haven't counted all the teeth.'

"You've got a line shaft, a drum shaft, and all in there, and they've all got different sprockets, and you've got several gears and you've got your main drive chain coming from your compound which is off of your engine up to the drawworks. So he told me, 'Sometime when you get time,' he says, 'I wish you would count the number of teeth on the sprockets for me,' and I did, but I can't remember them now.

"He was just trying to find out everything he could about these drilling rigs. Chris, by making the field and asking the questions and reading the books, well, he got it right quick. He was real brilliant. He didn't want a question asked him that he couldn't answer about a drilling rig. And he could keep it up there, once he found out."[8]

Employees consistently perceived him as fair. "The way I always figured him," said Jack Elkins, "[was that] if a man even acted like he was trying, Chris was on his side. If he didn't think you had any interest in your work and wasn't trying, he would unhook you pretty quick, I imagine."[9]

Chris Zeppa tended to maintain equanimity in the midst of unpleasant events. Once when he was at a rig, he instructed the toolpusher to make a telephone call. The toolpusher hurried to his car, put the gear in reverse, and backed out immediately—ramming his car right into the side of the Delta car Chris Zeppa had driven to the site. The toolpusher leaped out of his car, cursing and "raising sand," as Roger Choice recalled.

Chris told him, "You've run over it now. There's no need to talk now."

Then surveying the scene, he added, "Looks like I'm going to have to buy a donkey to ride."[10]

Sometimes Chris's solutions to problems were unconventional, but practical, as on an occasion when an employee was suspected of feathering his own nest at Delta's expense.

"They called him on the carpet," said Homer Lee Terry. "Mr. Joe wanted to get rid of him but Mr. Chris said, 'No, we don't want to get rid of him, because if we get rid of him we will have to fatten up somebody else. But he's got fat now; we can watch him.' "[11]

Chris also had his own way of determining whether someone knew what he was talking about. Bill Craig, who was with Continental-Emsco at the time, remembers when Delta was buying some new rigs for work up in Wyoming. Chris was working out the specifications for a rig with Craig.

"One day," said Craig, "I don't recall what the occasion was, but Chris said something and I said, 'No. Chris. It's this way.'

"He waited a little while and he said, 'No, Bill, it's this way.'

"And I knew that I couldn't let that go, because that wasn't right. That wasn't the way it should be and it was wrong. So I said, 'Now, Chris, that's not right. It's this way.'

"And I never will forget, Chris just stared. He sat back in his chair a little and looked at me. I thought, 'Oh, boy, I'm done.'

"He said, 'You and God just don't make mistakes, do you?' But he kinda had a smile on his face when he said it, so I knew that he wasn't really reprimanding me as much as it seemed. So I thought, 'Well, I'm right; I've gone that far, I'm going to go all the way.'

"I remember saying to him, 'Well, Chris, I'm going to let the Old Gentleman answer for himself, but as for me, I don't make very many.'

"I never will forget: Chris threw his hands up and laughed and said, 'Bill, I think you're right.'

"What he was wanting to know—I finally had sense enough to see it after he got through—he wanted to know if I was *sure* that I was right. He didn't know. That was his way of finding out. When I stuck with my guns, he knew that was it. It always made an impression on me. He didn't ever think he'd know everything, but he knew a way to find out [what he needed to know]."[12]

To most people, Chris was the watchdog of the treasury when it came to expenditures. "He was very conservative," said Thornton Tarvin, "and I remember he wrote me a letter one time. He said, 'Thornton, we've got some money to buy equipment with, but we don't have a nickel to waste.' "[13]

By all accounts, he was just as careful in his personal transactions.

"Chris and I and Mr. Mark Gardner flew from Tyler to Morgan City [Louisiana] in that company plane," said J. H. "John" Gilleland. "Old Mark, he bought some cards and said, 'Let's play pitch. Well, what shall we play for? Twenty-five cents a game?' Chris said, 'No, let's play for ten cents.' "[14]

Most of all, Chris Zeppa's memory is legend among people who have had close contact with him. Ed Gandy used to drop by the office early in the morning because Chris was always in by around 7:00 a.m.

"I used to ask him about so and so, a swivel or something on a rig, and without going to the files he could pretty well tell you every piece of equipment on each rig and he'd call it by name and give you the serial number on it. Now, that's the kind of memory Chris had. He could give you the numbers and all. At that time here in this [East Texas] district, we were operating 10 or 12 rigs."[15]

Hugh McKenzie, a retired toolpusher, cited another feat of memory. "One time we went over here to Rig 1—that old diesel. They had some bearings burned out and they couldn't find them. So we went around and got the numbers off the bearings, and the toolpusher did, too. The toolpusher wrote the numbers down and they drove over to Dallas to pick the bearings up. Well, the toolpusher had lost his slip when he got there. [Chris Zeppa] says, 'I declare. You can't remember those bearing numbers?' And he just spit them out just like that. He didn't forget nothing." [16]

"Chris has a memory of an *elephant*," said Keating Zeppa. "He is just incredible! Just incredible. Much to the consternation of many people who found out only after the fact that Chris could remember all sorts of things: when they were going to have a job done and how much it would cost." [17]

Warren Strahan summed it up. "He was a real good friend of everybody's but one thing about him: you never wanted to lie to him 'cause you didn't want to have to remember what you said, man, 'cause he'd sure remember. If he told you something this year, three years from now he'd come back and remember he told you that or ask you about it. It'd better be the truth as near as he knew it." [18]

When Chris Zeppa retired as executive vice president in 1968, his position was filled by his nephew, Keating V. Zeppa.

III

World War II came as suddenly to Delta personnel as it did to other Americans. Guy White, as a daylight driller, was drilling on the sixty-second well in the Amanta Meier lease near Griffin, Indiana, when the news of Pearl Harbor came. That Sunday afternoon, December 7, 1941, the evening-tour driller drove out to the well and asked White, "Are you ready to go to war?"

"What do you mean, go to war?" shot back White.

"Well, the Japs jumped on us . . ."

"Well, they got more sense than that," said White.

"They may have more sense than that," said the night man, "but they've done it anyway. You get in your car and turn your radio on."

"What station?" asked White, still unbelieving.

"Any of them! It don't make any difference. It's on all of them."

When they finished that well, they shut down the rigs there and stacked them. [19]

Relatively few were in on the "ground floor" of World War II as H. J. "Red" Magner was. After he received his degree in business administration in 1940, young Magner had gone back to work for Delta and was in Indiana working under drilling superintendent E. L. "Foots" Johnston when the peacetime draft began. Not wanting to be drafted as a Hoosier, Magner had his papers forwarded to his Texas draft board, which also delayed his report date. He was instructed to report at Dallas on December 8, 1941.

"And little did I know that they decided to start the war the same day I decided to get in," said Magner. "Well, they declared war on December 8, right when the doctor was looking up my rear end. Had the radio on and Roosevelt was declaring war."[20]

When the war began, Delta personnel, like other Americans, became wrapped up emotionally in the war. While many employees entered the armed services, others received deferments to keep the rigs running on the home front to supply the precious oil that moved the machines of war. Patriotism was a spark that motivated everyone.

Parts were scarce. Joe Zeppa came out to a rig one morning to find an old rotary chain worn out. The crew had tried to put it together. Zeppa asked the driller, Mutt Hays, "Mutt, what's the matter with it?"

"Mr. Zeppa, it's completely wore out!"

"No, it isn't," said Zeppa. "It can't be because there is no more to be had."[21]

With all forms of steel in short demand, a leak in a stock tank meant employees had to fix it themselves. Gus E. Jones recalled that the bottom was first cleaned out carefully and then hot asphalt was used to patch the hole; sometimes concrete was used. "Back in those days you had to do a lot of things. You couldn't buy the materials," said Jones.[22]

War scarcities continued to plague some rigs even after the war. "At that time our drill pipe was pretty well worn," said Bert Gauntt. "It was in bad shape and they couldn't replace it. I know out here at Opelika when I was still roughnecking, we had five fishing jobs out there in five days. The daylight driller twisted off just before we came on the job for five straight days out there. You know, them fishing jobs can eat you up."

Of all the shortages that faced Delta and other oil-field outfits during World War II, the manpower shortage was the most severe. Although many men obtained draft deferments because of the importance of the industry to the war effort, the rigs nonetheless frequently ran short-handed.

"I've gone out a number of times short-handed a man or two," said Ed Gandy, "and go out at midnight and go to work, say, with two men. I'd have some of the men that were out there double over or come back and help me pull out of the hole or this, that, and the other, and give them a tour for it to keep the rig going."[23]

As Red Magner recalled, "The problem was getting any type of hands to work at all. Sometimes you'd just have to shut down the rig till morning and wait till some guys came out there to do whatever you had to do."[24]

The war also had its impact at the executive level. In December 1942, J. W. "Joe" Beasley, who had been secretary and a director of Delta, resigned to enter the Navy. R. C. Johnston was elected to replace him on the board, and J. M. "Joe" Bevill became the new secretary.[25] The following year Johnston, who had been general superintendent, was elected vice president in charge of operations, at a salary of $8,100.[26] Johnston proved to be an interim officer, for when Chris Zeppa joined the company in 1944,

Johnston resigned to make way for the president's brother. In his letter of resignation as both vice president and director, Johnston gave as a reason his intention to enter business for himself, after 11 years with Delta.[27]

Eventually the war touched the Zeppa family more personally. In 1944, Trent enlisted in the Navy at age 17. He spent the remainder of the war at San Diego as a hospital corpsman. Another founder's son gave his life during World War II. Fred Sklar, Sam Sklar's son, died in 1944, a captain missing in action with Patton's Third Army during the Battle of the Bulge. A graduate of the University of Oklahoma in petroleum engineering, he was promoted to major posthumously.[28]

IV

The war and postwar years were a transitional period in many ways. For one thing, the old steam rigs were being replaced by power rigs. Red Magner and Guy White remember when they worked on their first power rig in the Illinois Basin in the 1940s. Before then it had been all steam rigs. The crews preferred the steam rigs in many ways, for the steam could be used to wash down the floor, to clean clothes, to keep warm by, and to keep hot coffee going.

"You could listen to that steam pump and tell whether everything was running all right," said Magner. "You had less repair and maintenance, obviously, than you did with gas or diesel engines." A power rig, however, didn't need a water supply to generate power, as a steam rig did. Rig-up time with power was only one-third of what it was with steam, because for one thing, water didn't need to be hauled.

W. B. "Jack" Kennedy, who started with Delta in the 1930s, liked steam but saw the disadvantages. "They cost so much to run, that steam rig. Sometimes I have laid water lines to them gas lines, five or six miles. Takes you longer to rig them up than it does power rigs. You can go out there now and in a couple of days these power rigs will be drilling.

"Down there at Haynesville [Louisiana], the last steam rig that we had, we went through the woods with a gas line. Old Herman Woods, he was kind of pushing tools. And we had to take a bulldozer and pull that pipe, see, all through them woods. Creek was up. It just took so long to rig them old steam rigs up. That's the only thing. A power rig is better."

Red Magner recalled: "I think we got rid of our last steam rig sometime in the '50s. But they were only the large ones that were over there in Louisiana where you could move in and maybe stay for 60 or 90 days on a job where the transportation costs and rig-up costs became lost in the overall costs. But like in Illinois we were doing these wells about every 12 days to 2500 or 3000 feet. These were days before we had drill collars and other things like that, and we would drill 12 days and be off three days while the rig builders would tear down the rig in one day and rebuild it in two days, and we'd have three days off to go honky-tonking or do what we wanted to do."[29]

V

In many ways the East Texas Division, headquartered in Tyler along with the main office, has been the center of Delta's constellation. Though not necessarily more lucrative than the other domestic divisions, it has remained an active, profitable operation and, perhaps most of all, is the place of Delta's origin, with all the emotional and historical meaning this might convey. The region gave birth to Delta and was crucial to its development: the East Texas field not only kept contract rigs busy, it also provided oil for Delta's own coffers. The East Texas field was the solid foundation on which Delta built and then grew. No matter where the company went from there, elsewhere in the United States or halfway around the world, this debt to East Texas would remain. As long as 50 years after the East Texas field had been discovered, Delta was still receiving revenues from production acquired in that field, as well as from other fields in the region.

Also, the East Texas Division provided a type of income only one other region did. For a number of years, Delta's gas plant at Chapel Hill brought in profits that were significant to the company's economic health. The only other Delta-operated gas plant has been at Ozona, in West Texas.

Important discoveries that added production to Delta's treasury were the Chapel Hill, Quitman, Como, and Birthright fields. The Chapel Hill and Quitman discoveries were Paluxy fields; Como and Birthright, Smackover fields. Chapel Hill produced a gas condensate and became the reason for building Delta's first gas-processing plant. Quitman yielded oil.

The Quitman field followed closely upon the Chapel Hill discovery. It was a profitable operation from the first, with Delta winding up with substantial production in the field that was to last for decades and into which Joe Zeppa brought some of his friends from earlier days, particularly George Keating, as investors.

"The thing that really pulled Delta out was the Quitman field discovery," said Nick Andretta. "That gave us a good big boost."

The Quitman discovery well was the J. B. Goldsmith No. 1, in December 1942, in Wood County, Texas. As was the case at Chapel Hill, the Quitman field was in the Paluxy zone around 6500 feet deep. The Quitman field also came as a farmout from a major oil company, Shell, and Nick Andretta and L. B. Benton from Fort Worth put the deal together that led to the drilling. In all, the two men spent a year securing commitments and farmouts from other companies that had leased there.

"The Quitman field has been a very rich field all these many years and is still producing an awful lot of oil, and Delta still owns a bunch of production up there," said Joe Bevill in 1980. "The Quitman field was a great big thing for Delta. It really kicked us up. We started buying rigs and equipment and leases and hiring people."

Bevill had joined Delta in 1938 as an accountant to audit some privately operated trusts for Joe Zeppa's friends, George Keating and E. P. Snyder, both of New York. These men's investments helped Zeppa keep Delta afloat in the early 1930s, while profiting themselves. Partially as an expression of his appreciation, he established the trust in which they invested money for Zeppa to manage. With the trust agreement, Zeppa had the authority to manage it as he saw fit.

Keating, Zeppa's boyhood friend in New York, and his associate Snyder became the beneficiaries of Zeppa's management skill, as the trust bought a lease in the East Texas field—the O. M. Victory lease in Rusk County in the southern end of the gigantic field. It was, as Bevill said, "a darn good lease," with several productive wells on it for a long period of time. "All of the partners made nice sums of money out of it," said Bevill, "until the lease went dry. It was still producing, I think, in 1942."

Delta had no interest in the trust, and Zeppa had only a small interest. "Primarily," as Bevill pointed out, "he set them up so his friends who knew nothing about the oil business could participate. And, for instance, he made Mr. Keating independently wealthy in the oil business."

When the Quitman opportunity came, Zeppa brought Keating once again into it. It worked to the advantage of both men, for Delta needed the additional investment and, as it turned out, the play produced substantial dividends for Keating over the years. As Nick Andretta said, "The Quitman field was one of the best strikes Delta had. It produced a high-gravity—42 gravity—oil." It came in the latter part of 1942.[30]

In a 1943 letter to Keating, Zeppa spelled out in considerably more detail than he usually did his mode of operation.

> My desire to sell producing properties is for the purpose of paying debts and accumulating a little cash. I generally buy royalties when they are still very speculative, and when they turn out productive I sell all, or part, to get back my money and pay for some of the luxuries which my salary does not permit. Royalties in the Quitman field are those I purchased long before the well reached a depth where it appeared productive.
>
> I conservatively figure that each acre of underlying royalties which I will sell will produce 40,000 barrels of oil. If the price per royalty acre is $1,500.00, then each acre will have to produce $12,000.00 of oil before the purchaser would get his money back from the one-eighth royalty. Figured at the current price of $1.20 per barrel, it would have to produce 10,000 barrels of oil for each acre. This would mean on the above price basis, the purchases of the royalty would buy four barrels of oil in the ground for each dollar, and these four barrels would only

be subject to taxes, as royalty is free of the cost of development and operation.

The above figures are based on the present producing sand. There are three deeper sands which are prolific producers of oil and gas in the same general area and which may also produce in this field. No well has so far been drilled deep enough to reach these formations. The price of oil is most likely to be much higher during the period in which the field is depleted. These are prospects which one acquires with the purchase and on which no specific value can be placed.

. . .

At the present the field is developed on the basis of one well for each 40 acres. The allowable production is 100 barrels per day; the price $1.20 per barrel. On this basis the current income is approximately $130.00 per year per royalty acre. The allowable is expected to be raised as the demand for oil continues to increase, as these wells are capable of producing without waste up to 1000 barrels per day flowing. I expect after the war is over the property will be developed on the basis of one well to 20 acres, which means that each 40 acres will have two wells instead of one, and this will, of course, considerably raise the rate of production and the rate of income per month or per year.

At the time, Zeppa was planning to sell $20,000 worth of royalties and offered them to Keating's family, to be applied toward Zeppa's indebtedness to them, or purchased, as they chose. And Zeppa added:

If you care to trust my judgment I shall try not to be unfair to either myself or to you in this transaction. Any little advantage you may receive in this transaction will be small compensation for the assistance you have given me from time to time. That I do not mention this very often is simply because I know that you know that I am conscious of it nevertheless.[31]

In 1980, geologist Jim Ewbank reported, "That discovery well [at Quitman] is flowing, which is damned unusual. That's 37 years of flowing oil."[32]

Growth was a continuing goal. At the end of 1961, Delta spent more than $1 million on lease purchases, development operations, and dry holes. The overall acquisitions, President Zeppa was to report, "have resulted in a substantial increase in the company's reserves." Most of the acquisitions had been in several East Texas counties (Freestone, Navarro, Hopkins), and Delta had held joint interests with Sklar Producing Company in DeSoto and Sabine Parishes in North Louisiana.[33]

Around 1949 Delta started looking for the Smackover lime in East Texas, though not a great deal was known about its features.

"Delta was a pioneer in discovering and developing Smackover production in East Texas," said Tommy J. Blackwell. "They had a very good safety record [with the hydrogen-sulphide gas found in Smackover formation wells]. They did a good job of it. Consequently, people who were investors would say to Delta Drilling Company, 'We'd like to buy into your deals. We'd like to have you as the operator because we don't want to manage sour gas.' That was the reputation they established."[34]

Two East Texas region fields proved to be particularly profitable to Delta in the second half of the 1960s. By 1965 Delta was negotiating a mortgage with the First National Bank in Dallas for its share of the plant-construction costs for the Gulf-operated Como plant (oil and gas), two counties north of Tyler. In 1966 Zeppa described production at Como in the Smackover lime field as "very profitable," and he expressed hopes that other Smackover fields in the region would increase rig activity. The following year, 1967, his optimism was justified when the Birthright oil field in north Hopkins County was opened. The Smackover discovery there was the Delta Drilling Company and M. B. Rudman No. 1-A Henderson unit, completed in March 1967, for 450 barrels of oil and 25 barrels of water daily.[35]

The improved techniques of operating in Smackover fields came over the years, as the result of experience. Ed Gandy's memories of 1949 when he was pushing tools while Delta was drilling the Chance No. 1 at Leesburg in East Texas are illustrative.

"It was quite an experience for me and the deepest well I had drilled. We drilled that well to 13,323 feet and we wound up in salt with some stuck drill pipe. Stuck it two or three times. And the last time we stuck we didn't get a-loose. So we finally backed off and left around 6100 feet of drill pipe in this hole. Which is still in it, as far as I know.

"That was quite an experience for me, and I used 169, I believe, rock bits drilling to 13,3-something. Now we drill the same well with about 18 to 22 bits. That's the improvement [that] was made in bits."[36]

VI

One of the more interesting technological improvements over the years was the jet sub, which Delta helped develop. Essentially the jet sub consisted of a piece—the "sub"—between the drill collar and the tri-cone bit. Three holes were drilled at an angle through the side of the sub to bring tubes down to each of the three cones on the drilling bit, and this tubing was braised into the sub, which was under the drill collar. The drilling fluid, jetted through the tubing, kept the hole clean. Originally the idea belonged to a Humble engineer, John Paul Nowley. As Marvin Williams, now president of Spencer Harris Machine Works at Gladewater, Texas, summed it up, "He had the idea and the design, and I took it [and made it]. The first one I ever ran was during World War II. This was the start of the jet sub, when Nowley designed it. We made three jet subs and then put them in bits, jetted them, and I carried them at night down at Sand Flat. We run them down there, and they did good. I watched it that night, the rate of penetration, and it was so much greater than the conventional bit, that in my mind I was convinced it would help a contractor and he would not have to run two pumps compounded, and [it wouldn't] cost that much extra for material and stuff.''

But with the war on, and the Humble staff overloaded with work because of personnel shortages, nothing more was done until the war ended. Then Williams pursued the matter once again and gained permission from one of Humble's petroleum engineers to jet five subs.

"Well, we did that,'' said Williams, "and they were running 54 bits to the well for the conventional bit. We cut it from 54 to 27 bits the first well—cut it in half—by the jet keeping the bottom of the hole clean while the bit would cut virgin material. It wouldn't be wearing on cutting particles down there. That is the principle today, to keep the bottom of that hole clean. Where the tooth is cutting on virgin ground all the time.

"Then it rocked on. In the mid-'50s I went to Mr. Chris Zeppa and I told him, I said, 'I've got something that will make you money. A jet sub that Humble's drilling.'

"He said, 'I've heard of it, but you got to have two pumps and compound them, and with all the rubber goods it will take, I don't think it will justify it.'

"I said, 'Chris, you won't have to buy but one pump. We won't make quite as good a time, but we'll use one pump and we'll do good.'

"He said, 'Well, we will try them on a well.'

"Humble was the only one that had run one. Nobody else. No independent. Delta was the first one to put one in the hole. The first one I had ever asked [to use it]. They had Gardner-Denver pumps, 20-inch pumps, which was good pumps in them days. But Mr. Chris said, 'If you go up there and you bust one of my pumps, I'm going to run your butt off.' It was a little stronger than that, but that's what he meant.

"I said, 'I'm not going to bust a pump.'

"He said, 'Well, who would you like to have?'

"I knew Ed Gandy ever since 1934, when I first met him, when he was a young driller, and I said, 'Can I have Ed?' He was up at Pickton and I wanted to run the bit in the area where Humble was running.

"Then he said, 'Okay.'

"Then to be exact, it was on the Minter No. 1 in the Pickton field, Ed told me to go and fix the bit for him, and I did, and they spudded that well on September 9, 1950, and I have the record. I was interested enough to and decided to work long hours. I would go to bed at eleven o'clock and then get up early the next morning when I was putting it on over. A lot of the time I was sleeping in the car. The first job that they ran up there, we cut the record at the total depth of 7986 feet and used 27 bits. That was half. Used one pump. We had been using two pumps. It took longer on the hours, but we didn't hurt the pump. On the first well, there was 456½ hours rotating time, to 7986 feet. Completed on October 7, 1950. Exactly 28 days. It would have taken much longer with the conventional bit. They would have been using 54 bits.''

At the time Delta was drilling for Mobil Oil.

Humble decided not to patent the jet sub but to let the industry have it.

"I was ready to put them out [to the industry],'' said Williams. "We'd moved out here [to Gladewater] and in reality the jet sub built this Spencer Harris Machine and Tool Company plant and paid for it. It is used 100 percent today. Everywhere. All the companies have them now. To my knowledge they could never get a patent on it then because it was too far gone, and they never did patent it."[37]

VII

The early forties saw Delta in a frenzy of activity close to home.

With larger rigs capable of drilling deeper than those used in the old East Texas field, the company drilled the discovery well—the Sally Walton—in the Chapel Hill field in East Texas, in the deeper Paluxy formation (the East Texas field had been produced from the Woodbine sand).

The Sally Walton was a Delta well, which Zeppa had taken as a farmout. As Joe Bevill put it, "We drilled a well and it was one heck of a big gas-condensate well—but gas in that period sold for three cents a thousand, if you could sell it. We produced that well wide open and flared the gas. For at least two years it lighted this whole countryside at night, to get the condensate out of it.''

The well brought in around $100,000 per month for a while from the condensate, he said. "And then the [Texas] Railroad Commission shut it in, for conservation. So when we couldn't find a market for the gas, we had to shut the well down for a long time."[38]

Dewey Strength worked on the lease early in its producing history. "They built four welded tanks and tied them together with a valve in-between them where they wouldn't run over. That gas well would fill those tanks full of distillate just as fast as it could run. That gauge kept the pumper running all of the time for the two years I stayed there, and, boy, it made them a fortune."[39]

A number of legal transactions followed the opening of the Chapel Hill field. Etexas Producers Gas Company was established complete with a pipeline system, with Delta the only stockholder. In 1942 Delta bought the pipeline system from Etexas for $175,000. At the same time Delta sold to Sinclair Prairie Oil Company an undivided half-interest in the pipeline system and certain oil and gas leases in the Chapel Hill area for $300,000.[40]

In 1942, Delta directors voted to dissolve Etexas Producers. However, slightly more than two years later the Etexas Producers Gas Company was revived when conditions changed. Under federal regulations, the Chapel Hill field was unitized and Delta's properties were taken into the Chapel Hill Paluxy Unit, with Shell Oil Company designated as the operator. The new Etexas Producers Gas Company would lease pipeline facilities from the Chapel Hill Unit to transmit gas to the old East Texas field for various uses, as well as to Camp Fannin, an army installation in the area. Delta would own all of the stock in Etexas Producers.[41] From that point on, Delta would retain Etexas, a profitable possession.

When Chris Zeppa hired him in 1945 in the Chapel Hill gas field, T. A. Everett said Zeppa said, "We guarantee 52 hours a week. No more, no less. If you can do your job in 10 hours, that's your good luck. If it takes you 100 hours, that's your hard luck."

That was fine with Everett, although he and Lee Browning, among others, were to spend "many, many 24-hour days without any sleep, keeping the lines open, keeping gas coming in," particularly in the coldest part of winter.

On one occasion, reported Browning, "The line came across a ditch out in the woods and it was exposed and it was frozen up, and we gathered us up a bunch of dead wood and built a fire on it, and it let us know when it broke through. You could hear it and feel it too, a jolt, sort of a crack. But, anyway, we didn't have any more trouble with that old line that night." In the summertime, the work evened out, however, to make up for the unusual winter schedules.[42]

To process the gas, Delta built a high-pressure absorption-type plant producing commercial natural gasoline (not the type used in automobiles), butane, and propane. With Chris Zeppa masterminding the construction from the office, Bob Waddell was on the site nearly all day. Delta employees built it, some of them coming over from the Kilgore yard.

"We did it all ourselves," said Waddell. "We had to use our own hired welders and we hired roughnecks and college boys during the summer."

Using blueprints and borrowing upon Chris Zeppa's experience, "we went and built it from scratch," said Waddell. An engineering firm in Tulsa, Oklahoma, designed the vessels. Some of the designs were sketched out by Chris Zeppa on a day-to-day basis. They were learning as they did it.[43]

The Chapel Hill gas plant was put into operation in October, 1949. The gas-gathering system had been in operation for several years by then, but a new gathering system was created that same year. There were from three to five gallons of liquid per 1,000 feet of gas, containing gasoline, butane, propane, ethane, and methane. The new plant made two "cuts" from the gas, first taking out the gasoline, and then the butane and propane mix, combined. In the beginning, the best production days brought forth around 18,000 gallons of gasoline, butane, and propane altogether, though the early range usually was only 7,000-8,000 gallons. When the Wright Mountain field in the area was drilled in the 1950s, additional gas from it was processed in the plant.

Was it a complicated matter, setting up the gas plant?

"Not really," said T. A. Everett, "when you had Mr. Chris doing all the thinking. He had been with Lion Oil Company and he had been construction superintendent for them at the gasoline plant. He had had it all. He knew it from the ground up."[44]

Among Delta's markets for gas from the Chapel Hill plant were customers at the town of Owentown, north of Tyler. This relatively small market included the state chest hospital there, residences, and industry. "And all at once," said Bob Waddell, "we woke up and Delta was in the public utility business. Changes the whole tax structure. So they had to separate it completely from Delta Drilling Company and form another company [Owentown Gas Company] to get Delta out of the public utilities."[45]

VIII

Delta went to the tri-states area with its office in Evansville, Indiana, in 1939, and Joe Zeppa acquired for both himself and Delta "a large number of prospective oil and gas leases and royalty interests" in Gallatin County, Illinois. At the time he had also acquired some leases for George T. Keating.[46] As had happened in other such ventures, the production proved to be profitable.

Conditions in the Illinois Basin were much as they were elsewhere. Leonard A. Bates, who joined Delta there, remembered those years as a time when "you couldn't buy a job." Long hours were simply the price of earning one's bread. There was never a slack period in the Evansville area, and when World War II came there was the manpower shortage that the drilling industry found everywhere. Guy B. White once pushed tools on three rigs in three different states. Later, Leonard A. Bates went him one better as a toolpusher on two wells in Kentucky, one in Illinois, and one in Indiana. Consequently, he stayed on the road most of his time, and he came to characterize the Evansville operation, as, most of all, "a workhouse." One driller, working on a rig close to his own home, told of his wife's cooking his meals and taking them to the rig: though he was so close, he was unable to go home to eat his meals.

There were other problems related to the formations and equipment as they drilled the relatively shallow wells in the Illinois Basin. One of the problems was "to get the mud right" and to make out with equipment that frequently was old. Sometimes, recalled Bates, the old drill pipe would twist off, which could be most frustrating.

Because of the location of some of the wells, the weather also provoked those who serviced the wells. One lease, drilled in the Wabash River bottom on the Illinois side, would be flooded when the river overflowed in the early spring; production men would have to service the well from a motorboat.

And if the early days in the Illinois Basin had some similarity to the earlier booms in Arkansas and East Texas, there was one other familiar sign for Delta hands. Louisiana Iron and Supply opened a yard in Evansville, Indiana, after the boom there and eventually had pipe yards in Salem and Crossville, Illinois, and in other tri-state locations. [47]

Evansville, Indiana, an air, rail, and highway center, was a well-situated central site for a tri-states headquarters. In southern Indiana by the Ohio River, it was near the other two states: Kentucky was just across the river, and Illinois lay on the other side of the Wabash to the west.

In 1940, Delta contracted to develop the Amanta Meier farm in Gibson County, Indiana, in an agreement with Hall Edwards, the receiver of the contested acreage. Delta drilled 62 wells on one lease, on the banks of the Wabash River near the town of Griffin. The Amanta Meier lease kept the rigs busy for a year and a half, drilling one producing well after another. Not only was the drilling a sure-fire matter, but there were three different pay sands.

Although the drilling on the large Amanta Meier lease was contract work, it was unusual, for while Continental Oil Company claimed to own the lease, Mrs. Meier disputed this. She contended that the Continental agent had not delivered the lease money until two days after the deadline, and she had refused it. At this point, Delta, contracted to drill the wells, offered to help Mrs. Meier with the lawsuit, with an opportunity to gain a half-interest in the lease.

Guy White, though only a daylight driller with a limited view of the situation, became a good friend of Mrs. Meier's son Paul, who visited the rigs frequently when he was not operating a huge corn plant. The family raised corn with which they fed cattle over the winter. Paul Meier kept White up to date on the lawsuit.

Since Delta had a contract to drill, operate, and produce the field with Continental and Mrs. Meier, drilling proceeded forward until wells filled the lease. Delta was paid as it did its work, out of income from the oil. But, meantime, other proceeds went into escrow until the suit was settled. When Continental finally won the lawsuit, Delta hadn't lost. [48]

Another lawsuit in the tri-states region involved Joe Zeppa more personally.

F. L. Carmichael and E. L. Johnston managed the Delta operation in Evansville. Carmichael was in charge. "Foots" Johnston looked after the rigs and production. In addition to their salaries, Zeppa had also permitted them to participate in deals. This was to lead to a spirited contention that ended up in litigation.

In 1945 a lease came to Delta in a rather unusual deal. Delta had drilled a well for a small independent oil company; as partial payment Delta was assigned 80 acres of the lease.

"Well," said Red Magner, "it just so happened that somebody drilled offsetting this lease and got a whale of a good well there. I'm talking about 500 or 600 barrels a day and a big sand thickness, and it looked like it was a real good property. And so Carmichael and Johnston felt that they were entitled to a percentage of that, even though they owned no interest in it and had nothing to do with acquiring the 80 acres. They just felt [that] because they were up there they were entitled to it because he had given them some previously. And Joe didn't think so. In any event, they sued."

In the midst of the early stages of the dispute, Guy White left for Texas on "a little vacation," during which he dropped by the Delta offices in Tyler to visit May Jones and Joe Bevill. As he walked in, May Jones announced, "Guy White, Joe Zeppa is looking for you!"

When the two men got together, Zeppa made White an attractive offer.

"That's the way I got to be division manager," said White. "After a little haggling I got a little working interest in the percentage of the profits and some interest in the production. He said he wanted to raise my salary. I said, 'No, just leave the salary like it is; what I want is a little working interest, and I want a chance to really make something.' "

White assumed his new duties on January 1, 1946. Three years later, as operations spread to the northeast, he became vice president.

In the lawsuit in Indiana, Carmichael and Johnston sued for a quarter interest on an 80-acre lease with four or five oil wells on it, quite a valuable property that was important in Delta's inventory of assets.

"The plantiffs' position, in effect, was that Delta had welshed on a deal," said Edward E. Kliewer, Jr., who handled the litigation for Delta. "And Mr. Zeppa was quite firm in what he knew he had done, and essentially the attack was one on his integrity. He just stood firm on it.

"They contended that under their arrangement, they had oral understandings and all that. But they had a written contract with Delta, and it very plainly stated that it did not include this. And so they sued up there [in Indiana] and the case went to trial, actually, and the defendant [Zeppa] moved for an instructed verdict. It was being tried by a jury, and the trial judge indicated that he was reluctant to grant that instructed verdict, but that if the jury didn't find for the defendant he'd vacate that verdict or something like that. I wasn't there. As a matter of fact, we engaged local counsel up there. But I was told that after the plantiffs got that message they dismissed their suit without prejudice.

"They then came back to Texas and sued on the same cause of action, and I handled that. We filed exceptions to the pleading for technical procedure, in effect saying, as it wound up, basically [that] the courts here had no jurisdiction because it involved Indiana land. And the trial court dismissed the [plantiffs'] petition and the intermediate court of appeals affirmed and the Supreme Court refused an application for *certiorari*. That case never got to trial because the court didn't have jurisdiction."

Joe Zeppa's personal approach to the case was characteristic of him. He dominated all discussions, including those with his attorneys. And the discussions didn't all go smoothly. Kliewer remembers the first trip he took to Evansville, Indiana, in connection with the Carmichael suit.

"We had a meeting in the hotel in Evansville," said Kliewer. "Joe gave us a lengthy dissertation on the facts. I guess he talked for an hour about the facts, and everybody listened respectfully. So when he finally ran down, why, I made the mistake, I guess, of then going back and beginning asking questions so that we could test all this. As a devil's advocate, because the other lawyer will do it, and plus the further fact [that] you've got to have your own education. He'd lived with this thing. He knew it like a book. Hell, we never even heard of it.

"He didn't like that worth a damn. And he said, 'Well, if you'd been listening, you would know what the answer to that question is.' And I told him, 'Well, Mr. Zeppa, I was listening, and I heard what you said, and I'm not sure I know what the answer is.'

"And so then we went forward. That's the only little bit of friction that ever occurred between us. We would find times when we were not in agreement. If I didn't think he was right, I'd tell him so. That he was entitled to. But I learned early on that the best way to handle him—and indeed the only way to even get along with him—[was that] if you have an opinion on the subject, you'd better be able to support it, and so I'd level with him as to exactly what I'd want and why I thought he was wrong. And then he would pass judgment on it and we'd go on from there."

After White became division manager at Evansville, Joe Zeppa sent Red Magner, then in Tyler, to Indiana to run the office.

"And so then he drilled, I think it was, ultimately 18 wells on that 80 acres," said Magner. "It was a multipay, had three or four horizons, but the Waltersburg sand was most prolific, and we had as much as 50 feet of pay sand there. And Joe made a deal with Guy, an eighth interest, he would get an eighth of the profits from the rig production. He allowed Guy to participate to the extent of his one-eighth interest in these other deals. So he messed around and had a whole bunch of deals he put together. We tried to keep the rig busy on contract, but you also had to have some deals going to fill in. And we must have drilled about 15 wells of our own that we participated in and [that] he took an interest.

"Mostly, what our game plan was, we would take farmouts from the various major companies like Sun, Carter, and Gulf. Carter [Oil Company] used to have a book of deals, as it were, and they'd go through an area and just buy up all the acreage they could get in this county. Then they'd come along and block it off and give it a name or a prospect number and say, 'We want to get a well drilled in here.'

"Well, they might have had a landowner that wouldn't lease to them or they'd have one that wouldn't let them build a location on, or there'd be some reason that they didn't want to do it themselves—geologically or personally or whatever. They called those their 'D' prospects. So we would go in there and thumb through their book and find one or two or three or four and build up a program and then we'd go out and sell it.

"One year we hit six out of eight wildcats. Now, keep in mind they didn't extend very far. Very small, amount to two or three wells. A big money-maker was one well, the Walker No. 3 well, which made 2900 and some-odd barrels a day. I never saw so much oil flowing into one little old tank from one little old well about 2500 feet deep."

On some wells, Delta took on the risk—and potential profit—to the tune of 100 percent.

"Joe was quite venturesome in risking, those days," said Magner. "They were cheap wells, by comparison, up there. At that time up in the Illinois area we were drilling wells only to 3000 feet, and the dry-hole cost was in the neighborhood of possibly $10,000 or $12,000. We were drilling for an average of $3.50 a foot up there.

"As a matter of fact, we were quite successful up there. At the time I left there in 1954 [and Guy White went to Tyler] we had operated over 100 wells, and Delta's net production was in the category of 700 to 800 barrels a day. Eight years. We had some good production in the tri-state area."[49]

In retrospect, one is amazed that no violent episode disturbed the smoothness of the tri-states work. Some of the Delta customers were gamblers who paid off in cash, peeling off $100 bills for the drilling done. It was not a comforting thought to have, as Guy White sometimes did, as much as $30,000 in one's pocket at a time, for there was always the possibility someone would slug him for the money before he could get it to a safe depository.

Joe Zeppa visited Evansville from time to time, and he enjoyed taking people out to dinner. Always well-informed, he invariably proved to be a versatile spellbinder and a sparkling dinner companion.

One morning after Zeppa had taken a number of people out to dinner in Evansville during one of his visits, Guy White and Red Magner scurried to his hotel to join him for breakfast. They ordered breakfast sent up. By then it was eight o'clock and Zeppa was shaving. All of a sudden, White and Magner were shaken by a booming sound from the bathroom.

Presently Joe Zeppa emerged from the bathroom with blood practically streaming down his face.

"That's a hell of a damn note," he said. "We were out last night and I spent $150 on you guys and your wives and all your guests, and here I am trying to get one more shave out of a damn penny razor blade!"[50]

IX

Although the Hawkins acquisition is a huge and dramatic symbol of Delta's growth, the company was actually spreading out before that, not only in Arkansas and the Illinois Basin, but in the Rocky Mountains states as well.

Early in 1947, Delta had been contemplating an operation in the Rocky Mountains. Upon investigating business conditions there, it was concluded that competition would not be as steep, with a better prospect of profit both from drilling contracts and from production. However, more and improved equipment would be required, costing perhaps more than $400,000.[51]

Early the next year Zeppa agreed to contract with Mountain Fuel Supply Company and the Atlantic Refining Company to drill in Wyoming. Soon afterward, drilling began on the contracts, and Delta opened bank accounts in Riverton and Rock Springs, Wyoming, to facilitate business transactions there.

By that summer an office had been established at Denver to handle transactions for the Rocky Mountain Division, and another rig was being moved into Wyoming. The Rocky Mountains operation highlighted the company's need to modernize its equipment, as the conditions could be expected to be more demanding for the deeper drilling, and Zeppa emphasized the necessity of converting some of the older steam rigs to internal combustion power.[52]

When Early F. Smith joined Delta's accounting department in June, 1948, the company had just dispatched two rigs to the Rockies, and the company was still small enough for Joe Zeppa to take a close personal hand in everything that occurred, as Smith soon learned.

"I feel like one of [those rigs] is mine," said Smith, "because I tore that thing down and put it together 1900 times on paper before I could get it to suit Mr. Joe Zeppa. I never saw it, but I had to account for the acquisition and then list the major components of the rigs and what we'd paid for them. He was a perfectionist."

Within the next few years, Delta had moved its operations also into Utah and Montana, and in late 1953, O. E. Mecham, who had been managing the Rocky Mountain District from the Denver office for several years, was named a vice president. The headquarters was later shifted to Casper, Wyoming, where the supporting facilities had been situated.[53]

The first contracts in the Overthrust Belt of Southwest Wyoming proved to be very difficult drilling, which is why Mountain Fuel had called in contractors instead of doing its own drilling. The transaction was packed with risk, and Zeppa's handling of the matter must have appeared naive to many outsiders. Yet, as William D. Craig observed, there was a method to Zeppa's apparent carelessness. The key to Zeppa's decision was his assessment of character.

Craig, who was with Continental Supply's Rocky Mountain Division at the time, recalled:

"I was calling on Mr. Nightingale [Mountain Fuel Company's president and chief executive officer] in his office at Salt Lake City. He knew that I was from Tyler, and when he found that I was as well acquainted with Joe Zeppa as I was, he wanted to talk about Joe, and he did.

"He said, 'Well, I'll tell you about that man. Joe's a smarter man than I am. When we made that deal, I didn't trust him, but he trusted me. The further that thing went along, the more I realized that the binding deal I tied him up in was totally unfair and Joe knew that something would be done. I was sitting at my desk one day and I buzzed my secretary and I said, "Get Joe Zeppa on the line for me." ' Then he said he asked Joe to meet him in Denver. Said, 'I gathered up all the contracts we had made and all this binding stuff and took it down to Denver. In the meeting, I said to Joe, "Now, Joe, let's get this contract where it ought to be." And I gave him the advantage that he should have had.'

"Well, to me, that was one of the finest statements I ever heard a man in his capacity make about a man in Joe's capacity. It meant to me that Joe decided that Mr. Nightingale was a man of character and that he would be a fair man. I don't know just what the details of the contract were, but he knew that he was making a rough contract, that it was close. He evidently also had faith enough that Mr. Nightingale would be the type of man that would come along with some changes in the situation. You can lose in that country with the best of contracts. It's hard. You never know exactly what you're going to run into."[54]

The conditions that mattered most in the Rockies had not changed at all by the late 1960s when Frank Dykstra encountered his old friend Bob Waddell. Dykstra, who had been Texas wildcatter Billy G. Byars's drilling superintendent before he went to Delta, had just arrived in Tyler after two years in Brazil for Delta.

"How'd you like to go to Casper, Wyoming?" asked Waddell.

"Where the hell is Casper?" retorted Dykstra.

"Oh," said Waddell, "it's a nice little town up in the mountains. Good hunting, good fishing."

"What am I going to do?"

"You go up there as assistant drilling superintendent."

Thereupon, Waddell took Dykstra in to see Keating Zeppa, who had just replaced his uncle Chris as executive vice president.

Waddell said, "Frank is going to go to Casper."

"How in the world did you talk him into that?" said Zeppa.

After several months in Wyoming, Dykstra was promoted to drilling superintendent, which he held for five years until he was named division manager of the Rocky Mountains operation. In the process he gained a detailed view of the region and its drilling problems.

Although the weather is not the only problem in Wyoming, it is a significant one.

"The second day they sent me to one of the rigs," said Dykstra. "This rig was on the tip top of this mountain. It was 40 below zero and the wind was blowing about 60 miles an hour. There was about two feet of snow. I had come out of that hot country down there [in Brazil], and it was so cold I'd get out of the car and I couldn't breathe. The snow was so deep it would pile up and knock your fan belt off. Just like mud. I didn't know snow would do that."

Once on the drilling rig, conditions were comfortable. Boilers fired with oil or butane provided steam heat, with heaters on the rotary floor, as the power rigs did the drilling.

"We [would] have big steam heaters with a fan in them that blows hot air on you," said Dykstra. "It's all walled in. The whole floor and all is warm in there. The main thing is getting to the rig, and rigging up. That's the biggest problem. Just learn to dress for it. They wear ski masks and all.

"But a major problem was help. People would go up there during the good months, then they would leave. And during hunting season they would take off and go hunting elk. There is good hunting up there."

When the snow comes, it arrives in a big way.

"On Easter Sunday of 1975," said Dykstra, "I thought somebody stole my car. I looked outside and thought it was gone. It was completely covered up with snow."

Weather and all, there were problems in the Rockies that one did not encounter in other regions.

"Casper has awful hard drilling on the side of those mountains," said Dykstra. "Deviation is bad. Lots of cracks in the formation. It was hard getting to and from. Most of the time now, they just slide them down with a helicopter. Set them down in a camp up in the mountains. We dug a well out in Medicine Bow and we had three snowplows working 24 hours a day. Just to keep the road open. They get about $25 an hour."

The length of time it takes to drill a well there depends on the location. "In Gillette, Wyoming, we dug two wells a month. You could go 50 miles anywhere from Gillette, and it would take you a month to drill one well at the same depth. That country is so broken up."

Delta drilled there for nearly all the major companies such as Texaco, Gulf, and Mobil, going to an average depth of 10,000 feet. The deepest well Delta drilled while Dykstra was there was 13,000 feet, though some wells later went to 18,000 feet. One company was a year drilling a well near Casper.

However, another major difference between drilling there and in, say, East Texas, is the loss of circulation. Lost circulation occurs when drilling mud pumped into the hole does not return to the mud pits, but is absorbed in the formations far below. In Wyoming, the broken formations would be responsible.

Though winter was a challenge, summer was pleasant. Mrs. Dykstra termed it "lovely, just lovely." There was no need for air-conditioning. "Had the prettiest flowers there I have ever had in my life," she said. Dykstra said only the difficulty of keeping employees caused him to retire when he did in the 1970s. "If I'd had some help, I'd stayed up there."[55]

X

Although Delta moved into the North Louisiana fields in the 1930s, it did not extend its operations into the southern part of that state until the Hawkins acquisition. Thus Delta inherited its South Louisiana rigs directly from Delta Gulf.

Some idea of the Gulf Coast District's ranking is suggested by the fact that the division has rarely appeared in the company's minute book, seemingly placing it between the extremes of a constant trouble spot and a spectacular bonanza. Part of the reason for this lack of formal attention, however, is that for many years the manager of the Gulf Coast District and Delta Marine was the same man. Since offshore drilling was the more glamorous of the two operations, it became easier to focus upon it, and the negative nature of Delta Marine's history helped ensure that the Gulf Coast Division was overshadowed by the problem child in its own backyard.

Both Delta's and Delta Marine's operating out of the same office added, as Gulf Coast Division secretary Flo Bonham put it, "a good bit of spice to the sauce." It produced problems and situations peculiar to the Gulf Coast, primarily because of Delta Marine's offshore operations. Sending a platform tender to South America, for instance, was never part of any other domestic region's experience. As a result of helping out with Delta Marine's chores, Bonham was on a first-name basis with the Coast Guard commandant in New Orleans.

At first the Gulf Coast headquarters, along with Delta Marine, was in New Orleans, with the maintenance shop and supportive facilities in Lake Charles, Louisiana, a considerable distance away. Having the operation divided and physically separated proved to be unsatisfactory, so when a location in Lafayette opened where the entire operation could be consolidated, the office and the yard were both moved. In 1969, Bob Waddell, vice president in charge of domestic drilling, negotiated with Coral Drilling Company to buy its yard and shop in Lafayette for $90,000. Subsequently the office was moved there from New Orleans, and the yard from Lake Charles. The site proved to be centrally located for Delta's operations in South Louisiana, and the company sold its property in Lake Charles.

As was the case in other regions, the condition of the drilling equipment was sometimes less than desirable. Jerry Hawkins's equipment in South Louisiana "wasn't exactly new," according to Garland Cox, a toolpusher, "but it was good equipment, in good shape." But as Delta expanded in number of rigs and operations, the existing equipment aged and was, more often than not, repaired rather than replaced.

When Warren Strahan went to work for Delta in South Louisiana in 1957 under shop foreman Joe Frazier, the division was still operating four steam rigs. To a certain extent that symbolized the state of much of the equipment there. "We limped along here with some obsolete equipment and some old rigs," said Strahan. "It was put together with baling wire and a lot of hope and prayer, I'll tell you for sure."

The upgrading of equipment began on a grand scale there, as in other regions, in the 1970s, allowing Strahan to append a coda to his assessment:

"We have some of the finest equipment now that money can buy. They added new rigs and upgraded the ones we kept, all at the same time. We have the finest equipment that's in this part of the country now."

But the condition of the equipment was not always the predominant factor in its level of activity. The capacity of the rigs is always pertinent, and in South Louisiana some of the rigs were designed for deeper drilling, which did not always coincide with customers' needs. This became particularly apparent in some of the slack times of the drilling industry.

"The marine unit [Delta Marine] stayed fairly active throughout, especially the barges," said Flo Bonham, "but some of the larger land rigs [weren't], and we had some fair-sized iron then. It would be totally impractical to put a rig with a drilling potential of 25,000 feet on a much shallower job. You could not be at all competitive by putting it on a well where the total depth was to be 8,500 feet. Most of the time you just might as well not even bother bidding, because there was no way."

Each locale has its distinctive characteristics that are reflected in its geography and in its people. Even after the eclipse of Delta Marine's activity, the Gulf Coast Division continued to feature circumstances peculiar to its region. It is the only division likely to have a land rig that must be serviced by crew boats, such as occurs when a well is drilled on an island or other sites made less accessible by rivers, lakes, or bayous. In such cases, the rig must be dismantled and floated to its destination by barge. Trucks also must be transported over water for rigging up. Once the rig is settled on solid ground, it operates like any other land drilling rig, but it still must be serviced by boats which take in the crews, fuel, drinking water, and everything else required to operate a rig.

"There's a little more variety, we think, than any other place," said Flo Bonham.

Other distinctions combine to give drillers cause for worry. "I know one thing," said Garland Cox, "that nearly anytime you take someone coming from West Texas down in here to drill, they get in trouble. On account of the [very high] gas pressure [at shallow depths]—they're not used to that. And they are not used to working the drill pipe whenever they break down. You know, when something is down they just want to shut everything down and go and repair it. Down in this part of the country, you let that pipe set too long, you're stuck. Because that shale will fall in, stick. They have to keep the pipe moving and watch for gas."

Furthermore, in South Louisiana, because the formations are so soft, it is not uncommon to drill 1,000-1,500 feet within a 24-hour period. "Lord," said Warren Strahan, "you can't do that anywhere else."

Just as hiring "hands" becomes quite difficult in Wyoming during elk season, a parallel problem exists in South Louisiana. East Texas driller J. W. "Turk" Hardin, after a three-month stint in South Louisiana, concluded: "Whenever it was hunting season or trapping season, you couldn't hire no roughnecks. You would think you had a crew that evening, and you get back out there the next day and you might not have a man."

Since the end of the Great Depression, every division at one time or another has experienced some difficulty in keeping well-trained workers. But among Delta's domestic divisions, the Gulf Coast's turnover rate for division managers has been the highest of all. The average tour of duty has been around two years. Marcus O. Black, a respected and popular manager, seemed well on his way to breaking the pattern after he was named manager in 1972. Then he was fatally injured in an automobile accident. As of 1981, however, division manager Kenneth Fowler was approaching his fifth year, at that point constituting a record tenure for the position.

The frequent change of managers could only have affected the division's efficiency over the years. When a new manager took over, as Flo Bonham said, "It was just about the same thing as me leaving Delta and going to work for another company, because each man that comes in has his methods to be followed. So not only I but other people in the office must adapt."

Despite the executive changes over the years, however, Flo Bonham has remained since 1969 as division secretary to become the steadfast personification of stability in the midst of flux. Like a bipartisan secretary of state serving under successive presidents—unquestionably loyal, most of all to the nation—she has provided a measure of continuity without which any business would soon suffer. Paper work, after all, is more than a modern affliction; it is a necessity, and its importance is most recognized when its flow is disrupted.

Despite the brief tenure of executives in the past and seasonal labor shortages, there has been more stability than meets the eye. Many members of a single family tend to hire on and stay with Delta. Warren Strahan, nearing a quarter-century with the company, for one, believes the concept of Delta as a family has been as powerful in South Louisiana as anywhere else.

"We're strongly family-type people here," he said. "We always have been. We all feel like we belong to each other. We're not a bunch of people; we're one group—just one family." [56]

The longevity of drilling superintendents and toolpushers like Sam Marceaux, Doris Darbonne, and L. A. "Gus" Mayton seems, at least partially, to document his contention.

XI

By the early 1940s Delta had completed many wells in the Schuler field in Union County, Arkansas, for Lion Oil Company. It was at this point that Delta first went to West Texas, several years before acquiring the West Texas properties that came with the Jerry Hawkins transaction. Nick Andretta was instrumental in making the contacts that led to Delta's drilling in the Slaughter field. Delta also developed its own production there.

"We were up there [in Arkansas] through '41, I think, looking for a place to use those rigs," said Joe Bevill. "We obtained some leases at Sundown, Texas, out in the country south of Lubbock. We moved one or more rigs out there and drilled a number of wells for the company during the war. There was a need for oil. We drilled on Delta's own lease, as an extension of the Slaughter field, and one of our leases is named Slaughter, the one we drilled on. Made a nice well. The only problem was, we did not have a pipeline connection, and they had to haul that oil a long way by truck to Sundown, Texas, and put it in a gathering system where it went into a pipeline. Lord only knows where it was refined.

"As the result of drilling these wells up there with borrowed money, we established relations with the First National Bank in Tulsa, Oklahoma, if I remember correctly. And kept adding to that mortgage at that bank, pledging these productive wells to pay off that mortgage. But the sale of oil from the leases was not consistent enough to really do a good job of servicing the mortgage and leaving Delta much.. We just couldn't sell oil. It was there and we had it. The result of which was, at some point in time during the war [1944] before pipeline facilities became available, we sold our properties in the Slaughter field to Stanolind Oil and Gas Company. I don't remember what the total sales price of the wells was, but the cash received was $575,000. We had a check for that amount of money, and that was the biggest bunch of money we had seen at any one time since I'd been with the company. It was big. We paid off our mortgage at the bank up there. Pretty nice wells."[57]

By early 1943, Joe Zeppa had been in negotiations with Stanolind Oil and Gas Company, but it was the middle of the following year before he had accepted Stanolind's offer to buy Delta's interest in the Della Slaughter Wright land in Cochran County, Texas, reserving a sixteenth as an overriding royalty.[58]

The Slaughter field deal was consummated just as Chris Zeppa joined the company as vice president and director.

Because of the distance involved in those "wide-open spaces" and the memory of the problems posed by inadequate pipeline facilities at Sundown, Joe Zeppa studiously avoided West Texas for the next several years—until, in fact, the Jerry Hawkins transaction virtually flung the company into the West Texas-New Mexico region once again, in better circumstances.

Distance, largely taken for granted by West Texas-New Mexico residents, made the automobile one of Delta's most important equipment items in that district. A supervisor driving from one locale to another will spend much of his time in transit. Ted Ferguson, Delta's long-time production superintendent, remembers the time Alton Blow, back in the Tyler office's bookkeeping department, told him how amazed he had been upon receiving Ferguson's car mileage report. "I'd say to myself," Blow told him, "he can't do anything, only ride around."

Ferguson explained, "I'd run 5,000 or 6,000 miles a month, every month. I was averaging better than 60,000 or 70,000 miles a year, every year. And it was company stuff too. It wasn't this running around. You don't ride just for the fun of it after you drive that much for a company. I was over there in [San] Angelo—this was '54, I guess—and we're drilling wildcats and so forth, one in Eldorado and up here at Rankin and one up here close to Carlsbad [New Mexico], and I could get in my car and make that round, and the shortest trip that I could possibly make was 732 miles from the time I left my house until I got back in front of it, usually that night, at one o'clock in the morning.

"We didn't have the 55 [m.p.h.] speed limit and didn't have near the highway patrol we've got now, and I certainly didn't drive no 60 or 70, either. I was driving 80 and 90 and sometimes more, most of the time, when the road would allow it."[59]

Distance inevitably has an impact upon personnel statistics.

"We probably have the highest turnover rate of all the divisions, because of our remote locations," said Lynna Jo Edwards.

The experience of Ed Doughty may not be typical, but it is illustrative.

"Did you know I worked on one well down there a year and nine months and it was, one way, 120 miles?" he said. "You can imagine the distance. I drove backwards and forwards from [Odessa]. Every day. It is unbelievable. We worked eight-hour shifts. I had about 10 to 12 hours at home."

Just as high-pressure gas is a familiar risk in South Louisiana drilling, lost circulation is a leading hazard in West Texas. "Out here is a porous formation," said Edward Doughty. "Some of it is real porous, and if it is not real bad you can put in some fiber or cottonseed hulls. You can just mix the cottonseed hulls or fiber with the fluid, and the pump will pump it around, and these particles will go into that porous formation and form a bridge. Now that is for some wells, but it could get so big you could pump trees in there and you couldn't stop it."

However, techniques which have proved to be of value in other regions have also been useful there. Air and mist drilling fits in nicely in many locales out West, as J. H. "John" Gilleland discovered while drilling a well in New Mexico. "Mist drilling, they called it," he said. "There were articles that Red Magner wrote in a magazine [about his experience in Pennsylvania], and that is what helped me out. I was up there by myself and trying to figure out how to drill it. Finally, I would read ol' Red's articles in that pamphlet. It was so simple, to read his description, when you get water, moisture, that I could just read about his drillings and go right ahead with it."[60]

Equipment was a continuing concern in all districts.

When E. G. "Eddie" Durrett, coming from Loffland Brothers, joined Delta Gulf in West Texas as a drilling superintendent in March, 1952, Chester Miller was the division manager. With Delta Gulf vice president Mark Gardner's encouragement they started improving the rigs.

"Believe me," said Durrett, "there was a long way to go, because the Hawkins operation had left a lot to be desired. All the other contractors were modernizing their equipment, and Hawkins had not done this, so to be quite frank about it, we inherited, as I recall, eight or nine rigs of the condition which was considerably less than desirable. We were faced with an immediate upgrading program.

"Well, we started on the upgrading program, and the whole bottom fell out of the business about mid-summer 1952 with what was later called the Great Steel Strike, and that strike had the effect of putting the brakes on the oil industry as the result of the industry's overdrilling in the late '40s and early '50s; [then] proration began to really take its effect and, as I recall, through the succeeding years we got down to as low as eight days a month of production.

"Boy, it was tough. Well, the updating program was interrupted at that point because rigs were just stacking. In fact, at the time I went to work for Delta Gulf there were almost 900 rigs running in West Texas and Southeast New Mexico. Well, the strike slowed a lot of that up, and rigs started coming down and the price structure began to decay and we thought, well, we're seeing the bad days. But, really, the bad days that were encountered in 1953 and '54 really were just hints of things to come. We didn't know how bad they could get. The late '60s, they got out to 900 rigs running nationally. Well, earlier we had had that many running right here in West Texas.

"In fact, on January 19, 1960, which I can remember quite well because of my fortieth birthday, I walked in and told my drilling superintendent, 'Lester, we got two firsts today. This is the first time I have ever been forty years old and this is the first time I ever waked up and hadn't had a rig running.'

"Two days later we had three going, but that's the way the ball bounced that day and I have never forgotten it."

Durrett succeeded to the title of division manager in Odessa in the mid-1950s, and by the end of the decade he had been named a vice president of Delta.

In the early days of Delta Gulf, there was "no such thing as a budget," according to Durrett. "You asked for what you needed. You either got it or you didn't get it. I think perhaps the first division budget in the entire operation was developed out here not so much for Tyler's use but for my own benefit, to know what I was doing in my own store."

The general policy of keeping expenditures low presented a challenge to keep pace with competitors. "I heard the expression made one time that in West Texas we were trying to compete with Model T's when everyone else had Model A's, but we did," said Lynna Jo Edwards. "In fact, Eddie Durrett has rigged up rigs out of auction sales."

When Delta Gulf was in the process of upgrading its equipment, new blowout preventers and closing units for the rigs were high on the list of priorities. After a half-dozen or so had been improved, Durrett sent in the last such requisition for a blowout preventer to Delta Gulf's Dallas headquarters, where Mark Gardner had his office. Gardner immediately phoned Durrett.

"Eddie," said Gardner, "what are you doing to these damn blowout preventers?"

"Gardner," said Durrett, "I'm putting them on these damn wells."

"I know that. That's what I'm talking about. Don't you know when you get through you are supposed to take them off?"

Chuckling over the conversation nearly three decades later, Durrett said, "Some wise old man said when you lose your sense of humor, you're dead. Well, I'll tell you, Gardner had one. He'll stay alive while we're gone."[61]

Maintaining the rigs kept welders like John E. Faustlin on the go. ("It's just long hours and hard work.") This is particularly true in the West Texas division where distance is more of a factor than it might be elsewhere.

"Some of them rigs then were 45 miles out of town," said Faustlin. "You would work up to midnight or two o'clock, whenever you got through. You would just go out and crawl in the toolpusher's bunkhouse and go to sleep. There wasn't any use in driving to town; they had everything you needed right there."

By 1961 when Edward Doughty went to work for Delta in West Texas, he recalled, the company's equipment was better than that of many competitors.

"There was very durn few that had better iron than Delta in 1961," said Doughty. "It's natural for roughnecks or drillers to bellyache about 'the damned old junk' and that kind of stuff. Anything is going to be good or bad by comparison, see. A rig would be down or something and I would go to work for another outfit and I could see that Delta had the best stuff. But some of them did have better iron that Delta did. I guess you would say Parker Drilling Company, just to name one. And maybe Delta had as good iron as Loffland Brothers, as long as I can remember. Delta was above average."

When T. C. Carlton, later to become division manager and then vice president, joined Delta in West Texas as its drilling superintendent in 1968, in comparison with the competitors there, "We were pretty far down on the list. Pretty bad reputation because of the type of rig. What really happened was that Delta was drilling a lot of wells for themselves in the early '60s and used all of their rigs to drill these wells. When they got ready to go back into the competitive field, they were a little bit behind. This was the early '60s when all of these West Texas people started updating their rigs, and Delta didn't start that then because they were drilling their own wells. Then they started updating their rigs at the time when the industry was slack during the '60s. It turned out to be real smart on their part.

"Delta got in with their upgrading program pretty heavy in 1974. We started putting on new engines—gas, diesel engines—back before Mr. [Joe] Zeppa died. That was the first step. The next step was the pumps. Then in 1977 they decided to completely change out some.

"We have a good reputation, now, but I don't mind saying that I think Keating would agree that back in the late '60s, early '70s, we were taking only those jobs in the West Texas area that nobody else wanted."[62]

The drilling that Delta did for itself more than made up for any lost drilling contracts. The company early acquired production in this fashion. In 1954, for example, Delta Gulf held a three-eighths interest in ten gas wells drilled in the San Juan Basin in New Mexico on a 4500-acre block, with 20 more wells to be drilled on 160-acre spacing. The wells cost around $20,000 each to drill. The shallow wells, around 2,000 feet deep, had low potential, ranging from 500 mcf-5,000 mcf, but with a long life expectancy. A contract had been made with El Paso Natural Gas Company for 12 cents per mcf.[63]

The importance of similar production grew evident at times when drilling activity lagged. One day in 1959, Chris Zeppa, Eddie Durrett, and Ted Ferguson were walking over the division's yard on University Avenue in Odessa, looking over the equipment. By then many contractors had been forced out of business. A number of drilling rigs were being auctioned off for a fraction of their actual value.

"Well," said Chris Zeppa, "we're no better than the rest of them. If it wasn't for Production, we'd probably have ours on the blocks too."

It was a small statement with a large insight, for on more than one occasion Delta's production, like a nest egg, pulled it through some shaky economic periods for drilling contractors.

Production is a department of an oil company that is often, to the chargin of those in it, taken for granted. Seeing that the oil moves from the well to the pipeline is not always a cut-and-dried process, however, and to a certain extent every well has its own "personality." As Ted Ferguson put it, "Every well is different. You drill 500 in the city of Odessa and there won't any two of them be exactly the same. Just little minor things, but there'll be a little bit of difference in every well. I don't think, in about 47 years of oil fielding, I've ever had two wells that I'd say, well, were identical. They'd be awful close, but they're not that close."

Ferguson was able to chart Delta's West Texas production growth personally. When he started out with Delta Gulf, the company was operating about 25 wells. By the time he had retired in 1978 after 30 years of employment, he was overseeing more than 200 wells operated by Delta, with more drilling in progress.[64]

"I'm sure that Production has carried them for years," said Lynna Jo Edwards. "Something kept this division running, because we went too many years where we didn't turn a profit."

Despite the high crew-turnover rate that sometimes affected the district, other employees built up respectable tenure. Lynna Jo Edwards went to work as Eddie Durrett's secretary in 1966 and ended up staying, eventually becoming administrative supervisor of the West Texas division. She was the first woman to be named a division clerk for Delta in that division.

The close personal supervision of top management in Tyler lent a special flavor to the work in Odessa.

"We turned in morning reports to Mr. Joe Zeppa and to Mr. Chris Zeppa, and don't ever kid yourself, they knew everything that went on," said Mrs. Edwards. "They both had fantastic memories. And you couldn't tell anything about any well they couldn't remember. So it was just like a little family operation."

At that time the West Texas office was in a crowded, uncomfortable wooden building.

"When people talk about having to share an office," said Mrs. Edwards, "I always laugh and think about the good ol' days when we worked out of this little narrow white barrack. We had Production, and Drilling, and ran the gasoline plant, Ozona Gas Processing Plant, out of this office.

"I have seen the curtains in the other room blow when the wind blew outside. Skunks under the building. It was bad. Froze in the winter and burned up in the summer. One time I had to be both receptionist and secretary. Had the Xerox machine within two feet of me. I could see T. C. [Carlton] all day long, because where his office was, I could look up from my typewriter and see him.

"There have been fights in the office with mad roughnecks. Years ago, we paid each man for showing up on the rig $3.00 a day. This one driller had to drive some of his hands and they were supposed to pay him their $3.00, and when it came payday, none of them would give this driller their money. They started swinging right here in the office and finally fell out the front door fighting. There were five of them.

"I've seen T. C. get out his shillelagh stick and lay it on his desk when two of them would get him cornered behind his desk. He had a wooden deal and he would just drag it out. This is how the roughnecks handled it when they didn't like the way we were handling our business back in those days."

In 1972 Eddie Durrett resigned to venture into business for himself. To look after the West Texas Division temporarily, Red Magner went out to Odessa. That winter of 1972-1973 he encountered a special problem in the field office.

"I was out there that winter and kept getting headaches and, oh, it was awful. I was getting a headache by seven or eight o'clock every night. We finally talked Keating into building a new building and finally tore the other one down and found out it had a gas leak right underneath where my office was."

When the new office building was completed in 1974, Joe Zeppa went to Odessa for the opening. By then, the division had grown, necessitating the separation of the production and drilling groups there. In 1977 the division's production and exploration departments moved to Midland. The drilling section remained at the 6.5-acre property in Odessa, where the Hawkins camp had been situated when Delta took over those holdings.[65]

XII

The Delta policy to acquire production wherever possible paid off handsomely in West Texas, particularly in the Ozona gas field out in the mountainous terrain of Crockett County in the southwestern sector of the state. Delta ended up not only with interests in scores of wells but also as the operator of the plant that processes gas.

Ozona, in the Pecos River country, technically is not a city or town. "It is all county," as Lonye Cain said, for it has only county government. Since one may interpret this as meaning its limits coincide with those of Crockett County itself, one of the huge counties of Texas, one can more readily understand the meaning of the sign along the highway before one enters Ozona: *The Biggest Little Town in the World.*

It was in this locale that Delta found one of its most valuable holdings. The discovery grew organically out of a systematic search for new production initiated by Division Manager E. G. "Eddie" Durrett and his associates.

"Joe [Zeppa] had sent a very fine young geologist to West Texas by the name of Clem Roberts," said Durrett. "Clem and I went to work trying to come up with drilling deals. Our success was so-so. We continued with the development of the Huldale field as the first real big block of production that Delta had gotten together in West Texas. A very fine little oil field. As I recall we wound up with perhaps 15 pretty good oil wells."

One lease for 80 acres was gained on a farmout from Amerada in the middle of an oil field, to the surprise of Durrett and Roberts.

They took another wildcat deal in Crane County to Joe Zeppa, who said, "Well, let's bet. That lease is probably one of the great discoveries in West Texas." The problem was that the leasehold was banded across the section. "We wound up with four wells; in the process, found Gulf 300 wells," said Durrett. "But those were four fine wells, believe me. It was a real support to us too, because that had good allowables. Those wells would just set there and make that allowable just like the metronome every month. It really helped our cash flow."

Meanwhile, as Durrett described it, "Things just began to get tougher and tougher in the drilling business, and we were really struggling and struggling to stay alive.

"Roberts had been looking into Crockett County. Well, Shell Oil had drilled two wells immediately east of Ozona that resulted in shut-in Ellenberger gas wells. They were big wells. For lack of gas market Shell plugged them, walked away from them. The closest gas market was some

50 to 100 miles away. I never will forget, Clem brought me a log in one day and said, 'Why don't you look at this sand, tell me what you think?' He said, 'Well, it is a Canyon sand.'

"Make a long story short, J. C. Williamson in Midland had about a 2500-acre block which was supposed to expire. I told Clem, 'Well, if you want to drill a well, let's see if we can put this block together.' So I went to Midland and talked to J. C."

After showing the deal to "some 18 to 25 people," Delta wound up with Pauley Petroleum and Felmont Oil Corporation as partners in the first well. They drilled the Friend No. 1 as a southwest extension to the plugged Shell Ellenberger wells. The Ellenberger was 8300 feet there, but they encountered the Canyon sand at around 6200-6300 feet. This resulted in a low yield, so with the gas markets 50 or more miles off, they plugged the well.

"Well," said Durrett, "Felmont said, 'We've had enough. We want to quit.' Roberts just threw his hands up and said, 'Oh, Lord, we are dead!' I said, 'No, we are never dead. You know, the thought occurs to me that when we drill a low well, Friend 1, to a high well, Shell Friend 1, that we got to go look on the other side [of the structure]. Well, Roberts gathered up all his stuff and said, 'Well, let me see if I can replace Felmont.' He beat the trails in Midland. He wore his shoes out. He came in and he was down. He said, 'I just can't sell it to anybody. Crockett County is a boneyard.' Well, [shortly after that] I stopped in front of the Midland Tower and I ran into an old friend of mine who was in the mud business, by the name of Hugh Munn. He was president of the Permian Mud Service and I found out later that there was another gentleman involved with him in Permian Mud by the name of Willard Johnson. I said, 'Munn, I'm looking for a partner to pick up an interest in Crockett County and drill another well.'"

This led to a connection with Willard Johnson, who brought in Johnson and Lindley.

"Well, to make a long story short, we staked the Friend A-1. Drilled a well to the Ellenberger; it ran high to the Shell well and we completed the Ellenberger after torture of eight days and nights when the temperature was no more than four above. For about a million and a half [cubic feet] of gas and 200 barrels of condensate in the Ellenberger. And dually completed in the Canyon. We perforated the Canyon. We picked up the same flare there that we did for Friend 1.

"Pauley had a vice president of the land department by the name of Larry Scott who was a pretty good kind of goer. Larry felt like I did, that we had something—maybe more than met the eye. In the meantime, Roberts had continued on with his method and we began to realize that Canyon sand was blanketed and could cover as much as 50,000 acres. Well, [with the acquiescence of Ed Pauley and Chris and Joe Zeppa] we acquired leaseholds, before the rest of the industry knew that we had a gas discovery, in the amount of about 35,000 acres."

This was in early 1961.

"We got our block together pretty well. As I say, everybody considered Crockett County the boneyard. I suggested that Pauley Petroleum run a 60-day flare test on one of the Canyon wells. We knew what the Ellenberger was. The Canyon was an unknown quantity.

"We reentered Friend No. 1 which was plugged. Went in and completed the Canyon sand, put a frac job on it, and it was pretty respectful. It came out with a million and a half, two million well. With Delta, the partners agreed to flare a 60-day worth of gas. Didn't have the market anyway. 'Let's find out if the reservoir is for real.' Well, we put that thing on a controlled flow of a million a day. Flowed it for 60 days. Shut the well in and ran a 1000-hour buildup and the thing was still building up at the end of 1000 hours. Sure, it was for real."

By that time they had acquired around 35,000 acres under leases from the ranchers. They drilled two more Ellenberger-Canyon wells, then took a bold step and drilled seven or eight miles southwest on another lease and made another well that was labeled a discovery. Then after months of persuading and negotiations, a contract was made with Northern Natural Gas, headquartered in Omaha, Nebraska, to take the gas. But a plant was needed to process the gas at Ozona, and Delta became the operator of the plant, since its partners did not have plant-operating capabilities and Delta, with one small plant already in East Texas, did. Chris Zeppa could contribute his own experience and advice in building the plant. In addition to Delta's pocketing operator fees for the plant, its men had drilled the wells and thus knew them well enough to maximize production. "And we did," said Durrett. "Pauley and Johnson-Lindley both were somewhat surprised at how much gas we could get out of that tight rock."

The entire team for developing the first 100 wells in the Ozona field was composed of Roberts doing the geology, Durrett and Roberts doing the fracturing design, and Ted Ferguson and Roy Bentley, assisted by a young production engineer Ray Rutledge, running the casing, pumping the frac jobs, putting the wells on, and hooking them up.

"We drilled—I really lost count of this, but it was something over 100 wells. I know at one point we had three rigs running the field. Each of those rigs was moving every seven days. So we had three wells logging and setting by two strings of pipe weekly, and we were pumping three frac jobs and completed three wells a week. We were on 320-acre spacing. Since then, it's 160 acres, and I think it'll go to 80 and perhaps even less than that."

Ultimately, Delta was able to sell its Ozona partners on drilling a wildcat at Eldorado, Texas, and at Sonora, Texas, both Canyon gas fields. This activity enhanced Delta's worth considerably.

Though in many quarters Joe Zeppa had the reputation of being conservative when it came to risk-taking, Durrett perceived him as quite bold at times. "I'll tell you, he could run very fast. Gosh, I could cite you a thousand cases where Joe's participating was considered by the outside industry to be a real high roller. And I think we would look that way when we drilled the discovery well in Ozona, because there had been 20 wells drilled through that Canyon sand that nobody recognized. They didn't know what they were looking for. But we did. Joe wasn't conservative—he was careful and he considered for us to gain."

Durrett, in his working agreement with Delta, had the privilege to secure overrides and purchase royalties, and in the Ozona strike he exercised that option to his subsequent reward.

"The first time I really went overboard I paid $30,000 for half of the royalty under the original block at Ozona," he said. "Immediately after we sold the Ellenberger. And I don't mind telling you that was more money than I'd ever heard of, and when I walked into the bank and borrowed it, it absolutely scared me to death. But it paid out pretty good after we got going straight."

With the rising price of natural gas and the projected longevity of the field, the Ozona field became a rich, ripe plum in Delta's treasury. Indirectly, we can measure it by Durrett's assessment of his own personal holdings there.

"I have told my wife that under no conditions, if I happen to go first, is she to ever divest my estate of Ozona. They could sell everything else, but don't sell Ozona, because the projected life of Ozona with its first production being in 1963, is 46 years. That goes down to the year 2009."[66]

Air drilling was a key to success in the Ozona field.

"Until they got air drilling they couldn't drill those wells," said J. H. "John" Gilleland. "It was too expensive to drill with mud. They drilled too slow. They were small wells. All of them. There were no big producing wells. But it got so we could drill them in about seven or eight days."

When the field came in, Ted Ferguson was the only production man there and, as he says, "I liked to kill my fool self down there, keeping up with the drillers," he said. "Just running after those wells and rigs. We had four rigs running in that area and they were drilling a hole on and off in about 12 days. And they got to where about eight or nine days they'd be on and off. J. C. Johnson drilled five wells down there and was on and off in 39 days. Course, he was skidding, he wasn't tearing down, but he drilled them in 39 days, five wells and run pipeline." One thing that facilitated the drilling was that they were drilling on 320-acre tracts and lost little rig-down and rig-up time as they skidded the rigs onto the adjoining tract.[67]

Construction on the new gas processing plant began in 1963, with completion in early 1964. Though the fruits of this operation looked promising, the outlay at this point was significant, for the plant and gathering lines were tagged at $2,880,000, with an additional $600,000 to be needed subsequently.

Delta was to bear 22.5 percent of the initial cost, or $648,000. The entire amount was borrowed by the partners from the First National Bank of Chicago at 5.5 percent interest on an eight-year note.

Originally the plant ownership was to have been divided among Pauley Petroleum with 50 percent, Delta with 30 percent, and Johnson & Lindley, Inc. with 20 percent, but the partners decided it would be to their interest to invite Northern Gas Products Company, a subsidiary of the gas purchaser Northern Natural Gas Company, to participate. Thus Northern Gas Products Company came to own 25 percent, in the process reducing the interests of the others by the same percentage. This transaction, it appears, was a kind of insurance to make certain the owners had a favorable market for the plant's products, perhaps even increasing the gas sales.

Their contract on the Ozona gas called for residue-gas deliveries of eight million feet per day over the first year; double that, or 16 million feet daily the second year; double *that*, or 32 million the third year; and 48 million the fourth year. The sales price was 16 cents per mcf. Although there would be little profit the first two years, the partners felt that once the plant was a going concern, Northern Natural would want to process gas from other producers in the plant, thereby increasing the volume.

While the four owners were building the plant, with Delta designated as the agent for construction of the plant and as operator once the plant was finished, Delta had already invested more than $1.5 million in West Texas leases and drilling, which included activities at Sonora and Eldorado as well as Ozona.[68]

The Ozona plant was not expected to start right in as a profit maker, but by early 1965 Joe Zeppa could report to the Delta board that the plant, now more efficient and with additional wells to supply it, had operated in the black for February and "it is anticipated that this source of revenue will continue to improve." By then, other gas production in West Texas, in Schleicher and Sutton Counties, was either in production or scheduled within a month or two. The Schleicher County properties, which Delta owned equally with Pauley Petroleum, "appear the best source of revenue from production in West Texas," Zeppa reported.[69]

Jody Conaway, superintendent of the plant since 1969, was one of 18 employees at the plant in 1980, with the product from 186 wells being piped into the plant. Of these, 85 wells belonged to Delta. The gathering system involved 48 miles of trunk line, not including lateral lines.

The plant extracts propane, butane, and the heavier gases, the latter all lumped together. In 1980 the plant was processing about 30 mmcf per day, with the capacity to handle 30 mmcf more a day. As for liquids, the plant has made as much as 75,000 gallons of propane a day, with 36,000 gallons of butane, and 1,200 barrels of condensate.

Any hydrocarbon processing plant has its high risks, and the Ozona plant is no exception.

"We work under extremely hazardous conditions," said Jody Conaway. "Everything overhead. Everything pressured up. Everybody has to know what they do at the time, whenever the problem arises. We may not get as dirty as everybody else or what-not, but we are working under extreme caution at all times.

"We have a great number of trucks coming in from Mexico. Most drivers are not English-speaking drivers. We are in constant threat at all times with whether or not they understand our procedures. Our employees load the truck and release the truck to be moved. We have five Mexican-Americans working here at the plant. We have a Spanish-speaking person on tour here at the plant at all times."

"One small mistake could develop into a large mistake. We're 16 years old. Each year the hazard increases. The wear of the equipment. And everything else. And it's just a constant watch."

Conaway himself was a burn victim in 1979 at a time when the plant had compiled the longest accident-free record in Delta history.

"We were working on an engine that had a leaky valve that accumulated gas over a period of a day," said Conaway, "and it flashed and I happened to be the lucky one standing there, and none of the good hands got hurt."

Conaway was trapped between two large cylinders in a two-foot space, and to escape he had to go through the fire. But he couldn't leave, and he shielded his face with his hands until he could run to safety. He was in the hospital for about a week, with his most serious injury the second-degree burns on his hands.[70]

The location of the Ozona field, in that rugged Southwest Texas terrain, contributed special problems.

"The distance problem and the complete different set-up we have here are two of the things that Mr. Chris never could really get into," said Lonye Cain, field operations manager. "I remember he asked Mr. [Roy] Bentley, who was the production field foreman out here at one time, asked him, 'Roy, do those boys have to drive all of those miles every day?' I will never forget, I was a real green hand at that time, and I thought it was quite amusing. Well, I can understand that, because expenses at that time, I imagine, were great. Gas was selling for 17 cents.

"The extra cost in tire wear, for instance. You can put a new set of tires on a unit and it is not uncommon to lose them all within the first two months or under 10,000 miles. They wear quicker, and naturally you are running over rocks, and they break and blow out.

"It is nothing for these switchers to drive 3,000 or 4,000 miles a month on a 21-day working period. They can drive 200 to 300 miles a day. One man would average around 45 or 50 wells. It is a lot of wells. I think one guy had 62 at one time, and he sometimes doubled back, two or three trips to one well, in the process of his making his rounds, for one reason or an-

"*I have been in the petroleum business since 1921 and have enjoyed every minute of it, including the time when I was released and turned loose to shift for myself. That was when we decided to form a drilling company with two assembled junk rigs, which we named Delta Drilling Company. Frankly, I did not expect that we would last very long, but here we are, years later, and still in business, expecting something good to turn up from day to day.*

"*I cannot say that I have anything to regret, except the fact that I could have accomplished a good deal more and didn't. However, my interest has been in the oil business and in the oil people, many of whom I have met and enjoyed and I am fully convinced that it has the best people in the business world today.*"

Joe Zeppa
July 2, 1970

L. E. "Red" Merritt, who had been in Arkansas during the El Dorado boom, was hired by Delta as their first drilling superintendent.

The staff which opened Delta's Tri-State office in Evansville, Indiana in 1939, included (from left) Virginia Buchanan, Ruby Effingham, H. J. "Red" Magner, and Guy White.

This crew drilled the Morgan No. 2 near El Dorado, Arkansas, in 1937. From left—Bill Williams, Jack Rodgers, Woodrow Hardige, Bill Phillips (seated), Vernon Hays, unknown, Claude Johnson, and L. V. Portier.

Traveling to the rigs to inspect their progress and the condition of the equipment was part of the daily routine for the management of Delta Drilling Company.

In 1941 on the Veal No. 2 in Hockley County, Texas, the Western Company performed an acidizing job, treating the Delta well with 14,000 gallons of acid. Eddie Chiles is standing third from left in the foreground. (Courtesy The Western Company.)

This 1933 photo was taken in front of the Delta's Longview office on the day of Sam Jones's funeral. Jones was the night watchman and pot fireman on "Ark" Carter's rig. Standing, from left: Bud Naul, B. W. Freeman, Pete Welch, Herman Murray, Jim Bradford, Guy White, unknown, Jesse Riggs, Frank Garlington, Homer Keel, "Ark" Carter, Elmer "Mutt" Hays, Woodrow Hardige, W. J. Baggett, _____ Harris, Bill Ferguson, May Jones, R. A. Stacy, and Evelyn Taylor (Reeves). Sitting: Nathon Bacle, "Frenchie" Portier, E. S. Young, Fred Burton, R. C. Merritt, Jimmy Lynch, A. C. Merritt, Paul Porter, W. N. Dailey, Claude Johnson, and Jim Ed Cook.

In 1949, Rig #6 was drilling the Knox No. 1 well in the East Texas area. Steam rigs were soon replaced by the power rig.

As Delta's operations expanded throughout the United States, rigs such as this one appeared in the Rocky Mountain Division, which was based in Casper, Wyoming.

The Etexas Producers Gas Company was organized in 1945 to process natural gas and make casinghead gasoline. The plant, designed by Chris Zeppa, was built in 1946 outside of Tyler on the Kilgore highway.

In 1938, employees gathered at the Hughey & Ross lease near Kilgore, Texas, for one of the first Delta picnics, a tradition which continues throughout the company.

other. If we can get about two, two and a half years out of a pumper's pickup out here, we are doing real good.

"We used to have to get out at night a lot of times and it would be cold and dark," said Lonye Cain, "and some of these guys are a little jumpy, or spooky. We used to kid each other quite a bit about snakes, and we still have mountain lions running around here in this part of the country, and you get an old kid out there that hasn't been raised out here and he is unfamiliar with it, you can give him a pretty hard time.

"We have aliens, wet Mexicans, walking through this country. I remember a guy working for the plant [who] was out in the middle of the night checking the line heaters. We use well heaters or direct-fire heaters in a lot of places just to heat up the gas. This one guy drove up beside this heater and stepped out to check it to see if it was hot and working all right, and this wet Mexican came crawling out from under that heater, and I think the old boy actually ran over his pickup two or three times trying to find a place to run. The Mexican went one way and the meterman went the other direction.

"We have had some close calls, but we have never had anyone bit by a snake. Some of the guys bring in some pretty big rattlesnakes. I had a pumper just a couple of weeks ago get one about five and a half feet long with 14 rattles. The little ones are just as dangerous. We have a little rock rattler out in this country, and he will only get 12-14 inches long, but he is just as poisonous as the big fellows, and he hides a whole lot easier."[71]

XIII

Delta moved into the Northeast region via Michigan. In 1950 Delta's Rig No. 10 was working in Michigan for the Brazos Oil and Gas Company, which was an operating arm of Dow Chemical. After drilling several wells, the last one was a deep, 7000-foot well which proved to be a dry hole.

Then on December 1, 1951, Delta assumed operation of large-scale gas properties in New York and Pennsylvania as a result of a deal for an undivided interest with Dr. Roland F. Beers and Dr. George C. McGhee. There was an assured market for the gas and a promising future, as some of the properties were producing and the nonproducing ones were "located on favorable structures." Beers was a professor at Rensselaer Polytechnic Institute at Troy, New York. McGhee, later to become ambassador to Turkey, was married to the daughter of E. L. DeGolyer, the Dallas oilman. Joe Zeppa knew DeGolyer, and it was probably through their friendship that Delta was approached to do the drilling.

Delta employed Charles E. Fralich of Bradford, Pennsylvania, to manage the district, and Rig No. 10 was shipped over from Michigan.

Red Magner, who handled the billing and other bookwork from his office in Evansville, remembered that crewmen began rigging up in Pennsylvania on January 1, 1952.

"And I remembered I got the freight bill for moving that rig all across the country there, and it weighed out about twice as much as when we set it up in Michigan. I guess I had forgotten about how much equipment had been added, but I complained to the railroad people. 'Hey, you charged me double on that stuff! We don't need all that snow hauled from Michigan to Pennsylvania, because they got enough in both places.' So they refigured their bill and found that they hadn't charged me enough. That hurt."

The first well drilled under the agreement with the partners turned out to be dry, but the second was a five-million-cubic-foot well, with the third also likely to be a small commercial gas well. Zeppa was optimistic.[72]

Approximately two years later, Delta's directors agreed to accept a sale of the joint holdings, of which Delta owned one-third, to John Fox for a total price of $1.2 million. A few gas wells and leases, the warehouse and yard at Driftwood, Pennsylvania, and drilling and other equipment were omitted from the sale, which would enable Delta to continue as a drilling contractor, for by that time the company was there to stay.[73]

When L. A. Bates transferred from the Illinois Basin in 1952, Delta already had two rigs in the Northeast and were sending another from Odessa, which would become Bates's outfit. Bates, as toolpusher, was told by Guy White to fly to Pittsburgh to join his rig. He had never been in an airplane before in his life, and he refused. White had to drive him up to the Northeast operation.

Pete Wolfe and Barney Owens were pushing tools on the two rigs already there, and Bates ramrodded the third one. By then, there was a blend of new and old in Pennsylvania. One of Bates's drillers was Ark Carter, then near the end of his career. When Bates subsequently became superintendent of drilling for the Northeast, he had eight rigs under him at one time, which kept him on the road steadily.

In 1953, William H. Young replaced C. E. Fralich as head of the Bradford, Pennsylvania, district, which still came under Guy White's supervision in the Evansville office. In 1954 Young resigned, and Red Magner, who had been handling the paper work, replaced him as manager.

In 1955, when the Evansville office also closed, Delta moved its offices from Bradford to Pittsburgh, Pennsylvania, to be closer to its customers. The division's office remained there until 1970 when it was shifted to Indiana, Pennsylvania, 50 miles away, where the supporting facilities, shop, and yard had been maintained for years.[74]

Delta had entered the oldest petroleum-producing region of the United States, where, in fact, the domestic industry began with the historic Drake well in 1859 at Titusville, Pennsylvania. Some of the old equipment, in the form of antiques, were still around when Delta moved into the region in the 1950s. Don Carman remembered seeing relics in West Virginia that traced back to the nineteenth century.

"You'd see in the older fields the old wooden surface pipe," said Carman. "It was octagonal shaped. You'd drive down the road, and an old hillside farmer might have a bunch of old wooden sucker rods along there for a fence, kind of an antique." [75]

The Northeast at that time was cable-tool country, and rotary rigs were rarities. A cable-tool rig uses a cable with a 40-foot piece of solid steel on the end as a bit, which pounds into the ground to make a hole.

The cable-tool rigs enjoyed certain advantages. "A cable tool didn't have much expense," said Auburn Bryant. "They could spend a lot of time on a well and still come out on it. The cable tool just has one engine and they usually worked the men 12 hours and they just had four men [two per shift] and they didn't pay them rotary wages. If you were going to run a rotary, you had to have about three engines and 15 men at that time. A lot of cable tools would shut down on weekends to cut down on their overtime. In bad weather you can't afford to shut a rotary down. You have to keep going. At that time a rotary bit would cost you $1,000, and a cable tool has a little old bit they can sharpen and it don't cost them much of anything." [76]

At first, cable toolers had a clear field in Pennsylvania. "It was a bad crooked-hole country," said Dallas Bryant. "Everybody swears you couldn't drill a well with a rotary because of the deviation. We couldn't compete with the cable tool. It took us about 90 days to drill a well with fluid and it cost so much more than it would have with a cable tool, we couldn't have ever stayed if we hadn't started air drilling." [77]

When Melburn Miser went to the Northeast division with the first rigs in 1952, there were 100 cable-tool rigs drilling in the region, and rotary drilling was almost unknown there. "Cable toolers kind of thought of the rotary people as outsiders," said Miser. "It took a while for them to get used to it. After we drilled a few holes they saw we were here to stay. In a year and a half they were moving out." [78]

Soon Delta was drilling in Ohio, New York, Pennsylvania, West Virginia, and Maryland, all mountainous country.

The key to Delta's success in the Northeast was, as Dallas Bryant said, air drilling. In a few years Delta's rig count there had soared to eight, as the result of this major innovation. Air drilling, spearheaded by Red Magner, revolutionized the industry in the old oil province.

"The prime problem was that the surface formation was fractured," said Magner. "The country obviously is hilly and rolling, and it's been distorted pretty well. You don't have water and sand like you find around [East Texas]. So you go to drill and you lose circulation. You can put all sorts of mud and circulation material in it. Sometimes you'd stop it and sometimes you wouldn't. It's a slow process. It wears the bit out and it takes a lot of fluid."

He and others had read that some air drilling had been attempted in California because of lost circulation.

"Guy White said, 'Go out there and rent some compressors and let's try it.' We had the compressors out there because we'd paid a month's rental on them, so we said, 'Well, okay, we'll try them.' It was just fantastic. It was the first well we ever tried [that] we drilled in about half the time and half the bits. So we were off to the races.

"The difference, basically, is instead of pumping fluid down through the drill pipe and out through the bit and up through the lines, you pumped air. With the fluid, for instance, your pumping rate would be in the category of maybe 120 feet per minute. That means that the amount of fluid you're pumping would be moving and traveling up the hole at approximately 120 feet a minute, so that if you were 1,200 feet deep it would take you ten minutes for the cuttings to come out. And the amount of *air* volume that you put down there, it traveled at a rate of 3,000 feet per minute. So if you're drilling at 3,000 feet you'd get your cuttings back in one minute. Put another way, you're always having a clean bottom. Like your dentist, now they've got those fast, high-speed things with a jet on them and it just swooshes—well, that's the same basic principle."

"It'll work, wherever, in the absence of water."

Delta, following the early air drilling, also introduced gas drilling in the Northeast. It was based on the general principle of substituting air for fluid to ensure better circulation; in this instance, gas was substituted. The first air drilling on March 18, 1954, had been a success. Soon afterward, while Magner and his colleagues were still patting themselves on the back, they faced another problem that was similar to one others had experienced in the San Juan Basin in New Mexico.

"El Paso Natural Gas had a problem with drilling into these fractured-type formations and having damage done to the formation and not getting gas out of them," said Magner. "So they tried gas drilling, setting pipe on top of the pay and drilling into it with gas. Make gas wells. So in Pennsylvania we had a situation—same thing—except we had pressures that were in excess of normal. They were about 4,000 pounds there at 6,000- or 7,000-foot depths. And we'd have to drill in the heavy mud and that would only accentuate the fracturing, sealing, by the heavy muds. So this particular guy that was president of New York State Natural Gas says, 'Let's drill this one in with gas.' He knew what he was saying, but he didn't know how to do it, and I didn't know how to do it. No one knew how to do it.

"We'd never done it before, so I read something in a magazine about it and I called Shaeffer Tool Company and said, 'Have you got an arrowhead or a rotating head?'—that's what they called it—and talked to their engineer and I told him the specifications and the pressure and all that and he says, 'I'll be right out.' And he put his arrowhead on motor freight and shipped it on, and without him we couldn't have done it. He just rigged up the whole thing for us. Because he'd done it before in the San Juan Basin.

"We didn't know what was going to happen. We didn't know what size well we'd get or anything else. Well, we drilled it in and it was a five-million [cubic foot] well. It took us a long time to do it, though; we had conditions that were not like West Texas and Four Corners. It was cold and wet and we had to dry the gas out and all that, but we achieved it, and it was a success, both the air drilling and the drilling in with gas on the very first well.

"It was the first time anybody had done that in Pennsylvania. Never had even thought about it. Matter of fact, I gave my first paper and my first talk on it at Marietta College, Marietta, Ohio, as it was going down."

This did not mean that all drilling problems were solved for Delta and those who adopted air drilling in that part of the country, but it simplified matters. As Magner pointed out, unlike fluid drilling, every aspect of rotary drilling with air requires a different approach. A basic difference is that, using air, there is a totally open hole and the hole itself can fall in. There is no bouyancy as there is with fluid drilling. Greater care must be taken not to twist off drill pipe, since fishing it out is next to impossible. Special strategies must be worked out to combat high-volume gas pressures. Any technique carries its own risks and problems, but the advantages of air drilling became so immediately obvious that Delta's reputation in the Northeast was made for good.

The introduction of air drilling necessitated an overnight reevaluation of rotary drilling. To drill a 6,000-7,000 foot well, a cable-tool rig would require six months, though it could drill cheaper. At that time the cable toolers could drill for $6 a foot, while a rotary drilling job would cost $9 a foot. But the difference was that a rotary could finish the assignment in about 45 days. But since the "law of capture" existed in Pennsylvania—meaning that whoever reached the gas or oil first could take it—the man who "got there fustest" won the race. This enabled a company to build up reserves faster by using rotary rigs. However, when air drilling came along, rotary rigs became competitive overnight, for then they were as low in cost as cable tools. As a result, some of the old cable-tool rigs were stacked when they finished a project and were never moved away.

The Benezette gas field Delta was working in was a large one, four or five miles wide and perhaps 20 miles long.

"So it provided drilling for us for four or five years," said Magner. "By that time we had about eight rigs and were spreading out all over the Appalachians, and for almost a year we didn't even have a competitor in the rotary field. And we were going down into southern West Virginia and over into Ohio and all around, so we were spreading the air gospel, as it were. It was a flamboyant-type era, and economically, I think, our division up there was making more money than all the rest of the rigs put together. One year we had 92 percent activity ratio, and that's most difficult to do in the wintertime. Meaning we had as many as six months' work ahead for every rig. Well, in Pennsylvania we had a bird's nest on the ground, as it were. Because we had people that would guarantee us a year's work and pay for our rig up there and back, so we couldn't lose."[79]

Melburn Miser, who was the first one to drill on Delta's first air-drilled well in the Northeast, said, "I don't know how we lived through it. Dust was so bad around the rig, I don't know how we kept things running. I will never forget, we pulled the bit out and I made 300 feet of hole that night. We hadn't made that in three days—about 100 feet a day. So I called the toolpusher and told him. He said, 'You didn't make that much hole.' He just wasn't expecting to make that much the first eight hours. That was an evening tour. They started it that afternoon."

Once they were able to control the dust, air drilling caught on, and eventually all the companies operating in the Northeast were air drilling.

Not only did air drilling return cuttings from the bottom in record time, there were other pleasant side effects. "It's not hard on your drill pipe," said Malcolm Eakins. "Expensive drill pipe wears out. I worked up there on one rig eight years and still had that same stack of drill pipe."

Foam drilling, or mist drilling, also proved to be useful in the Northeast. "When they first started, when the hole would get wet, then you would have to go with fluid," said Auburn Bryant, "and then they came up with detergent and mixed it with water and injected it down your drill pipe with the air to keep the hole clean and keep it from mudding up. On nearly all the surface holes you have to have it." Don Carman described the detergent concoction as "very foamy, or lathery, which would be more like shaving cream. The primary purpose of it was to lighten the intrusion water. You would break it up easier with soap and return the water to the surface in larger quantities."

This technique also simplified many drilling procedures.

As Don Carman said: "For a long time there we would take cable tools, drill a surface hole, and then they would move them off and move the rotary rig on and it would be straight air. But after we got involved in foam drilling or mist drilling it was more economical to move the one rig on and drill the surface hole and the whole thing."[80]

Delta drilled a great many gas storage wells in the Northeast, particularly for Consolidated Gas in some of the old fields that originally had been drilled by cable tools. The most prolific portion of the field would be saturated with wells and the gas piped there in the summers; in the winters the gas would be pulled out and delivered to customers. But with storage fields, one can't have potential leaks, which meant the whole storage area had to be combed thoroughly for old abandoned holes through which gas could escape. "That kept us busy for many, many years with various rigs," said Red Magner.

The gas-storage concept also led to what was in the 1950s a revolutionary idea on oil storage, which Joe Zeppa proposed to his fellow oilmen.

Guy White had spent a great deal of time visiting the Northeastern region while Delta was drilling deep gas wells there. He learned there were old oil fields, long abandoned, near the banks of the Ohio River. One, called the Seven Sisters field, covered more area than did the East Texas field, and there were others similar to it. White eventually mentioned this to Joe Zeppa.

At the time, the Independent Petroleum Association of America (IPAA) was actively opposing the importation of cheap foreign oil which was driving down the price of domestic crude. Foreign crude from the Middle East could be brought to the East Coast for $1.00 a barrel, too cheap to enable domestic oilmen to compete.

By that time gas-storage fields were already in use in the Northeast, and, in fact, one Delta rig worked practically all the time for New York Natural Gas Company, drilling and reconditioning old holes where they were storing gas.

Why not do the same with cheap foreign crude oil? The $1.00 a barrel oil could be brought up the Ohio River at a total cost of $1.50 and pumped into the nearby abandoned oil fields in West Virginia to be stored until needed. With large companies or perhaps the federal government sponsoring the project, the United States would then be prepared for a petroleum shortage or international crisis.

Zeppa presented the idea to the IPAA. To say that the idea fell on unreceptive ears would be an understatement. Almost all of his fellow members present voiced their enthusiasm—against the proposal. The idea's time obviously had not come yet.[81]

As in other locales, the weather and the terrain influence men's attitudes toward work. Men native to one region tend to prefer that one to others, as East Texans have learned upon being sent to the Northeast. Rarely was one tolerant of the cold winters. As Melburn Miser said, "When they looked up and saw those geese fly south, look out. I saw a lot of people get up and go."

Hunting seasons seem to affect the availability of crew nearly everywhere. As for the Northeast, Miser said, "The first day of deer season up there, the hands—well, it is just like they all want to get sick, but we do manage to operate."

Though drilling primarily in Pennsylvania and adjoining states, rigs went up to the Canadian border and as far south as Georgia. In the time

Dallas Bryant was with the company, Delta crews had drilled in New York, Ohio, Pennsylvania, Maryland, West Virginia, North Carolina, and Georgia. The Rome, Georgia well was 812 miles away from headquarters. Dallas Bryant was toolpusher on it, relieving Z. L. "Nig" Spraggins from the East Texas division.

The company didn't restrict its activities to petroleum. Delta even drilled some salt wells in West Virginia for a chemical company. The pure salt was loaded on barges and shipped to Charleston where a commercial grade of salt was made from it.

Through the years, Delta concentrated on contract drilling in the region and accumulated very little production. Production there primarily is gas. Drillers and toolpusher like L. A. Bates have never drilled an oil well in Pennsylvania.

Until the latter part of the 1970s, Delta's production in the Northeast amounted to perhaps 20 gas wells. In 1977 the division initiated plans to drill 20 wells a year for itself. By 1980, the production figure had jumped to 80 wells, which kept three pumpers covering from 25 to 30 each.

By then, Delta was selling its Northeast gas production from the 80 wells to three companies—Lafayette Gas, Peoples Natural Gas, and Consolidated Gas.

By 1980 Delta, under the management of Frank Whyte, had ten rigs in the Northeast, all of them purchased new within a six- or seven-year period. The upgrading had started before Joe Zeppa's death and then had continued apace. All but one were portable rigs.[82]

XIV

As deeper drilling picked up in Mississippi and Alabama, the Southeastern division was opened in Hattiesburg, Mississippi, in 1971.

In what may be seen as a precursor to the Southeastern division, Delta had sent a rig in the late 1960s to Rome, Georgia, to drill a storage well. The rig was moved from Pennsylvania. Searching for a porous zone to store gas for an Atlanta company, the drill reached its destination, but the zone lacked the porosity that was sought.

Operations soon spread into Florida, with a lot of contracts in the Jay field there. Subsequently, Delta was offered a contract by the Louisiana Land and Exploration Company to drill a 16,000-foot well in Florida.

By 1975, the Southeastern division had closed. It became the shortest-lived division and was subsequently merged back into the Gulf Coast division, headquartered in Lafayette, Louisiana, from which it had sprung.[83]

XV

The latest division opened was at Victoria, Texas, covering the South Texas region stretching from Port Arthur to the Rio Grande Valley. This division got under way in late summer, 1978, with Russ Hudeck as division manager and Herman Gordon as drilling superintendent. The district began with four rigs—one coming from Lafayette, three from the East Texas division—with a fifth rig following later from East Texas. By 1981 the new South Texas division boasted ten rigs.[84]

8
UPS AND DOWNS

I

In addition to the regional divisions, two corporate subsidiaries have played roles—not always positive ones—in the parent company's economic history: Delta Marine Drilling Company and the Tyler Hotel Company. But even these do not complete the survey of Delta's domestic activities. What was it like to operate during periods when drilling slackened off, particularly in the late 1950s and early 1960s? Many independent drilling contractors were forced out of business in that era. How did Delta survive?

Then there is, always, the human factor. What were working conditions like? Who were some of the men and women whose labor helped Delta succeed? What were some of the risks they faced? What were some of the contributions made?

Finally, using the financial benchmarks available, what has been Delta's pattern of growth from the 1930s to the 1970s?

II

Delta Marine was born in optimism, at a time when offshore drilling was in its early phases. In late 1955 Joe Zeppa outlined to his fellow directors, Chris Zeppa and Sam Sklar, his plans for a new and separate corporate entity to be concerned with contract offshore drilling.

This operation, Delta Marine Drilling Company, a Delaware corporation, was to be capitalized at a total of $3.4 million, with Delta owning 51 percent of the common stock and two-thirds of the preferred stock at a total cost of $500,000, and purchasing $100,000 of the ten-year debentures (total $1.5 million), which would carry two percent of the common stock. This would give Delta a total 53 percent control of the company. Alexander Shipyards had agreed to purchase 12 percent of the common and one-third of the preferred stock, and would build one submersible barge and convert one LST into a drilling tender. Delta would furnish the drilling equipment.[1]

Within a few months, the organizational plans had been completed, with a total capitalization of $3,677,900. Delta owned 51 percent of the voting stock and Zeppa expected the new company's financial obligations to pay out in about three years.[2] J. M. "Mark" Gardner, who was to run Delta Marine, was also elected a vice president of Delta.[3]

Thornton Tarvin went to work as vice president of Delta Marine in 1956, and then a year later he worked also for Delta, so he was responsible for Delta Marine's two offshore rigs and the 15 rigs which Delta had in the Gulf Coast Division in Louisiana. He remained at the New Orleans offices until 1962, when he went with another company.

Delta Marine at that time was in the process of building an offshore tender, the *Joseph Zeppa* and a barge, the *Chris Zeppa*. The *Joseph Zeppa* was ready first, operating in the Gulf of Mexico off the Louisiana coast, and then in 1957 the *Chris Zeppa* went out also. Later another tender was added, though it eventually went on to Brazil where it went to work for Petrobras, the government oil company there.

What were some of the problems in drilling offshore?

"The big problem was my offshore rigs," said Tarvin, "because they were breaking down so often. We had a lot of trouble with those, and we spent a lot of money fixing them up and finally got them to running pretty good, and after I left 'em, they got around with them for several years.

"It was all new to all of us, because offshore was just started then. Nobody knew a whole lot about it. You look at it today and see the improvements—my God, it was kinda like pioneering then.

"The land rigs, of course, you can drive up to them. You have to tear them down and move them and go from one location to another; and the offshore rigs, you're already rigged up and all you do, on the barge, you pump it out and raise it and move to a new location. The tender ties up to a platform, so you finish the work on the platform and then you take the rig, put it on a floating barge, move it to the next platform [where] you have kind of a crane to unload it and load it, and untie your tender, and you're ready to go again."

At that time, in the beginning, there was only one well to a platform. Later, multiple wells would be drilled from one platform.[4]

Offshore drilling had many advantages over land drilling.

"It was altogether different from land drilling," said W. H. "Pop" Whitfield. "If I had it all to do over, I would make all of mine on the water. I wouldn't make it on the land. You don't have all that driving. All your room and board is free."

Some locations were as far as 90 miles out in the Gulf of Mexico, with a work schedule of seven days on the job, seven days off.

"When I first went to work [in 1960] they were working 15-7," said Whitfield. "Then they went to 10-5. Then after a couple of years they went to 7-7. It is like living in a hotel, on the barge. They had ten bathrooms and

38 bedrooms. Anything in the world you wanted to eat, recreation room. You could look at television, play dominoes, cards, just anything you wanted to do. Everything was air-conditioned.''

He worked on the barge *Chris Zeppa* for ten years.

"You just don't have no mud, no ditches to dig. It stays set up all of the time. All they do is let the derrick down and move and raise it back and start again. They just pump the water off and it floats, and a tug boat pulls it then to wherever you want to go. When you get there, you sink back and raise the derrick and go to drilling. That derrick is a jack-knife, that is what it is. That rig was made for 25,000 feet.''

Of course, there were serious weather problems, such as hurricanes.

"We always came in ahead of time, so we never were on the rig. The Coast Guard would give you a certain time to leave. You lay the pipe down and fill your hole up with mud and close all your blowout preventers and disconnect all your turns.

"I wasn't working on it at the time when [Hurricane] Audrey came, the one that hit Cameron [Louisiana] over there. They had tin on all of the outside—they eventually took all that off and put on metal—but they didn't even lose a sheet of tin over there in Audrey. I think they were seven miles off the beach."[5]

By perusing Delta's minute books for the period from 1956 into the 1970s, one can gain some sense of the ups and downs of the Delta Marine venture. A big factor in its economic status in the 1950s was the state-federal controversy over ownership of the tidelands. While in the late 1950s there was generally reduced rig activity, the dispute between, in this instance, Louisiana and the federal government contributed considerably toward the decline of offshore activity. Nevertheless, Joe Zeppa reported in 1957 that Delta Marine had concluded "a very successful year" and he hoped to put the tender *Joseph Zeppa* in foreign waters, since domestic offshore contracts with Magnolia Petroleum Company and Ocean Drilling and Exploration Company were expiring in early 1958.[6]

Joe Zeppa continued to be optimistic about Delta Marine's prospects in the face of decreased rig activity. He recognized that the company's deepwater equipment needed to be supplemented with shallow-water barges, and purchased one called *Doc* for $335,000, approximately half its original cost. A lease-purchase contract was signed for a barge from Marine Drilling Company for $950,000, and Zeppa expected the two new properties to practically earn their payments.[7] By the next year he could report that Delta Marine had been "very fortunate" in keeping both the *Joseph Zeppa* and the *Chris Zeppa* at work most of the time, along with the two inland barges.[8] In 1964 Zeppa noted encouraging rig activity, with $434,000 left in Delta Marine's long-term debt. The year's rig activity, he said, had been "relatively high," with realized net earnings of $384,854 after taxes and depreciation.[9]

But the strained optimism of earlier years gave way to the reality of a losing proposition in 1966 when Zeppa reviewed the year's profits and losses. Delta Marine had incurred "heavy expenses" for revamping the *Joseph Zeppa*, resulting in losses after depreciation. The immediate prospects, he noted, "are not too encouraging," even though the tender was working for Tenneco, and the submersible barge was working for Chevron. Though two of the inland barges were busy, one tender was stacked at Morgan City, Louisiana, and its future prospects were not promising. The Tenneco contract would soon expire.[10]

From that point on, Delta Marine had no more ups, only downs. It continued on "a marginal basis" in "an intensely competitive market," and did not have the revenues with which to meet its obligations. Then a storm damaged the *Chris Zeppa* to the tune of $291,000. By 1969 the company was in "very serious financial difficulty," and by that time Delta had advanced Delta Marine a total of $1,075,000, without which it would not have been able to pay its debts.

In the fall of that year, Joe Zeppa approached Mark Gardner, who had by then resigned from Delta Marine and was president of High Seas, Inc. in Houston, with the possibility of some sort of arrangement between the two companies whereby High Seas might utilize Delta Marine's tax loss. But after checking the matter out, Gardner reported that under existing laws such a thing was not possible.

By the latter part of 1970, Delta Marine owed its controlling stockholder, Delta, $1,480,000 for cash advances. The following year, 1971, as the operating losses mounted, it was recognized that not only were maintenance and operating costs extremely high but that "the equipment is old and can no longer be considered efficient." The submersible barge, no longer operable or in demand, was to be salvaged.[11]

At the 1972 annual stockholders meeting, the problems of the marine drilling firm were examined in some detail. The equipment on the barges was old and could not be operated competitively because of the low horsepower rating of the motors, pumps, and drawworks; thus it was unacceptable to the larger companies that might contract it. Because of the equipment's age, much time was lost in breakdowns, and repair costs were excessive. Nor could it drill as deep as the work frequently required. At the board meeting that same day, the minority stockholder directors advocated liquidating Delta Marine. Dr. Sam Y. Dorfman, Jr., suggested that "every effort be exerted to dispose of Delta Marine Drilling Company as quickly as possible on any terms which might be available and acceptable to the directors of Delta Drilling Company."[12]

But instead of coming to a head, the Delta Marine crisis rocked along. Delta acquired additional stock in Delta Marine, raising its ownership to 83.32 percent. From that point, there were occasional profits for the company, and losses of various sizes for different reasons. Drilling under contract with Petrobras in Brazil brought income, as Delta Marine No. 8 was employed there, but its refurbishing created additional expense.

By the end of 1973, the book value of all fixed assets of Delta Marine was $1.5 million, with indebtedness to Delta amounting to $3.288 million, leaving the stockholders' equity a negative $1,720,949. By then Delta Marine had been insolvent for a great many years and continued to exist only because of the financial backing of Delta. An exploration of the legal aspects of the company's dissolution was under way. Soon afterward, Delta Marine sold its barge and Rig No. 3 for $100,000, the tender *Joseph Zeppa* was sold for $127,500, and other equipment for $152,000. Rig No. 6 was stacked at Morgan City.

By early 1975, Delta Marine Rig No. 8 had finished a Venezuelan contract and, subject to an expensive repair, was to go to Brazil. But the hull was in such bad condition it was not economically feasible to repair it. Instead, another vessel was purchased at Mobile, Alabama, and converted to a drilling tender, Delta 9, by using some of the old tender's equipment. The total cost of the project was estimated at $2.5 million, but the payoff was a three-year contract with the Brazilian government.[13]

Delta Marine, for most of this time, had staggered along, a continuing burden to Delta, but the end was obviously coming, despite the Brazilian contract. The minority stockholders were disgruntled with the exasperating drain of capital, and the firm was clearly in no position to compete with the large offshore drilling outfits which had grown into giants while Delta Marine faltered.

Mark Gardner, who ran Delta Marine at the beginning, assessed the company's major difficulties as being the use of secondhand equipment that could not compete favorably with other companies, and Joe Zeppa's personal reluctance to share his investment, and control, with a larger body of investors.

"This is one time where Joe took in some stockholders other than Sklar and Dorfman," said Gardner, "but the reason for Delta Marine's absolute failure was that Joe built his company on junk. When we started Delta Marine Drilling Company, we decided to build a tender. In those days a tender-platform combination was the way a lot of the offshore drilling was done. We got an old LST that belonged to a company then called American Marine Corporation. Oddly enough, Joe talked to the president of American Marine Corporation one time. I mean, we were down there and he said, 'This is the damnedest junkyard I ever saw,' and it was. This old LST was not worth a damn, and they pawned off a bunch of other junk on us and we put secondhand drilling equipment on it.

"We also had the same company build a submersible barge for thirty feet of water. This was in the early days and we weren't getting out into the deep water. The tender was named the *Joseph Zeppa* and the submersible, the *Chris Zeppa*. Everything on it was mediocre at best.

"This was when everything was in a great rush to get started. I mean, the thing was just snowballing out there [offshore], so you didn't have to make any excuses. In fact, I made a contract for both those units—long-term contracts—long before they were ever done. Even though they were inferior in many, many ways to other equipment. It was just like a rig today. You can hire out anything you have got. But then there came a time when there was a big shut-down occasionally because of controversy between the federal and the state governments on where the hell the line was out there, and everything shut down. Everything in the Gulf of Mexico, offshore or elsewhere, and there were some terrible times. We didn't operate too long, maybe two years, and then everybody shut down.

"This inactivity existed for a long time. Companies weren't willing to go on until this controversy was settled. They didn't know where in the hell they stood. Titles weren't clear and all that. After that, then, things really took off, and the necessity to have really good equipment and a lot of it, if you were going to be in that business, was on you.

"Well, the only way to do it was with public money. That is the way that all of these big offshore companies did it. You can look at all of them. That was public money. Now there is a lot of it being done with these limited partnerships, but Joe would not consider making Delta Marine Drilling Company a public company, and there were no funds to upgrade the equipment to get more equipment, and there came times when there were slack periods—even though the general trend was up, there were always little dips—and, of course, equipment like Delta Marine had would be the first laid off. It was just old, sorry equipment."[14]

Delta owned more than two-thirds of Delta Marine, with the rest of the ownership spread among 30 other stockholders. The company remained in the doldrums because Zeppa refused to make a public offering of stock and therefore was unable to generate the funds to upgrade the equipment, yet he refused to go out of the business. Actually, he could have gone public with Delta Marine and still kept Delta privately owned, but he apparently didn't want any part of the public telling him how to run his business—not an unusual turn of character among self-made entrepreneurs such as he.

At one point he did offer to let employees purchase stock in Delta Marine for about 50 cents a share, but when it became obvious the firm wasn't going to soar, he bought the stock back for about $1.50 so that they could reap some profit on it. But, according to Gardner, Zeppa would not attempt to buy up the stock from the other original stockholders in order to consolidate Delta Marine with Delta for tax purposes. This move would have been helpful to the parent company, for Delta was paying taxes and Delta Marine was losing. Delta would have benefitted from credit for the losses.

"Because of his pride he would not go to these other people and buy their stock," said Gardner. "It certainly went against every business reasoning there ever was, and he was a hell of a businessman. So his pride and ego overrode his business [sense]."[15]

Although economic conditions played their part, Joe Zeppa's personality, as Mark Gardner indicated, was probably the most crucial factor in Delta Marine's vicissitudes.

"Dad was not one to change quickly," said Keating Zeppa of his father. "Once a course of action was established, he pursued that course of action. He was not one to easily change from a committed course of action. And I can site you several business instances where it acted to the detriment of the business and to the eventual demise of it.

"Probably the classical example was the Delta Marine Drilling Company. Dad perceived a certain trend. His perception was correct. That was going to be a new area of drilling contracting. He also perceived that it was possible to publicly finance such an effort. But to my knowledge it was the first and only time any of his business ventures had public investment. Delta Marine was a publicly held company, and stock was sold to the public. [But a public offering of stock was not held, and stock was not traded on the stock exchange.]

"Delta Marine started with what were then current types of drilling equipment—drilling tenders, fixed platforms, inland barges, this sort of thing. But Dad brought to Delta Marine the same techniques that he had used in Delta Drilling Company, and that was to go out and buy some junk iron, patch it up, and put it to work. And he did not do his homework.

"I say this, if I can, in a dispassionate business sense, using hindsight, which is 20-20, but it's clear to see what happened: We went out with old patched-up equipment. I'm quite certain that during those days several of the people that he had involved in it as managers and people like this in the Gulf Coast at the time tried to convince him otherwise, but he was gonna follow his traditional pattern. He went out with old equipment that had been refurbished, but that technique would not work in a marine environment. Just too demanding, and you couldn't afford to break down the way you could on land, and patch up and fix and make do.

"Furthermore, the industry quickly moved into the early jack-ups and, eventually, semisubmersibles, this sort of thing. Dad simply was not prepared to undertake the financing necessary to purchase what at that time was just a horribly expensive device. Those first jack-ups were—gosh, the jack-up itself was $1.5 million. That was unheard of.

"And Dad could not see the trend, and consequently Delta Marine, throughout its corporate life, grew, peaked, and declined, all on old equipment—nothing new and no new-style equipment or different equipment. So, frankly, his inability to change course or to depart from something tried and true was the downfall of Delta Marine. It has had a very precarious existence and [1981] will probably mark its end. We have negotiations going on now and plans are to dispose of Delta Marine's last remaining assets and liquidate the company.

"The total loss to Delta, we'll never know, because all we can do is look at Delta Marine's financial statements and consider that this will be the write-off. But the cost to Delta in terms of agony and preoccupation, loss of use of funds, and people tied up on Delta Marine projects who could have been doing something else . . .

"He didn't perceive the different, absolute distinct nature of the marine operations. But the marine environment was not one with which he was familiar, either by training or background. And apparently he chose not to take the advice of people who had marine experience. So that, I would say, would be a classic example of his unwillingness to change course. Once it was set in stone, he never changed it."[16]

In 1980 Delta Marine had one rig left, a tender-type rig operating in Brazilian waters, the only offshore rig that Delta or its companies had. Known as Delta No. 9, it was the same type of rig as the *Joseph Zeppa* had been, a tender worth from $10-$15 million.

In 1975, Joe Zeppa himself, a few months before his death, summarized the Delta Marine story succinctly and accurately:

"We've got Delta Marine, but we're about to go out of it. The company has not been successful. We actually got into it with the wrong equipment at the wrong time. We own about 82 percent of the stock."[17]

III

The Tyler Hotel Company—the corporate label for the Blackstone Hotel—was in many ways the most persistent problem Delta faced during the period following World War II. It never made money. In fact, it consistently lost money. And though minority stockholders grumbled over the money constantly disappearing down the rathole, Joe Zeppa clung tenaciously to the property. Financially, the Tyler Hotel Company can only be viewed as a burden which Delta endured for 20 years.

The Blackstone Hotel, the result of an idea fostered by Tyler businessmen who formed the Tyler Hotel Company for the purpose of building the five-story, 100-room brick landmark on Broadway, downtown, opened in

1922. E. P. McKenna became president of the company. R. E. Pellow, a Waco hotelman, leased the Blackstone. By the time the East Texas oil boom arrived, ensuring full occupancy, Pellow had sold his interest to McKenna, and McKenna, resigning the presidency of the company, formed a partnership which leased the hotel. With Tyler packed as a result of the oil boom, the Blackstone flourished enough to justify adding rooms to the structure.

Trent Zeppa summarized the economic and social environment:

"The lessee made a *hell* of a lot of money because the East Texas oil field came in just at the right time. The hotel was the greatest cathouse in East Texas for all the old oil wowsers for years and years. For high-priced hanky-panky, the Blackstone was it. *Quality* folks did their business at the Blackstone."

As the oil boom settled down to a steady production of reserves, the hotel business slackened, only to spurt up anew during World War II. The Army contracted with the Blackstone operators to house soldiers training in Tyler, and soldiers from Camp Fannin also patronized the hotel on weekends. With the movement of people and the upsurge in business, the Blackstone received its share of prosperity.

E. P. McKenna died in 1944, and his partners sold their leasing rights to H. Fuller Stevens, who had operated Dallas's Adolphus Hotel. But with the end of the war, the Blackstone reached another slack period.

The Blackstone, a block off the square in downtown Tyler, had been operating for nearly 30 years in 1949 when it was acquired by Joe Zeppa, independent oilman Louis A. Grelling, and Henry Bell, Sr., president of the Citizens First National Bank.

Trent Zeppa summarized the transaction:

"The Peoples Bank crowd was gonna build a new hotel called the Carlton, and Mr. Henry Bell, Sr., did not want to let the old Blackstone go down the drain, so he talked Mr. Zeppa and Mr. Lou Grelling into putting up the money to buy the Blackstone and re-do that, and Mr. Bell was gonna do all the work. Mr. Grelling and Mr. Zeppa were oilmen and couldn't care less.

"Well, Mr. Bell dropped dead of a heart attack. That left Mr. Grelling and Mr. Zeppa. Mr. Grelling's health was not that good, but Mr. Grelling was gonna do such remodeling of the hotel as possible, but he got sick and couldn't, and then he and his wife died in an airplane accident [four years after Bell's death].

"So that left Mr. Zeppa. He set about to re-do the hotel and go into the hotel business. It never worked, from beginning to end."

Upon Bell's death, his two partners purchased his stock from his widow. This left Delta and L. A. Grelling each with 49.45 percent of the outstanding capital stock, with the minute remainder in other individual's hands. After Grelling's death, Zeppa eventually bought, for Delta, the Grelling stock from the estate.

Following Bell's death, Zeppa engaged Henry Bell, Jr. to refurbish the hotel, which resulted in a $1 million overhaul. Adding 30 rooms and a garage for guests, the hotel assumed an air of elegance with a New Orleans motif. In 1954, Zeppa announced to his Delta directors that the hotel was "now, by far, the best in this area and the accommodations compare favorably with any in Dallas." Although he acknowledged no profits should be expected for a few years, he thought the future could be approached optimistically.

At the time he became involved in the Tyler Hotel Company, Henry Bell, Jr. had just finished college and returned to Tyler.

"Mr. Zeppa and Mr. Grelling bought my family out of it, but I served as vice president of Tyler Hotel Company to help Mr. Zeppa with the enterprise, and actually I managed the Blackstone Office Building, as a moonlighter for him.

"Then we went through several years of construction, adding onto the hotel and remodeling it. I had many long evenings with Mr. Joe. He would like to meet with architects at about five o'clock or five-thirty in the afternoon and we'd go on to midnight.

"He was quite a guy. He called one day and he said that he was flying down to New Orleans the next day to the christening of a barge rig, and asked my wife and me to go with him so that we could look for some furniture—antiques—for the hotel. And we flew to New Orleans, went to the barge christening, and then we went looking for big marble mantel pieces to put in the lobby of the hotel."

When neither the Blackstone nor the Carlton proved to be profitable as anticipated, it was suggested that owners of both hotels consolidate their interests and close one of the competing hotels.

As Mark Gardner recalled, "Joe said, 'That's a great idea. There is no reason for us to be losing money. There is enough business for one hotel, but not two.' This was great until the decision had to be reached which hotel they were going to close. Were they going to close the Carlton, which was probably at that time under 15 years old, or were they going to close the Blackstone that was many years older? As long as Joe owned the Blackstone they weren't going to board that son-of-a-bitch up, I guarantee that, even if it cost $20,000 a month."

Around this time Zeppa negotiated for the acquisition of the Blackstone Building, built in 1939, next to the hotel, which was to become Delta's main office building. The board went along with him on accepting the $300,000 price tag on the building, provided a working interest or oil payment could be exchanged for it. However, it was 1963 before the office building had been repaired and remodeled so that Delta could move its office there from the Citizens National Bank Building. The Tyler Hotel Company owned the building, now renamed the Delta Building, and Delta owned the Tyler Hotel Company. Delta leased the Delta Building and used it for offices until 1981.

Throughout the Delta minutes from the 1960s on, the thrust of recurring phrases is evident: "The chairman reviewed the financial situation of the company's interest in the Tyler Hotel Company and the difficulty encountered in disposing of the property." ". . . a substantial operating loss." "All were in agreement as to the desirability of disposing of the property if at all possible, since there appears little prospect of eliminating operating losses in the immediate future." "The operation continues to lose money and there is little prospect of improvement . . . ." A new manager reduced losses but did not halt them by any means, and the trend continued.

The only bright spot was the expectation of incorporating the losses into Delta's income-tax return. By the late 1960s, motels on the traffic loop north of Tyler had cut severely into the hotel's occupancy rate. The future for the Tyler Hotel Company appeared more dismal than ever.

The constant drain upon Delta's coffers periodically provoked strong objection from minority stockholders, who thought that its subsidization should be halted. In one 1973 board meeting, for instance, Dr. Sam Dorfman, Jr., noting a $93,355 loss that year from the Tyler Hotel Company, coupled with an even greater loss for Delta Marine, recommended that no further cash advances be made to either. Dorfman couldn't understand why Delta's management, which had proven so effective in its other operations, continued to allow assets to be eroded by "the historically non-productive operations of Delta Marine Drilling Company and the Tyler Hotel Company."

It was an intriguing point which Dorfman raised. There seems to have been general agreement, even among Zeppa's critics, that he was a financial genius. Yet he kept throwing good money after bad, as his expectations failed to materialize. It was inexplicable behavior on the part of one so skilled in assessing the bottom line potential of a property. One might speculate that, after all, he knew nothing about the hotel business and, having wandered into a field with which he was unfamiliar, made a habit of losing money while hoping to turn the situation around. The greater probability, however, is the factor of personal pride: To have disposed of the property might have constituted an admission of failure, which all his life he had labored to avoid. In the absence of a precise and certain explanation, his behavior in this matter is one of the most puzzling of his carreer.

Delta held on to the Blackstone as long as Joe Zeppa lived.

Trent Zeppa summed up the situation as succinctly as anyone:

"It was a horror story, from beginning to end, except for the lessees. It was just terrible, up until Keating [Zeppa] came along."[18]

IV

While Delta Marine and the Tyler Hotel Company, by virtue of their continuing deficits, assured themselves regular entries in the company's minutes, other holdings, though important, barely received official notice. Among these were Algord Oil Company, Sklar Oil Company, L. T. Campbell Trucking Company, and stock in Premier Oil Refining Company of Texas.

Zeppa purchased controlling interest in L. T. Campbell Trucking Company during World War II. Campbell Trucking had been contracting its services to Delta, and Zeppa perceived ownership of the truck firm as both an economical and practical move. However, within a few years it became evident that contracting for such services was preferable to owning a fleet of trucks. When Joe Beasley returned from the service, Zeppa hired him in 1946 to handle the sale of Campbell Trucking, after which Beasley entered law school. [19]

Other transactions affected Delta's daily operations less, but represented considerable money. Zeppa and his partners on the board of Premier Oil Refining Company—J. R. Parten and Sylvester Dayson—accepted an offer to sell that company to Farmer's Co-op in 1948. The offer was $25 per share for the stock—"a very advantageous sale"—and Zeppa urged the Delta directors at a special board meeting to accept. They agreed and Delta sold its 16,800 shares for $420,000. [20]

The transactions involving Algord Oil and Sklar Oil came in the early 1950s in response to current economic conditions. When the steel strike of 1952 precipitated a shortage of tubular goods, drilling was affected deleteriously. Although the company expected to expand as development programs proceeded in Pennsylvania, Kentucky, West Texas, and the Rockies, Joe Zeppa anticipated that, nonetheless, many Delta rigs would be stacked as a result of the steel shortage. At the same time, risks grew larger. There was the danger that gas blowouts in deep wells with high pressures might produce damage suits. For these reasons, Zeppa found it "highly desirable to separate the company's valuable oil- and gas-producing properties from its drilling operations, thereby removing the assets represented by these properties from the possibility of attachment by judgment."

One step in this direction was the purchase of all the common stock of Algord Oil Company, which was a production company only. Zeppa anticipated that Delta might sell its and other producing properties to Algord, which would carry on as the exploratory and development arm of Delta. He arranged with Douglas Whitaker, a Shreveport oilman who often turned deals for Delta, to acquire certain properties for purchase from Sklar Oil Corporation, for which Delta would pay Whitaker more than $650,000. The end result would be a wholly owned subsidiary, which would lead eventually to separating the drilling and production facets of Delta's operation.

In June, 1952, Zeppa executed a transaction with Allen Guiberson to buy all 40 shares of common capital stock in Algord for $5,000. Thus the corporate vehicle was in place. Delta then sold to Algord properties in Smith, Wood, Gregg, and Hopkins Counties in Texas, along with others in the Lisbon field in Claiborne Parish, Louisiana, as well as assuming a mortgage in connection with them.

The purchase of Sklar Oil proved to be one beneficial to both buyer and seller.

"I transacted that sale to Joe," said Leonard Phillips, Sam Sklar's son-in-law and business associate. "It came about because Sklar Oil Company, a corporation, could only deduct its depletion and pass it down to the stockholders through a taxable dividend, which became uneconomical to do. And having been a lawyer, it occurred to me it was a useless thing—you must remember, oil was $1.65 to $2.65 at that point; were it in the present I might rethink it. Mr. Dorfman and Mr. Sklar decided perhaps we'd better sell it. Thereupon I contacted Mr. Zeppa with the thought in mind that if Delta bought it, it would give Delta additional income to utilize for the purchase of new rigs or the acquisition of new properties. And Mr. Zeppa was very receptive and we consummated the deal."

The properties of Sklar Oil, a production company, were scattered over East Texas, Louisiana, and South Arkansas. Some of the Arkansas wells were still producing from the El Dorado boom. Once these holdings were redistributed, Zeppa dissolved Sklar Oil.

The original plans for Algord Oil failed to materialize, however, and its assets were liquidated in December, 1956, at a time when Delta Gulf was being absorbed into Delta. Algord was not legally dissolved until 1973.

V

As Delta expanded into a variety of operations, the executive hierarchy grew more complex. In 1956 a number of district managers were elected vice presidents in charge of their districts. At that time Guy White and Mark Gardner already were vice presidents, and three division managers were elevated to that rank: E. G. "Eddie" Durrett, West Texas-New Mexico; Thornton Tarvin, Gulf Coast; and H. J. "Red" Magner, Northeast.

After leaving Delta for a few years, Tarvin returned to Delta in 1965 as head of an office in Houston, where his job as a vice president, basically, was calling on the trade to generate new customers and drilling contracts. The Houston office had been inherited from Delta Gulf, whose staff there was moved to Tyler following the Hawkins deal. Personable Tarvin called on representatives of the major companies such as Exxon, Shell, and Continental, as well as a great many independents, including some who drilled only a few wells each year.

But while Tarvin's office might initiate a job for Delta, the actual bidding was left to the managers of the regional divisions in which the actual drilling would be done.

As Tarvin told Joe Zeppa, the division managers "know what they can do, they know the formation, they know how the drilling is, which I wouldn't, so if I had to get out and get the information it'd take me a little longer."

During much of the 1940s and into the 1960s, company planes provided transportation for a great deal of the executives' travel. Jimmy Rider was the pilot for many years. The company started off with a two-passenger plane, later used a four-place Howard, and eventually went to Twin Beeches. Ultimately, however, in 1975, the company sold its last plane and instituted the policy of chartering a plane when business required a trip by, say, eight or ten persons.[22]

VI

During the years Delta was maturing and expanding, Joe Zeppa's personal family—his two sons—were following paths that would take them toward different destinations. After attending grammar school in Tyler, Trent Zeppa left for boarding school—the Midland School for Boys in the mountains of southern California. He returned to Tyler for a year of high school before going to McCallie School in Chattanooga, Tennessee, from whence he graduated in 1944. At 17, with his parents' permission, he joined the Navy and served the remainder of World War II at San Diego as a hospital corpsman.

After release from the Navy in 1946, Trent enrolled in Tulane University in a pre-med program, into which he incorporated study of the classics. But before he completed his degree requirements in 1949, he and Majorie Barr married. That summer Trent worked for Delta on a drilling rig in Wyoming, under one of the two senior drillers of the company, Ark Carter, who was no stranger to him.

"He had known me as a child," said Trent Zeppa, "so he was very careful that I didn't hurt myself nor—he had a brand new rig—that I didn't do anything that could damage that.

"It was nice to get out of the New Orleans heat," said Trent Zeppa. "But I was of slight frame and the wind would blow very hard, and a time or two in pulling pipe, I'd get pulled off my feet and hauled across the derrick. But that was about all. I never had a job where I displaced a regular crewman. Because Mr. Zeppa wouldn't have me displace a regular wage earner on one of the drilling rigs. Mostly I scraped tin and cleaned machinery because Mr. Zeppa was supposed to make an inspection up there and Ark wanted to make sure that brand new Emsco rig was very clean. And it was covered with Black Magic, which is a kind of drilling mud using diesel oil in it in frozen climates. So everything was covered with oil and he wanted it all off before Mr. Joe showed up. So I spent a lot of time that summer chipping Black Magic off the rig, derrick, tin, that sort of thing."

That fall Trent entered law school, the University of Colordo at Boulder. The following summer he once again worked for Delta to supplement the VA allowance for married veterans attending college on the GI Bill of Rights. But when the Korean War started that summer, Trent was called back into the Navy and attached to the Marine Corps at El Toro, California, as a hospital corpsman. When he got out of the Navy this time, he went to Southern Methodist University and completed his law degree there.

In 1954 he went to work first as a law clerk for Blalock, Clower, and Kliewer in Dallas. Then on April 1, 1955, he went to work for Delta.

As Trent Zeppa recounted, "Well, when I came over to Tyler for Thanksgiving in '54, [Mr. Zeppa] said I needed to work to repay the company for its expenditures of money in my behalf, to reimburse him for my care and maintenance since I was young. It was always 'the company,' some inanimate entity that exists, and so I said, 'Fine,' and I did. I agreed that ten years would be long enough, and he did too.

"I didn't have to go, and I did it against Mr. Kliewer's advice and my mother's advice and Jack Blalock's advice, as I knew it was gonna be an utter fiasco for me to try to go to work for him. The thing they didn't understand was, Mr. Zeppa wanted to be paid back for the money he had advanced me for my welfare, and I wanted to make damned sure that he had been paid back and then have nothing more to do with any of it."

As Trent Zeppa explained it, his "reimbursing the company" consisted of working at a lower salary—"about that of a second-class secretary"—so that what he wasn't being paid consituted the repayments. By obtaining a state securities dealer's license and practicing law on the side, he was able to supplement his pay from Delta.

The relationship between Joe Zeppa and his adopted son was an uneasy one, as Trent remembered it. The conflict appears to have been between two highly intelligent, strong-minded men with very definite opinions.

Frequently, during his years with Delta, Trent was the one who ended up breaking the company's bad news to Joe Zeppa.

"My uncle [Chris] would say that all of the company junk was always dumped on my desk," said Trent Zeppa. "And foreign divisions or anywhere else, if they really got ripe, I would end up getting them. Because I wasn't afraid of him. The company politics was such that sometimes when something would go wrong, they would bring it to me, make sure I took it to him because they didn't want to. So I usually got all the guff. Oh, he'd fire me in the morning, and my uncle would hire me back in the afternoon. Not being a masochist, I had enough of it."

Throughout this period, Trent maintained a formal relationship with his stepfather. "Oh, I would call him Pop occasionally," he said, "but he would prefer me to call him Mr. Zeppa, particularly when I went to work for him. It was formal, and he liked it better, and it made it easier for me. I just tried to conduct myself, and did, in the proper deferential manner. But once my obligation was over, I departed. And that was that.

"Finally after the ten-year period, I asked for a pay raise commensurate with what he said I should be getting, but of course he wouldn't give it to me, and so I set about leaving."[23]

Joe Zeppa, as others would attest, was not always an easy man to work for, whether one were related to him or not. However, despite Trent Zeppa's view of his father by adoption, a close reading of Joe Zeppa's personal correspondence during this period produces no evidence that Joe Zeppa held any antagonism toward Trent or considered him differently from a natural son. And, whatever may have occurred face to face between the two men in 1965 when Trent and his family moved to Dallas, it did not interfere with Joe and Gertrude Zeppa's driving over to spend Christmas day that year.[24]

When Trent Zeppa left Delta in 1965, Keating had just returned from Argentina.

"Keating had always had visions that he and I would run the company," said Trent, "and I told him, 'No, never would,' because there was no way I could. And when he suggested that I go back to work for them again, before Mr. Zeppa died, there was no point in my doing it, didn't want to."[25]

Keating Zeppa acknowledged that his brother seemed not to have felt a need to become involved in the company. "If he did feel it at one time, I think it sort of withered, because in later years Trent and Dad never did get along. They really didn't. They were two entirely different personalities. I don't think one *could* understand the other. And so if Trent ever did have any desires to be a part of Delta, I suspect that it was just battered to death. I don't mean that literally, but figuratively."[26]

After leaving the company permanently in 1965, Trent Zeppa worked as a trust officer for the bank about a year, then moved to the West Coast. Following a divorce from his wife Marjorie, he eventually married Nancy James, a lawyer and a daughter of old-time East Texas oil-field lawyer L. L. James, and they moved to Port Townsend, Washington, where Trent died unexpectedly, in his sleep, of a heart attack in 1981 at the age of 54.

VII

If Trent Zeppa found no reason to identify with Delta Drilling Company, it was an altogether different story with his brother, Keating Zeppa.

Though Trent was six and a half years older than Keating—which meant that Trent was already in school when Keating was born, providing Keating with a status similar to that of an only child—the boys did have a number of parallel experiences. Both grew up, for instance, in the midst of a busy household.

"My recollections of my parents tend to be tinged with a feeling that it seemed like they were always going somewhere," said Keating Zeppa. "And that I was left at home with the servants. They were off traveling in Mexico or going here or there. In *fact*, I know that they weren't gone that much. But apparently their absences from home, whether it was for two or three weeks or a month or something occasionally, left an impact on me, because my recollections—they were always gone. I know now they weren't, because it's documented. But I'm left with that feeling."[27]

There is at least one piece of documentation, in a 1943 letter from Joe Zeppa to his old friend George Keating, that young Keating Zeppa, then ten, accompanied his parents on a trip to Mexico. "My trip to Mexico was partly business," Zeppa wrote, "and at the last moment I decided to have Gertrude and Penny [sic] go along. They followed me two days later intending to go all the way by plane, but they were grounded at Laredo and had to continue on by train. We all enjoyed the trip and particularly Penny. My purpose for having him along was to stimulate his desire to learn Spanish and to be given an object lesson as to the need for it." The most memorable sight of the trip was the volcano, then active, at Paracutín; they drove by auto to within two miles of "this spouting furnace," then rode horseback to within a mile of the base of the cone: "It is a sight which not many have the opportunity of witnessing during their lifetime."[28]

Before World War II, Gertrude Zeppa took her children to southern California for the summers, leasing a house for three months at Balboa Island or some other site south of Los Angeles. Joe Zeppa would join them for a few weeks near the end of the season, and then they would return to Tyler.

"Those were enjoyable days," said Keating Zeppa. "Consequently, I grew up, in many respects, around salt water. And that's where I developed my love of the water, of the ocean, sailing, the sort of thing you normally don't associate with East Texas."

When World War II came, travel difficulties as well as the West Coast's being a potential target for Japanese attack caused Mrs. Zeppa to shift the family vacation site to Santa Fe, New Mexico. "It was the nearest cool spot Mother could find," said Keating Zeppa. "So I feel very comfortable in and around northern New Mexico and also in southern California, because of each particular experience."

Keating Zeppa's childhood relationship with his Uncle Chris was based on infrequent visits. He did not get to know Chris well until he was older.

To Keating, Uncle Chris was one who supervised some fascinating machinery in Arkansas.

"Occasionally Dad would go up to El Dorado," said Keating Zeppa, "and once or twice I think I tagged along. He'd go up and visit Colonel Barton. I do recall specifically the one visit to El Dorado. I don't remember much about the trip except the awe with which I was struck as my Uncle Chris took me around the refinery there, the refinery he was running for Lion Oil. I guess I was nine or ten. It's the only thing I remember about the visit to El Dorado—going up there and Chris taking me all around.

"Chris was always very patient with me. There weren't many occasions, but he took time to take me along and explain things. And, really, Dad very seldom, if ever, did that. He was too busy.

"Actually, Chris and Dad, for all their likenesses, were very different in their approaches. And so it was not as though, for example, Chris and [his wife] Gustine were over at our house a lot, or we over at theirs. They were both sort of independent people who went their own ways. There wasn't a lot of this, you know, Italian togetherness and back slapping and singing and stomping grapes. They were two hard-working men, doing their thing. They were *close*, yes—I mean, they were very fond of each other. To this day, I think Chris probably respects his brother Joe as much as any person he's ever known. Dad had a very deep and abiding affection for him. Dad would just eat Uncle Chris out something fierce sometimes, but it was always with a fondness and affection, and I think he always looked upon Chris as Little Brother—you know how big brothers treat little brothers."

Keating Zeppa attended elementary and junior high school in Tyler, and was generally a good student who had no difficulty making As.

In 1946, at the age of 13, he went off to McCallie School in Chattanooga, where his brother Trent had gone before him. A friend, Richard Morris, the son of "Buck" Morris who was running the East Texas Salt Water Company, was his roommate. Academic standards at McCallie were much tougher than young Zeppa had been used to in Tyler.

"I would not say I was a particularly brilliant student. I think I probably for three years maybe had a B average or low B average, but I can guarantee you that I learned a lot that I never would have, here, in the public school system, if I had been a straight A student. Just more demanding academically."[29]

Joe Zeppa was equally impressed with the school. "It is the type of school which makes a dollar go a long way," he wrote George Keating. "It does not cater to the rich, and takes in a great many boys at greatly reduced rates who cannot afford the modest board and tuition of $1,000.00 per year. This is done in a manner which does not permit the students to know which ones are there on part scholarship. The school stands very high academically and is run by men who are not ashamed of the Christian faith, and it still believes that the Bible has not gone out of style."[30]

In the summer of 1945, by which time Keating was 12, he had a Social Security card and worked for the first time as an office boy for Delta, every afternoon, doing routine filing, running errands, sharpening pencils, and general "go-for" chores. Since Joe Zeppa had once been an office boy at Shearman and Sterling, he apparently deemed this job a logical place for his sons Trent and Keating to begin.

After two summers in the office, Keating worked under B. W. Freeman in the transportation department at Kilgore one summer. He stayed with the Freemans, sleeping at their house, and took his meals with Homer Lee Terry's family, who lived at the Kilgore yard. The following summer, 1948, he worked in the welding department for Joe Blasingame and stayed part of the time, and took some of his meals, with production supervisor Gus Jones and his family at the Delta camp.

As a summer employee, Keating had a small role in building the gasoline plant at Chapel Hill. Delta supervisors had been instructed to treat Keating like any other hand, and most of the time they complied. By this time, the compressors were already running in the gas plant, and Keating and a fellow summer employee had been assigned the chore of painting inside the compressor building.

"It was the middle of summer and it was just hotter than the hinges of hell up in the top of that building, painting those big mufflers and this sort of thing," said Keating Zeppa. "And both of us, I think, were probably red in the face and about ready to pass out. So we climbed down out of there for a few minutes to cool off. And *just* the minute our butts had hit the concrete and we were beginning to fan ourselves, Bob Waddell comes around the corner. 'Corking off, huh? Well, we'll see if we can't find something easier for you to do.' So he got us a couple of 'yo-yos' [sickles] and he sent us out to cutting Johnson grass on the flare line right-of-way out through the jungles to where the flare pit was. Well, the temperature was about the same, so the only difference was the smell of Johnson grass and a few extra ticks."

Upon completion of high school in 1949, he attended a 12-week summer session at the University of Colorado, instead of working. That fall, at the age of 16, he enrolled at Rice Institute. Though his intellectual capacity could handle the workload at Rice, emotionally and psychologically—he realized later—he was too young to face the challenge. The summer of 1950 came as a welcome respite; he went to Europe on a student tour and, at 17, for the first time visited the old Zeppa hometown of Fubine.

"I remember my Aunt Pauline, or Paola, Dad's sister, met me at the train in Milano and took me down to the farm for a couple of days. Generally speaking, I remember sort of being fussed over and made to feel that, you know, I had three heads and six legs. But it was a very pleasant experience. Then I rejoined the group in Rome.''

As he struggled into his sophomore year at Rice, he came down with pneumonia, and after a few weeks in Hermann Hospital in Houston, falling behind in his studies, he dropped out of school upon release from the hospital. In January 1951, he enrolled at The University of Texas at Austin, in petroleum engineering. By that time the Korean War was in progress, with college students increasingly under pressure of the draft.

"In evaluating the risk of being drafted into the Army," said Keating Zeppa, "I decided that maybe the better part of valor was not discretion but to charge ahead, so I volunteered for the Navy, for aviation cadet training."

The decision to join the Navy in March, 1953, represents a turning point in Keating Zeppa's life. It was his first opportunity to stand or fall on his own ability, and the reassuring memories of his years as a cadet and naval officer were to nourish him emotionally from then on.

His choice of the Navy was perhaps influenced by his brother Trent's having served in that branch, and his own love of the ocean since those summers in California when he was a small child. But his desire to enter aviation cadet training was merely the logical extension of his interest in flying as an early teenager.

"I started flying in 1946," said Keating Zeppa. "I was 13 years old. Jimmy Rider, who was the company pilot at that time, had just come to work for us; he had been an instructor during World War II, and I had expressed an interest in flying. And Dad, of course, in his younger years [in California] had taken flying lessons, and apparently he didn't object to Jimmy teaching me how to fly. Which I shall ever thank him for, because as it turned out I got my private pilot's license when I was 14, which was legal in those days. But the problem was, in Texas you couldn't get a driver's license till you were 16. So I had to have someone drive me to the airport so I could fly.''[31]

Within a few years he had become a skilled enough pilot to fly considerable distances, all without mishap. In January 1951, a few days before his son's eighteenth birthday, Joe Zeppa wrote George Keating: "Pinny flew off a couple of hours ago by himself in our BT plane for Austin and, weather permitting, then to Houston and home. He seems to be very confident in his ability to fly and navigate, and I hope everything goes well. Our pilot, who has been training and coaching him, seems to feel that he is qualified to handle the plane.''[32]

By the time Keating Zeppa entered the Navy he not only had a pilot's license but had logged perhaps 1,000 hours of flying time before beginning flight training.

"*But*, for some reason or other, I wanted to get some experience in water flying, in the flying boats," he said. "Granted, that's sort of like learning to ride a dinosaur just before their extinction. So I guess it's a bit of almost archaic knowledge that I have that no one will ever have anymore, and that's flying big flying boats on water. A lot of fun! Enjoyed it and it probably—that four years in the Navy—gave me the chance to develop a self-confidence that I had never had at any time of my life prior to that, because I found that there was one thing, at least, that I could do better than anybody else, around me—that I could fly.

"I was not in my father's shadow. Nobody knew a Zeppa from anybody else. When I finished training, the 18 months of flight training and everything else, the day before our class was awarded our wings and our commissions as ensigns in the Navy, my training squadron commander said, 'Zeppa, I'm sorry to tell you, but you were aced out by less than two points in the overall training scores, from beginning to end, in taking honors for the highest scores ever made in the Naval Air Training Program'—by a lieutenant (j.g.) in the Coast Guard who had come back for flight training. He was a heck of a fine guy, really fine. And you know what is funny? I didn't feel a bit bad about that, because he was really a super guy. By then, you know, I didn't need that score to tell me I knew how to do what was being done.

"I think probably that was a real turning point in my life, the four years in the Navy. It gave me self-confidence. It gave me a tremendous opportunity to work with people, that you don't get in college. And with people all up and down, and in an atmosphere, or ambience, in which my background and my name didn't mean a damned thing."

Following basic training at Pensacola, Florida, he took advanced training in sea planes at Corpus Christi. Thereafter, he was based in San Diego, but the squadron was periodically deployed to Manila, so he saw duty in much of the Pacific. In 1955, for instance, he telephoned his parents from Hong Kong and Tokyo.

Looking back over his time spent as a pilot, he sees it as a positive experience.

"I think my life-long love of flying has provided me the security of knowing that, while I might bump my toe here and there, my experience in the Navy gave me, without question, a sense of self-worth that I had not obtained in school up to that point in time, and it had never particularly been heaped on me by my parents. It was always 'Strive higher.' But I found that in flying I could excel, that I'm more at home in an airplane than I am in an automobile. First time I'd been out from under the Zeppa name. I was just another person. And I found, as just another person, I was acceptable as a human being and could actually do something pretty well.

"But when the four years were up, I also knew I didn't want to make a career of the Navy. I was discharged in March of '57 as lieutenant (junior grade) and came back to UT—Austin—in September '57."

While at The University of Texas he married a fellow student, Eleanor Packard, from Dallas, in January 1958, approximately 16 months before his graduation. Eleanor, or Ellie, was studying scene and costume design in the drama department. Her father was a geologist with DeGolyer and MacNaughton. Their first child, Christopher, was born in December that year.

In June, 1959, he graduated with a degree in petroleum engineering and "showed up for work," as he put it, with Delta Drilling Company.

"By the time I got out of school there sure wasn't much demand for petroleum engineers, I can tell you!" said Keating Zeppa. "The domestic industry was starting down. My job offer from Delta was $500 a month. That was a very handsome salary. Because I recall of our graduating class at UT, I think the top offer that anybody received was $585 from Humble, and there weren't that many job offers for anybody. In '59 nobody wanted petroleum engineers. So John Vinzant, who was a classmate of mine, came to work for Delta at the same time. We both felt pretty good about getting $500 a month."

As far as Keating was concerned, there had never been a question as to what he would do upon graduation.

"It just never occurred to me that I'd do anything else, other than work for Dad. I never realized I had any options. And I don't say that with any bitterness. It just never occurred to me. I mean, part of it was, I had the feeling, right or wrong, that Dad would just have been horrified if I had gone to work for Humble or some other organization—he probably would have considered I was an ingrate. I don't know if it was *true*. My perception was probably false, but that's my recollecton of the way I felt."

Although there was no agreement or discussion with his father on these matters, as the son said, "I wasn't told *not* to. And, again, I never was that close to Dad, really. It was just sort of an assumed thing."

Keating and Ellie moved to a small rented house in Tyler, and he went to work for Carl Haskett, the senior engineer in the production department. His career commenced with a great deal of field work, carrying out routine engineering chores such as well completions and well stimulations.[33]

He subsequently went to Argentina to represent Delta there in the 1960s, returning to Tyler in 1965 to become a vice president and to rise, when Chris Zeppa retired in 1968, to executive vice president.

VIII

In the late 1950s and early 1960s, domestic drilling lost much of its promise, as there were fewer contracts to divide among the competing drilling firms. Rig counts went down in the industry. By then, most of the action had moved overseas in search of more plentiful, cheaper oil. This slack period, ironically, followed close on the heels of a high point in drilling in the middle 1950s.

"That broke a lot of people, because they went into the drilling business[when it was booming], and it didn't last long enough for them to pay for their rigs," said Early F. Smith.

It also had unfortunate repercussions for Delta and its personnel.

"I've seen the time when the company would put rigs to work on a losing contract, just so that their crews could make some time and keep their seniority with the company—their longevity so they could still be eligible for their group insurance," said Smith. "The reason I would know that they weren't making anything is because my main function for a great deal of the time was accounting for contracts and doing the billings, and I knew what the rigs cost to operate, and I knew they couldn't make any money.

"I said to Chris Zeppa, 'We can't make any money on this.' He said, 'We *know* we're not going to make any money on it, but that iron's going to set out there and rust and those boys are hungry, so we'll just let them make a few weeks' pay.'

"In East Texas we'd gotten real bad there for a while. I've seen the time in this area we wouldn't have a rig running out of eight or ten rigs. Course, they'd drill enough to keep their leases up, active, and if it was a pretty good looking prospect they wouldn't let the lease expire."

"I can remember Mr. Joe telling Eddie Durrett to take jobs just to put the rigs to work," said Lynna Jo Edwards, in the Odessa office. "I mean, to make your overhead and depreciation, but work the rigs. Break even, but keep the men working and keep those rigs running. So it's been real tough. I mean, we have seen very hard times in West Texas, and it's a very competitive area."

During one four-month period in the early 1970s only one rig—No. 36—was running in the entire West Texas division.

The company had a policy of keeping the toolpushers on the payroll. When times were tight, they might have an option of overseeing more than their normal run of rigs or of taking a job as a driller until things brightened up. Drillers, bumped by the toolpushers, might continue working as roughnecks.

"I remember one time," said Ed Gandy, "out here at Wright Mountain, all three of my drillers were ex-pushers. Most of the men, if Delta didn't have a rig running, why, they'd try to hire out to somebody else if they could and come back with Delta when Delta was hiring. The reason I didn't leave and go back to the company I had worked for, Delta had much better benefits. At that time they were paying us a bonus at the end of the year, and others weren't doing that. Delta had come up with the paid vacation. That was a problem with independent drilling contractors."

"I've seen the time around here when we had a solid crew of drillers," said retired paymaster L. D. Hoppers. "Keep the drillers working so they'd come back."

All the employees personally felt the pinch in the drilling industry.

"We had a celebration when we'd get a job," said Warren Strahan. "I mean, we'd all holler and cheer, you know. We were going to work. During the hard times we had eight or nine rigs, I guess, [in the South Louisiana division]. Most of them stayed stacked all the time. But when we'd get a job, that was something to celebrate. One job. Hell, that put beans on the table. Jobs were hard to come by."

At one point the drilling industry's future looked so dire that Warren Hirsch, the company's safety director, suggested to Joe Bevill that the salaried people at Delta "plow back" into the company, perhaps as a loan, a portion of their salaries until conditions improved. "And Joe double-talked me out of it," said Hirsch. "He said, 'Warren, you'd break Mr. Joe's heart if you mentioned that to him. He wouldn't have it because he knows all the people on the payroll need their salaries for sustenance.' I'm sure Bevill was right; had it ever gotten to Mr. Joe, he would have objected strenuously. Because his credit was always good, but, you know, you just respond in kind to good treatment."

But while many people were forced to sell out and leave the drilling industry during these years, Delta took advantage of the opportunity to buy good used equipment at bargain prices. Chris Zeppa did most of the buying, often purchasing equipment that the company had no immediate need for. Most of the top executives attended their share of auctions.

"From 1957 up until I retired in '63," said Guy White, "I spent just about half of my time going to auctions where contractors were going broke, and we were buying that equipment, and it was good equipment."

Warren Strahan, who worked over a lot of the equipment in South Louisiana, was bemused by the policy—and impressed by Chris Zeppa's trading skills. "We had 'em stacked and they were buying more rigs. I couldn't understand why they did that. But I do now. They were planning on the future. In the long run, it proved itself out.

"I saw Mr. Chris, now, he traded a great big old pump in and then turned right around and bought the damned old thing back from the same people that he traded it in to, for half the money he had traded it in for. Now I saw him do that. He made money on it. He was a shrewd trader. He did all kinds of things that I consider ordinary people wouldn't never thought of."

There were advantages to having the Delta operation spread over a number of divisions in different regions of the country. "Business would be slow here," said L. D. Hoppers, "but it would be real good up in the Pennsylvania area or in West Texas. They never were all down at the same time." At one point, Delta bought two rigs that cost less than $100,000, both of which were put to work in the busy Northeastern district. Subsequently another one was shipped there from Louisiana.

Although Delta was large enough to have operations in many regions and foreign countries, which helped the company survive, its domestic oil and gas properties deserve most of the credit.

"I think what sustained Delta Drilling Company back then was the fact that we had production," said geologist Jim Ewbank. "We had income at all times, as compared to other companies that went broke because they had no income from production. If all their rigs were stacked, why, they had a zero cash flow, as compared to Delta, which had some real good production. Mr. Zeppa was quite intelligent in getting production as a 'back-up cushion' for Delta."[34]

IX

"Drilling is kind of the glamorous part of the oil business," said Roger Choice, a pumper; "a lot of people don't know much about production, and yet it's very important and it's complicated."

The actual production of oil and gas—the pay-off, after all—continues to be overshadowed by the search for those minerals. As anticlimactic as the production stage may be, it is the goal and the justification for all that precedes it. And though it may appear to be a cut-and-dried procedure, there is much more to it than meets the eye.

Most people never see the equipment that moves the petroleum. A tourist driving through a field sees only the visible symbols: wells, like gigantic metallic mosquitoes, rhythmically sucking hydrocarbons from the earth like perpetual-motion machines—which oilmen wish they were.

Probably the most unsung worker in the oil patch is the pumper, who sees after the wells. As essential as his job is, he is likely to go unnoticed as long as things go smoothly. But, invariably, cold weather brings him special problems, wherever he is situated. In the early days a pumper might have to thaw a frozen line with hot water from a teakettle. Or, if a forecast of cold weather arrived in time, he might, depending on the monthly schedule, shut the wells in until warmer days.

"There were lots of times that I had to be around at all times," said Joe Craddock, who worked in East Texas. "I'd go from one to the other—thawing out and keeping it going. I don't know how I managed it right now, with the equipment we had then. I'd check my wells all hours of the night."

Today, though a heater may be on each line, a night visit to the field is often necessary in extremely cold weather. Chemicals may have to be injected into the line to prevent freezing, and the expensive chemical, unlike the antifreeze in one's radiator, eventually must be replaced. In the Northeast, Melburn Miser instituted a system of placing alcohol bottles on the well so that the alcohol would drip into the lines and prevent freezing. The alcohol bottles go onto the lines around the first of December and remain until March.

Although water moving through the lines along with oil and gas causes most of the concern in the winter for a pumper, paraffin can also produce serious trouble.

"It'll plug your line plumb up," said John Robert Gunn. "And then you have to get a hot-oil truck and pump the hot oil down your line. It's heated oil, up to about 300 degrees. After it's heated, the paraffin will go to your tank. Over a period of time you will probably have to get the tank cleaned."

Oil and gas wells have individual personalities.

"There's not any two wells that would flow alike," said Joe Craddock. "They was just like an old mule. They have different habits, and if you just stayed with it enough, you'd learn the habits. How the formation gives it up, you know; you could determine about what size choke you could flow it on to the best advantage of the well."

In production, as in drilling, improved equipment has benefitted the employees. "Years ago," said Gunn, "you didn't have the facilities you have now to work with. We didn't have tongs to break pipe with or anything. We had to do it by hand. It didn't get too bad or too cold to pull one of those old wells. You had to beat the ice off a ladder and go on around."

W. L. "Buster" Medlin remembers a few decades back when he was in charge of 12-14 wells in the Quitman field. "We'd have to look after that lease seven days a week, ten hours a day. We might get off 30 or 40 minutes early, but no more. Most of the time they'd have you loaded with so many wells it would take a full eight hours or more. If you didn't have no problems it would be a real easy eight hours. It was just the way you had to work back in those days. In those days you would have to do it or somebody else would be waiting on your job.

"I noticed a lot of changes," he continued. "Engineers came in and brought things out where it was made easy on the working class of people. Out there where you really have to dig the ditches and lay pipe by hand, that has changed. New inventions came along and it improved by a hundred percent."[35]

X

A great deal of the production Delta acquired over the years came as the result of taking interests in the wells the company was drilling. Usually the deals came from independent operators who were primarily interested in keeping their dry-hole costs down if they failed to find oil. So the main figure these independents were interested in was their dry-hole cost.

"They didn't care too much about what it was going to cost them after that," said Bob Waddell. "If they had a good well, they could afford to run pipe. The main thing is dry-hole costs. Then you'd tell them an estimate of what it would cost, then they'd go out and sell it on that basis, probably add some to it. They'd say, 'Well, do you want part of it?' Then we'd come back and say, 'Well, we'll drill for x number of dollars and take a quarter interest in it.'

"A quarter was just about the most. We took an eighth or a quarter, get enough out of it to cover our costs, you know, cover our quarter. If it was dry, we didn't make anything, but we didn't lose anything. We didn't have the risk. Hedging our bets. But that's where we got into quite a bit of production.

"Usually that was handled through Joe Zeppa. Most of the time they talked to Joe Zeppa, because Chris was more conservative than Joe. I'd take the guy in there with his maps and all these pretty pictures and [he'd] show exactly what it was, and Joe Zeppa'd look at it and ask a few questions. 'Bob, what do you think about it?' It didn't take him long. He'd either say No or Yes. That's one thing you could get out of him. Joe had pretty good judgment about it. Most of the time he had."

Trent Zeppa described his father's policy as a no-lose approach. "He would always try to make the deal pay for itself in dry-hole contributions, plus return a profit to the drilling. Very seldom would he scavenge his profit as a drilling contractor in order to participate in a deal. And he was a superb salesman. He could talk people into it."

Was he honest? "Yes!" said Trent Zeppa. "Yes, oh, very much so."

And shrewd? "Very much so! Very difficult."

Did he still maintain the respect of people? "As far as I know," said Trent Zeppa, "though he would make people awfully mad."[36]

XI

As a company that had started out with secondhand equipment, Delta continued to lean heavily on used materials through World War II.

"When I went to work for them [in 1945]," said Bob Waddell, "the first [company] car I drove was an old two-door Ford they bought in New York from some old lady that never drove it any further than the grocery store. They'd go up in the East and buy those Eastern used cars and bring them down here. And that was all we had. A year or so later, well, we started to buy new equipment and new materials and new cars."

Although more new equipment was purchased after the war, used goods remained common for years. The main thing that held Delta back from the top ranks of drilling contractors, in the opinion of Mark Gardner, a former vice president, was that the company used "pretty much poor equipment," which hampered its operation, "in spite of anything Joe Zeppa would say." Gardner added: "I had one of the old hands tell me one time, 'Mark, I have been working for Delta Drilling Company for 25 years, and the biggest single new piece of equipment that I had at any time was a three-foot length of rotary chain.' "

While this experience may not be typical, it does dramatize the problem that many employees faced.

"We had a nickname here, Delta did," said Dallas Bryant, referring to the tri-state district. "Scrap Iron and Drilling Company. It was a joke with everybody. But they were a good company to work for. They treated us good. Pay was always on time. That's what we were looking for."

When Guy White took over in the tri-states region, World War II had just ended, and parts and equipment were especially scarce. White described the problem up there to Joe Zeppa.

"Now I know that you don't know about this, but Delta Drilling Company is known to have the best men in that area, and the sorriest rigs, and it's not Delta Drilling Company's policy to have sorry rigs. And I want to have the authority to buy anything I need to run the organization up there." Zeppa, he said, gave him what he asked for. At that point, operations in the tri-state district were to continue only a few more years.

Warren Strahan in 1980 recalled the situation in the 1960s in South Louisiana. "The lowest point in my whole Delta career—now maybe I shouldn't tell this, but it happened—we were running casing on one of our rigs, and I heard a casing outfit say, 'Well, we had to come back out to Junk Drilling Company.' Now that hurt my feelings. We worked hard keeping that stuff running. I'd admit it was in bad shape, but, hell, we couldn't do any better at that time. I thought about it many a time. Calling my company Junk Drilling Company."

Almost everyone who has worked for Delta for many years has an anecdote about the condition of the old equipment. "When I went to work for 'em, they had some wore-out tools," said driller James Wylie Smith, who worked mostly in the East Texas district. "We was setting up the boiler and had an old wrench that wasn't no 'count. It slipped. Mr. Joe's standin' up back here behind. We didn'y know he's out there. This old boy let out a big cuss word and said, 'You Joe-Zeppa-looking so-and-so!' Only he didn't use those words. He was talking to his wrench. Mr. Joe didn't say a word. Next day we had some new wrenches out there."

Though the workers of each district tended to perceive themselves worse off than the other divisions, apparently old equipment was a part of every operation. "I always felt like West Texas and the Rocky Mountain divisions was kinda like the black sheep of the family," said Charlie Martin.

"We kinda got whatever everybody else had worn out. We had to put it back together and run it some more."

"We didn't have too good drill pipe, for some reason," said J. C. Johnson, a retired West Texas toolpusher. "It seemed like it wore out or petered out on the deeper-hole rigs. We usually had quite a bit of problems with the holes, and drill pipe would twist off and what-not. We did manage to get 'em down."

As noted earlier, Delta bought a lot of used equipment at auctions during the slack period of the late 1950s and early 1960s with an eye to the future. Hi Cole, purchasing agent for the company then, recalls that Delta representatives attended a great many sales. Out of it might come a good rotary table for a specific rig or a bigger or a newer mud pump for another rig. The auction purchases were always aimed at upgrading the equipment, at economical prices.

"We bought whole rigs as well as individual pieces of equipment," said Chris Zeppa. "I bought several whole rigs. Some of them were old, some of them were younger. A whole rig in those days meant the derrick, the drawworks, the engine, the pumps. Some of them were almost new.

"We were changing the old rigs we had. We were improving our machinery. And we were taking advantage, cost-wise. We bought a lot of rigs we knew we didn't have any work for, but we figured sooner or later this oil business was gonna move, and so it did."

Coincidentally with the advent of the Arab oil squeeze of 1973, Delta began upgrading its equipment. Though not in direct reaction to the crisis, since the process started before the international event, the program proved to be timely.

J. C. Johnson, a toolpusher at Odessa, summed it up:

"I noticed over the years [that] for a long time you didn't see a lot of improvement in the equipment, but the last three or four years I worked for them [early 1970s], you could really see the changes then. We were converting all the engines and everything down to Cat diesels, getting new drawworks and what-not. It was really looking up."

By the latter part of the 1970s, Delta's equipment situation had changed dramatically. Warren Strahan, recalling the days when some people called Delta "Junk Drilling Company" in South Louisiana, said, "But they don't tell us that any more, friend, I guarantee you. They're privileged to come on one of our rigs, because, boy, we got some good ones." And Charlie Martin, in Odessa, added: "As far as the equipment now, we've got as good equipment as anybody in West Texas." The pattern could be applied to the other divisions as well.[37]

XII

Ultimately, any company—whatever its financial statements reveal—is only as good as the people in it, at all levels of the organization.

Bob Waddell put it plainly when he said, "If you don't have the people, you're just lost. Your people can make you or ruin you. You've got to give them the credit. If they're good, they make you look good. If they're not good, you can't possibly look good."

This effect continues all the way from the highest echelon of management to the roughnecks and roustabouts, for they ultimately determine the quality of the service the company provides.

"You stop to think," said Bob Waddell, "today you've got a rig out there that cost you $6.5 million. Of course, they didn't cost that much years ago, but it was relatively the same. A rig that cost you $1 million 10 or 15 years ago is close to one that costs $6 million today. You've got, say, $6.5 million invested in that equipment; you might have another $2 million invested in that hole in the ground, and you put a man on there and turn it over to him, especially from midnight till six or seven o'clock, eight o'clock in the morning, by himself now, and you've got to have confidence in a man like that to handle that job. It's a high risk. And that's what goes on every day, today."

An eagerness to do the job assigned was an integral part of life to those employees. "We were products of the Depression, and all we'd known was hard work," said Eddie Durrett, who headed up Delta's West Texas division in the 1950s until the early 1970s. Many of those hands came off the farm. Willis Howard's father-in-law told him, "Well, you've swapped the cotton patch for the oil patch." Bert Gauntt chopped cotton for 50 cents a day before starting out roughnecking at $6 a day. "That was more money than I'd ever seen," he said.

Some came to the oil patch, and Delta, by indirect routes. Glen Koonce as a boy of three left Mount Vernon, Texas, with his family for Oklahoma in a covered wagon, with a cow tied behind. They crossed Red River on a ferry. Koonce's first oil field experience was as a trucker. Later he talked a driller into hiring him as a "weevil" hand, and sometime later got his first drilling job when a toolpusher mistook him for another man who was an experienced driller. "Boy, I was so thrilled, I never slept a wink all night," he said. "I was wondering what in the hell I was going to do with a damn drilling job. I never had as much as made a connection on a drilling rig." But with the support of his crew he made a go of it. Years later, in 1958 when Frankfort Oil Company sold the rig he was on to Delta, Koonce, a toolpusher by then, elected to go with the rig.

A lot of Delta's hands came from backgrounds similar to that of Hugh McKenzie, who ended up spending 35 years with the company before he retired in January, 1978. A farmer who had also strip-mined coal near his hometown of Alba, in East Texas, McKenzie joined Delta as a roughneck after a variety of jobs: driving a road grader for a county commissioner, working in an airplane defense plant in Fort Worth, and two other brief jobs on rigs. Two weeks before he went to work for Delta, he hired out to another drilling outfit—"I was fresh off the farm; I smelled like crabgrass"—but soon quit. A friend recommended McKenzie to Nathon Bacle, who was drilling for Delta in the Quitman field in East Texas. The friend told Bacle, "I know an old boy [who] chews the right kind of tobacco." McKenzie and Bacle chewed Day's Work tobacco and worked together for the next 15 years.

It was a happy moment for McKenzie. He had always wanted a roughnecking job since the time, as a boy, he had gone to a rig with his father. "That's the most fascinating thing," he had thought at the time, "and it's right here in Alba." The rig had an old coal-fired boiler.

McKenzie was 31 years old and Delta ran 15 rigs, all steam, when he went to work for Bacle. He soon received an unexpected initiation.

"We'd been there about four days," said McKenzie, "and we was shorthanded, of course. And Nathon got ready to come out of the hole, and he said, 'You get on them blocks there [in the derrick].'

"And I said, 'Who, me?' Because I never had been no higher than gathering corn and no lower down than digging potatoes.

"I said, 'Man, I can't make it up there.' He said, 'Yeah, you'll make it.' And I think we pulled out 48 strands [of pipe], and it didn't take us over three hours to get them out. And when I came out of that derrick, the sweat was running out the top of my boots. I was scared. I couldn't turn nothing loose. I lacked one hand having enough to work with."

Oil-field work soon "got in the blood" of most employees. Claude N. Clark, who gained the sobriquet, "Bollie," from having pulled so many cotton bolls in West Texas, experienced an early fascination for the oil fields and soon departed the farm for jobs with Sinclair, then Texaco. When he returned to farming for three years, he yearned to see the oil fields. "When I was farming I would get so hungry to see an oil well I would drive my tractor up in the corner and get in my car and I would drive over there about 15 miles to those oil wells and stay over there and look at them." He left the farm that time to go to work for Delta Gulf's production department, which later merged with Delta.

Hard work was the first thing a man learned around a drilling outfit, particularly in the 1930s when almost everything was done by hand and sheer brute strength. Red Magner, as a college boy working as a roughneck in the summers, had a severe test during a cementing job on a well, in which he found himself, as a weevil or novice, doing the hardest part of the work. Sore and exhausted, he worked his way through the acre of cement sacks, conveying them to the hopper. He hadn't shaved that morning and sweat was pouring off him, with cement caked on his face. The superintendent, Claude Johnston, took one look at him and said, "Red, by God, you'd better get an education, 'cause you'll never get anywhere on your looks!"

Even education did not exempt one from heavy work—nor from the jocular taunts of one's co-workers. Bobby Payne started out on a rig in West Texas. "He came out of college and roughnecked on my rig, with an engineer degree," said J. H. "John" Gilleland. "I used to walk by him and say, 'You just can't beat that college education!' He would be covered with coal oil. I knew he was going up, though." Payne eventually became manager of the division.

Delta attracted dutiful practitioners of the work ethic. "I liked a lot of work," said Claude N. Clark, who pumped in the West Texas division. "If I wasn't working, I wasn't satisfied. So I was working seven days a week." He had joined up with the right outfit, for, as Raymond Lewis, a roustabout mechanic in Kilgore said, "Delta doesn't run out of anything [to do]."

Some men were so habituated to physical labor that they found it hard to slow down when they were promoted to supervisory positions. "It didn't make any difference what I was going to do, I went after it and worked," said Nig Spraggins. "Well, I got my drilling job over there on Rig No. 17 on the morning tour and Hugh McKenzie was [the driller] on the evening tour. When we got ready to start rigging down, he got me out to one side and said, 'Now, listen, boy, you are a driller now and you ain't supposed to work like that.' He was giving me some good advice. He didn't want me getting out there working and showing him sitting around chewing tobacco."

Loyalty was a constant among the long-time employees. W. O. "Shorty" Meyer might be seen as a prototype of the loyal Delta hand. He and his wife Georgia Belle reared a family of 11 children on his wages. "Delta raised them," said Meyer. "I worked 29 years and I didn't miss a day from being sick. I have worked a few days when I thought I might have had the flu, but I went ahead and worked. One time I might have had to have lost time because I hurt my back. I don't know how I hurt it, but I got where I could hardly walk. But in place of laying off I took my vacation, so that way I didn't lose no time working. They told me I didn't have to do it that way. I told them I had rather do it that way, 'cause something else could happen. I didn't believe in taking advantage of anybody. I never was late on the job, but you have to give my wife credit for that."

The loyalty was based on faith in the men who ran Delta, as Fax C. Marshall illustrates. Marshall had been offered a job with another company during a slack period when his Delta rig was down. He went to see Chris Zeppa, who said, "Well, Marshall, we know you, and other people don't, and I think you'd have a better chance with Delta." Marshall decided to take Zeppa's word for it. Two months later he was a driller.

Invariably, more than loyalty to the company was at work, as the men tended the drilling rigs in all kinds of weather. Sandy-haired, red-faced Eddie Chafin, a chunky fellow who claimed he'd been in the oil fields all his life, was one of the best roughnecks that ever hit a rotary. "He could take what we call a weevil roughneck," said Edward M. Gandy, "and you could put [that weevil] on the floor with Eddie Chafin, and Eddie would make a roughneck out of him in a week's time. He could work anywhere on a drilling rig, but he wouldn't drill. He didn't want the responsibility." Chafin was roughnecking in the Opelika field in Henderson County just west of Tyler in the dreary, cold, wet winter of 1944. Heavy rain had transformed the oil field into sheets of cold mud, and mere rubber boots didn't do much of a job of protecting the drilling crewmen. Out of frustration one day, Chafin said, "They say a mother's love is stronger than a father's love. But I'd like to see the mother that would come wade through this mud to feed those kids!"

Inclement weather is not the only complication in life on a rig. On a drilling rig, the work is close, and what one man does affects the others—perhaps even a life; thus personal relationships are of top importance. Usually crews work smoothly together, as Donald Peters indicated in describing likeable driller Turk Hardin: "He was one of those that was so nice to work for that nobody else ever got a chance to work for him unless they were just lucky." But human relationships are never perfect, and often one has to make an effort to assure an effective shift. As Peters said, "You can't whip them all, so you got to try to get along with them. It's a crew deal, kinda like football. You've got to work together."

Competition between the shifts on a drilling job sometimes leads to social friction. As Bob Waddell said, "There are three crews on a rig, and every crew feels like the other one is not keeping up. If you'd listen to them, you'd think they were the only ones that do anything on a rig."

At times, the friction becomes physical. "They tie up every once in a while in a fight," said Waddell. "I'd have to give them a two-week vacation without pay when they'd fight on a rig. We never did have too much of a problem with it, but it would happen. Sometimes it's over whose time it is to drive back and forth to work. Little minor things. They'd fight today, then be good friends tomorrow.

"I remember one day, an old boy—he was pretty rough—got a hatchet after the driller and ran him off the floor, and the driller ran out to his car, and he had a .22 rifle in his car. So the driller gets the rifle, and the other old boy with the hatchet, he got behind a steel drum out there, and this driller waved that rifle, and he'd duck behind there. And finally—they had a big old roughneck worked there with them, Red Martin, he was about six feet six inches, weighed about 250—[Red] got up there and he took the hatchet and took the rifle away. But, you know, they worked together, good friends, for Delta for years."[38]

XIII

Though a toolpusher's job is a promotion for a driller, the raise in status and salary carries with it a high degree of responsibility, and frequently, stressful day-and-night duty. Until recent years a toolpusher might supervise several wells at the same time, and he has traditionally resolved all problems the driller has chosen not to tackle. In the past, because the wells he was overseeing might be spread many miles apart, he spent a large percentage of his working time in the company car, sometimes even sleeping there. He might go a month without seeing his family, and when he finally got home, he might hardly sit down when the phone would ring—and he would drive off again, to wrestle with the latest crisis.

Many men approached such a job with mixed emotions. "I was on duty 24 hours a day," said Willis S. Howard. "If I wasn't on the [rotary] floor, I was in the bunkhouse, or around, or on the way. I never stayed at home much. I lived them rigs."

Hugh McKenzie, who retired in 1978, added, "Well, it's just a headache. You've got to be where they can put their finger on you."

The main responsibility of the toolpusher is to keep the rig running, for "down-time" means no money is being earned by the very expensive machinery, while costs continue whether revenue is being generated or not. If a repair job or a part replacement is needed on the rig, it is the toolpusher's responsibility to see that it is done—without loss of time.

"Just say a pump would break a rod or something," said McKenzie. "[You] gotta go get a rod. If it's the middle of the night, you just go to a telephone and keep calling. You got to get it fixed. Everything's a-running.

"I remember down there at Livingston [in Southeast Texas] one night, I called seven welders, and it was just a-raining and a-lightning and a-storming. Oh, my God, it was a-raining and a-thundering. Not a one of them would come. And I wound up calling one of Delta's welders out of the yard at Kilgore, Texas, to drive to Livingston. They drove all the way down there. That old boy got done that next morning about 8:00.

"About the last 12 years I worked, I was a toolpusher. The last five years, I didn't know what home looked like. See, the boom had got back home [in the 1970s], and everybody was trying to drill. Maybe you'd be a-tearing down in Alba, and they'd want you to be a-rigging up in Mount Pleasant in the morning. The toolpusher would be there in charge of tearing down and rigging it back up."

Recently, the toolpushers' life has changed radically. The responsibility is the same. He still works 24-hour shifts, but he has twice as much free time.

"There's two men working the same rig that I was working," said Mc-Kenzie. "Seven days on, seven days off. They're drawing about $500 a month more than I was at that time."[39]

XIV

Injuries around a drilling rig can be serious, even fatal. Even today, there are few veteran roughnecks or drillers who have all of their fingers or are free of scars. In earlier days, injuries were more common. One of the risks that no longer exists in the oil fields is an exploding boiler. A boiler explosion on a Delta steam rig in the Schuler field of southern Arkansas, in August, 1938, killed one man and seriously injured three.

Dewey Strength was a survivor. "I was standing over there by the driller and [the explosion] picked me up. It threw me 150 feet down next to the rig. I don't know how high I went. [The driller] Mr. Teel's body—there was one dead snag out there on that location and it wrapped him around that snag. He was killed instantly.

"It picked Woodrow Tillery up. It skidded those pumps, and that fellow Leon Davis, that skid caught his foot and threw him down in that hot water. It threw Woodrow Tillery in the boiler pit. It broke his arm. It scalded me all over, took all the hide off my face."

Strength, along with the others, was rushed to a hospital in El Dorado, 13 miles away. "They didn't call an ambulance," said Strength. "Somebody threw me in a pickup. I was unconscious all the way. Was for a long time after I got there. Even the doctors pronounced me dead. It busted this shoulder, broke my neck, cracked my back. The first thing I remember about it was, whiskers came out when my face started healing up and scabbing, and my nurse came in and was picking those scabs off. I was there for almost a year before I went to do any kind of work."

When the doctor pronounced him dead, a news reporter picked it up and published it.

"A driller I had worked for was down in South Texas, and he read it in the paper and sent a funeral wreath to the hospital. My wife got a $15 funeral wreath."

He and his second wife, Nancy, had just married the preceding February. Strength was 40 and his wife was 18, and she was taking care of his two daughters from his first marriage, who were 14 and 13. That night, the toolpusher drove to Bernice, Louisiana, and notified Nancy Strength that her husband was injured; at the time he didn't know the extent of the injuries.

"It was a pretty big load," said Mrs. Strength. "I felt scared mostly. I had lived in the oil field all of my life. That was my life. Of course, it was a shock not knowing what had happened and all. No money. Delta took care of us. They got us an apartment in El Dorado."

Strength expected doctors would find him 100 percent disabled, but he wanted to work anyway, and he told the insurance claims adjustor he didn't want any compensation, but he wanted a job. When he got out of the hospital with his arm in a cast and a sling, he worked as a checker in a warehouse yard at Schuler. Later he went to Sundown, Texas, in the western part of the state, before moving to East Texas where, most of the time, he worked as a pumper until he retired.

Of all the unpleasant things that can happen on a drilling rig, blowouts and fires lead the list in unpopularity among both roughnecks and stockholders. Anyone who has ever witnessed one, not to mention having been a participant, has never forgotten it. The fire on a well on a Delta lease in the Como field near Sulphur Springs in East Texas is an example.

Originally the hole had been drilled down to the Smackover, a deep formation, but didn't make a well. The hole had been plugged. Subsequently another well had been drilled a quarter-mile away, inspiring the idea of going back into the plugged well in hopes of making a shallow well. This was about six months after the original hole had been plugged.

Bert Gauntt was pushing tools over the project, and Hugh McKenzie, though drilling on another rig, had been called over to help out. McKenzie was suspicious of the way the plug was set and warned that the hole might blow out. The rig had blowout preventers, but he remained wary. And when the trouble began, he made plans to shut off the existing gas with a closing valve.

"I never did get it on," he said. "It blowed that kelly right up through the derrick. Well, naturally, if you ever run a rig, you'll turn your back to it to keep that mud and stuff out of your eyes. It was in March and I had on an old slicker, and when I opened my eyes back, it wasn't nothing but fire, boy. Just all over, boiling out on you and everywhere else."

How did the fire start?

"Well, I figure a light bulb busted under the floor or somewheres. They didn't have safety lights on it like they do now. Well, I started to the doghouse. But I could see Bert standing back there. Looked to me like he was afire all over. And I said, 'My God, that ain't no way to go.' And I turned around and went back in front of the motor and down the back steps.

"[Gauntt wasn't really on fire], but looking through that fire, it looked just like he was a burning torch. And I got to them back steps and I was out of the fire, but I was afire all over.

"And I thought, 'Well, I'll just run and jump in them mud tanks there.' Well, I jumped off of the damn steps, or fell off of them; I don't know which it was. But I thought, well, that tank will be afire in a minute. That live oil'd come right on around the fire. And I run and jumped in the water pit. Then I got put out and everything, and I had this old slicker on, and all I had left was the collar and the string where the sleeve was. I guess it was hot enough it stuck to my neck, and I thought I was still afire.

"You won't believe it, but I had third-degree burns from there to there [gesturing over a wide area of his body]. I've not got a scar on me. It just grew back. The way they treated me.

"But Bert had the forethought enough that he moved the car. That's the first thing he thought of—'Get the car out; we're going to have to have it.' "

Then the other men came, bringing W. D. "Kilroy" McKenzie, a cousin, from the reserve pit.

"It had blowed him [Kilroy] off the [rotary] floor and he was afire," continued Hugh McKenzie. "And it burnt every rag of clothes that boy had on him off. I mean every rag. And he went into shock.

"And so we finally got him in there [to the hospital in Quitman]. The doctor was there in five minutes, 'cause he was our family doctor. And he didn't know neither one of us. That's how black we was from the oil."

All three men were lucky to escape with their lives. The drilling rig burned up. But all three men stayed in the oil fields, though "Kilroy" McKenzie carried scars with him thereafter.[40]

XV

Fortunately, all was not hazard in the oil fields. There was also time for recreation, which in the early days was often as wild as the work was hard.

"It was a way of life," said Red Magner. "They worked hard, they drank hard, and they played hard."

In southern Arkansas in the late 1930s, some things hadn't changed a great deal since that earlier wild El Dorado boom.

"In those days the girls did follow booms," said Red Magner, "and usually they'd just have a little old honky-tonk out there set up on some pine stumps. It'd be just a one- or two-room shack, but they'd sell booze, and nobody worried about their license or whatever they were selling—homebrew or beer or white lightning or straight grain alcohol. They'd have a bunch of these little old gals hanging around there, and one was a peg-legged madam. We called her a peg-legged whore. She kinda ran that joint, keeping up with all these little old chippies around there."

The jerry-built businesses, like the Schuler Wildcat Cafe, were little more than a tent with boards around the sides. At least one establishment was more temporary than the proprietor realized.

"I'll give you an idea as to how flimsly it was," said Magner. "This Duren Rogers, he loved to drink, and when he was drinking he just loved to have fun. So I was single there and of course I'd honky-tonk with all of them. So old Duren got drunk and they threw him out of this honky-tonk. Keep in mind that it probably wasn't more than about 40 feet by 40 feet upon these old pine stumps, about two feet off the ground. And he said, 'Well, I'll get even with those son-of-a guns.' So he and a couple of his buddies got out there and he trimmed them off some old pine boughs and fixed it up on them old stumps out there and got on the end of the pine bough and started rocking that dern thing, and those guys came a-pouring out of there. They thought the earthquake had already hit. The bar was rattling.

"Oh, boy, I tell you, I've been around roughnecks all my life, and there's no way you can outsmart them."

On one occasion Guy White took a group of roughnecks out honky-tonking to "break in" a new crewman. Many hours later they took the young man home, with two men holding him up. The poor fellow had passed out.

His mother came to the door.

"What hit him?" she said as she, with great surprise, observed her son's condition.

"Seagram's Seven!" someone in the crowd replied, as they deposited him inside the house and fled.

Guy White and Red Magner used to visit some of the hot spots in Henderson, Kentucky, when they had time off, drinking beer and playing the slot machines. One night they were particularly active. The next morning Magner asked his boss, "Hey, Guy, you got a headache?"

"Yeah, I got one," said White unenthusiastically.

"You don't suppose it's that beer we drank?"

"Hell, no," responded White. "It couldn't be that, because we just had 19 bottles apiece, which is our usual."

Others made their amusement on the job. Horseplay on drilling rigs has never been recommended by safety experts, but at one time it was a part of rig life. Hugh McKenzie and Nathon Bacle had a roughneck on their rig who refused to work derricks. He just would not go that far off the ground.

"But me and Nathon tied him on the block one night and hauled him up to the crown, and tied the brake down and went off and left him. You could hear him screaming all over Alba [where the drilling was]. Me and Nathon talked about that—what if he'd died with a heart attack? We'd never [have] got that explained."[41]

XVI

Sobriquets are endemic to the oil patch. Many roughnecks have become known by their nicknames—Slim, Whitey, Bones, Mutt, Tex—and not remembered by their actual names. The practice seems to be contagious around oil booms. When J. M. Bevill went to Tyler, his name was James Milton Bevill, but the man in whose house he stayed, and who gave him job leads, said, in effect, "Well, we can't call you James because I've got a son named James, and then we'd have two." So he settled on "Joe," and thereafter, wherever Bevill went, he was Joe Bevill. "Red" Magner gained his nickname earlier in life, and more logically, with his hair, "red as a fox's tail." Certainly "Red" was easier to say than "Harold Joseph," his legal name, and it was infinitely more acceptable than some labels which might have been pinned on him in the oil fields. But he did harbor serious thoughts of dropping his nickname.

"I thought that if I came down to Texas, I'd be more dignified and just use my middle name, Joseph, Joe," Magner said. "But we already had Joe Bevill down here, even though his name was James, and we already had Joe Zeppa. There was too much competition. So I said 'Well, all right, I'll be undignified and just be Red.' "

L. D. Hoppers, as Delta's paymaster, came to realize that nicknames were the most reliable means of identifying an oil-field worker.

"Mr. Chris gave me a title before he learned my name: Timekeeper," said Hoppers. "I hadn't been there but about a week and he called. He said, 'Timekeeper, what rig is Red Martin on?' D. H. Martin. Then we had two or three Martins on the payroll, and I didn't know Red Martin from anyone else. From then on I learned the guys' nicknames."

One man's nickname, Addie Go, had nothing to do with his real name. William Bryant Kennedy was known as "Jack" Kennedy, J. W. Parks as "Blackie," and Jack Hill was really Orville Hill. One man was known as "Tangle Eye" because his eyes were crossed. Sometimes the acquisition of a sobriquet had a logical history; sometimes it did not. Garland Cox's nickname was "Big Knuckle." Why? "I don't know," he said. Z. L. "Nig" Spraggins's first name is Zelphin, but he has never used it. "One of my great aunts from Arkansas sent that name for my mother," he said. "My dad didn't like that Zelphin, so he named me 'Nig.' " James H. Gilleland is known familiarly as "John," because when he was roughnecking he had the initials "J. H." on his lunch pail and someone started calling him "John Henry," which was then shortened to John. Even the company pilot had a nickname. Garrett Paccio Rider was known as just plain "Jimmy" Rider.

At least one worker named himself. Red Magner, because he was a college boy at the time, was assigned much of the paper work in the field office in the Schuler field. One hand came to him for help in filling out his Social Security form. The man only had two names.

"You've got to have an initial," said Magner.

"I don't have one," said the worker.

"Well," said Magner, "think of one."

"All right," said the man, "put M down there."

Magner printed an M for the man's middle initial.

"What the hell does that stand for?" asked Magner.

"Merridy."

"Merridy?"

"Yeah, I always did like that name."

On occasion, nicknames could complicate man's life, as Lee Browning learned in a little oil-field town near El Dorado. Browning was known as "Brownie." There was also another man called "Brownie." One morning Browning walked into a cafe there for breakfast, and the waiter greeted him, "Morning, Brownie."

This set off a reaction in the cafe that puzzled Browning. As he recalled it, "I saw this man step out from the kitchen with a shotgun in his hand, and then he saw me and turned around and went on back. I wondered why he did that, and I come in that night and they told me that this other Brownie had gone down there the night before and walked up in the doorway of the cafe and let them have a load of buckshot.

"I went on over in Louisiana, and I'd meet somebody on the street there ever' once in a while, and they'd say, 'I thought you got killed over at El Dorado.' "[42]

XVII

A number of Delta employees have become, over the years, legendary figures in company lore. One of these was tall, strong, bulky Joe Blasingame, who has been described as a self-trained metallurgist, machinist, and inventor. Blasingame took over the warehouse and welding at the Kilgore yard when A. B. Owens transferred to South Louisiana; when B. W. Freeman died, Blasingame also took over the maintenance and repair of the company's vehicles until he died of a heart attack in 1970.

As a boy, Keating Zeppa worked briefly for Joe Blasingame. "Working for Joe Blasingame was not, in that day and time, particularly unique, because in the '40s the work ethic was still pretty strong in this country. People weren't that far away from the depression years of the '30s. Joe expected everyone to—you know, if it was a ten-hour day, he'd want a full 12 hours! And that was the way he was about himself. Joe was definitely a man of action. He was a man of very positive opinions and statements, and Joe was never one to say that something couldn't be done. He was a very creative person with his hands, with his mind, with iron and steel and machinery. There was really nothing that Joe couldn't do with machinery.

"He was not the easiest of taskmasters, but he didn't see it as his job to coddle a bunch of kids. His job was to turn out work."

Marvin Williams, president of Spencer Harris Machine Company, was a close friend of Blasingame.

"Joe was one of the best welders, and could take nothing and make something out of it, in the days when you had to do that. The greatest piece of work I ever saw Joe do was at Opelika [field]. A driller was raising a crown and it dropped, and when it fell it caught the engine and busted the back end of it out, and Joe took the pieces and put the engine back together and ran it. It was cast iron. One of the greatest pieces of work I've ever seen. It was just unbelievable what he could do with stuff he had to save.''

"He was the type of fellow that if you told him that he couldn't do something, he would do it or kill himself," said J. W. "Turk" Hardin. "I liked Joe. A lot of people didn't. He was pretty strong-headed and set in his ways, but he treated me real nice."

Hardin had a special reason for liking Blasingame. He suffered an accident in 1968 that jeopardized his career as a driller, but the ingenuity of Joe Blasingame extended his working tenure. Hardin was 16 feet off the ground on an elevator when the cable broke.

"I was standing straight up in it and hit the ground and it crushed the heel bone in my foot," Hardin said. "I don't have a heel bone in my foot at all. My ankle looks like a road map. The doctor told me that I would never be able to work on a rig again. They had a mechanic out in the yard in Kilgore and they put me out there driving a hotshot pickup.

"I was working with Joe Blasingame. Joe asked me one day could I run a rig off a seat. I told him that I could and that was all [that was] keeping me from going back to drilling. So he built me a seat. Rig 21 was the only one that had a place to put it. He built it out of a bicycle seat. That's the only reason why I still work on a rig today. I just can't stand for no length of time."

Blasingame also used his imagination on projects unrelated to his work. "He had a greenhouse out there in the [Kilgore] yard," said Jack Elkins. "He had it central heated, and thermostat-controlled doors on it. When it would get so warm in there, those doors would raise up and let the fresh air come through. He raised tomatoes year-around out there. I mean big nice ones too. He kept those tomatoes at the right temperature."

He also maintained a regular outside garden there during spring and summer.

"He had a big cantaloupe down there one time," said Elkins, "and a driller and a roughneck came over there working in the yard and [Blasingame] would go down and check on that cantaloupe ever' two or three minutes, and he went down there one time and the cantaloupe was gone. Now he was ready to run somebody off for that cantaloupe. The old driller finally told him what he did. [Blasingame] didn't do nothing, but he didn't like it worth a damn."

Blasingame's eccentricities were conversation pieces at Delta.

When young Charlie Martin, then roughnecking, wanted to go to work in the Kilgore yard as a rig mechanic, Blasingame was hostile to the idea at first.

"So Bob Bateman got aggravated at him one day and told him, 'Well, just give the boy a chance, and if he don't work out, we'll send him back out on the rigs,' " said Martin. "So Mr. Joe [Blasingame] said, 'Charles,

I'm going to put you on temporary, and if you work out, fine, and if you don't we'll run your ass off,' and that's what he said. He was pretty well plain-spoken. Rough as hell.

"He was kinda strange in a way. He knew what he was doing, and if he liked you, you were fine, but if he didn't, you didn't work very long. You might make a first paycheck and then again you might not.

"We had one boy to come in there, a trainee mechanic, who was working in the shop one day. This trainee mechanic took out some Copenhagen and took a dip, and he walked off and Mr. Joe looked at me and said, 'If I had known that boy chewed that mess I would never have hired him.' "

Work never ceased. As Jack Elkins recalled, "When you went through that cattle guard at the Kilgore yard it might be twelve or one o'clock in the morning, but you'd have the feeling somebody was standing up there waiting on you to turn you around and send you right back somewhere else. Many a night you would come in there and the mechanic would be working on the drawworks or something, and Mr. Blasingame—it didn't bother him whether you were a truck driver, mechanic, or what—would say, 'Let's jump in there and let's get this thing out,' and you were liable to work there in the shop the rest of the night. The time didn't seem to bother Mr. Blasingame.

"We came in there one Saturday night and we had been to Houston, bringing a load of new drill pipe back to the yard, and he met us. He said, 'How many hours ya'll got in?' and we told him, which was up about 70 or 75 already. He said, 'I don't give a damn if you have got 100 hours, we are going out to the lake at Mr. Zeppa's farm in the morning.' "

Bright and early the following morning, a Sunday, they were at Pinehurst Farm to install a device in the lake to regulate the water level.

"He was a fine fellow," said Elkins. "It had taken a long time to get to know him, but after you got to know him, he was all right. He expected you to work, I guarantee that, and if you didn't work you didn't stay with Mr. Blasingame. Joe was a little bit stern. He demanded a little more perfection."[43]

XVIII

When Jack Elkins went to work for Delta, old-timers like Ark Carter and Mutt Hays were still around, nearing the end of their working days, and they still exercised their traditional authority and privileges.

"I was running the gin truck," said Elkins, "and, of course, back then I thought the world was made in two days, and I thought we had to get that rig strung out in a day, and I backed up there and Ark Carter and old Mutt Hays was leaned up against one of them hauling racks where you put this stuff in, and they were talking. They were right in my way. I had some steam lines on the gin post, and I backed up there and sat and sat, and they just leaned up against the rack talking, and after a while I tooted my horn, and I shouldn't have done that, because both of them old drillers came around there and got right square on top of me. They said, 'Who are you honking at?' and I said, 'Well, I want to get this stuff on that rack.' 'Well, by God, you can set there until we get through talking.' They don't operate like that now.''

In the early 1950s Ark Carter still insisted on doing the same work he had done as a younger man. "Mr. Chris found out that he was going on top of the derrick when they were rigging up," said Homer Lee Terry. "When Mr. Chris found out, he told him, 'No, Ark. We've got younger men.' 'Well, Chris, I can still do it.' 'I know you can, Ark. You just stop it.' "

He continued to be a colorful character until he retired.

"The old rascal wore them glasses, and they usually had one lens out and the other one covered with mud," said Jack Elkins, "and I don't know how he saw anything." Shortly before he retired, an accident in Pennsylvania sent Carter to the hospital. The old driller's hands were so tough he never wore gloves; nonetheless he decided there was a limit to the number of injections he needed. Don Carman recalled the scene in that medical ward: "They came in, first thing, and gave him a shot. So Ark gets wise to that. He knows when they're coming, so he switches his name tag over on the other bed and gets that name tag over there, and they'd come in there and shoot that guy. That one guy looked like a pin cushion, and Ark wasn't getting a shot."

Carter retired in 1967 and returned to a home he had already built near El Dorado, Arkansas.

In his sixties, L. V. "Frenchy" Portier, who had been drilling for Delta from the early '30s, was beginning to feel the impact of age. His reflexes weren't as fast as they had been, and he wasn't up to the old-time physical labor required of roughnecks. But he was to young to retire, and not only did he still want to work, but Joe Zeppa said, as many employees would remember, that Portier had a job with the company as long as he wanted to work.

"As long as he could get on the rig floor," said Herman Gordon, who worked derricks for Portier, "and if he still wanted to work, the roughnecks would carry him on the floor. Now that was the way I understand it. They said Joe Zeppa said he could work till he was 100 or however long he wanted to."

In the latter part of 1959, Portier's crew was drilling a well at Omen, 25 miles from Tyler. It was cold.

"We shut down to do something," said Gordon, "so I told him, 'Frenchy, I'm going to grease that crown block today, because it's supposed to be in the low twenties tomorrow, and I'm going to have to go up there tomorrow.' So he said, 'Fine.' Back then we rode the elevators, rode that block up and down, so I got up on the elevator and I got off the board, and he slacked the blocks back down to where I could reach the grease from the board. I put my belt on and ran out there and greased the fittings, and I unshucked my belt, went to the crown and greased it, and came back down and put on my belt."

While Gordon performed his chores in the derrick, Portier was talking with a salesman from Continental Supply. Portier, apparently thinking the blocks were on the rotary floor instead of up in the derrick, responded to Gordon's yell that he was through by clutching the rig into high gear. He never looked up.

"He just plowed right into the crown and cut the line and the blocks fell down," said Turk Hardin. "That block fell right down on that center pipe in the hole; when it did, it knocked the slips out and all of that fell in the hole. Just as Frenchy and the salesman went into the doghouse, that board fell up against the door and shut it to, behind them."

Herman Gordon watched, horrified, from the derrick. "Well, there were three men on the floor besides Frenchy, and they seen what was happening, and they got out of the way. It did make a wreck. So as soon as everything quit falling, I unshucked that belt and I come down that ladder, taking about five steps at a time. I was okay, but I didn't know if anybody else was. Got down there and not one soul hurt."

Turk Hardin was in the crew that relieved Portier's men. "He hit the crown about noon. We got there at three o'clock, and Frenchy come down the steps and everything was piled up there and he had his clothes rolled up under his arms. He said, 'Me too old, John. Me too old.' He called everyone John. Whether he knew you or not. So I took his crew. He never did go back to drilling."

Herman Gordon rode home with Portier that day. "Bless his heart, rest his soul, Frenchy never did break down till he got home. We got off that afternoon and we drove in front of his house and he got to crying like a whupped baby. So the next day we went by to pick him up, and he said, no, he wasn't going back to the rig. He felt like the Good Lord had warned him; he never killed nobody, but he came close yesterday, and he never did go back to the rig. So then Mr. Joe Zeppa and them put him out at the lake, overseeing Delta's fishing part of Lake Tyler, and he stayed out there until he retired in 1963."

In an interview in 1971, Portier summed up the experience: "I quit working on a rig in 1959 when the traveling block fell four feet from where I was standing. Right then I decided somebody up there was trying to tell me something, and I never worked on a rig no more."

Jack Elkins expressed the sentiments of many Delta employees when he said, "They didn't make but one Frenchy."[44]

XIX

May Jones was another living legend at Delta who aged along with the company. She knew how to talk to roughnecks in their own language, and though she was almost invariably gruff-toned, they seemed to rejoice in whatever attention she gave them. She seemed to know how they thought. Otis Wilson once went to the office in Tyler the day before payday. He asked May Jones if Tommie Smart, Jones's assistant, was in.

"She won't give you your check early!" said Jones.

"I don't care," said Wilson. "I just want to talk to her."

He went on into the office where Smart was working. While they were chatting, Smart slipped him his check so Jones wouldn't see it. A few minutes later, Wilson walked out.

"She give you your check?" asked Jones.

"No, she wouldn't give it to me," said Wilson.

"You're a damned liar!" shot back Miss May.

As Marguerite Marshall described her, "She had a sharp tongue, used foul language sometimes, but she had a heart of gold."

L. D. Hoppers classified her as "salty, but the best person you ever saw." When he had been at Delta about a month, Hoppers was off sick for most of one week and had failed to turn in a time sheet. When he returned to work, he told May Jones of this oversight.

"Well, I made one out for you," she said.

When he received his check, Hoppers learned he had not been docked for the time he had missed from work.

"That's all right," she said. "You'll make up for it if you stay here long enough."

Red Magner remembered a time or two when he took May Jones "honky-tonking." Joe Zeppa had tickets to a dance which he passed on to Magner. May Jones was about 20 years older than Magner, but proved to be entertaining company.

"I had my own Model A Ford," said Magner. "We went out to the American Legion, and May got drunker than Cooter Brown, and so did I, and we got outside there and it was cold. And you wore vests in those days, too, and had an overcoat, and I couldn't find the dern keys to my car. Boy, I'm telling you, I finally counted out I had 21 pockets, looking for that key to get the car started.

"May was just *good*. She ran the business and she kept up with the money. She was a pretty salty old gal herself and knew how to take care of the toolpushers and the roughnecks, and she was great."

"She pretty much worked the hours Mr. Zeppa did," said Alton Blow. "I know on occasion I would stay late doing something, and you'd hear her. She'd go in there and argue with him. She was doing a lot of the pay-roll-report filing then. And she just griped about the state requiring all this, or the federal government requiring this, and she said it was just to give some S.O.B. a job. She was really a character."

"May Jones was kind of a different person," said John Justice. "She was quite outspoken. You'd have to be careful not to set her off. She was a true friend. And a loyal friend. But she didn't hesitate to let off a string of cuss words around there. She'd fly off and let go. She and Joe Bevill, I guess, used to have a lot of set-tos. They were right there together, and of course Joe Bevill was Mr. Zeppa's right-hand man, so their paths would cross and they would tangle. I could hear at a distance, and I just would not get any closer."

Everyone seems to have been aware of the give and take between Joe Zeppa and May Jones. "They would just yell at each other sometimes," said C. L. Vickers. "She could talk back to him." In the early days of Delta Gulf, Joe Zeppa was in his office discussing a matter with Red Nixon. Presently he needed to dictate a letter. He buzzed for Marguerite Marshall. No response. He buzzed for Tommie Smart, and she didn't appear. Finally he rang May Jones. When she came in, the president asked, "May, where are Tommie and Marguerite?"

"Hell, I don't know Joe. I've got work to do. I don't have time to keep up with them!"

"Well, you ought to," he said.

"Well, I'm not," she retorted.

"Wait a minute, May," Zeppa said. "I want you to meet Red Nixon. He came over with the Jerry Hawkins deal."

Then Zeppa turned to Nixon and said, "Red, this is May Jones. She's been with us a long time. She does about as she pleases around here."

In declining health for the last several years of her life, May Jones died in 1964. At the time, she was the assistant corporate secretary, the highest position then held in Delta by a woman.

Richard J. McDonough may have voiced her most enduring epitaph when he said: "She was one of a kind. The only one that I know of that could order Joe Zeppa around."[45]

XX

From May Jones on, Delta has been blessed with competent, loyal women in its offices around the country. One probably could obtain a detailed and accurate history of the company by interviewing a half-dozen or so of these women: Grace Baker who has been in charge of the central files for about half the life of the company; Marguerite Marshall and Joann Blalock who have served as executive secretaries; and division secretaries such as Flo Bonham in the Gulf Coast division in South Louisiana and Lynna Jo Edwards in the West Texas-New Mexico division. Though unheralded, these and other such women have been essential in the company's operations.

Like Annie May Jones, Flo Bonham went to work for Delta on a temporary basis in 1969, as a Kelly girl. "But it was sort of like the man who came to dinner," she said. Delta was in the process of moving its Gulf Coast yard from Lake Charles to a consolidated office in Lafayette, Louisiana, and when she was asked to accompany the move, she was delighted.

Tall, blonde, attractive Tommie Smart joined the Delta office staff when she was 19. Hired by corporate secretary Joe Beasley in 1941, she was one of the first employees to use the new payroll machines purchased soon after she arrived. Later she became secretary to Joe Bevill.

She became widely known for her information programs in the East Texas area on both the economics and the products of the petroleum industry. She put on the "Magic Suitcase" talk for schools, civic organizations, and professional groups, discussing the variety of items produced from petrochemicals.

"We remember her as being one of the most versatile speakers we had," said Jack Rolf, of the Texas Mid-Continent Oil and Gas Association's Oil Information Committee, which sponsored the programs. "She did just about everything. She could handle about every subject that we dealt with."

In the process she won several awards from the committee, including its Outstanding Performance award for having put on 100 programs.

Like May Jones, Tommie Smart rose to become assistant corporate secretary. When she died unexpectedly in December, 1979, she had been with the company 38 years. Had she lived until retirement, she would have been the woman with the most years of service in the company.[46]

XXI

Employee loyalty has been forged largely because the employees have perceived the Zeppas to be interested in them.

Ed Gandy explained, "Like [the Zeppas] say, they didn't make the company altogether. The people that worked for them helped them make it, and they'll let you know that."

To many employees, particularly the long-time ones, this perception has been translated into the image of Delta as a family.

"Mr. Zeppa always felt about his people as a family," said Alton Blow. "Things were tight back in the late '50s when all the other contractors were going out of business, and he didn't have that much business—he was still trying to buy these rigs—and he started off at the Christmas party by talking about how bad things were, and then he wound up by telling people how to manage their personal finances, not to be going into crippling debt. Everybody got a kick out of that."

Although Delta did not always pay top wages, it has always been re- membered by older employees for meeting the payroll. High on the prior- ity list was paying the men on time. One Christmas the time sheets were picked up in South Louisiana and West Texas by the company airplane to ensure that the paychecks went out on time. This, no doubt, enhanced the feeling of personal concern for the workers.

Some employees left for jobs with major oil companies because of more attractive salaries, but many remained at Delta. "Back in those days, peo- ple stayed because of the people you worked with," said Sarah Clark. "It was a well-known fact that Delta paid less all these many years. I had sev- eral chances to go with majors, but I chose to stay with Delta because of the people. It's nicer to come to work and want to come to work. I don't ever remember a day back in those years that I did not want to come to work."

Over the years the company improved its fringe benefits, moving ahead of many independent contractors. Drilling superintendent Doris Darbonne remembered when insurance, hospitalization benefits, and vacations were not common with Delta's competitors in South Louisiana.

"I found that amazing [that Delta had such benefits]," he said. "I worked for other contractors and never had anybody say I could have a week's vacation on the end of a year, or two weeks on the end of two years. Never had that before."

The Christmas bonus had become a fixture since the early days. In the 1960s the money usually allotted to this fund was diverted to an employee retirement program, which went into affect in December, 1964. Lee Browning was the first man to retire under the pension plan, and Chris Zeppa wrote him a letter, apprising him of that fact.[47]

By the late 1970s, salaries had been increased to levels that competed favorably with major companies.

The pattern over the years has been one of evolving complexity, from a time when the firm was small and the president could personally know ev- ery employee, to a time when he could not possibly know everyone. The Christmas bonus is a personally distributed gift. A retirement fund is less personal, but a more reliable symbol of security. The feeling of close fam- ily-like ties, one to the other, might persist, but with the increasing growth and complexity, the old image could not help but erode to some extent.

XXII

As Delta grew, its internal needs became more complex. Tighter control of supplies and expenditures became desirable. One day Joe Zeppa asked Bill Craig, with Continental-Emsco, to drop by his office when he had the opportunity. Zeppa pointed out that Delta, indeed, was growing, and in need of more organization. Purchasing, particularly, was a field in which some central control was needed.

"Every toolpusher, every field man, what-not, does his own buying and so forth," said Zeppa. "Of course, as the company grows, you know better than I do, that's not the way to do it."

The company had reached a stage where it needed a separate purchasing department.

Zeppa continued, "Now the reason I'm telling you, I'd like to ask you if you would take your time and pick from your company someone that you'd want to recommend to us as a purchasing agent. Whatever you do, just tell Chris. Bring him to see Chris, and let Chris go on from there."

Zeppa, pointed out Craig, knew what he was doing when he asked someone to select the kind of man he was looking for. Zeppa had trust in the man and his judgment. The man Craig sent to Delta was Hiram S. "Hi" Cole, who became Delta's first purchasing agent.

Sometime later, Zeppa told Craig, "Well, I see you sold your man to us."

When Hi Cole went to work for Delta in 1950, he modeled the new purchasing department on Continental Supply's procedures. But all the major purchases flowed through Chris Zeppa for his approval.

"Buying is very important," said Hi Cole. "You don't want to ever overbuy. I learned that over and over from Chris. That was constantly in their minds."[48]

In the thirties this conservative policy had been essential to survival. A lot of companies, after all, hadn't come out of those years.

XXIII

The ultimate scorecard in any business is the tallying of profit and loss. What has the scorecard revealed about Delta's economic health over the past few decades? To answer this question, we will examine selected sectors of the company's activities, particularly its production pattern, and then the concrete end results of the company's efforts as reflected in net income and rig count/evaluation.

A survey of production statistics in the 1960s may not be representative or average, but does suggest patterns. By the 1960s, Delta's policy on taking interests in drilling wells was well established, with the final decisions still depending upon Joe Zeppa. This sampling will examine both ends of that decade, the early 1960s and the late 1960s.

In one compilation of exploratory and development wells for a period of 1961 and through March, 1962, a 15-month period, Delta listed 66 wells in which it had an interest. Delta's interest in some of them amounted to 100 percent—but very few, it might be noted, and those in the East Texas district. Of the 66 wells, 42 were in East Texas; 8 were in West Texas-New Mexico; 1 in the Rocky Mountains; 8 in South Louisiana; and 7 in the Northeast. Because the East Texas district sometimes drilled beyond its immediate region, this labeling may be slightly misleading. Though most of the wells were within a reasonable driving distance from Tyler, at least one was in South Texas (Jim Wells County), three were in Mississippi, and ten were in northern Louisiana, primarily the Many field in Sabine Parish. As may be seen, the farther the wells stretched from East Texas, the fewer interests Delta took in them and the less likely that working interest was to be more than a third. Of the eight wells in West Texas-New Mexico, Delta had a half-interest in only one, and this was at an early stage in that well's history—meaning a part of it could yet be sold; furthermore, that well, with a total depth of 5491 feet, was the most shallow of the eight wells, so that its total cost would likely be less than the others. In the Rocky Mountain district, Delta had taken only a one-eighth interest in one well in Wyoming. In South Louisiana, Delta had 50 percent of two wells, around 25 percent of the others. The Northeast interests hovered around 25 percent per well of the seven.[49]

At the end of the decade—using 1968 as the sample year—the basic overall pattern had not changed, though there were some modifications. The heaviest concentration of working interests in wells remained in East Texas, where there were 29 such wells, plus five in Arkansas, Mississippi, and North Louisiana which were administered from the East Texas district for a total of 34. These individual interests ranged from 0.992 percent in one well to nearly half (7/16) of another, with most of the range from five percent to one-fourth. Elsewhere, Delta took interests in the Northeast, 5 wells; the Rocky Mountain district, 4; and West Texas district, 14.[50]

A recapitulation of Delta's 1968 operations in which a working interest or override was involved showed the following breakdown by districts.[51]

District	Number Drilled	Dry Holes	Producers
Northeast	4	2	2
Rocky Mountain	4	3	1
West Texas	12	1	11
East Texas	22	16	6

As can be seen, the most profitable Delta interest that year was in West Texas, where only one dry hole marred the records. However, more than half of the producers came in the Ozona field, where six wells came in that year. East Texas, which previously had been a mainstay, that year yielded 16 dry holes to only six producers.

Other assets were acquired more directly. From time to time Delta purchased stock in other companies upon the recommendation of Joe Zeppa. A case in point is the investment in Texas Illinois Natural Gas Pipe Line Company in 1950. As president, he had subscribed to 21,871.5 shares at $10 per share, but recommended to the board that Delta limit its holdings to $100,000. This apparently was no problem, as the other stockholders were interested in whatever amount of the stock the company did not want, and key Delta employees were given an opportunity to buy the rest of the stock.[52]

On another occasion, Zeppa's friend and frequent business associate, Sylvester Dayson, turned over to Delta his rights to subscribe to 12,865 shares of Pan American Sulphur Company stock at $7 per share. In turn, Delta gave Dayson the option to purchase 60 percent of this new stock at the subscription price within three years. By the time Zeppa reported his agreement with Dayson to the board in early 1953, the stock was selling for $11 over the counter. Zeppa waxed optimistic over the prospects for Pan American Sulphur—a view Sam Sklar failed to accept fully, though he went along with the deal.[53]

Although the bottom line of financial records may not tell the whole story of a company's operations, it is useful to plot general patterns. The following recapitulation of selected net income figures from the middle 1960s to early 1970s will give some indications of Delta's general condition for the period. To simplify the examination, the figures will be restricted to the net income or loss. Although this chapter is concerned primarily with domestic activity, all of the Delta-owned companies, including foreign operations, will be represented in the grand total.

Delta's land-drilling operations continued to be *the* reliable, steady moneymaker over the years. Delta Marine operated habitually in the red, as did the Tyler Hotel Corporation, a domestic but non-drilling enterprise. With few exceptions, the foreign operations added relatively little to Delta's coffers. To gain some idea of the strength of Delta's domestic operations, let us compare the net income of Delta Drilling Company and Etexas Producers Gas Company with the overall net income of Delta and her subsidiary companies together, both foreign and domestic, over a representative seven-year period.

	Delta and Etexas Net Income	Grand Total Net Income All Companies (Including Tyler Hotel, Delta Marine, and foreign)
1966	$1,642,625	$1,959,259
1967	1,968,876	2,225,500
1968	1,305,678	1,395,402
1969	1,547,684	1,791,140
1970	1,145,106	388,693
1971	1,356,879	1,475,280
1972	1,154,595	1,014,673

As can be seen, all of Delta's subsidiary companies together never raised the grand total more than a few hundred thousand dollars in any one year, and at least in two of these seven years, the other companies actually lowered the overall total income. This pattern of Delta's domestic land operations generating the bulk of the profit held true over the long run.

Rig counts, though not a conclusive means of measuring success, nonetheless provide a handy index for evaluating a drilling company. By 1975, near the end of Joe Zeppa's career as chief executive officer of Delta, the domestic rig count was at 38, ranging in value from a $285,000 Bethlehem S-60 (Rig No. 48) working in the Rocky Mountains, to a $1,615,000 Emsco Electrohoist III (Rig No. 68) in the Southeastern division. Total value of the 38 rigs was estimated at $26,218,000.[54]

A breakdown by division:

Division	Number of Rigs	Value
East Texas	8	$4,482,000
Gulf Coast	7	5,654,500
West Texas	9	5,428,500
Rocky Mountain	4	1,850,000
Northeast	6	3,323,000
Southeast	4	5,120,000
Total	38	$26,218,000

In number, there was little change from the early 1950s, but the value of the more sophisticated machinery was another thing: a far cry from the two rusty old second-hand rigs with which Delta had first started in 1931.

9

ON FOREIGN SHORES

I

Delta's expansion into contract drilling outside the United States came—at first tentatively—during World War II, grew in pace with the American oil industry's foreign activity, and then began to retrench in the 1970s. Since 1944, Delta has drilled in 12 countries other than the United States. By 1981, the company's rigs were active in only three foreign nations—Italy, Mexico, and Brazil. The pattern of foreign activity—growth and expansion followed by retrenchment—has been much in line with that of the overall American petroleum industry.

For the most part, the foreign drilling was done by the company's foreign subsidiaries or associates, in which Delta's ownership ranged from 49 percent to 100 percent.

Delta's foreign operations have been concentrated in Venezuela, Italy, Mexico, Argentina, and Brazil. Contract drilling for shorter periods was carried out in Canada, Spain, Libya, Australia, and the Philippines, and crewmen worked in South West Africa (now Namibia). Delta's Italian company drilled one well in Turkey. Negotiations in three other countries fell short of agreement: Saudi Arabia and Bolivia in 1957, Iran in 1973.

By 1956, Delta's overseas operations were in full swing, described by Joe Bevill as a "wholesale expansion of foreign operations" which had generated a massive volume of extra work back in the corporate headquarters in Tyler.[1]

This was a propitious moment for an expansion, whether domestic or foreign, for by that point Delta had fully digested the Hawkins gains and had both rigs and capital with which to pursue opportunities it might otherwise have been forced to forgo.

A summary Joe Zeppa made of the future of the drilling business in 1957 also provides a partial rationale for Delta's entering the foreign drilling business in the 1950s. The domestic business was not quite on a breakeven basis, with no significant improvement in sight. Drilling activity in the States wasn't expected to climb, because petroleum consumption hadn't increased and cheap foreign oil was being imported at a high rate. Furthermore, new oil and gas discoveries in the States were becoming fewer and more expensive, and producing days had been reduced in Texas and other states, bringing decreased net income from oil operations here. Foreign operations, Zeppa said, offered "the best prospects for profits, although they entail increases in capital commitments and present personnel problems of a serious nature."[2]

At that time, Zeppa was highly optimistic about the foreign ventures he already had, as well as those to come.

II

It might be argued that Delta's first foreign venture was in Mexico during World War II, though one would have to stretch the facts to do so. More properly, this first episode might be called an adventure. Joe Zeppa, as an individual, went to Mexico in 1943, apparently in hopes of developing mercury and tin deposits there for the American war effort. We are able to piece this matter together partially by reconstructing fragments of records.

In a letter to his old friend George T. Keating, Zeppa mentioned a Mexican trip that was "partly business," and his wife Gertrude and son Keating, or "Pinny," followed him by two days so they could enjoy sightseeing. Zeppa did not volunteer any details at all of the "partly business" facet of his trip, though he went into considerable detail about the volcano which had recently erupted at Paracutín. We do not even know for certain that he had mercury and tin on his mind when he left for Mexico, but in light of what he later told his fellow directors, it seems a safe assumption.

Joe Zeppa's associations, personal knowledge, and reputation attracted men to him who, in turn, led him into a variety of business propositions. One such connection occurred during World War II, when there was a scarcity of strategic minerals needed for the war effort. A group of men—including officers of Humble and Standard Oil Companies—invited him to try to develop mercury and tin deposits in Mexico. The mercury was in the state of Guanajuato, the tin in the Coneto area near the city of Durango. Guanajuato is in the central region of Mexico, Durango in the north.

Zeppa, reporting the matter to his directors, asserted he had participated for two reasons, one patriotic and the other because of the advantages for Delta of his association with these men. The group acquired—as individuals, in the names of Mexican citizens, since foreign corporations were restricted by Mexican laws—mercury and tin mining concessions and built a furnace with which to smelt mercury. However, the furnace proved to be uneconomical when the price of mercury dropped abruptly. By this time Zeppa had personally spent $14,270.23 of his own funds. After he had detailed the experience, however, the directors—he, Chris Zeppa, and

Sam Sklar—agreed to approve his Mexican mining venture and to reimburse him for his expenses in connection with it.[3]

III

Delta first went beyond its own country to Canada when the company contracted with Lion Oil Refining Company to drill "one or more wells" on Cape Breton Island in Nova Scotia in 1944.[4]

It proved to be a short-lived drilling program, out of which Delta gained experience more than anything else.

"Drilled a dry hole and came back home!" Chris Zeppa summed it up. "It was a contract job for Lion Oil Company of El Dorado—our good friend, Colonel [T. H.] Barton. That was just about the time that I landed here, and that thing landed right in my lap. I had to put a rig together right here in Overton [in East Texas]. In fact, we bought a little rig to send over there. That's the first rig acquisition we made after I joined Delta. We loaded everything in flatcars in Overton, Texas. The reason we did it there was, the rig was there. We bought it from a fellow from the East Texas field and went up to Nova Scotia, drilled one well, was a dry hole, and that rig returned to Mississippi."

The rig which Delta sent to Nova Scotia was owned jointly with Watson W. Wise, an independent in Tyler. Delta employees manned the rig. Herman Woods went as the toolpusher, and Bill Phillips, one of the company's old, old hands, was a driller.

Wise recalled the venture: "Joe was an enthusiastic pioneer, and Colonel Barton, who was president of Lion Oil and a great friend of his, called him up and said, 'Joe, I think there's some oil in Nova Scotia. Now do you have a rig you could move up there?' And Joe said, 'Yes,' and after some negotiations he called me and told me what sort of an arrangement could be made, so we sent a rig to Nova Scotia and drilled for three years without finding any oil. And the upshot of our adventure in Nova Scotia was momentarily not advantageous to either one of us, even though we had a rig on contract.' "[5]

IV

In July 1952, Joe Zeppa informed his stockholders and board of Delta's participation in the exploration of concessions in northern Spain. A group of American oil operators, primarily Texas-based, was contracting with the Spanish National Institute of Industry to explore the Milagro concession in the provinces of Navarra and Logroño. The group contributed a fund of $1 million. Members of the group included General American Oil Company (Algur H. Meadows's company), Delhi Oil Corporation (the Murchison outfit), DeGolyer and MacNaughton (the Dallas consulting firm), E. E. "Buddy" Fogelson, and Sol Brockman. On the basis of studies by DeGolyer and MacNaughton and other geologists, the concession's commercial possibilities seemed to be excellent. Meadows was handling negotiations with the Spanish government. Delta took a ten percent, or $100,000, interest in the venture. The contract was duly signed with the Spanish government, and by the end of the year Delta had agreed to rent its Rig No. 18 and supervise the drilling operations with Investigaciones Petroliferas Valdebro. The Valdebro expenses would be advanced to Delta from the group's funds.[6]

Delta was thereby launched upon what was truly its first overseas drilling assignment.

The Spanish venture with General American Oil and its partners was very much a learning experience. Hi Cole was the purchasing agent for Delta at the time and he experienced as many of the problems as did anyone.

"I believe it was the first rig to go abroad after I joined the company," said Cole, "and it was quite a deal too. You wouldn't believe how much red tape is involved in that. That rig had to be overhauled, and when it went over there it was in tiptop shape, and then we bought supplies for, I guess, a year's operation. That was one of our functions, to work out all the supplies and parts that would be needed for the operation of the rig during that time. Fortunately, I was very familiar with drilling over there, and it really stood me in good there.

"We didn't know what was available over there. Chris went to Spain and kind of surveyed the situation and found out there wasn't too much available. We had to send nearly everything. There was no end to the detail. In export there was more red tape and paper, but we eventually got it over there and went to work.

"You had to transfer everything to kilos. I spent many a night working on this. I would carry work home with me, but looking back now, it was fun, it was a challenge, and we did it. We did it and it worked. We had a wonderful guy behind us, pushing us all at the time and leading the way, and that was Chris. That man was something; Joe Zeppa was too."

What was it about Chris that stood out in this operation?

"I would say his unbounded energy and enthusiasm. He tackled everything with a spirit that was going to get it done. He was a good teacher."[7]

The crews went over in 1953.

Morris Hayes became the manager of the Spanish operations for Delta. Although ensuring that equipment and supplies were collected and delivered was a major problem, the key to the entire operation proved to be personnel. Many had been working in their own backyard, more or less, in the East Texas region, and foreign work was completely new to them. It was the first time that practically every one of the crewmen had been on an airplane. Some adjusted, while others experienced great difficulty.

Nolan R. "Buddy" Jones went over as a driller and, as it wound up, liked his Spanish experience. But the problems he encountered were also encountered by the other men.

"Somebody asked me where was my first foreign job," Jones said, "and I told them it was in Illinois. While I was in Louisiana, I got a letter from the company asking me if I would be interested in going to Spain, and they said, 'If you have some good men that you would recommend, bring them along and we'll talk some business.' So I came and I had four men working for me and three of them went to Spain with me.

"I got on the plane, and I had never been on a plane before in my life. I was dreading it. They sent us on the train to New York and [we] got on the plane and I said, 'Lord, you just open that door and I'll get out of this plane!' I looked out there and the manifold was red hot. I thought that was just bad. They set there and roared the motor and I shut my eyes and when I opened them again we were still setting there. When it finally did take off, I looked down and we were a thousand feet up. We got way out over the ocean, and one of my men—Elmer May, we called him Duke—said to me, 'I bet if this thing was to fall, them sharks would eat a man up.' I said, 'Boy, quit talking to me like that.' I said, 'Man, I'm scared to death right now.' "[8]

Jack Elkins, who went over temporarily to drive a truck until the equipment was unloaded and set up, was a rarity among the Delta contingent going to Spain. He had been on an airplane before, in the Army during World War II. Because he had already been in the Philippines, New Guinea, and Australia, he was prepared for the difference between Spain and the States. "In fact," he said, "it was better than I thought it was going to be, but some of them boys had never been away from home and I felt sorry for Mr. [Chris] Zeppa because we were all out there on location and they got on him: 'This ain't this and this ain't that.' Well, I knew what I was getting into, because there ain't no place like the United States. They were all expecting vacations down on the Riviera, and it didn't turn out that way. They were all getting on Mr. Zeppa and I told them, 'Hell, I knew what I was getting into when I came over here.' "

Unloading was a major task.

"They set that drawworks off the ship and had it lined up with that pier," said Elkins. "I mean, the skid of the drawworks was just even with the edge of that pier, and I was backing up, fixing to load that thing. Mr. Chris came by and he said, 'How are you going to load that?' and I said, 'I am just going to back up there and pull up on it.' He looked down there and he said, 'I don't know what to tell you to do, but I am going to tell you one damn thing: if you lose it in the water, your ass is mine.' he wasn't smiling when he said that. He meant that."

Carrying a big load of equipment such as pumps and drawworks through the Spanish countryside required the services of a police motorcycle escort. In Spain this complicated, rather than expedited, the trip.

"You would pay the escort expenses plus your swamper—he was a Spaniard—and, of course, didn't none of us actually ever know what a peseta was worth or nothing like that," said Elkins. "We would just get us a sack full and leave and spend it. We would go in that bank and those people standing in those banks, their eyes would bug way out and we would get a pocket full of those pesetas—and that was a lot of money to those people. We were doing a lot of spending over there. We were living high on the hog.

"Those motorcycle patrolmen, about 45 miles an hour is as fast they will go, and sometimes you would have to hold your breath going down one of those hills to keep from running over one of them, but every little town that had a beer joint, that is the first place those jokers would stop. They would shut her down and they would eat and drink that beer and wine like it was going out of style, and you had to pay all that. Those motorcycle cops were having more fun than anybody.

"We were staying in this nice hotel—and I mean it was out of this world. They had this live music, them boys playing those fiddles and stuff while you eat, and we would go in there every night and eat beef steak and potatoes; that is what we would order. That is all we knew what to order, and Mr. [Chris] Zeppa came in one night and he said, 'How come every night ya'll order beef steak and potatoes?' I said, 'Mr. Zeppa, that is all in the hell I know how to say, just beef steak and potatoes.' We were over there eating cheese and drinking wine and eating beef steak and potatoes.

"You have to park those trucks at a garage. If you didn't, those people would strip them at night. We got a pump loaded one night and Mr. Zeppa followed me to the car and I pulled down in this garage there in town and parked and I had a big suitcase and he had a big suitcase—and those taxis you could ride all over town for a dime—and I drug that old suitcase out and Mr. Zeppa got his. I set mine on the sidewalk and started looking for a taxi. Chris took off up the street. I said, 'You going to walk?' 'Oh, yeah,' he said, 'the walk will do us good.' I didn't feel like it would do me any good, though. I had to agree with him and went right along with him."[9]

Despite his trepidations while airborne, once Buddy Jones landed in Spain and got acquainted with his new surroundings, he found a lot to like. His wife, Ethel Marie, joined him, and that aided the adjustment.

"I was the first American to take an all-Spanish crew and make a well," said Jones. "It took us nine months to make the well. They took my two American men, the May brothers [Elmer and Jack], and put them on another crew, and I had an all-Spanish crew and they had an all-American crew. I was kind of mad about it. I said, 'I don't understand why you would give that to me when you got one man who is studying Spanish.' All I had was a little handbook that I used. Morris Hayes [the Delta manager] said, 'Buddy, I want you to take a crew and make a well in Spain.' I said, 'You mean, I am going to have to take them and I can't even speak Spanish real good?' He said, 'Well, that's not it. They like you and that's what they want.'

"I drew a many pictures. You wouldn't believe how many pictures and signs I made. They had a Spanish engineer who spoke English; he said, 'Hey, Jones, you have got to learn Spanish.' 'Well, how am I going to learn it? I've never studied it.' He said, 'You pick you out one of these men here and they'll help you, one that speaks good Spanish.'

"We finished the well in 1955. When I left there I had four men that could run the rig and I had five who could work on the derrick."

Delta had sent three rigs over at first. When Jones drilled his well in the country east of Madrid, he used a German rig that Spain had bought before World War II, though the labor was Delta's responsibility.

The Spanish operation, generally speaking, had to be self-contained. When equipment turned out to be inadequate, then more primitive local means had to be used.

"I can remember when we got over there on our second well and tore up the only two trucks that we had to rig-up with," said Jones. "So, to put this rig together, we had to use oxens and mostly cows. We hauled pipe down the side of these mountains, and a girl drove them. She was 17 years old, and she said that she stayed out of school to help her father when he had something to do. When we got things running on the rig, we would get jacks and jack them up. We had one truck and it usually takes four or five trucks to rig something up."

Jones and his wife came to know the Spanish people during his off-duty hours.

"I used to go and eat lunch with them in their homes sometimes," said Jones. "They were very poor people. One day a boy came to us and said that his grandmother wanted me and my wife to come to dinner on Sunday. So we went and my wife had a little trouble eating, but I didn't. She [the grandmother] would tell her boy to ask us something and have us say it in English, and she would just laugh. She hadn't ever heard English before. I guess I had as little education as any driller there, but I bet I could speak Spanish better." [10]

With others, the Spanish adventure did not run as happily.

"When we first sent a group to Spain," remembered Marguerite Marshall, who was secretary to both Joe and Chris Zeppa, "they were a motley crew. They really were. And we found out some really weird things about some of the people we sent over there.

"We had one roughneck who had worked for us for *years*, and he was solid, he was a steady worker, and when he got overseas we had trouble with him from the very start. And finally discovered that it was the first time he had been away from his wife and four children, and he could not read nor write! And he couldn't hear from them, he couldn't communicate with them. And he developed into a malingerer! He wouldn't work, he wouldn't do anything. We had to finally bring him home."

Bill Davis, a clerk, wrote "very colorful letters" from Spain that kept Mrs. Marshall and others in touch with the intriguing turns of human character which developed there.

One of these reports involved a young mechanic who was dispatched, in a company truck, to buy some parts. Three days passed without word of him. A search party began to trace his steps, expecting to find the company truck which, in turn, would help locate him. He had made a "detour." They found the truck parked outside of a whorehouse. He had been there three days. When they went in after him they discovered that he hadn't paid his bill. The charge was 100 pesetas per endeavor, and his bill was for 1,090 pesetas. They couldn't decide whether he had secured a cut-rate price after the first ten, or couldn't complete.

Bill Davis, who reported the anecdotes, had to settle the man's accounts in order to rescue him.

"There also was one young man who came from near here," said Mrs. Marshall, "and we thought, 'Well, that would be one we wouldn't have any problems with at all.' He was so clean-cut a young man. And in Spain tobacco is a government monopoly, and you couldn't take any in with you. And he dipped snuff. And he wouldn't work without his snuff. He came home."[11]

John Murrell, now chief executive officer of DeGolyer and Mac-Naughton and a partner in that firm at the time of the Spanish venture, summed up Delta's experience:

"Joe Zeppa took the bull by the horns and said he would put two rigs over there, which he did, and it was most difficult at that time, due to the fact that the Spanish are a prideful people, and they thought that everything in Spain should be utilized before anything was brought from the United States. And the Spanish national should be taught as soon as possible to operate any machinery that was brought over there. It was a very difficult situation. Joe sent Chris over, who did an excellent job of pacifying those people. First, because he could talk Italian and Spanish. And, second, because they liked him. From a standpoint of amicable relationships with the Spanish people, there is no question but that Joe and Chris did a marvelous job, as well as all the people they sent over there. There was never any doubt of that. However, it was an unsuccessful search for oil, except they did find one field—a field that produced oil which had arsenic in it. No place in the world have we found it.

"Spain was buying oil and importing it at that time, and it was a drain on the economy, but they connected up this field and they used maybe a tenth of their oil mixed with 90 percent of other oil, which made it usable. That's the Spanish venture in a nutshell: Joe carried forward. They must've drilled 40 wells over there, and only two of them produced oil. That was in the arsenic field."[12]

V

The Venezuelan operation may be characterized as one that began with high hopes, seemed on the verge of becoming a flowing success, and then began to dribble into doubt and, finally, into a losing situation. The boom which had initially lured Delta into that country gradually faded in the face of threatening nationalization of the foreign oil companies. Generally, Delta's equipment was not first rate, and the cost of improving it added another burden to an already difficult problem.

The Venezuelan story began in 1954. Within several years the operation was in trouble, and in serious difficulty by the 1960s. By the 1970s, a sensible way out was being sought.

The Venezuelan venture began at a time when the American oil industry was in the heyday of foreign exploration. Following a trend of the industry as a whole, Joe Zeppa kept his eyes and ears open for a way to exploit this trend.

In 1954 Zeppa orally briefed partners Sam Sklar and Sam Dorfman on negotiations that would take the company into South America for the first time. At the end of the year he spelled out the details in a formal board meeting.

By then, Delta had acquired 80 percent of the stock in the new Perforaciones Delta C.A. with capital of $30,000 (100,000 bolivares), under Venezuelan laws. Perforaciones Delta then acquired from John R. Peddy of New York an option to purchase all of the stock of Perforaciones Guarico C.A. of Venezuela, and then actually purchased the stock. In purchasing the company, evaluated at $1 million, $279.686.20 was paid initially, followed by the balance in equal installments over the next four years. Delta lent Perforaciones Delta $225,000 for its share. Peddy received five percent of the stock in the new company and $25,000 cash for his services in making the purchase possible. When the dust had settled, Delta owned 75 percent of the Perforaciones Delta stock.

Along with stock control of the new company, Delta gained from Perforaciones Guarico one small and six medium rigs, along with a number of trucks, cars, Caterpillars, a large warehouse and yard, and "very substantial inventories" of drill pipe, machinery, bits, and operating supplies.

Zeppa pronounced it "an excellent purchase," for the new company, as he saw it, "has adequate working capital, good prospects for profitable work, and an established business and reputation in Venezuela, where drilling operations should be profitable."[13]

In one of the first transactions after Perforaciones Delta was in business in 1955, the new firm bought all of the drilling equipment from Compañía Anónima Maracaibo Towing Company—including four rigs—and a contract with Richmond Exploration Company to drill a number of wells in the Boscán field near Maracaibo. This cost $200,000 cash and $450,000 in installments over a three-year period. In turn, this new equipment was leased to Perforaciones Delta's wholly-owned subsidiary, Perforaciones Guarico, to operate it, and the contract was assigned to the subsidiary. Delta's new Venezuelan company then purchased two rigs and other equipment from Richmond Exploration Company for about $345,000 cash and work that was to be performed.[14]

Watson W. Wise, the Tyler independent operator, remembered the day Joe Zeppa, at the onset of the deal, called him and said, "Can you go to New York? There's a company available in Venezuela, and I'd like to check it out."

Wise said, "So I went with him and we got in the old DC-3, Jimmy Rider the pilot, and believe it or not, we flew from Tyler to New York in a DC-3 and took Ed Kliewer, Joe's lawyer, along, a very capable man, and visited the people in New York who had a group of four drilling rigs. None of them were very modern, but Joe was particularly interested, I remember, because as he examined the assets of the company he found that they had $475,000 in the bank in cash, and this could countermand a lot of old depreciated drilling equipment. I felt personally that the equipment was not modern enough to be utilized for a very long time. Joe was interested in what he could do with that $475,000 in replenishing, and he bought the company and operated it for several years."

Bob Waddell said, "Joe Zeppa had told me about a month or so before that 'I may want you to run down to Venezuela.' He said, 'A contractor down there has got some rigs for sale, and we may want to buy them, and I may want you to go down there and look at them.' Well, he called me in and asked me to run down in the morning. He always wanted you to go in the morning. I said, 'May be day after tomorrow before I go.' I said, 'You want me to go down and look at those rigs? You still thinking about buying those rigs?' He said, 'No, I've already bought them, and I want you to go down there and look and see what I bought.' He bought some fairly good equipment, some not too good."

Mark Gardner, a Delta vice president, remembered: "I didn't know anything about it, and Joe bought that company without anybody ever seeing the equipment; without any question about their operation, we bought it, and we ended up with about seven or eight drilling heaps over in eastern Venezuela that really were a pain in the neck."[15]

Most of Delta's early months in Venezuela were spent acquiring additional equipment and fixing it up. In 1956 a barge was bought in Venezuela for $40,000, and $124,500 was spent to convert the barge for drilling operations. Through its financial connections, Delta arranged financing for the subsidiary.

Meanwhile, the drilling subsidiary, Perforaciones Guarico C.A., gained a contract with Venezuelan Sun Oil Company to drill wells in Lake Maracaibo, but this necessitated constructing a drilling barge for that purpose, plus equipment—the total tab coming to $685,000.

As it turned out, Delta's Venezuelan outfit drilled many of Sun Oil's wells on Lake Maracaibo, primarily because Sun Oil executives knew Zeppa and his reputation. John G. Pew, a retired vice president of Sun Oil, said, "One well we drilled down there was 7,000 feet and we drilled it in ten days, which was pretty much of a record, as far as my experience. It turned out that that property that we bought at Lake Maracaibo produced over one billion barrels of oil before the Venezuelan government confiscated it."[16]

"Venezuela was in a boom there on Lake Maracaibo," said John D. Hall, "and we sent three lake tenders with all the necessary equipment. Made good money for a while. These tenders were all built specifically for Lake Maracaibo, for platform drilling."

One barge was built in Pascagoula, Mississippi, and two near Houston. Then they were towed down to Lake Maracaibo, according to Walter L. McElroy, then in the purchasing department.[17]

By the end of 1956, the future for the Venezuela operation appeared rosy, with a $450,000 profit expected; as a result, Zeppa believed that "foreign work and Gulf offshore work would offer the best chances for profit," making it "highly desirable to concentrate on the expansion of these operations rather than in the continental United States land operations." Accordingly, he was in process of purchasing controlling stock—perhaps up to 70 percent—of Compañía de Perforaciones y Servicios S.A. of Caracas, which he considered "very cheap" at $500,000. That company had five drilling rigs and other properties, which would bring Delta's Venezuelan subsidiaries to a rig count of 18 drilling rigs and three workover rigs. As Zeppa remarked to the board, "This will make it the most important outfit in Venezuela, with the possible exception of Loffland Brothers, who may have a few more rigs."[18]

"Joe's ambition was to be the biggest drilling contractor in the world, in number of rigs," said Mark Gardner. Zeppa's comments to the board do provide support for Gardner's contention, for at that time Venezuela was a leading world site for drilling, and Delta was making a run for the top ranking.

In January 1957, Perforaciones Guarico C.A. was liquidated and dissolved into Perforaciones Delta C.A. With its drilling operations consolidated into one company, the Venezuelan company was operating ten land rigs, one workover barge rig, one heavy-duty shallow-water barge rig, one heavy-duty rig on barge tender for Lake Maracaibo, with two additional ones under construction—the last three representing a cost of about $3.3 million. Zeppa saw "an unprecedented demand for drilling equipment," especially in Lake Maracaibo. By then Perforaciones Delta had three contracts—with Sun Oil, San Jacinto Petroleum Company, and Mene Grande Oil Company. Despite this $3 million worth of business, it was necessary for Delta to advance money to its subsidiary to keep it going by acquiring additional equipment.[19]

Financing was the biggest problem in foreign operations, complicated by Delta's being a private company. The money provided by Delta for part of the operations came from bank borrowings, and the company couldn't always get the money back from foreign operations as rapidly as it was needed to meet commitments. "We had to meet it from other sources," said Joe Bevill. "This created problems. Logistics were pretty bad. We had the materials and equipment shipped from the United States. Most of it was bought in the name of Delta. We didn't have to tell them who Perforaciones Delta in Venezuela was.

"Continental-Emsco was our principal supplier, and they knew our business real well. We laid all this information before them. They knew what we were doing."

C. L. Vickers, who was with Continental at the time, remembers: "[Joe Zeppa] bought several new rigs and some barges from Continental which were sold to the company in Venezuela. His expenditures at that time were very great, and at one time collectively the Delta company in Venezuela and the barge company down in New Orleans [Delta Marine] were all indebted to Continental for equipment, and the total added up to about $4 million dollars. That would be in the late '50s."[20]

In the late 1950s, though the situation still appeared "quite favorable" in Venezuela, a decline was in the offing. In 1958, while drilling was down both domestically and abroad, Delta enjoyed long-term contracts on Lake Maracaibo with four companies—Sun Oil, San Jacinto Venezuelano, Mene Grande, and Phillips Petroleum Company. The future seemed safe.

The optimism was short-lived. A year later, Zeppa reported that prospects for Perforaciones Delta in Venezuela had been "less favorable than expected." A few months later still, Zeppa reported the Venezuelan subsidiary was suffering a loss of $113,934.78 through May 31, with drilling operations there "at the lowest level in many years," with "no signs of improvement." Among the factors were stiffer competition, closer profit margins, increased costs. With domestic drilling in a depressed condition and low oil allowables in Texas, this was hardly good news.

The Venezuela operations never really picked up significantly enough to recover. By the end of 1959, Zeppa reported operations there at "a low ebb throughout the entire year." Only one of its four Lake Maracaibo drilling tenders and one land rig were operating. Rig activity overall in Venezuela was down to 40 percent, with only 50 rigs operating. Previously there had been 120 rigs running in the country. Consequently, one of the Venezuelan rigs was sold to the parent company for removal to Mexico. By 1960 Zeppa was reporting that contracts for land rigs in Venezuela were "virtually non-existant." He continued to expect improvement, while acknowledging that Perforaciones Delta "continues in a pitiful condition." The subsidiary had operated from one to three rigs during the year, and a barge and rig were to be shipped to the States. A strong sign of the unhappy conditions there could be seen in the fact that Shell Oil had stacked 16 rigs in excellent condition and had them up for sale at "a remarkably low price."

But Zeppa clearly was not ready to give up. We can almost read his surge of optimism in the minutes when it was recorded, "There is a general feeling in the industry that the drilling industry has about reached bottom and a turn for the better should not be too far off. Many contractors have gone out of business and many others are barely keeping their heads above water from day to day."[21]

As if economic conditions were not wretched enough already, Delta's Venezuelan subsidiary suffered yet another blow when a derrick inexplicably fell into Lake Maracaibo, killing one person. As for the equipment, the crown block and the top part of the derrick were lost, while the remainder of the rig stayed on the barge, which suffered no damage. "The loss of the man is the most tragic part," Zeppa wrote to a friend. Noting that the friend had disposed of drilling equipment in that country, Zeppa commented: "You really did sell your rigs in Venezuela at the peak of the market. You couldn't give them away today."[22]

By the next year—1962—hope seemed to have been abandoned for Venezuela, and rigs were being shipped to the United States. By the end of the year, however, one barge rig had been working for Sun Oil, and there was guarded optimism that things might pick up. By 1964, a small profit from Perforaciones Delta could be seen; another rig was being shipped from there to Brazil. But the long-term prospect was dim, because the company would need to overhaul and modernize much of its equipment, a further financial drain.[23]

While the general picture in Venezuela appeared almost uniformly dismal, one bright spot pierced through the gloom: Perforaciones Delta's 50 percent ownership of Petroleum Machine Works there, which, though small, continued over the years to generate a respectable profit for its size. A pertinent comparison of the two companies may be seen in the fact that in one three-month period in 1968, Perforaciones Delta lost $35,000, while Petroleum Machine Works *made*, before taxes, $100,271.

By the end of 1969, however, Zeppa expressed renewed optimism when Perforaciones Delta showed net income of $339,269 for a nine-month period ending September 30. Drilling activity seemed to be up throughout Venezuela, and especially on Lake Maracaibo. The optimism soon faded, for by 1971 the company was back in the red, one reason being the high cost of maintaining the old equipment. Added to this, "an unfortunate series of fishing jobs and mechanical failures" during the drilling of five wells made the losses greater.

As if this were not enough, Leonard Coats, the manager and 49-percent owner (by now Perforaciones Delta owned 51 percent) of Petroleum Machine Works C.A., died of a heart attack during a flight intended to take him to Anaco in eastern Venezuela, which a hijacker had diverted to Havana. The Delta subsidiary was some time acquiring the remaining interest, finally buying the remaining 49 percent of the stock in Petroleum Machine Works from his widow for $125,000.

Though Zeppa had not given up in Venezuela, he acknowledged in 1972 that the equipment was "outdated and inadequate" at the time when operations in that country were becoming oriented toward the government-owned company, CVP, which would eventually be the only customer there.[24]

That year Joe Zeppa admitted to his old boyhood friend, George Keating, that "conditions in Venezuela went from bad to worse" and that Delta would reimburse Keating for his $51,657.10 invested in Perforaciones Delta, despite the fact that over the years the value of the bolivar had dropped from 3.30 to 4.50 to the dollar. But he repaid Keating in dollars, not bolivares, and urged him to "put the checks in the bank and forget Venezuela."

No better summary of the Venezuelan venture can be gained than Zeppa's comment to Keating:

"PDCA had excellent prospects and was making money, but, unfortunately, it had to reinvest it all in drilling equipment, and about the time it should have been producing earnings the bottom fell out of the drilling business in Venezuela."[25]

By 1974 Perforaciones Delta could report a small profit for the first time in quite a while, but a severe storm on Lake Maracaibo in September turned over and sank Drilling Barge Tender No. 4, causing a total loss. The insurance company was prepared to settle on its insured value, 6 million bolivares. No Delta personnel were lost in the accident, though an employee of a service company was killed.[26]

In October that year, Jack Robinson went to Venezuela as Delta's manager, shortly before the government nationalized the oil industry in January, 1975. "Before I got on the airplane, the first newspaper I picked up, I knew this was going to happen, and we did start taking steps to reduce our staff. So we went from a six-rig operation down to a one-rig operation in about six months."

With the nationalization of the oil companies in Venezuela, the contracts were canceled on the Delta subsidiary's rigs, though the Delta operation itself was considered a service company and not affected by the new laws. The rigs were shipped back to the States and sold at auction in the fall of 1976. Because the rigs had not been maintained well, they were not economical to use in the States.

The last barge in Venezuela was sold in 1980, closing out a 26-year story.[27]

VI

While other movements abroad were dictated by the needs of the industry at the time, attorney Ed Kliewer feels that Joe Zeppa's going to Italy was "very close to his heart."

"There were wells to be drilled in these foreign countries, and he wasn't the only one there, nor was he the first one. The American companies were going foreign more and more," said Kliewer.[28] However, Americans were not stampeding to drill in Italy, which suggests Zeppa may not have chosen to operate in that country had he not had emotional ties to it.

The Italian operation, while exhibiting an occasional low spot, has been consistently one of the happier results of "going foreign."

The operation began in 1955, a time when foreign exploration was already catching the eye of American entrepreneurs and when Delta was looking for profitable opportunities overseas. We don't have Joe Zeppa's personal account of how the venture came about, except as it has filtered through the pages of the minute book, but we do have the testimony of Chris Zeppa, who kept a close, personal hand in the matter from the beginning.

He said, "The Italian operation started this way: You know, wherever you were born, it never leaves your body. And by that time we were already pretty well foreign-drilling operators anyway, in different places.

"But the actual impetus to the thing was this: In 1955 the oil people had their conventions [World Oil Congress] in Rome. So Joe Zeppa went to the convention, and Rome had just laid it out for them. Every convention has a certain movement, whether it be a convention for apple growers, or whether it's a convention of people buying bicycles or automobiles. On the outskirts of that convention are people that [have] got something to sell to these conventioneers. And, of course, the minute the convention gets together they begin to get acquainted with this guy and that guy.

"Well, Joe Zeppa, still pretty well knowing his language, he was able to do a lot of contacting, more than anybody else. Some people—three partners—had a little drilling company over there. At one time they were oil engineers working for the state. In fact, they were early oil engineers in Turkey—for Italy—and in Yugoslavia, and across to the Middle East. But somehow or other they didn't make it go real good to suit them, and so they decided to sell their little company. And as this convention gathered, they began to run around and see who in the hell would want to buy them out.

"And so they started making connections, meeting people, and knowing people, and most of them speak some English, but all of them could speak Italian damned well—they were all graduates from the finest university in Italy! So who would they talk to first? They'd say, 'That's old Joe, that's Joe Zeppa; he's an Italian, isn't he?' So there was the connection. They buckled him up. And they got to talking.

"Now the next thing they proposed: 'Look, Joe, we have a nice little company here. We got a little drilling rig, working for the state, and so forth and so on. We don't want to get bigger and if we don't get bigger we just wouldn't be sufficient. We certainly would like to sell that little company. Want to buy it?'

"Joe says, 'No.' I'm just guessing—I wasn't with him. But I'm sure he had lots of gaff, and then he asked what they had and this and that and how much money they wanted, where they were working, that sort of thing. And they were really three fine partners, finest engineers you could possibly [see], from the best university in Italy. One was a Tuscan, one was a Venetian, and one was a Sicilian who had it on all the rest of them—he had a big orange farm in Sicily!

"Anyway, [Joe] came home, the convention was over, and the first thing he said to me, he says, 'Chris, you know what?' And he began to recount this story about these three fellows and one thing and another. At that time, just a little bit before that, Gulf Oil drilled a little well on the Adriatic coast and found a small shallow field. Well, these boys knew that that'd be the time to [sell]."[29]

So when Joe Zeppa returned from Italy in August, 1955, he formally reported to his board that he had conducted negotiations to buy Compagnia Industrie Metanifere e Affini (CIMA), which he described as an established drilling company with one medium rig. Delta agreed to pay the three owners—Mariano Amico, Italo Veneziani, and Antonio Andalo—a total of $208,000, or 130 million lire, in a combination of cash and notes, some of the proceeds to go toward liquidating CIMA's debt, which had been one of the reasons for the partners' wishing to sell. This transaction left Delta in possession of all the capital stock.[30]

"So I went to Italy for something or other, not direct to that, but while I was over there, would I look into it?" said Chris Zeppa. "I looked into it and talked to them, so it looks like the shallow field on the Adriatic coast was going to be good one. So [Joe Zeppa] says, 'Look, if you like what you see, send me a cable and we'll see what we can do about it.' And that's how we got into Italy. We bought that rig out and we immediately sent two or three good rigs and we started."[31]

Chris Zeppa executed a contract of purchase on August 23, 1955, which provided for closing the deal by September 30. The contract refined the terms, providing for the purchase of all the stock—face value, 100 million lire—for 40 million lire, subject to adjustment which depended upon the current position of the company as determined by an audit. This audit report showed the capital stock was closed for 39,155,875 lire, amounting to $62,865.66. Because CIMA did not have sufficient operating capital, provision was made for the previous owners to lend CIMA 30 million lire, which was secured by notes to be repaid in three or four years. At the same time, additional capital stock was provided, and purchased, by Delta.[32]

Before the year was out, Delta had a $25,000 interest in a $400,000 pool contributed by American companies and individuals to acquire an old concession of about 7,500 acres between Rome and Naples, which had been in shallow stripper production for a number of years. The group planned to drill a deep well on the land, with CIMA hired for the drilling.[33]

Back in Tyler, the transaction had its impact. When Walter McElroy started as a materials clerk in the purchasing department in 1956, Delta had just gone into the Italian operation, and he was soon in the midst of getting two rigs ready to go to Italy. Because parts might not be available locally, they had to outfit a self-contained operation. "For continuity of operation you have to send with that equipment things you think would keep it running," said McElroy. "You've got to have a warehouse capable of keeping that rig running. Going to Italy or Sicily we took hand tools and even operating supplies that we had no assurance at that time would be available to us locally. It was quite a chore getting a project like that off and on its way."

Among the moves to better equip CIMA in 1956, Delta supplied, through Continental Supply Company, two rigs, one a Delta rig and the other one a new Emsco A-800 power drilling rig.

The Italian acquisition began to look good by the end of 1956, with a profit of about $100,000 expected.

Soon CIMA was operating four rigs, with capability ranging from 4,000 to 12,000 feet. CIMA kept its same personnel. "The organization was there," said Joe Bevill. "Those people were well-qualified, well-educated people, engineering wise. When we enlarged the operation in Sicily, we did send a few Americans, but not many. When the Sicilian operation was over, the people came home."

The Sicilian venture came about early in 1956 when Zeppa added to Delta's Italian holdings by purchasing all of the stock of L'Aquilea

Mineraria S.p.A. from Franco F. Fenzi for 1 million lire. The purchased company controlled three concessions in southern Sicily and had two permits to obtain concessions of about 60,000 acres. In addition to the stock purchase, Delta paid Fenzi about 20 cents an acre, or $12,000, which had been the cost to Fenzi.

By 1957, the fully owned subsidiary was operating four rigs in Italy and Sicily, with L'Aquilea Mineraria S.p.A., the Sicilian company, housed in Palermo. By that time CIMA had long-term contracts with the Italian Edison Company and Monsanto Chemical Company for drilling in Sicily, which Zeppa termed "very profitable."

"We bought two additional rigs to send to Sicily and they were operating for the Italian company," said Bevill. "Operated for two years or more in Sicily. Didn't find any oil, as I remember, and those rigs went on to the mainland."

James Wylie Smith served in Sicily as a driller for 19 months, during which he saw Joe Zeppa twice and also Chris. He had an all-Italian crew.

"We drilled one well 12,000-and-something right off the side of a mountain," said Smith. "Looking to the east, that Mt. Etna. I always wondered what would have happened if we'd a drilled into that volcano."

Smith faced a problem other Delta drillers experienced on occasions while on foreign assignments. "They couldn't speak English and I couldn't speak Italian," he laughed. How did he get by? "Just by motioning. Motions."

They drilled east of Palermo, the Sicilian capital where they had entered the country. But as the rig moved about Sicily, Smith found the language barrier even more complicated than he had realized.

"I was stationed in four different towns," he said. "They didn't speak the same lingo."

That fact probably discouraged him from undertaking a systematic study of the language.

All dry holes inflict pain upon an oilman, but the Sicilian results probably brought Joe Zeppa a bit more disappointment than a similar experience elsewhere would have. As C.L. Vickers said, "I think that it would have been a great satisfaction to Mr. Zeppa if he could have discovered a real oil field in his native country."[34]

In the long run, one of the most important events in the Italian company's history was the decision of bespectacled, stocky Vinicio DiCocco and several other key employees to remain with the operation. DiCocco, a 1946 engineering graduate from the University of Pisa in northern Italy where he was born, had joined CIMA in Bologna four years before the Delta purchase. Zeppa appointed DiCocco general manager of the newly acquired firm, and in 1956 CIMA's headquarters were moved to Rome.

DiCocco was to remain with the company, and in 1974 was named *consigliere delegato*, or managing director.[35]

By 1959 CIMA had been reorganized as Delta Overseas Drilling Company, remaining as Delta's fully owned subsidiary.[36]

By the end of the year, however, the Italian company was exibiting some of the same distressing signs seen in Delta Marine and the Venezuela firm—it seemed likely to lose money. It was operating at "a very low ratio," with only two of its four rigs keeping partially busy. Though there was hope for increased drilling in Italy, it was also recognized that if business did not pick up within a reasonable period of time, some of the equipment would have to be transferred elsewhere. The following year brought more unwelcome news, for in the early part of 1960, Delta Overseas Drilling Company had nearly all of its equipment stacked; by the end of the year, however, three rigs were working, albeit at low rates.

Unlike the story in Venezuela, though, prospects in Italy brightened so much that by late 1962 Chris Zeppa could complain, in a letter to his sister Paola, now back in the old home at Fubine, that he wouldn't get to visit her in Italy for another year. "Things with Delta Overseas in Rome are going along so well it gives me no excuse," he wrote.

Then Chris revealed an expectant bit of business news to his unmarried sister, who had never seen one of Delta's Italian rigs: "If DiCocco gets the contract he is working on at this time, he will move the largest rig to the little town of Villa Deati, which is about 25 or 40 miles from home—halfway between Alessandria and Chivasso and quite convenient for you to make a nice little trip . . . . So, you see, we are getting closer to home. Who knows, maybe there is an oil field in the valley where the meadows are."[37]

Delta Overseas stayed "fairly active" in 1963, despite sluggishness near the end of the year. Its activity was enough to expect a net profit of $120,000 after depreciation—a forecast that was revised to "very small earnings" by the end of the year, then to an actual loss before depreciation. Joe Zeppa was looking into the possibility of applying air drilling in Italy. The shales there were giving trouble—they disintegrated upon contact with water. The firm searched for a less costly way to penetrate the formations there.[38]

By 1965 the Rome-based firm's financial picture brightened considerably when Chris Zeppa negotiated a job in Turkey with Mobil Oil Company. Although Delta Overseas was continuing without financial support from the parent Delta, the work in Turkey came at a most propitious moment, for activity in Italy remained low. With the boost from the Mobil job in Turkey, by the end of November that year, earnings of $119,455 had already been posted. This can be seen as the hint of an upsurge, slow in coming, but by early 1967 being interpreted favorably as "a full schedule of work" that had brought "rather handsome profits." By the end of 1967, three of its five rigs were operating. Although the profits generally were small, the Italian operation appeared to be causing no concern, unlike the one in Venezuela. By the end of 1968, Zeppa was even considering moving a rig from Australia—at a cost of $160,000—to Italy for deep-well drilling. Since such equipment was unavailable in Italy, the customer had agreed to absorb the expenses of ocean freight and duty.[39]

Although prospects appeared "reasonably good" for the Italian company at the end of 1969, it owed Delta Drilling Company, the parent company, $614,812—a debt of $273,895 on the original drilling equipment plus the sales price of the rig moved from Australia. For the year 1959, Delta Overseas earned $50,000 after taxes, but was unable to make any significant reduction of its debt. The following year the company lost $50,720 after depreciation because reduced operations failed to produce enough income to cover overhead. But Joe Zeppa persisted in backing the project: "The company is well-managed and fairly well-equipped for the drilling in Italy, and it is hoped that conditions will improve." In mid-year, Joe Zeppa flew to Italy following a meeting of the Seventh International Petroleum Congress in Moscow, visited Milano and Fubine, then conferred with DiCocco about the Italian business.[40]

While Delta Overseas stayed in debt to the parent company, profits continued to be nominal or nonexistent. An attorney suggested that the Italian firm be recapitalized, increasing the capital to 165 million lire, which Delta subscribed to, thereby reducing the amount of the debt to Delta. This left Delta owning stock in Delta Overseas with par value of 315 million lire, with the Delta Overseas debt to the parent coming to 305,845,258 lire.[41]

The next month—April, 1972—Chris Zeppa visited Italy and reported from Fubine to his brother, "In Rome they are busy for the near future. They [have] more work than they have men for three rigs and maybe four." His optimism seemed to have been justified in early 1973 when Delta Overseas reported a net income of $247,680 after depreciation. By then, one rig, No. 6, held a long-term contract, and No. 5 was to be put into service.[42]

But by late 1974 the debt to Delta had risen to $413,848.71, of which $142,000 represented the National 130 rig moved there from Australia. At the time, Delta was being repaid $15,000 per quarter toward retiring the rig debt; once that goal was reached, the installments were to be applied to Delta Overseas' debt for a Gardner Denver PZ-9 pump which had cost $115,397. The remainder of the debt—$155,646.96—was for supplies and materials shipped.[43]

In early 1975, prospects were picking up in Italy, as Delta Overseas secured a contract with AGIP, the state agency, for two years, with an option to extend it for two additional one-year periods. The site of activity was a newly discovered high-pressure gas field east of Milano, for deep wells. This represented the most lucrative event in Delta Overseas' history, for the rigs would be working on a day rate, earning around $8,400 per day. In order to facilitate the deep-drilling contract, the parent company prepared its Rig No. 68, a Continental-Emsco Electrohoist III in Lafayette, Louisiana, to drill to 20,000 feet, after which it would be sold to the overseas affiliate. Out of the day-rate income, Delta Overseas would pay Delta the sum of $91,250 per month for 48 months to repay the sales price of the rig, which came to $4,380,000.[44]

Much of Delta Overseas' activity was along the Adriatic coast, where it drilled for the government agency AGIP, the firm ELF, and Montecatini-Edison, looking for deep gas and condensate. By this time Delta Overseas had five rigs out of an estimated total of eight or nine contract rigs available in the whole country. By then Italians were customarily paying $1.25 per gallon for gasoline, a strong motivating factor in an accelerated search for oil and gas.[45]

Though the Italian operation had gone through some troubled times, it had managed generally to hold its own or generate a small profit. Now as the search for fossil fuels intensified, it appeared to be in a strong position to take advantage of what lay ahead, for it had effective management, capable crews, and proper equipment—the traditional ingredients for success, as long as economic conditions are favorable.

Joe Zeppa himself summed up the Italian operation in 1975, shortly before his death: "We have six rigs there. We are shipping a new one. Italy is good. Italy is pretty good to work in. And you get paid, that's the main thing."

Five years later, Chris Zeppa provided an updated report. "Now our Italian company is a going jenny," he said. "It has drilled the deepest wells in that country. Actually, the last few years it has drilled for practically nobody else but the state-owned company. Strictly contract drilling, no production at all [for Delta]. Well, at one time we had an interest in a well south of Rome a-ways, but I don't even know if it's producing any more. We drilled one of these for our own account, and it produced a while.

"So the company now has a beautiful office. They moved out of Rome, because Rome has become impossible for traffic. So moved to Fiumicino—you go out to the Adriatic straight out of Rome. And we bought an old glass factory that was not being used any more; we converted that to an office.

"The company is very successful, doing real good. Now they have four drilling rigs, all working for the state. They all are on land. One of them is between Milan and the big airport north of Milan—that's the deepest rig, drills 15,000 or 16,000 feet. One is now drilling south of Milan, not too far from where I was born, by the way.''[46]

VII

Delta Drilling Company pioneered the exploration of Libya for petroleum, drilling the first well in that nation in 1956, under contract to Libyan American Oil Company.

In July 1955, Libya's new petroleum law became effective, making large concessions available to foreign oil operators. Libyan American, the Libyan operating firm of Houston-based Texas Gulf Producing Company, was one of 12 American, British, and French companies to file applications for oil concessions. Libyan American was awarded a concession in one of the large surface anticlinal structures which reconnaissance studies had indicated might be particularly promising—the extreme northern Cyrenaica region, or Zone II, as it was categorized. Among the stipulations, Libyan American was obligated to drill at an early date.

Libyan American, with Delta doing the drilling, spudded in the first exploratory well in Libya on April 30, 1956, in Concession 18 of Zone II in northern Cyrenaica. This was near Benghazi.[47]

Delta drilled three wells on contract in Libya during a period from 1956 to 1958. Since drilling was in super-arid country where centers of population were far apart, the operation had to be as nearly self-sufficient as possible. Chris Zeppa saw to it that there would be a full supply of spare parts. As one Delta hand put it, ''You had nearly two of everything but a drawworks, because we had extra motors.'' The parts inventory came to around $100,000, a princely sum in those days. Having those extra parts on hand, according to Hugh McKenzie, who pushed tools in Libya, was ''just like walking into Continental Supply or National Supply—you go buy what you wanted and go in there and get it.''

Because of the isolation and climate, the effort in Libya was a challenge to the crews there. The relative newness of foreign work to Delta crews added to the difficulties which might ordinarily be found there.

Soon after the first contingent of personnel had gone to Libya, complaints from the operator reached the attention of Tyler, pertaining to the efficiency of the work. Bob Waddell was dispatched to the scene as troubleshooter.

''Being in the Tyler office, right under the gun, when anything like that came up, Joe Zeppa'd tell me to pack my clothes and go,'' said Waddell, ''and sometimes he'd want you to leave that afternoon, but you couldn't. So I went into headquarters of the company we were working for in Tripoli. The rig was quite some distance from Tripoli; in fact, it was east of Benghazi over toward Egypt in the area where Rommel and Montgomery fought back and forth. There were still land mines, and you could see remnants of all the fighting back in there when I was in there about '56. There weren't any roads. You had to run land mine detectors out to any location before you could go into an area, and we were finding mines all the time. Once in a while you'd hear one go off over there in the desert. Camel would step on one or something.''

When Waddell checked out the manager of Delta's operation, he decided then and there to dismiss him, and he wired Chris Zeppa to send over a replacement. But he had to remain until a new man arrived. Meanwhile, Joe Zeppa, in Europe at the time, decided to fly down to Tripoli. Waddell took a commuter plane into Tripoli to meet him. Zeppa wanted to go to the location, so they flew back to Benghazi and drove 75 miles out to the rig over the rough terrain. Zeppa spent a few days at the camp, where the crews lived in Quonset huts.

"This manager that I had fired, he was very apologetic,'' said Waddell. ''Joe Zeppa had a soft heart. He got in the room there and talked with him. Joe Zeppa felt sorry for him, and thought maybe if I'd relent he might be able to make it. I finally told him, 'Now, listen, if you're going to keep him, I'm going home.' ''

Finally Joe Zeppa acceded to Waddell's decision, but invited the discharged manager to take his wife to Cairo, then on a tour of Europe, at Delta's expense. Meanwhile, Waddell waited nearly two months for his replacement to arrive.

While on location there, Joe Zeppa wanted to see all of the local sights. Over Waddell's protests he insisted on being driven to a watering site for Arab tribesmen.

"We had a water well for that rig. We were pumping water out of an old dirt well about six feet in diameter, right through that old limestone rock, and it was an underground stream that the Arabs had found probably 1000 years before that, and we made a deal with those Arab tribesmen, with the sheik, to pump water out of that well to furnish the rig. And that old well furnished all the water for the rig, camp, everything. We had agreed to pump water for their camels and goats and sheep that they brought in there to water, and we poured some concrete troughs down each side of the well. We pumped water into those troughs, and they'd come in there sometime with 600 camels and they'd drink 30 gallons a piece, and we pumped water for them.

"It was about three miles from the camp. Joe Zeppa decided he wanted to go down there and see the camels and goats that were watering at the well, and I tried to talk him out of it, but nothing would do, and there he was. He went down. We got in a jeep, and the sand was running-board deep on that jeep, and the dust just rolled all over you, and we got about halfway down there, and he was bareheaded and that dust just covered him, and then he started on me and he said, 'I know why you wanted me to come down here. You wanted me to get this dust on me!' I said, 'Wait a minute. This is your idea, not mine. I told you it was rough.' "

When they arrived, as it happened, the Arabs also came in with a large flock of camels.

"He got out there fooling around, and this old Arab sheik was on a gray horse with an old homemade Arab saddle and with his head all wrapped up and all those families—see, they run in tribes, the whole family, and they'd come in there and water those camels, goats, and sheep. The first thing they did, they'd get over there to those concrete troughs and they'd wash their hot feet in that water. And then they'd get down there between those camels and start drinking.

"Joe Zeppa wanted some pictures made, so he got this old sheik with his horse and he wanted a picture made with him, so they started taking pictures of him. I tried to tell him, 'No, no,' but nothing would do but he would get some pictures. Well, they just swarmed him. I could see his bald head, and he made matters worse: When he got some pictures he reached into his pocket and started handing some piasters. When he started handing that money, I mean they would swarm him, and I didn't think I was ever going to get him out of there!"

Unfortunately, there is no photographic record of the event, for Zeppa had a new camera and the film apparently was improperly loaded. Later, Waddell—again over protests—took Zeppa to the old Arab market in Tripoli, where Zeppa wanted to mingle with the local people. He insisted on getting out of a horse-drawn buggy to take pictures.

"He was hard-headed, you know, and he wanted to walk around," said Waddell. "I said, 'You walk through there, that old Arab market, and you going to have fleas and everything else all over you. Don't get out.' Naw, he got out and walked and I don't know how many fleas he got on him, but I wouldn't get out of the buggy. But he was that way. He enjoyed it."

The first well, Gerdes el Abid A1-18, was abandoned in January 1957, at 9,372 feet. The second well, at Ras Marawa, B1-18, was spudded the following month.

Nathon Bacle, Hugh McKenzie, and Bernie Wolford were among the Delta men from East Texas who were in Libya for the concluding phase of the contract. Bacle went over in early 1957, followed a month later by McKenzie. Wolford, just 21 at the time, arrived that summer.

"That was the first time I ever walked on a plane," said McKenzie. "Went by myself. Never had been no farther away from home than the mailbox, hardly. Louisiana was the farthest I'd been, and drove back and forth [from home] then."

He flew from Dallas to New York and then to Benghazi, Libya, on the Mediterranean coast, via London. He soon joined the Delta drilling crew.

"We were drilling about 50 miles out on that road that Rommel retreated out of, there in World War II," said McKenzie. "Well, when me and Nathon got there, well, it was just another East Texas job to us. We thought you was supposed to work, and we went at it."

He soon found out it wasn't "just another East Texas job" in certain ways. The hard work was fitting, but when he started circulating one night, bringing up the drill hole cuttings to the surface with fluid, he learned that wasn't the way it was done in that desert country where water was so scare. Unaware of the local policy, he was surprised when he was chastised for circulating without authority from one of the supervisors. Water had to be hauled 40 miles across the desert to the rig.

Living conditions were different from any McKenzie had known.

"They lived in camp on the first job," said McKenzie. "It was just made. They went out there in the desert and built a camp. Most of them made out of stone. And a lot of fabricated houses, you know, that they had shipped in there.

"But when I got there they'd moved and was within 50 miles of this little old town. And they rented a big old building that was cut up in rooms just like a hotel. Somebody said it was going to be a macaroni factory. See, the Italians were in there before World War II. Anyhow, we was in this little old town and we drove back and forth. One toolpusher stayed three days and the other one three days. And you'd come in and get your rest. And, if we needed anything we just loaded up, supplies and everything."

The second well, like the first, proved to be dry and was abandoned at 8,652 feet. The crew moved about eight miles to the northwest—"a little old place right on the cliff of the Mediterranean Sea," said McKenzie—to begin drilling the third hole.

That summer a young man who was to see a great deal of foreign service for Delta made his first trip overseas to join the crewmen already in Libya.

Bernie Wolford had just turned 20 when he heard Delta was looking for mechanics to go overseas. He didn't land that job, but he did accept a job as trainee rig mechanic in the Kilgore yard under B. W. Freeman in spring, 1957. That July, he went to Libya for them; they had made him wait until he was 21. Guaranteed $650 a month, he went to Benghazi for ten months.

When young Wolford got off the plane in Benghazi and went through customs, Kenneth Ward, the man in charge of the Delta operation by then, came for him. Ward took one look and shook his head. "Are you the diesel mechanic they sent over?" Wolford replied, "I sure am." Ward said, "Well, I'll be damned. They're sending them over here before they get them out of diapers now!"

The next day they went on to location. As they drove out, Ward told Wolford not to worry; if he couldn't make it in mechanics, he would let him stay over in some other capacity. Wolford replied testily, "I came over for mechanics. I can make it, and I'm not planning on doing anything else.''

The rig had six engines on the drawworks, four of which were out of order. He began overhauling two of the engines while Ward watched with dubiety. After both engines started up as Wolford knew they would, the two men became good friends, with Ward's doubts about Wolford's youth positively resolved.

The third hole also was dry, this time at 12,000 feet. There had been no significant shows of oil or gas in any of the drilling, to the disappointment of all concerned. It was 1958 by the time the contract was completed. Libyan American had the option of either shipping the rig to the States or buying it from Delta, but Joe Zeppa considered keeping it if more work could be obtained in that country. Accordingly, bids were prepared on other work, but when nothing materialized, Delta sold the rig and pulled out of Libya.

Delta's drilling efforts for Libyan American were in advance of the exciting strikes that were to come in Libya. Esso completed a 500-barrels-a-day well in December, 1957, a production not considered commercial in that region because of distance. The first major discovery was Esso's Zelten-1, completed in June, 1959, for 17,500 barrels daily. Others followed, including the gigantic British Petroleum-Bunker Hunt Sarir field in 1961.

Libya subsequently was to develop into a petroleum giant. From 1967 through 1972, Libya produced a total of 7,309,824,495 barrels of oil.[48]

VIII

The Australian contract-drilling activities carried out in the Adelaide and Brisbane regions, like the Libyan operations, were transacted through Delta Drilling Company in Tyler and not through a subsidiary, despite their being halfway around the world.

In 1958 Delta committed itself to a drilling contract with Delhi-Taylor Oil Corporation to drill wells in Australia.

The biggest break on the Australian operation was the fact that the home office could deal with the government there in English, for, consistently, one of the salient problems in foreign ventures had been the language barrier.[49]

Robert Waddell, who had a hand in the operation, described how it began.

"Delta had a deal working to drill some deep wells in the state of South Australia, out of Adelaide, and for a Dallas company—at that time it was Delhi-Taylor, the Murchisons. Anyway, [Delhi-Taylor] had bought some new rigs, and they were drilling way up there in the northeast corner of Australia, and they drilled one well. They got a pretty good well, and they built a camp and air strip and went on all the way out to drill some more, and they didn't find any more oil up there. And they had a big rig for sale, nearly new equipment. Delta was negotiating with Delhi-Taylor to drill down there in South Australia. We could buy that rig over there and move it around and have a very inside track getting that work."

After an on-the-spot inspection of the Australian rig, the company bought it and took on a contract.

Waddell said: "They made a deal to buy the rig and then for the people over there to load it out. It all had to be barged out and then loaded on a ship and brought around to Adelaide and then hauled overland about 500 or 600 miles inland, over the cattle trails, and they carried those road graders and all up there to make a road in front. When they got the rig on location and getting ready to drill, then I went over there and helped in getting the job started and seeing that everything was going all right.

"And we got things pretty well going and got the office set up, and about the time I was getting ready to come back I got a wire from Tyler to stay, that the office man we had over there had decided to come home. He didn't tell me. He wired Tyler he was going to come home.

"I got stuck again, so wound up three months over there, and they drilled the first well and moved over to another one."[50]

Hugh McKenzie went to Australia as a driller on that first rig. He stayed 17 months there as they drilled for Delhi-Taylor. The location was 900 miles from Adelaide, the capital of South Australia, in the Simpson Desert. It took five hours to fly from Adelaide to camp in a two-engine World War II English plane. The crews would stay 28 days in the self-contained desert camp, with full facilities that enabled the men to exist in the boondocks, then go into the city for four days.

Chris Zeppa made inspection tours to some of the overseas sites. McKenzie remembers when the executive vice president visited the crews in Australia.

"He'd stay right out there at camp with us," said McKenzie. "He didn't want to go into town. And we all would sit around there and we drank beer and played poker at night. And Mr. Chris would just have him a big time when he was out there with us. Oh, he'd stay about two or three weeks at a time."[51]

By the end of 1959 the rig had drilled a dry hole at 12,637 feet, and had moved to another location. This was at a time when Delta was operating at a loss, drilling was slack generally, and all were hoping the rig would keep busy in Australia. But those hopes fizzled, as the National 130 rig lay idle for most of 1960, and then flared again at "good prospects" of an extended contract with Phillips Petroleum Company beginning early in 1961. The importance of putting this rig to work cannot be overemphasized at that low point in both domestic and foreign operations.[52]

As anticipated, the Phillips contract did materialize. Although the 18-month contract promised a nice profit and cash flow, it necessitated a $180,000 outlay for a camp, electric generating plant, and transportation facilities.[53]

Bob Waddell said: "We moved our headquarters around to Brisbane and drilled a well for some people in that area and got a pretty good well there, and again they drilled and made some production. Phillips Petroleum was the operator, and they thought they had a big field discovery and they moved, went in there and built a camp and made roads and did all that and sent another rig or two in there, and never did get anything."[54]

By 1962 other contractors had moved into Australia, increasing competition. Joe Zeppa felt Delta needed diversified equipment to meet these challenges, but wanted to watch future developments before shifting additional rigs there. After all, it would cost $85,000 to ship a medium size rig from Italy, or $100,000 from Houston.

On November 1, 1962, Delta completed its Phillips contract and stacked the rig at Phillips's last location. It had been profitable, but since any future drilling there was likely to be to depths of less than 7,000 feet, different rigs would be needed to operate profitably at that depth.

In early 1963, Zeppa called a special board meeting—at which Sam Sklar and Sam Y. Dorfman, Jr., of the five directors were absent—to discuss the formation of Delta-Palmer Drilling Company, an Australian corporation, which would be owned 55 percent by Delta, 45 percent by Can-American Drilling Corporation, Limited. Under the agreement, Delta's National 130 rig would go into the new company, along with a National 55A and T32 rig belonging to Can-American. This deal would provide what Zeppa had been seeking in Australia—a capability to drill both shallow and deep wells.[55]

Ed Kliewer, Delta's attorney, was sent on a mission to investigate the Australian deal.

"Well," said Kliewer, "I concluded that this deal just wouldn't fly. One of my tasks was to look into it and see about it and tell [Joe Zeppa] what I thought about it and what I thought he ought to do. So when the conclusion was reached that this wasn't good, I just cabled him. I don't remember what I said. The effect of it was, 'The deal's off. Report follows.' I was either in Sydney or Brisbane. Of course, Delta had an operation over there, but he was negotiating a deal with the fellow Palmer who also had some rigs there."[56]

As Kliewer said, the proposal failed to jell. Instead, Delta leased a rig from Can-American, kept it busy most of the year while the National 130 rig lay stacked, and brought home a profit. Though several gas discoveries in Australia brought encouragement for future prospects, the one rig Delta owned there was, after all, the largest one in Australia and not likely to keep busy as long as the discoveries remained in shallow fields. The slow pace continued. For all of 1963 the National 130 rig remained stacked, while the leased rig drilled two wells, earning enough to pay the overhead. A new Phillips contract was in the offing which was expected to take five months and net a $85,000 profit. At that moment Zeppa's game plan was to wait for an opportunity, which he felt would come.

"Doubtless there will be signficant discoveries in Australia in addition to those already made, and by maintaining a small organization on the ground, we hope to benefit from developments at a later date."[57]

The break he had been expecting came late in the year when Phillips drilled a promising well.

"They thought they'd really hit something," said John D. Hall, "and Mr. Zeppa hauled off and sent two rigs out there without a firm term contract, just gambling. He was a gambler on things like that."[58]

With Phillips needing two more rigs, Delta purchased a used National 130 rig in California and sent Rig 37 from the Louisiana Gulf Coast, which along with Rig 41 gave the company three rigs in Australia. The Phillips work was expected to bring in a profit of $754 per rig day. Ed Arnold, recently hired as a drilling engineer, would go to Australia to help set up the rigs and get the operation going.[59]

Although three rigs earned satisfactory profits for the next year, problems arose. Costs rose because of distance between rigs, necessitating additional camp equipment. Labor turnover increased. New wage schedules for Australian nationals were expected. And by the end of 1966, a sad, familiar story seemed to be repeating itself halfway around the world: all three rigs which had been hired by Phillips were stacked. Drilling activity had distinctly dipped in Australia. Delta was preparing bids on offshore platform work. One of the rigs went to work briefly for Amerada Petroleum Company of Australia, but no hopes were raised for steady work, as it was noted that "land operations have come to a virtual standstill."[60]

With the rigs stacked, for a time attention turned toward possible large-diameter ventilation-shaft drilling for mining companies, although Delta's equipment would require some modification. While preparing bid proposals, Delta was discussing a joint venture with Camay Drilling Company, as Camay possessed the necessary big-hole tools and the experience in that type of work. By then Delta was down to two rigs in Australia. The outlook was poor. The Delta rigs were too big and heavy for most of the work available. The large-hole drilling proposals found no takers.[61]

In 1971 plans became final for closing the Australian operations. By that time Rig 37 had already been dispatched to the Philippines for short-term duty with Fil-Am Resources. Delta now brought it and Rig 41, still in Australia, back to the States. Happily, demobilization funds from Fil-Am Resources helped pay for the ocean shipment. By 1972 the rigs and other material had been returned, and the Australian story was over.[62]

IX

By 1970, Delta's rigs in Australia had been stacked for several years, and the company gained a contract to drill wells in the Philippines.

Bernie Wolford recalled, "Keating [Zeppa] asked me if I'd go over to Australia and start getting this rig ready to bring in to port, which was about 700 miles out back of the base, called Charleville. Charleville was a little town where Delta had their warehouse and camp, kind of northeast of Brisbane about 700 miles.

"It was an Emsco J-1100 rig, complete. Keating asked me if I would go over there and see what needed rebuilding and overhauling. When it got to the Philippines they wanted to drill a well with it.

"So I flew out and went into Brisbane. I caught a plane the next morning and went out to Charleville, and I started getting this rig together, and when I landed in Charleville it was the first time they ever had any measurable rainfall in several years. The whole town was having a party, I guarantee you, having a party. I hired a bunch of sheepherders and everybody I could get to help me, and we started overhauling equipment right there in Charleville, and shipping the rest of it into Brisbane to the shipyard. I stayed about a week out there and then flew on back into Brisbane and we overhauled the whole rig."

By then one rig had been shipped from Australia to Italy.

Once Wolford had the rig overhauled at Brisbane, he returned to the States. Meanwhile, Delta was drilling in the Phillipines for Fil-Am.

"A guy named Bingham was their superintendent, and he was down looking at the rig, and he was a Canadian. After we put the rig on the ship there at Brisbane I came on home. Well, about a month later Keating called me and said Bingham wanted me to come back over there and be present during rig-up back in the Philippines. So I went back over there and we had an American crew, drillers and all. We moved it in on a southern island, Mindanao. We moved it in and rigged it up and I stayed there until we got to drilling and then I left and came back home."

The Phillipines operation was short-lived.

"They drilled into the basement less than 5000 feet. Drilling into granite. Then they moved it across the island somewhere and they drilled into basement shallower than that, maybe 3000 feet. They then moved it across somewhere and maybe drilled in the basement about 5000 feet. That shut the operation down. The rig was shipped back to Beaumont."[63]

10
SOUTH OF THE BORDER

I

There is at least symbolic significance in the fact that the two foreign nations where Delta has fared well have in common their flag colors—green, white, and red—which also are the colors of Delta's logo. The similarities, however, go one step further: the operations in Mexico and Italy are characterized by nationals' running the subsidiaries and manning the rigs.

Delta's Mexican operation began in 1959 when Joe Zeppa came to an agreement with Guillermo Álvarez Morphy, an independently wealthy engineer who had attended Notre Dame University and was well-known in Mexican business circles, and Ciro Gonzales, representing Construcciones Alph S.A. de C.V. of Mexico City, to form a Mexican drilling company. Álvarez Morphy ("Bill" Morphy, as Delta executives were to call him) had negotiated a contract with Pemex, the government oil company, to drill 20 wells near Tampico, and the new company would carry out that obligation as well as perform other contract drilling in Mexico.

The name of the new company was Perforadora Central S.A. de C.V., capitalized at 1 million pesos, subscribed 50 percent by Álvarez Morphy, 50 percent by Delta. Delta would sell two rigs, capable of drilling to 9,000 feet, to the new company. By mid-summer, 1959, one rig was being assembled for delivery at Tampico, with the second one not far behind. The new company would be obligated to pay Delta $600,000 for the equipment.

Álvarez Morphy was to be the president and general manager of Perforadora Central.[1]

Accountant John Hall was dispatched to Mexico from Tyler to set up bookkeeping procedures; Bernie Wolford, to get the equipment delivered and in working order. Problems abounded, with the language barrier leading the way. John Hall summed up his trepidations in a letter to the Tyler office: "It is going to be very easy for things to get in a mess without detailed instructions [to Mexican employees in the field] as transactions are processed." Hall found his assignment delayed by the unavailability of Mexican officials when papers had to be signed, and he concluded: "This is not a 'mañana' country. 'Siguiente semana—quizás [next week—perhaps]' describes it better." Nonetheless, he was bustling along, recording equipment transactions, and posting materials to warehouse cards.[2]

One of Keating Zeppa's first assignments with Delta was to assist in moving two drilling rigs into Mexico for the Mexican affiliate that summer. While the rigs lay stacked in a field on the north side of the Rio Grande at Progreso, Texas, downstream from McAllen, the paper work was being processed on the Mexican side. This site was selected because it boasted the closest bridge strong enough to carry the heavy truck loads. About six weeks passed before the Mexican trucks could cross the Rio Grande to load the three rigs' various components and move them to Tampico. The young Zeppa left for Tampico for several weeks to supervise the unloading and to be present during rig-up of the small rigs.

"I think probably the one thing that sticks in my mind most about that summer," he said, "is what happened when I finally got back from Mexico. I had accumulated substantial reimbursement for expenses—at least, what appeared to me to be substantial—and I remember working up my expense accounts and routinely submitting them to Mr. Chris Zeppa for approval. A couple of days later I got a call. Chris wanted to see me and so I came up to the office, and he said, 'Pinny, you know, I see here where you've shown the expense of $12 to $13 for laundry.'

" 'Well, yes sir, of course; why not? Been down there for six weeks and I had just so many work clothes.'

"He said, 'Well, it doesn't seem to me that laundry is necessarily a reimbursable item.'

"I was flabbergasted. I said, 'Well, Chris, what am I supposed to do, work in my skivvies? If I was living at *home*, I wouldn't have that expense. Ellie would do the laundry.'

"And I remember him looking up at me with a twinkle in his eye and he said, 'Well, Pinny, I'll tell you. There are just several things that you get for free at home that you have to pay for on the road—and none of them reimbursable.'

'Well, as a matter of fact, he did allow that reimbursement for laundry expense. After a hassle."[3]

By the end of the year, Perforadora Central was off to an exciting start. With the first two rigs shipped, Álvarez Morphy had obtained a contract for 30 wells which would provide work for two more rigs. One of the rigs would be shipped from Venezuela, where an economic headache was growing, and another was purchased from Rimrock Tidelands. The total cost to Delta was pegged at $1.4 million, but Joe Zeppa estimated that the money should be paid back within 36 months.[4] Within a year all four rigs were operating, and the debt to Delta, being repaid at the rate of $45,000 per month, had been reduced to $1.175 million.

Not even the gray, practical language of the minute book could muffle the excitement over the Mexican company: ". . . Were it not for the Mexican income taxes which take about 50 percent of the net earnings, this operation would be incredibly profitable, but even with this deduction the investment should be liquidated in much less than three years."[5] Mexico was about the only bright spot in Delta's picture as the 1960s began. During its first year of existence, Perforadora Central had repaid Delta about $400,000, and, for once, it appeared that events were exceeding predictions. By the end of 1962, at the rate things were going, Perforadora Central would liquidate its debts to Delta, with net assets of its own reaching $500,000.[6]

Although operations slowed down in late 1961, with only one rig running, full work resumed early in the new year. In addition, the company had secured a contract to drill 40 wells with a barge rig then being assembled on the Gulf Coast. Subsequently, Perforadora Central purchased a clamshell dredge for $86,525 to aid in its inland barge drilling. The barge rig was soon at work at Yucateco.[7]

Compared to most of the other Delta enterprises at that slack period, the Mexican operation was an economic marvel. By May 31, the end of the 1963 fiscal year, Perforadora Central had paid off all its obligations to Delta, was free of all debt, and had declared a dividend amounting to $216,000, payable in 1964—half of it going to Delta.[8]

The following year, with three land rigs and two barge rigs, Perforadora Central obtained a three-year contract from Pemex for a large barge rig. For this purpose, Delta, through Continental-Emsco, sold its Barge Rig 57 to Central. Toward the end of 1965, Central was on a profitable basis, even though its three land rigs were stacked and not expected to be working "at an early date." Its two barge rigs, however, were hard at work and making monthly payments to Continental-Emsco for Barge Rig No. 2 which Central had bought in the summer of 1964.[9]

Though the land rigs remained idle, inspiring the idea of selling them to Pemex, the fact that they were already paid out meant they were not a burden. The barge work, on the other hand, continued to be profitable, so much so that a second dredge barge had been added to fulfill a long-term contract. Net income after taxes for an 11-month period ending April 30, 1967, was more than $400,000.[10]

The Mexican connection was a profitable one, with Delta receiving "substantial payments annually" and regularly collecting a technical-service fee each month.

By early 1969, Central's net earnings after taxes had soared to a record $1,039,288 for *eight* months ending January 31 of that year. From this "exceptionally good year," Delta's 50 percent came to $519,644.[11]

Mexican capitalism obviously was a going concern. With two barge rigs and two dredges operating, but with the land rigs still stacked, Central formed a subsidiary company to perform pipeline construction work for Pemex in 1969. The subsidiary, Lineas de Producción, S.A. de C.V. (LIPSA), immediately proved to be profitable, earning a net income of 6.5 million pesos ($520,000) in its initial ten-month period. Delta owned 37.7 percent of LIPSA. The only apparent problem in Mexico was that both companies were forced to carry "extremely large" accounts receivable with Pemex. Toward the end of 1969, Central was carrying 19 million pesos; LIPSA, 25 million. While payments were made partially in two-year interest-bearing notes, settlements were made only once or twice a year. In 1971, the stockholders of Central purchased the stock of LIPSA. Because no contract work had been completed at that time, LIPSA showed a loss of $21,447—the first sour note in the Mexican story—but that was expected to be only temporary. [12]

There were a few hints of potential complications: Pemex announced plans to discontinue operating one of the barge rigs. And by this time Central's equipment had been in operation for several years without the cost of major overhaul or dry docking; but since no equipment can run forever without significant repairs, expenses could be expected to increase in the near future.

The 1970s, nonetheless, were profitable ones for Central. In 1972, Delta's half of the net income came to $501,909. [13]

By early 1975, Perforadora Central was flourishing enough to lend aid to the ailing Delta Marine. As Keating Zeppa, then executive vice president of Delta, reported to the stockholders, arrangements had been made to sell Delta Marine's Inland Barge 6 to Perforadora Central for $700,000, against which a prepayment of $400,000 had been made, along with a $800,000 prepayment against the cost of rebuilding it and modernizing the rig equipment. [14]

Undoubtedly, the Mexican venture had been a happy one for both the North American and the Mexican partners. Álvarez Morphy and associates—his son "Pat" Morphy was later to run the company—had the advantage of capital investment from the United States, plus equipment and access to technical know-how which Delta employees had acquired through years of experience. On the other hand, it was an excellent arrangement for Zeppa's outfit, for Delta was dealing with a man whose personality, background, and connections could only benefit Perforadora Central. Central was effectively managed and in a position to stay attuned to the Mexican political and economic environments.

But in addition to having effective management by a Mexican national thoroughly in touch with his country, Perforadora Central was half-owned by Mexicans, further motivating them to keep profits high. Just as significantly, the association in Mexico came at the right time in history: Mexico was enjoying an unprecedented economic boom, and with the growing energy crisis, the state-owned search for petroleum was likely to be rewarded in a country where oil and gas had been found since the early years of the century and in which undiscovered reserves still lay waiting for the bit.

II

When domestic drilling went slack in the late 1950s and early 1960s, Delta, like many other companies, went overseas, as Keating Zeppa put it, "in search of revenues." Because there was little, if any, political risk, and the operator usually did most of the work, a drilling contractor's job was appealing.

Delta's Argentine adventure, however, illustrated practically all of the pitfalls that could be encountered, short of expropriation.

The promise which Argentina represented to Delta was almost completely eroded by the fluctuating economic trends within that country. Caught at a time when devaluation of the peso and the national oil company's sluggishness in paying its contractors added to the uncertainties of galloping inflation, Delta experienced disappointments from the outset.

Delta's interest in Argentina can be traced at least back to 1959 when a "serious effort" was made to obtain a contract to drill 300 wells there. When Delta was underbid, no further steps were taken to match those terms, which Zeppa and his executives considered "unfavorable and imprudent considering the risk of the capital involved." [15]

According to Mark Gardner, a number of companies were seeking drilling contracts in Argentina. William C. "Bill" Clements, representing Texas-based SEDCO, had close connections with powerful men in the Argentine government. Meanwhile, Joe Zeppa had become acquainted with an Argentine who he thought had the necessary "clout" to help secure contracts. Zeppa dispatched Gardner to discuss the matter with the Argentine.

"The first time I ever talked with this guy," said Gardner, "I knew he didn't have any damn stroke at all, but Joe makes you stay down there and battle this damn thing out. Kerr-McGee had a former undersecretary of state for Latin American affairs that would voice their rights, and the Italians were over there. The net result was, Kerr-McGee got 500 wells, the Italians 500, and Bill Clements 1,000. That's what made SEDCO. But, anyway, Delta didn't get any. There was a deal floating around down there for some workover rigs, and Joe cabled me to look into this matter, and I did, and I cabled him that it wasn't anything that we should mess with and that I was coming home. And I had learned enough to know that I had better send Joe a cable and get the hell out of there before he could get a cable back, because for some kind of perverse reason, if he thought you wanted to come home, he cabled you to stay there. So I would send him a cable, and I would get on an airplane. I told him I was coming home, that I wouldn't have anything to do with this, and damned if he didn't send another fellow down there and made a deal with these characters. For seven rigs. There was no protection against devaluation of currency. It damned near ate us alive."[16]

Following Gardner's fact-finding trip, attorney Ed Kliewer flew down to assess the situation and the men Delta would be dealing with.

"I went down there and I looked at the damn thing and I made a mistake," said Kliewer. "I called Joe up, talked to him. I didn't make that mistake the next time. I told him I didn't think it would fly. I told him I didn't think Delta needed to be down there, and I told him I didn't like the attitude and reaction of these two local businessmen. My recommendation to Joe was to forget it. Well, that didn't do anything but just get him up to the point where he was determined to make the deal. And he did."[17]

Thus in 1960 Delta plunged into its Argentina operation. Two Argentine subsidiaries were established—Con-Tra-Pet (Contratistas de Trabajos Petroleros, S.A.I.C.) and Quitral-Co, S.A.I.C. The companies had a contract with the government company, YPF, for oil-well servicing for four years, with the probability that the work would continue and even increase. By the end of 1960, five well-servicing units were operating in Argentina. Delta's total investment, including working capital, came to $1,250,000, which Zeppa expected to be returned during the first three years.[18]

Seen against the background of depressed conditions for the drilling industry, the Argentine effort can be interpreted as searching out, rather than waiting for opportunity.

The Argentine companies—Quitral-Co and Con-Tra-Pet—were wholly owned by Perforaciones Delta, the Venezuelan company, in order to take advantage of nontaxable dividends received in Venezuela. Later in the operation, this arrangement was to be changed, as Delta purchased the stock from the Venezuelan company and resold some of it to Argentine nominees, who held Delta stock in their names, in order to comply with changing Argentine laws.

Keating Zeppa was assigned the task of getting the operation kicked off. Equipment and materials were ordered and loaded on a ship at Beaumont to go to Argentina. This included a portable camp, well-servicing equipment, vehicles, and a warehouse. Delta's purchasing department sent all the necessary equipment and spare parts. The equipment arrived before the warehouse and facilities were ready to receive the various items. Then it was discovered how bad pilferage could be.

"They cleared through customs," said Walter McElroy, "and took it out to the site where it was going to stay and stacked it on the ground until they got their warehouse building ready to put it in. When they started unpacking these automobile parts to put them in the warehouse, they discovered there was nothing in the boxes. So we had to duplicate that original order for automobile parts, which was in excess of $100,000 worth."[19]

Keating Zeppa flew to Argentina to meet the boat when it arrived at Puerto Deseado in southern Argentina. The port was six or eight kilometers inland, and consisted of a concrete quay with a rail spur alongside it. Other than sheep, there was not much else there.

"So I flew down there," said Keating Zeppa. "We got the boat unloaded, in spite of a dock-workers strike. There were no dock-workers resident there. To get dock-workers you had to bring them from Buenos Aires or elsewhere. And since they were all on strike, there was no one to be brought; so, literally, the ship's crew and the Delta personnel unloaded the vessel. There was no engine available to switch the flatcars and what not. We were using some rail cars to move with. So what we did, we unloaded one of the big tandem trucks and used it for a switch engine. Pushed cars around with it. It was a narrow-gauge railway, small line, and European-sized cars, so it was pretty easy to do."

In the latter part of November, these preliminary chores having been done, Keating Zeppa flew back to the States where his and Ellie's second child, Cathy, had just been born. In December, the family packed up and left for Argentina, where they were to remain for four years—at least twice as long as was originally expected.

Though Joe Zeppa never explained why he sent his son to Argentina, Keating Zeppa believes it was to gain administrative experience.

"I think he saw it more as a good training ground for me. And figured it was a small enough operation that if I just screwed it up entirely, it wouldn't hurt that badly. I would say, practice without too much risk on the company's part. Well, there was a fair amount of risk. Our initial investment was about a million and three [$1.3 million], which at that time was a lot of money to Delta. But those were not big contracts. They were not drilling contracts; they were well-servicing, this sort of thing. And they were just sort of ham and eggs. As it turned out, they were underpriced anyway.

"I don't know what his thinking was. He never conveyed it to me. But his idea was that we'd go down there for 18 months or so and get things kicked off and then probably come home. But things didn't go as they were planned to."[20]

A hint of trouble appeared in 1961 when Joe Zeppa reported to the stockholders, "We know that we are deriving a substantial profit, but it is not yet possible to state exactly the amount because of the high expenses in initiating the operation, the trouble we have had with personnel, and the fact that it is not, as yet, performing all the work it is capable of. We still feel that it will pay out the investment in about three years, although it was considerably higher than originally anticipated." He nonetheless believed prospects there to be bright. Later that year it was necessary to make cash advances to the two Argentine companies and to arrange for credit purchases from

IDECO, a Dresser Industries subsidiary. Zeppa expected that even further advances might be needed, for if Con-Tra-Pet and Quitral-Co were to provide four more servicing units, as requested by YPF, additional expense would be needed to enlarge the camp and the service facilities. Despite the unexpected higher initial costs, personnel difficulties, bad weather, and a lack of skilled men, Zeppa remained optimistic on a long-term basis and believed additional investment was called for.[21]

The first note of what was to become the gravest problem in Argentina came at the end of the year when Zeppa announced to his directors that, among the other problems besetting the Delta subsidiaries in Argentina, the government company YPF "has not been meeting its financial obligations, either to our companies or to other companies operating in the Comodoro [Rivadavia] area."[22]

The Comodoro Rivadavia field, which was really a series of fields, had been discovered in 1907 during the drilling of a water well. It gave Argentina its first recorded commercial production, amounting to 101 barrels a year. By 1960 it was producing more than 34 million barrels, over half the country's total for a year.[23]

As Keating Zeppa recalled it, "First thing that happened was our customer, YPF, which was the government oil company, stopped paying its bills, and, of course, our contracts were all in pesos, Argentine pesos, while our obligations, our debts, were in dollars because we'd purchased all our equipment in the States. We had a rapidly declining peso and an insolvent customer, and the combination of the two ultimately cost this company $1.5 million or $2 million in exchange losses. We couldn't collect our bills on time and therefore buy dollars to pay our suppliers. So Delta Drilling Company, essentially, financed the two subsidiaries for the equivalent of about a year and a half or so, invoicing. A settlement was finally made with YPF on all past-due items. We took a hell of a beating. Plus the fact that [those Delta representatives who] had worked up the pricing on the contract and had made the deals with YPF—so much an hour for the units and what not—were very optimistic. The pricing was way out of line. We lost a ton down there, part of it due to my inexperience, but part of it due to circumstances beyond anyone's control.

"In fact, I remember one time Jack Harbin, who is now the CEO of Halliburton, and Dad decided to come down to Argentina and visit with these people and tell 'em how the cow ate the cabbage and, 'By God, they were gonna pay their bills!' All service companies were having trouble; they couldn't collect their bills—Halliburton, Schlumberger, everybody. We weren't by ourselves.

"Ken Abel, who was resident manager of Halliburton down there at that time, and I met 'em at the airport, greeted our respective bosses and what not. These two Texas oil types went charging over to YPF and visited over there and then went over and talked to the American ambassador about the situation. The upshot was, they both left after seven days or so, and I don't know what Jack Harbin's comment to Ken Abel was, but I know that JZ looked at me finally before he got on the plane and he said, 'Hell, I don't know what you do with these sons of bitches—they won't pay!'[24]

Mark Gardner summed up this aspect of the venture: "We got involved in this workover business down there. Sent all of the equipment down there. This was all new equipment from the United States, and, boy, it was a fiasco, and this is one time that Joe was real worried. He got him something there that was real rough. I think finally it worked out, but it had Delta Drilling Company in a tight squeeze for a while. Now they weren't the only ones. Clements and Kerr-McGee and, I guess, the Italians had a guaranty with the central bank. They got their money. It didn't matter what happened to the peso. This is one time when Joe was really doggone sick with worry."[25]

In 1962 Joe Zeppa reported that though the Argentine operation was "now on a profitable basis, YPF is behind in its payments to the companies to the extent of about $1,250,000." It was not a problem unique to Delta's subsidiaries, but extended to other North American companies. He foresaw no solution until the end of troubled Argentine political conditions—which that year climaxed in the arrest of President Arturo Frondizi by military leaders, in the imposition of José María Guido as president, and in the dissolution of Congress and political parties. Political and economic unrest was to continue for years.

By mid-1962 the Argentine situation had grown so serious that Zeppa called a special board meeting of Delta in Tyler to discuss it. The failure of Yacimientos Petrolíferos Fiscales (YPF) to pay Quitral-Co and Con-Tra-Pet had imposed so severe a cash strain that Delta, in effect, had to bail them out of the crisis, which included paying off a debt of $500,000 to the Republic National Bank of Dallas, which Delta paid out of a loan for which it mortgaged some of its leases and production.

Another blow came late in the year when Quitral-Co and Con-Tra-Pet sustained "severe losses" as a result of devaluation of the Argentine currency. Zeppa felt that eventually the subsidiaries would collect the entire amount of their invoices to YPF, and steps were being taken to try to recoup, as much as possible, the losses resulting from the devaluation.[26]

The contract with YPF provided for losses related to peso devaluation between the time work was performed and payment was made, but in 1963 this provision was being viewed with hope rather than with certainty. In order to minimize the shock of fluctuations in the exchange rate, Delta accepted promissory notes from YPF in dollars converted at the rate of exchange on date of issue, but payable in pesos at the rate when paid. This would preclude the severe losses the subsidiaries had suffered in the first year. Zeppa pointed out that YPF had paid its notes on time and that YPF "has never defaulted on such obligations." By then the Delta companies had $1.6 million of such notes. Of that figure, Delta had discounted $900,000 with the Republic National Bank of Dallas and the rest with Continental-Emsco, toward retiring the company's debts. The Argentine companies were in their fourth year of operation and could be expected to pay out the original investment in October, 1964—if YPF honored the claims on currency exchange losses.[27]

But there were other problems. As Keating Zeppa remembers, "We had American supervisors and Argentine and Chilean laborers. Largest problem there was a matter of finances. General area of personnel was the second largest problem. [There was] a very high turnover rate among the Americans, many of whom came out of Venezuela simply because things were beginning to slow down there, and we mistakenly presumed that since they had worked in Venezuela for several years they'd be fluent in Spanish and come down and fit right in. [But it did not work out that way.] They had been spoiled in Venezuela. Times were good. Pay was big. Nobody had to work very hard. They were accustomed to this sort of thing. The living was easy. Climate was nice. Well, they got sent into Argentina, and that's about like hitting Montana. There is nothing around for 100 miles, and the winters are brutal. And they lived in a camp. They didn't stay very long. The Argentine personnel by and large were competent enough if they had had good enough supervisors. And, of course, we gradually evolved down there to the point where we had no Americans. But that took some time."

The weather in southern Argentina, where operations went on, was memorable. John Hall was there one July Fourth. "I never will forget," he said, "I was down there at the camp, and it was snowing like everything. Cold, windy."

Jack D. Robinson went to work for Quitral-Co in Argentina as a toolpusher in 1963, and worked his way up to manager by 1972, when he left for Brazil as director and general manager for Perbras, Delta's company there.

"When I went to work we were operating at that time eight service rigs," said Robinson. "We had a camp where we provided room and board for about 200 workers. We worked around the clock, 24 hours a day, seven days a week.

"We did have a strong labor union. We did have a few strikes. We had one 24-hour strike that I remember quite well. The night shift had come off duty, and they were having breakfast in the mess hall, and one of the workers ordered his seventh T-bone steak with two eggs, and [a representative of] the catering service came over and told me, 'This is his seventh steak that he has ordered, and all he is doing is eating the filet side and then ordering another steak.' I said, 'Well, serve him one more and if he orders another one, then I will refuse it.' So he did order the next one, we refused to serve it, so we had a 24-hour strike. We lost 24 hours of rig time for eight rigs, but anyway we won our point. But generally speaking we did enjoy good relations with them.

"We were in the southern part of Argentina, around the forty-sixth to forty-seventh parallel south; that is the extreme south. It is very cold down there. Most of the Argentines didn't want to live in that area at all. They wanted to live in nicer climate areas like Buenos Aires.

"We had all nationals on the rigs—about 70 percent were Chileans. We had some American toolpushers. We had some American mechanics, American radio men, one American electrician. That was in the earlier years. Finally we went all national. We were able to train our people in each department where they were experienced enough and capable enough to run the operation. It was to our advantage to train the nationals. So for approximately three years there I was the only American in the company. All the rest were nationals, in 1965, 1966, something like that. And in 1967 we brought in two new drilling rigs, and at that time we brought in some expatriates—an American drilling superintendent, American toolpushers, and American mechanics, and they stayed there until about 1969. By that time we had trained enough nationals that could take over operations."

Robinson cited one scene, relating to the inclement weather, that has never left his memory.

"We did move down and drill a well on a small ranch. They had about 140,000 head of sheep on this ranch. Around 10,000 head of cattle, around 1,000 head of horses, and their gaucho complement was something like 75 to 80 cowboys. We moved a rig in there and you could see all four seasons in one day. You would have a little sun, a little sleet, a little snow, and lots of wind. One thing that does stand out in my mind; I think it was December 11, 1968, we recorded winds up to 255 kilometers per hour in Comodoro Rivadavia. Telephone poles just popped, windshields, back glasses out of buses. Just the force of the wind. No vacuum or anything, just the force of the wind. That was a day that I really remember. I guess during a normal season you would probably have a month of warm weather. It would get up to maybe 80 degrees. December and January were the warm months, but you had wind down there 365 days a year. A calm day was winds in the neighborhood of 20 to 50 kilometers per hour. It is windy all the time. We used wind lines on our traveling blocks to make trips or to ensure that our blocks would travel in the center of the derrick. We had to anchor all of our trailer houses down. Sometimes you would have to wait a day or two to remove a trailer house because of high winds."[28]

Although operations were in the southern part of the country, the offices were situated in Buenos Aires, where the weather and living conditions were more hospitable. The first visit back to the States for Keating and Ellie Zeppa and their children came at Christmas of 1961 when they returned to Tyler for a short respite at Pinehurst.[29] And there were occasional Delta visitors to Argentina. Joe and Gertrude Zeppa flew to Buenos Aires in 1963 in time for the Fourth of July celebration for the American colony there. A luncheon was held in the hotel where the Zeppas were staying. The American ambassador spoke. Charles Delton, office manager for the Delta operation in Argentina, remembered most of all his visit with Mrs. Zeppa. "Mrs. Zeppa and I had the best conversation you ever saw, and you know what it was about, mainly? Baseball. And she was telling me, 'Why in the world did they ever put Willie McCovey in the clean-up spot [for the San Francisco Giants in the 1962 World Series]? The very words she used. 'That's what lost the Series.' She knew the players and their qualifications in detail. That's particularly unusual for a lady in that social standing. She was precious to the last breath. I never will forget that social affair, one of the nicest I've ever been to on foreign soil, and I've been to many of them."[30]

In December, 1964, Keating Zeppa turned over the Argentina operation to William Triest, and he and Ellie returned to Tyler with three children (Joe had been born in the States during the Argentine stint); their fourth child, Marie, was born the following year. He had devoted almost four years to managing the Argentine operation.

"For my personal development, certainly, the stay in Argentina was a good one," said Keating Zeppa. "If nothing else, I at least got a very good grip on the Spanish language as it is spoken in Argentina, which is very helpful."

Joe Zeppa was able to tell the Delta board, according to the minute book, that his son had carried out his assignment "quite successfully," and upon his return to Tyler, Keating Zeppa became assistant to the president. "I was on the road, just a hell of a lot, as soon as I got back from Argentina," Keating Zeppa said. "We had opened up business in Brazil. We were still operating in Argentina, and I ended up spending a lot of time in Brazil and also back in Argentina. As I recall, a number of trips to Venezuela. Some of the trips to Brazil in that period were rather protracted, because they involved contract negotiations."

In 1966 Keating Zeppa became vice president in charge of foreign operations, and in 1969, when Chris Zeppa retired, he replaced him as executive vice president.

As 1964 ended, Joe Zeppa could report net earnings in Argentina of $1,044,836.33 for 11 months ending September 30, with many of the old problems solved and the organization more efficient. The companies were operating on an extension of the original contract by then. With a revised contract, though, Argentine taxes would probably have to be paid. And the currency exchange problem lingered.[31]

A year later the companies were working under a new two-year contract and trying to resolve the exchange crisis. The new contract left the companies subject to Argentine income taxes, but Zeppa expected the operation to yield "substantial profit." The major problem continued to be currency exchange. By now the Argentine government had put controls on transfer of funds outside the country, and had issued dollar bonds bearing five percent interest toward settlement of Argentina's foreign debt. Delta's two companies purchased $458,000 of these bearer bonds which were to be brought to the States by Keating Zeppa. Permission was also granted for purchase of Argentine dollar bonds for transfer of note collections, involving the Republic National Bank of Dallas and the First National City Bank, for $273,000, on which there would be an exchange loss of $18,000. Only the exchange controls seemed to be in the way of the companies' prosperity; even so, their debts were substantially liquidated. [32]

By early 1966 the exchange problems persisted, but Zeppa was able to report to stockholders that "The original dollar investment in Argentina has been returned to the parent company almost in entirety, and when dividends presently declared and for which exchange has been granted are paid during the next six weeks, the Argentine companies will be free of dollar debt." [33]

In 1966 a new contract was made by Quitral-Co with YPF to drill 35 wells in the Comodoro Rivadavia region "under extremely favorable prices." The projected income justified the purchase of two new Electrohoist rigs and other equipment for $2.5 million from Continental-Emsco. Purchased through Delta, the parent company realized $130,000 for its services. Dollar exchange was to be granted for note payments for the equipment, and thus the operation seemed to be headed for profit, although there would be a four-month gap from the first drilling until the first cash payments would be received, because YPF was paying its drilling invoices with 90-day notes. By October 1966, the two new rigs were at work and optimism was in the air. [34]

By the end of 1967 the Argentine companies owned and operated ten well-servicing rigs, three drive-in workover rigs, and the two Electrohoist II drilling rigs. Although their principal services were to YPF, substantial work also was being performed for Pan American Oil Company and Cities Service. Since the profits from these operations were "very substantial," Quitral-Co was expected to retire the balance of its debts to Continental-Emsco for the Electrohoist equipment by the end of 1969. [35] This was not, however, to be.

By early 1968 the two companies had suffered a loss for a four-month period. Drilling activity was down, and it was anticipated that for the year the companies might merely break even, earning only the very heavy depreciation. As a matter of fact, they operated at an actual loss over the fiscal year ending October 31, 1968. By then the workover projects for YPF had declined by half, and both of the Electrohoist rigs had been stacked since April. By early 1969 one of the rigs was under contract to Signal Oil Company, and hopes were that it would continue for six months or more.[36]

This turn of events left the Argentine companies in an unexpected plight. The 35-well contract, on the basis of which the $2.5 million rigs had been purchased, had been expected to be extended, but it had not been. To compound this, the well-servicing operations by then were spread into three districts instead of one, which increased the overhead.[37]

In fiscal year 1969 the companies lost $128,114, as well-servicing work continued to be "very slow." Although bids were going to Cities Service of Argentina and YPF, additional capital outlays would be necessary if the new work were obtained. A year later, the companies earned a small "disappointing" profit of $48,516 after depreciation. Work was lined up with Sun Oil and Amoco, and again hopes were up.[38]

By 1971 it had become obvious that the companies were not going to be able to repay their obligations as expected, and that Delta, which had been making substantial cash advances and providing supplies and paying salaries of the Americans working there, would face more of the same in the future unless changes were made. Since the Venezuelan company could not assume the debt owed to Delta, an "awkward situation" had developed. To resolve it, Delta purchased all of the stock in the Argentine companies from the Venezuela company, thereby transferring ownership to Delta, the parent company.

However, by December 1970, Argentine laws further complicated the matter by requiring that companies working for YPF and other governmental agencies be classified as Argentine companies, with 60 percent or more of the stock Argentine-owned. Thus Con-Tra-Pet purchased 156,500 shares of Quitral-Co stock, Delta the remaining 243,500 shares. These bearer shares were to be held for the benefit of Delta by its Argentine nominees. Delta then purchased all of the shares in Con-Tra-Pet. Quitral-Co operated Con-Tra-Pet's small inventory of material and equipment under a lease agreement.[39]

By the following year, the Argentine companies had shown a net income of $457,838, but the problems there seemed "almost insurmountable." Joe Zeppa commented that "Inflation and currency devaluation consume the profits earned before they can be converted to U.S. currency." By the end of that year, Delta's advances to Quitral-Co added up to $1,771,546, and by then, as the political situation changed, it was becoming more and more difficult to transfer money to Delta in the States, even when it was available.[40]

By the end of 1972, Delta divested itself of ownership in Quitral-Co to comply with Argentine laws which, by then, penalized foreign companies doing work for the Argentine government and gave preference to Argentine-controlled companies. To become an Argentine company, 60 percent of the stock must be held by Argentines, with at least 60 percent of the directors also Argentine. Under a new law, Con-Tra-Pet would be required to publish its balance sheet, which would reveal that Con-Tra-Pet, wholly owned by Delta, owned 39 percent of the Quitral-Co stock. This would brand Quitral-Co as a foreign-owned corporation, it was felt, and place the company in a difficult position financially. To overcome this, it was contemplated that the Quitral-Co stock owned by Con-Tra-Pet might be transferred to those individuals owning the remaining stock. A likely next step would be to sell Quitral-Co's rigs and heavy trucks to Con-Tra-Pet, which then would lease them to Quitral-Co. A rental agreement then could ensure that Con-Tra-Pet effectively controlled Quitral-Co's finances. Though Quitral-Co had had a profitable year in 1972 and had repaid Delta $800,000, the future was acknowledged as being cloudy. With unstable political and economic conditions, characterized by strikes and protests, there was concern as to whether it might become difficult to get the rigs and trucks out of Argentina in the future.[41]

To comply with the Argentine laws, Delta transferred the necessary stock to Quitral-Co's president and other officers and associates, with each agreeing to surrender the stock on demand as long as Quitral-Co owed Delta. Con-Tra-Pet then sold its stock in Quitral-Co to the operating officers of the company who owned the other 60.875 percent. Then the two Electrohoist rigs were sold by Quitral-Co to Con-Tra-Pet for $950,000. Meanwhile, Quitral-Co arranged funds to buy dollar exchange bonds with which to pay Delta for technical services. These various transactions left Con-Tra-Pet owing Delta $600,000, and this was expected to be paid to Delta from the sale of 39.125 percent of the Quitral-Co stock and drilling equipment rental to Quitral-Co. The Electrohoist rigs were leased to Quitral-Co for a year, with options on two annual extensions, at a rental price of $1,200 per day, whether the rigs were working or stacked. Then, to ice the arrangement, Delta had the opinion whereby within a ten-year period it could purchase 100 percent of the Quitral-Co stock from its present owners, in a separate option agreement.[42]

By late 1974, the often untidy Argentina operation had become more orderly. Quitral-Co owed Delta nearly $500,000 in all. In order to secure this indebtedness, the Argentine company had borrowed locally and purchased Argentine dollar bonds which were pledged to Delta. Meanwhile, Con-Tra-Pet, still wholly owned by Delta, owed the parent $266,800 on its $600,000 February, 1973, loan, and Con-Tra-Pet was repaying Quitral-Co the remaining $63,750 for the two rigs in installments, the last one due March 15, 1975.[43]

Joe Zeppa was clearly disturbed about the Argentine complications, and he said in 1975, "They've got some damned good equipment down there too. Well, I don't know what the hell we're gonna do with it, to tell you the truth. You get to the point where you're fighting the government all the time."[44]

Clearly, the handwriting was on the wall for all to see. Delta's days in Argentina were numbered—even if the minority stockholders raised no objection to continued operations there.

The problems in Argentina had been somewhat different from those in other countries. The companies there had made money—quite good money—but the unsettled political and economic conditions there obviously were not conducive to any long-term operation. No matter what the companies did to solve their internal problems—which they were able to do over a period of time—they could not overcome the external difficulties which plagued the nation and all those, native and foreign, who did business there.

In 1980, looking back, Ed Kliewer summed up the frustrating years. "For quite a period of time it was just nothing but problems—things that nobody could anticipate. Certainly I didn't. Inflation, inability of the government to pay its debts and all that.

"That Argentine operation took a lot out of Joe. It was a worrisome son-of-a-bitch. It was such a drain on his energy. And as it's turned out, it's gradually working out to a point where Delta can move out."[45]

III

By 1960 Zeppa had made an effort to develop a drilling operation in Brazil. When it finally materialized, it was to have its ups and downs, with both promise and risk, for two decades. It was one of the three foreign ventures still under way as Delta entered the 1980s.

Initial negotiations involved two rigs, with the corporate vehicle to be a new company which Delta and the Brazilian government would own jointly. However, these plans were delayed by a change in administration in both the government and Petrobras, the Brazilian government's oil company.[46]

It was not until 1963 that a drilling contract was finally negotiated with Petroleo Brasileiro S.A., or Petrobras, the mixed-economy corporation of Brazil which held jurisdiction over the mineral development. The contract provided for Delta to drill a minimum of 25 wells over a period of 730 days in the state of Bahía in Brazil. The contract called for three rigs; two were sent from the States, the other one was leased from Perforaciones Delta in Venezuela. Despite certain built-in protective clauses for Delta, including substantial advances by Petrobras to cover some purchases, and transportation, bits, muds, and chemicals being furnished by Petrobras on a reimbursable basis, there remained risks, particularly in light of Petrobras's payment history. By now Zeppa had no doubt benefitted from the education in Argentina, and he felt the risks had been reduced sufficiently to justify the new foreign plunge. There was a large profit to be had, and with Brazil striving to make itself independent of petroleum imports, Zeppa felt the association had a good chance of continuing far enough into the future to ensure a repeating profit.[47]

Instead of a jointly-owned company as had been first envisioned, Delta operated in Brazil under its own name, but without registration as a foreign corporation because of the type of contract it had. But since some of the payments would be in Brazilian cruzeiros, bank accounts were established in Rio de Janeiro and Salvador, carried in the names of Delta's local manager H. Clyde Johnson, drilling superintendent Albert Jenson, and office manager Charles C. Delton. It was anticipated that the receipts in cruzeiros from Petrobras would come to about the same as Delta's disbursements in cruzeiros.[48] In December 1964, Ardle V. Hill, a veteran at foreign work, replaced Johnson as manager of operations.

Delton, an accountant who had already worked for Delta in Argentina, went to Brazil as "the trail blazer" in May, 1964, to act as office manager in Salvador and set up the accounting and financial operations. Salvador is about two hours north of Rio de Janeiro in the state of Bahía.

One of the first obstacles he encountered was one that sometimes seems endemic to the oil patch—heavy rain. "It had rained like it had never rained before in the history of Brazil," said Delton. "I think they said three or six months straight. One man said he had to close his transportation business. The rain ruined it. He kept a record of the rainfall, and it rained 252 inches in one month."

Salvador is a port city, and the Delta equipment was shipped directly to the docks there.

"Well, when the iron came," said Delton, "I remember very well it was still raining, and Bernie Wolford, who became manager of maintenance, went into the hold [of the ship], connected the batteries and all, and I saw that guy with my own eyes, as the boom went onto the ship and brought up one of those trucks, he was still on the damn [truck], working on it, and the boom was coming right on down and setting it on the dock. By the time he got through connecting everything, the guy drove it off. Tractors and all."

The warehouse was situated about 16-18 kilometers, or 10-12 miles, from Salvador.[49]

At that time Delta had three drilling rigs and acquired a fourth one on a labor contract. Wolford, as the head mechanic, was the only American mechanic on the operation, with experienced nationals assisting. The nationals had had most of their experience working on cars and trucks rather than on rigs, however.

"I hired me a couple of machinists out of the machine shop in town," Wolford said. "They were real good machinists. Building everything, lots of parts for Delta. You name it, and in the oil field, we built it. We would build shafts for the different compounds and drawworks and all, rebuilt all our rotary tools, slips, tongs, and elevators. We really got together a good maintenance crew."[50]

When Buddy Jones arrived in 1964, three rigs were being sent there, and one was being rigged up. "Most of the stuff had not got there yet, and was still on the boat. The first rig that I got on, we were on it 18 months and all we drilled was nine months. The rest of the time was fishing. We had as many as four different items in the hole at one time, and got every piece out."

Problems seemed to never cease for him. "Just as soon as I would get to sleep, they would come up there with some problem," said Jones. Then there were drilling problems related to the geology of the region. There was a great deal of serious trouble with cave-ins, along with deviation difficulties.[51]

Then there was geography. The jungles of Brazil were a far cry from the pine trees of East Texas.

"I was about half scared of it at night because of these big snakes, animals," said Jones. "The most poisonous snake I ever seen was this little striped coral snake. They usually lay under boards and we were careful at night about picking up a board and laying it around a rig. Snakes were bad about coming to the rigs at night. They had a lot of monkeys, but I think they had more wild animals in Venezuela."[52]

"Everywhere it's a jungle," said Charles Delton. "Where you get out of the city, you've got nothing but dense forest. It's hilly, not flat by any means, and transportation—it's a major problem. And, of course, the elements: there you have a real problem. The number one is always your health."[53]

Early in the operation, the North Americans learned to work with natives. "We had all Brazilian roughnecks," said Frank Dykstra. "We had an American driller; all the rest were Brazilian. I was American, and all the rest were natives, and these natives would work barefooted, and when they made some money they would buy rubber boots, and when their feet would get hot they would fill their rubber boots full of water to keep their feet cool.

"Man, it was awful hot in Salvador, and they are such poor people down there. They had to buy one cigarette at a time.

"We lived on a camp job. Delta leased a ranch house from some Brazilian down there, and we made it into a camp for the crews. I ran this camp and all, besides pushing tools. I'd have to get the cook and we'd go in every week, about 80 miles in, to get groceries, and then come back. Not too much exciting about it."[54]

Salvador represented a foreignness that many of the Americans had never seen before.

"We had a world of wives that had never even left the country, and when they walked into an apartment that had the wires sticking out, a lot of them didn't stay long enough to let the water get hot," said Charles Delton. "Big old cockroaches and scorpions. I don't know what they expected. We had turnover just tremendous."[55]

"When I first got there," chuckled Frank Dykstra, "I had a fly in my coffee, and I called that waiter over and said, 'There is a fly in here.' He said, 'No problem.' He just takes a spoon and takes the fly out! 'Go ahead and drink it.'"[56]

Charlie Delton, the office manager in Brazil, found that the most difficult part of his job was the human relations, not only smoothing over problems with the local Brazilians but, most of all, helping to work out the behavioral problems of the Americans.

"When I wasn't taking one out of the tank," he said, "it was out of the hospital or in there, so, you see, you've got a job like mine, you're not just office manager, you're father confessor, you straighten out all the conflicts, everything."[57]

Inevitably, adventurers turned up in the foreign assignments, but probably few were as colorful as—we'll call him—Jimmy Lincoln, who was one of Frank Dykstra's drillers in Brazil.

"He was a character," said Dykstra. "He would get just wallowing drunk, but one of his roughnecks would get hurt and he would take his shirt off and put it on his hand, get it all bloody—just give you the shirt off his back."

He was from Canada, and had been all over the world. At one of Delta's Christmas parties in Brazil, Lincoln turned up with a streetwalker instead of his wife, to the horror of the other personnel.

"Well, he took her in there, and the people running the food table tried to stop this gal from eating, and she took a turkey leg and hit one of these women with the turkey leg," said Dykstra. "The manager told Jimmy, 'I ought to send you back to the States.' Jimmy said, 'You S.O.B., that would be a favor to send me back!'

"They put us all in a hotel there for a week to kind of let us get climatized before we went to work—charged everything to the company. Boy, Jimmy just sat down there in that bar and charged everything to Delta. He had a terrible bar bill. That manager liked to have a fit when he got that bar bill. Jimmy said, 'Hell, you said charge everything to them.' That guy was something else.

"Delta would give them a case of beer every time they made so many feet of hole, and Jimmy had about six cases of beer coming, and they decided they would just pay him, give him the money instead. So he takes his whole crew and stops at a bar. We waited supper on him at the staff house where all of our Americans stayed—my drillers, myself. He didn't show up, so I got in my pickup and went to hunt him. I thought he broke down somewhere. We found them all in a saloon over there, just drunker than skunks, the whole bunch of them. They wouldn't even come to eat."[58]

Some returned to the States with reminders of their stay in Brazil. When Bernie and Maidie Wolford left for Brazil in May, 1964, their oldest son, Bernie, Jr., was four, the younger, Jimmy, two. Their third son, David, was born in Brazil in late 1965. When they left in 1969 to return to the States, the boys all spoke Portuguese well, but David, slightly over three then, could speak no English whatsoever. "He couldn't say, 'Good morning,' " said Wolford. "We communicated with him in Portuguese. It took him a while to break over when he came to the States. When he'd go spend a weekends with his grandparents, one of the other boys would have to go and translate for him."[59]

Operating in a foreign country involves many considerations taken for granted in this country. "Over there, you've got everything [to worry about]," said Charles Delton. "You've got to find out about their local taxes. One main worry is when are the damn banks going to close or stay open. That could hurt you badly. Little items like that are not little. You've got to foresee all that.

"Pilferage is one thing that will cost you something wild if you don't watch it. Down there they have it—consistently. How things disappear is just uncanny. Because there's a market for any tool or anything. Payroll abuse, if not padding, is wild. Personnel problems are greater than your working problems, let me put it that way.

"I found out in Brazil that we could not possibly be using as many tires as we were buying, and I put a big long $200 chain and lock on [them]. That ended that.

"The accounting for that venture was one of the most demanding of my entire experience, because the contract between Delta and the government there involved advance payments from the Brazilians to Delta, and those payments had to be credited against invoices to them with variable limitations, and if you didn't watch those limitations you would wind up paying for some of their goods. And then you'd have to do it in two currencies at variable rates, and everything to tie in and conform to their laws—and don't forget ours. It took one hell of a lot of work."[60]

By early 1965 Petrobras had acquired eight rigs, capable of a range of depths, and was requesting bids on labor contracts to operate them. Delta bid on the contracts because the company needed more work, with little additional investment, to compensate for the payments in cruzeiros that had been insufficient to pay all the local expenses.[61]

Though Delta became committed to operate three heavy rigs owned by Petrobras on a labor contract basis, this work would not begin until early 1966. Meanwhile, the payments by Petrobras in cruzeiros frequently were insufficient for Delta to keep all of its Brazilian commitments, as only 25 percent of the payments was drawn in local currency. This occasionally required exchanging the dollars sent to the U.S. for Brazilian currency.[62]

Nonetheless, by early 1966 the work was breaking even before depreciation. The operation was well-organized and stabilized. Petrobras seemed pleased and Keating Zeppa had flown to Rio de Janeiro to negotiate a price increase under the current contract. The younger Zeppa succeeded in securing better rates, but awaiting the beginning of the labor contract on the three Petrobras rigs, bought in Romania, earnings were "unexpectedly small," as the company ran into inflated costs and unforseen drilling problems.

It was at this point, in 1966, that it was discovered that by operating in Brazil without registering the company as a foreign outfit, Delta had been operating contrary to law. On October 1, 1966, the problem was corrected by forming Perbras, a Brazilian limited partnership—Empresa Brasileira de Perfuracoes Ltd., wholly controlled by Delta—which would be jointly responsible for the 1963 contract.

One of the three Romanian rigs bought by Petrobras went into operation in 1967. As with several other foreign ventures, the operation started out "disappointing," but hope still existed for improvement.[63]

Frank Dykstra worked on a rig that was built in Romania. The Brazilian government had traded coffee for three drilling rigs. "And dug an 18,000-foot well. Took over a year to drill it, and the rig wasn't any good at all. They don't know how to heat-treat in that country. The sprockets and chains wore out real quick."[64]

By the end of 1967, Perbras' deficit, financed by Delta, had reached $100,000. Under the agreement, Delta received 75 percent of the income from its three rigs in dollars in the United States, with 25 percent going to Perbras in cruzeiros, along with all of the income from the labor contract on the Petrobras (Romanian) rig in cruzeiros. It was anticipated that the operation would continue to lose until the contract could be renegotiated to provide for Petrobras to pay 45 percent in cruzeiros instead of 25 percent.[65]

By the end of 1968, many of the bugs had been worked out of the Brazilian venture. On August 1 of that year, currency-exchange problems were resolved when Delta agreed to divide the dollar payment, with Delta receiving 55 percent and Perbras, its Brazilian subsidiary, 45 percent, payable in cruzeiros. Amounting to a modification of revenue division, the revision would permit Perbras to begin paying off the large advances made to it. By then Delta's operation ran four rigs—Delta's two, the one leased from Perforaciones Delta, and one owned by Petrobras. At last, "a substantial profit" before depreciation could be reported, as a result of improved organization and drilling procedures, and the elimination of some of its earlier costly errors.[66]

Libya was the location for a Delta rig in 1956. The company already had foreign operations in Spain, Italy, and Venezuela, with Australia, Argentina, and Brazil yet to come.

In 1960, Delta began operations in Argentina.

Morris Hayes, standing second from right, managed Delta's operations in Spain which began in 1952. Hayes used hands from the States and local Spaniards (front row) to man the rigs.

Working on Rig #40 in East Texas in the early 1960s were (front row from left) Randall Holley, Shorty Lewis, Jack Kennedy, Gordon Wilson, James Caldwell, and Jim McNew; (back row from left) unknown, Jim Jackson, and Marion "Foots" Taylor.

In 1954 at the Benezette Gas Field near Driftwood, Pennsylvania, Delta became the first drilling contractor to successfully air drill a well in the Appalachians, revolutionizing the industry in the region.

DELTA DRILLING COMPANY
25th Anniversary Dinner
Blackstone Hotel - Nov. 19, 1956
PHOTO BY JAY OLSTAD

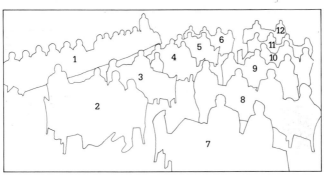

The Twenty-fifth Anniversary Dinner was held in the Ballroom of the Blackstone Hotel on November 19, 1956, in Tyler, Texas. Table 1 from left: The Rev. Jack Bentley, Annie May Jones, Nick Andretta, Mrs. Nick Andretta, Joe Zeppa, Mrs. Joe Zeppa, Mrs. Chris Zeppa, Chris Zeppa, Mrs. Guy White, Guy White, Mrs. Joe Bevill, Joe Bevill, Trent Zeppa, (standing). Counterclockwise, Table 2: Douglas Whitaker, Mrs. J. C. Stacy, J. C. Stacy, Warren Hirsch, Mrs. Gus Jones, Gus Jones; Table 3: Mrs. J. T. Wilson, J. T. Wilson, Mrs. John Parrish, John Parrish; Table 4: E. L. Young, Mrs. E. L. Young, Bolie Williams, Mrs. Bolie Williams; Table 5: Pat Donahy, Mrs. Pat Donahy, A. C. Smart, Tommie Smart; Table 6: Frenchie Portier, Mrs. Frenchie Portier, Mrs. R. C. Rhoades, R. C. Rhoades; Table 7: R. T. Mothershed, Mrs. R. T. Mothershed, Mrs. C. L. Bishop, C. L. Bishop; Table 8: Ed Gandy, Joe Blasingame, Mrs. Joe Blasingame, Mrs. Nathon Bacle, Nathon Bacle, B. W. Freeman; Table 9: H. H. Craver, Joe White, Mrs. Joe White, Irene Wood, Herman Wood; Table 10: W. N. Dailey, Mrs. W. N. Dailey, Mrs. D. H. Strength, D. H. Strength; Table 11: H. J. "Red" Magner, W. B. "Jack" Kennedy, Mrs. Jack Kennedy, W. M. Williams, Mrs. Bill Phillips, Bill Phillips; 12: Ken Bennett on piano and Danny Steuber on drums.

The Ozona Gas Processing Plant in West Texas came on line in 1964 to process the natural gas produced by Delta wells in the area.

Although drilling activity was low in the 1960s, Delta used this slack time to repair and rebuild rigs and equipment so the company would be ready when the activity rate picked up again, which it did in the 1970s.

SHAPLEY' ROUGHNECK — Mrs. Claudine Shapley of Tyler, Tex. has brought women's lib to the oil industry, taking a job as the world's first floorwoman on an oil rig operated by Tyler's Delta Drilling Co. for Phillips Petroleum Co. in Northern Louisiana. Mrs. Shapley is shown on a rig tour with Tool Pusher Ray Young, left, and Motorman Don Gladden. Until taking the "roughneck" job on an oil rig this week with Delta, Mrs. Shapley was a legal stenographer for 21 years.

Oil Industry Getting
First Lady Roughneck

Safety and training have long been priorities of Delta Drilling Company. The development of such devices as the derrick climber have benefitted the entire industry. Delta also led the way in developing and distributing information concerning the dangers of hydrogen sulphide gas.

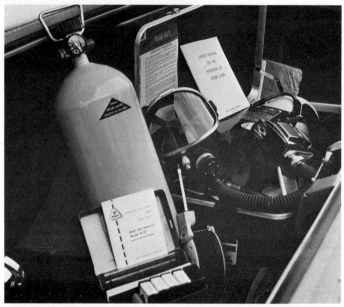

In 1969, Brazil's president Arthur da Costa e Silva, ruling by decree after recessing Congress, suffered a stroke, and military leaders imposed General Emilio Garrastazu Medici in the presidency. With a new president and a new slate of officers and directors at the helm of Petrobras, Delta's venture there took on a wait-and-see nature. It was expected that the new leaders were likely to curtail land operations in favor of offshore development. Accordingly, Delta had proposed that its three rigs be placed under a contract whereby they could be operated anywhere in Brazil, with compensation for work outside its original contract area. Though such an arrangement would provide for payment in cruzeiros, Zeppa reasoned that with the $1.8 million advance made to Perbras by its parent company registered with the Central Bank, Delta would have the right to repatriate the amount as the funds became available.[67]

Despite the problems of operating, plus inflation, Delta continued to make money in Brazil and sought to diversify its services to meet changing conditions in that country. As the earlier contracts neared expiration, Perbras successfully bid on two water wells in the state of Piaui with equipment rented from Petrobras. The program, drilled for the Brazilian government but involving United States aid funds, offered the possibility of drilling 200 more wells if the first two were successful.[68]

Within the year, a contract had been signed for Perbras to operate a tender offshore. Delta Marine's Tender Rig No. 8 was moved to Brazil—a godsend for Delta Marine's sagging fortunes—but repairs and mobilization would run close to $400,000. However, the day-rate payment to Delta Marine in dollars was expected to bring in a profit that would not be taxed because of loss carry-forwards.[69]

Delta Marine Tender Rig. No. 8 arrived in Salvador, on June 20, 1971, but adverse weather delayed its rigging up. By that time the Romanian rig was still being operated on a labor contract, and Perbras was operating four workover rigs under a four-year contract.[70] It remained to be seen how profitable the operations would be.

When Joe Zeppa reported on the Brazilian matters to the shareholders in 1973, the Brazilian partnership—Perbras—had suffered a $684,016 loss, composed of a direct operating loss on a cruzeiro basis, a dollar expenditure loss, and an unrealized exchange loss.

When minority stockholders inquired as to what was being done to resolve the losing operations—Perbras was included with Delta Marine and the Tyler Hotel Corporation—Zeppa replied that Edwin D. Arnold, vice president in charge of international operations, had been sent to Brazil to see what could be done.

By the end of October 1973, the Delta Marine 8 had completed its contract with Petrobras and had been moved to Venezuela. Activity in Brazil had reached a low point.[71]

By October 1974, Perbras owed the parent company a total $3,571,022.72. Of that amount, $3,018,903.14 represented dollar transfers which were registered with the Central Bank as loans for working capital with specific maturity dates—capital which could be repatriated. However, the Central Bank of Brazil then issued regulations which froze repatriation of foreign loans until December 31, 1978. Interest, however, could be remitted, and interest for 1974 amounted to $266,418, which Zeppa hoped would be earned and remitted by Perbras.[72]

Management changes occurred in Perbras in the early 1970s. Frank J. Semper, who had been in charge there, resigned in December 1973, and was replaced by Jack Robinson, who had been in Argentina for Delta. Robinson was then transferred to Venezuela to manage Perforaciones Delta, and L. L. "Bud" Worsham, who had worked for Delta in Venezuela and Mexico, became manager of the Brazilian operation.[73]

The turnover in managers during the time Bernie Wolford was in Brazil stood out in his memory. "The old saying was, if the boss called, be sure to get his name," he chuckled.

Even though Delta was facing losses, there was the sum of $3 million in Delta's accounts receivable from Perbras, the debt registered with the Central Bank so that it could be transferred to the United States as funds became available, beginning in 1978. Thus, there was a $3 million motive for continuing operations in Brazil. By then a long-term contract had been negotiated with Petrobras for offshore work. Delta Marine's Tender Rig No. 8 had been expected to perform the work, operated through Perbras, but a dry-dock inspection of the tender in Curaçao revealed it to be in such bad condition that it would take $2.5 million in repairs for it to meet certification from the American Bureau of Shipping and the U.S. Coast Guard. But if Delta failed to meet the offshore contract with Petrobras, it might signal an end to its business in Brazil, thus jeopardizing its interest already there. A substitute tender was sought in a 30-year-old tender owned by Atlantic & Pacific Marine. Despite its age, the tender was virtually new, having been in storage practically all of its existence. By salvaging Delta Marine 8 and chartering the Atlantic & Pacific Marine tender, Delta could continue its operations for Petrobras; otherwise, the entire Brazilian activities might be shut down. But with the tender operating, Perbras would be able to start a cash flow of $700 to $1,000 per day. At a time when the annual loss by Perbras had been recorded at $349,857, this opportunity for profit had a definite appeal.[74]

As 1975 got under way, Delta found its Brazilian operation to be in a rather precarious situation—losing money, but risking to lose much more, especially money it had already earned, if it did not keep providing services for Petrobras. The status of the Brazilian operation was somewhere between the dismal story of Venezuela and the success of Mexico, possibly a notch above the money-making operation, cash-flow problem in Argentina. The reality of the situation justified continued activity, but with careful scrutiny of the risks involved.

IV

In late 1969, Delta sent a rig to South West Africa, just north of the Republic of South Africa. That area is now independent and called Namibia. Delta's involvement consisted of purchasing, installing, and testing equipment for the operation. Bernie Wolford, a long-time Delta employee now in the drilling business for himself, went to South West Africa in late 1969 to set up the rig. He had just returned from Brazil, where he had been in charge of maintenance.

"I took a few weeks off and Joe Blasingame was in the process of building a rig in the Kilgore yard, a 500 Gardner. It's a rig capable of drilling 8,000 or 10,000 feet, and I helped complete that rig. It was going to South West Africa."

Delta was dealing with Emanuel Rosenblat, head of Etosha Petroleum Company.

"He was from New York," said Wolford. "He had a Dallas office. Delta was gonna be partners with him, and he had a large leaseholding or concession in Etosha Game Preserve in South West Africa. I didn't know until I later went to South West Africa that it was the largest game preserve in the world. I think they said about 24,000 square miles.

"As we got the rig further completed, Delta had just decided they would enter into a labor contract with them and operate the rig. Rosenblat's rig and Delta labor. About the time we got it all loaded on the ship to go over there, Rosenblat canceled the labor contract with Delta.

"Rosenblat owned the rig. Delta rigged it up for him, and he paid Delta for the rig-up. We loaded it in Beaumont on the Gulf Coast and picked up a tandem truck. We were going to ship it to Walvis Bay, South West Africa. He didn't cancel the labor contract until it was on the high seas. It was a trying thing. Keating [Zeppa] wanted me to go to South West Africa and unload the rig and carry it to the location, rig it up and stay there until they drilled service, where the people couldn't say, 'Well, you didn't rig-up properly' or something.

"So they had hired a drilling superintendent by the name of Philoughby.* He and I left Dallas together to go to South West Africa. Philoughby didn't sober up from that day forward. We went to New York and flew on over. He started out in Dallas and didn't draw a sober breath from there on. He drank on the plane. We flew on over to Greece and on down to the East Coast of Africa. Into Johannesburg and flew on in to South West

---

*Not his real name.

Africa. We didn't fly into Walvis Bay; we flew into some large town there in South West Africa and took a private plane, flew up to Grootfontein and that's where Rosenblat had a mining operation. There were several minerals over there, and I don't know what all he was engaged in, but he had a pretty good size exploration and geology department. He was still in the exploration stage.

"We beat the rig there two or three days. We went up to Grootfontein there in a private plane and some of his people picked us up. They had a saloon that opened at six in the morning in a hotel, but they didn't take us to the hotel. First they took us out of town a couple of miles to a little rock house that didn't have any doors or windows and told us that was our quarters. No electricity. I wouldn't get out of the car; I was afraid it might leave. I said I wasn't going to stay there and they said, well, that's where Rosenblat wanted us to stay. We said, 'Well, we want to go back to the hotel.' Well, he wasn't going to pay for it, but it didn't make any difference to me whether he paid it or not, I was ready to go back to the hotel.

"So we went back to the hotel and stayed there a couple or three days. When the rig started to come in, well, we flew down to Walvis Bay. The location was about 50 miles over inside the Etosha Game Preserve, and from Walvis Bay it was approximately 400 miles up there. About 20 miles of paved road, and we loaded part of the rig on a railroad car and the railroad ran that way for about 180 miles or 200 miles and we were going to pick it up there and haul it to the location, and the rest of it we transported all the way there by truck. We hired dump trucks and cattle trucks and everything to get the rig up there. We had this one oil-field tandem to unload everything.

"After I stayed in Walvis Bay for about a week, well, I left Philoughby there and I went on up to start rigging up the rig, and it was really like living in a game preserve. There was every kind of animal in the world up there: elephants, lions, giraffes, zebras, gazelles, sternbucks, and you just name it. Some of them you couldn't name. We drilled a water well at the rig, and at night there would be a lot of tracks around.

"For some reason, they had decided to make the camp in one location and the drill site in another location. Which was about 35 miles apart, and this was some of Etosha's people. Rosenblat's company was Etosha, and he got the name from the game preserve.

"The second drill site was going to be where we made camp. [It was an all-German camp.] Why they didn't make the first one there I don't know. But we had a water well there already drilled, and none of these Germans had any experience rigging up anything. None of them had any experience rigging up anything but a small truck-mounted water-well rig.

"If I hadn't carried some pictures over there I don't think I'd had it rigged up *yet*. Philoughby—what days he would go out to the drill site—

would get to laughing and a-kidding, and he'd make them mad because they didn't know what they were doing. They would get mad and they would quit. It is very hot over there in the daytime. It'd get up to 130 degrees and, of course, you don't do too much in that kind of weather anyway.

"We finally got the rig rigged up and got a water well drilled and got spudded in and set the conductor pipe. That fulfilled Delta's obligation.

"While we were rigging up there, close to the end of rig-up, well, we got the derrick all put together and strung up. I told this German drilling superintendent that we wanted him to raise the derrick, and he said that he wouldn't take the responsibility of raising it. And I said, 'Well, you need to, so that you will know how to do it, because after you get it raised I'm not going to be here when you lower it.' It took about a day to talk him into raising it. We got everything rigged up, and during this course of 30 days or so, two or three times the jeeps would break down between location and camp, and a couple days we didn't even make it. We were broke down all day, and we weren't about to start out on foot across that game preserve—I wasn't and no one else was, either.

"We got spudded in and drilled two or three collars down, and that fulfilled Delta's obligation. During all this rig-up time, I had a good relationship with all these Germans. I had one guy who could speak a little broken English, and I kind of worked through him and showed them how everything goes. At night we would have classes in camp about how to work on mud pumps and different things, what it was all about. It was a kind of school I conducted. Philoughby would sometime attend and sometime he wouldn't. He would go down every few days to the railhead over in town. He'd get low on booze, and he would make it a point that he would have to go down there and get somebody to drive him there."

Ordinarily it would take a couple of days to rig up. This job, because of the special problems, took about a month. They spudded in, and Wolford and Philoughby left for Johannesburg where they met with some people from Etosha, then flew home. Philoughby had completed his agreement with Delta.

Back in Africa, the well proved to be a slow job.

"I think after about a year they still didn't have that well completed," said Wolford. "They still weren't but about 7000 feet."[75]

V

What might be labeled "close calls"—those countries into which Delta almost went, but didn't—are almost as interesting as the actual foreign operations. Although Joe Zeppa probably discussed more foreign deals than those which finally made their way to the minute book, negotiations al-

most reached the contract stage for operations in three other countries—
Saudi Arabia, Bolivia, and Iran.

The discussions over Saudi Arabia and Bolivia began in 1957, but did
not survive the next year. At the end of 1957, Zeppa announced to his
board of directors that Delta was "about to conclude" an agreement with
Arabian American Oil Company to operate one of Aramco's rigs in Saudi
Arabia. Delta had been selected over several other competitors who had
submitted proposals. "This may, later," said Zeppa, "develop into a much
larger operation."

Exactly why Delta did not go into the Aramco deal is not altogether
clear. Not only did the minority stockholders tend to view askance most
foreign operations, but the fact of anti-Jewish feeling by the Arabs must
have had some impact.

At the same time, Delta was negotiating with TULM (Tennessee Gas
Transmission, Union, Lion, and Murphy) to send drilling equipment to
Bolivia. "If agreement can be reached on this initial contract," Joe Zeppa
told the board, "the prospects are excellent that it will turn into a large
operation."[76]

It never did. Lion Oil and Murphy Oil, it might be noted, came from the
same place Joe Zeppa had started his own career in oil—Arkansas.

In 1973 Delta seriously approached the idea of going into Iran. The Iranian Oil Exploration and Producing Company—"The Consortium, made
up of interests from British Petroleum, Shell, Standard of California, Gulf,
Texaco, and Mobil"—invited Delta, as well as others, to submit a drilling
program for South Iran. H. J. "Red" Magner and Jack Robinson, then stationed in Argentina, attended a meeting of interested contractors in
Ahwaz, Iran, in May, and reported back to Zeppa.

First-class used rigs would be acceptable to the Consortium, and Delta
had four diesel electric rigs which it could use—two from the Southeastern
division, two from Con-Tra-Pet in Argentina. Computing its possible proposal for the three-year contract, Delta's valuation of the four basic rigs
was set at $4.5 million. The total long-term investment would be
$13,979,370, which would make the long-term cash requirement
$9,279,370. The cash flow for the three years, amounting to 4,216 con-
tract days, was estimated at $11,070,696, which would mean $1,791,326
more than the long-term cash requirements. There were, of course, the
prospects of further work beyond the three-year contractual period.

Because the Shah of Iran was exerting "extreme pressure" to increase
oil production, and because of the prolific reserves of that country at a time
of an increasing worldwide demand for petroleum, steady future work
seemed likely. The report to the board concluded, "There is every reason
to believe that the successful contractor on this tender can look forward to

several years beyond the initial three-year contract of continuous uninter-rupted activity."

When presented to the board of directors at a special meeting in June, 1973, the proposal met immediate resistance. Director Albert Sklar argued that Delta's financial resources were insufficient to take a $9 million risk in the Middle East, an area that to him was "politically explosive and subject to extreme changes which are impossible to predict." Keating Zeppa, ex-ecutive vice president and director, pointed out that the opportunity for profit on land drilling was much greater there than anywhere else in the world. Sklar countered that only a much larger company should consider such a risk, ideally a publicly-owned company which would spread the risk among more stockholders than would Delta, where "the total investment would represent approximately one-half of the net value of this company." Dr. Sam Y. Dorfman, Jr., took this opportunity to suggest bringing the Argentine rigs to the States, improve their capacity, and operate them do-mestically where conditions were more predictable.

Joe Zeppa, noting Dorfman's point, reported that "large capital expen-ditures" were being considered in order to improve domestic equipment. New engines—Caterpillar engines or their equivalent—would necessitate spending $2 million in a 12-15 month span.

Once the discussion had ended, a poll of the directors revealed strong opposition to submitting a bid proposal: four against, one for, one abstain-ing.

"The bid was, therefore, not made in the name of Delta Drilling Com-pany."[77]

In retrospect, with the hindsight that a few years of history have already provided, the decision against tendering a proposal to the Consortium on Iran was an extremely sound one, even though the stability of the Shah's government did extend until late in the decade.

The minority stockholders in Delta generated little or no enthusiasm for foreign operations generally. "I can vouch for it," said Seymour Florsheim, who was closely associated with Sam Dorfman, Sr. before his death and subsequently went on the board himself; "the rest of the minority owners did not advocate the foreign ventures."

"Insofar as after I got on the board," said Dr. Sam Y. Dorfman, Jr., "the only foreign venture we really had an opportunity to stop was the Iranian thing. And the other ones, the only thing that I have seen is retreat, retreat from those ventures. The Italian and Mexican properties have been the only ones that have really been prosperous."

Though foreign drilling has had an aura of glamour about it, Dr. Dorf-man notes, with a chuckle: "It's like having a pet cobra at home. It sounds great, but it'll bite you in the butt! You know, no matter where you are, you're gonna get hit. And, of course, there were numerous companies that lost their rear end in Iran.''[78]

VI

Delta's only two presidents viewed the foreign divisions from the vantage points of two different generations—Joe Zeppa in 1975, Keating Zeppa in 1980.

"Of course," said Joe Zeppa, "lot of these [foreign] things have not all been successful. Not because of their planning, but because of the way the exchange rate went. We went to Argentina, for instance, and we lost the first year close to $1 million in the decline of rate of exchange. And we shouldn't have lost it, because it was all money due us from the government company! They wouldn't pay. And while we were waiting, why, the peso kept going down. It's fantastic, you know, when you see your money disappearing because of quotations on the market."

Keating Zeppa looked back in 1980: "Taken on the whole, everything outside the United States, from the early, tentative outreach in Nova Scotia to the current time, and all that has transpired in between, I don't know that Delta Drilling Company has made much money overseas. We either stayed too long or we didn't commit enough capital or for one reason or another; but, generally speaking, Dad was not too successful in any venture in which he could not exercise his own brand of business acumen, if you will.

"When he had to rely on a hired manager that was not easily reached, even by telephone many times, [he was less effective]. Dad, as most entrepreneurs, ran his business on the basis of personal eye-to-eye contact and verbal instructions and verbal expositions on what he wanted to accomplish. I don't think he was inclined to put down on paper a structure and say, 'I want this and this and this, and here's how we're going to accomplish it.' Not that he didn't make the analysis, or didn't know what he wanted, but it was generally in the realm of oral communications, and I think we probably over the years have not been as successful and as perspicacious in the hiring of managers of foreign operations as we have been domestically.

"But taken all in all, if we did a summary of Delta's foreign operations, Delta would not come out too well. For that reason, and wanting to concentrate more particularly in the United States on exploration and production, we started about three years ago a program of shedding our foreign operations and converting those investments into a transportable form of capital."

At the time in 1977 when Delta began simplifying its foreign holdings, Keating Zeppa recommended to the board members that they keep Mexico and Italy for two reasons: they had the best outlook for profitability in the next five to ten years, and there was a high probability that they could remit those profits. "There's no good in making a profit in a foreign country if you can't bring it back home." The board went along with it.

The first to go was the Argentine operation, which was concluded in 1979. The sale of the Venezuelan properties was concluded in 1980. With the sale of Delta Marine's last remaining asset, a drilling tender in Brazilian waters, Delta Marine was liquidated in 1981.

That left the remaining operation in Brazil mainly workover land rigs, which were attractive to several potential buyers. Perbras was sold on December 29, 1980.

These actions left Delta with its interests in Mexico and Italy: a 24.5 percent interest in Mexico, beginning in May, 1980, with a 37 percent interest in LIPSA, the pipeline company there; and 100 percent interest in the Italian company, which had become more profitable by the end of the decade.[79]

11

MEANWHILE, OUT AT THE FARM

I

Joe Zeppa's two great loves, as almost anyone who knew him well would say, were Delta Drilling Company and his large farm near Winona, north of Tyler.

Pinehurst Farm was more than just another farm or ranch that an oilman had spruced up and transformed into a showplace. All this Joe Zeppa had done, but for a farm boy from northern Italy his identification with Pinehurst went much deeper.

The farm, in fact, was one of the major reasons Delta remained in Tyler instead of being moved to a large city such as Dallas or Houston. An Italian peasant's love for the land seems to have been deeply ingrained in his psyche. The farm also gave him an opportunity to show his fellow landowners and citizens how to operate a model agricultural project.

Pinehurst, a few minutes' drive from his downtown office, was near enough that he could spend his weekends at the farm. He could not have done this so easily had he been headquartered in Dallas or Houston.

Yet the farm was even more than that. It was, in a large and deep sense, an extension of Joe Zeppa's office downtown, just as it was an extension of his own personality. When he left his desk for the weekend, he repaired to his retreat for relaxation and a change of pace, but often he also was conducting business in another, informal setting.

Old friends and business associates were entertained at Pinehurst. Close friend and philosophical debater Sylvester Dayson was a regular weekend visitor. The Sunday morning get-togethers between the two men usually ended abruptly in shouting but were resumed the following weekend.

As in much of his business, Zeppa was forever tinkering on his farm, undertaking agricultural improvement projects to make Pinehurst Farm like no other. Like some of the business ventures, not all of his plans for the farm worked out. Some, in fact, came a cropper. But no setback seemed ever to curb his prevailing optimism. He would just take a new approach and try it again. There seems to have been little difference in the way he handled the farm from how he operated his business. Though Delta and

Pinehurst were two different entities, in Joe Zeppa's mind they were virtually inseparable.

This attachment to Pinehurst was never better symbolized than in a statement he made one day to an employee—a desire he apparently never translated into definite language to his heirs. That day Roger Choice and some other employees were drilling a water well at the farm when Joe Zeppa came by.

"He told me way back there that when he died he wanted to be buried up there on that hill," said Choice. "And I said, 'You mean right out here on the farm?' He said, 'That's right.' "[1]

II

Joe Zeppa bought the original acreage—about 800 acres—of Pinehurst Farm in the early 1940s. At the time he purchased it, the land was known as the Simpson Peach Plantation. Large quantities of fruit were shipped by rail from a sidetrack at the farm. Many peach-growing farms still operate in the area. Some of the old buildings that came with the farm when Zeppa bought it were built shortly after the Civil War.

In 1946, Zeppa inspected the Smith County prison farm, which the county had decided to sell. It was at Lindale, in the Tyler area. At the time, A. D. Winston was managing the "pea patch," as the county farm was called, and his son, Milton Winston, was working for him, guarding prisoners and farming.

Joe Zeppa, while looking over the farm one day, asked Milton Winston, "If I buy this place, will you go to work for me?"

Winston told him he might.

A few days later Zeppa purchased the 860-acre farm from the county, and Winston went to work for him. Zeppa soon afterward named the old county prison farm Oakhurst. The other one, near Winona, was already named Pinehurst. Why did he name one Pinehurst and the other Oakhurst? "I asked him the same thing one time," said Winston, "and he said, 'Well, oak trees over there and pine trees over here.' "

When he got married, Milton Winston moved out to Oakhurst Farm, and his father moved back to his old homeplace at Whitehouse, Texas, south of Tyler, though he still worked as a consultant to Joe Zeppa on the farm. Meanwhile, Milton Winston's cousin, Leon Winston, managed Pinehurst Farm, and Milton, Oakhurst. Leon had been there about five years when he was killed in an automobile accident in Dallas. At that point Milton's duties were expanded to see after both farms.

"I was coming back and forth from Lindale to Winona, from Oakhurst to Pinehurst, every day, and it was getting pretty rough on me," said Milton Winston. "Even on weekends Mr. Zeppa wanted me around. If I wasn't here [at Pinehurst] on Saturday and Sunday, he wanted to know why. He just liked for me to be around. And he asked me one day if I wouldn't like to move over here, and we did so."

Despite the fact that Joe Zeppa's interest in the farms was the only reason the Zeppa family or Delta had any connection at all with them, the two farms were altogether different operations. Oakhurst Farm was owned by Delta Drilling Company and was operated commercially for the benefit of Delta. Pinehurst Farm, on the other hand, was Joe Zeppa's personal property and had no affiliation with Delta. Zeppa, however, did operate Pinehurst as a commercial agricultural property.[2]

III

Oakhurst Farm eventually grew to around 2,100 acres. The pattern of acquiring land to add to the original 860 acres was the same one Zeppa was to use in expanding Pinehurst Farm and, in fact, Delta itself.

There was a similarity in the operation of the two farms, although it was clear that Pinehurst was Zeppa's more cherished of the two properties. At Oakhurst, Milton Winston said, "It was at first general farming, cow and calf operation. We had row crops and we raised some cotton at first, corn, sweet potatoes, and peas. Then we got away from the row crop and went strictly to cow and calf. Ready for the markets. Most of our calves we shipped to Iowa and Nebraska—we have been for 30 years. They go to wheat field stubble, corn stubble fields, silage. They go there before they go to the feed lot."

The same pattern was still being followed at Pinehurst Farm in 1980 when Winston was interviewed. The calves would be shipped when they were six to eight months old.

Pinehurst Farm and his care for it symbolized Zeppa's love of the land that he never lost until the day he died.

"Mr. Zeppa loved the earth; he loved the farm," said Winston. "He was a farm boy from his own country, in Italy, and I've heard him tell people, and tell myself, that he made his money out of the earth. That's the way he had of putting some of it back. And it is just hard to envision and hard to realize how much money he put into soil and water conservation."

Pinehurst, at first roughly the size of Oakhurst, grew to 3500 acres through the process of buying adjoining tracts as they became available over the years, the same kind of steady accumulation that can be seen in much of his Delta transactions, particularly in his building up production reserves.

Many of the existing buildings at Pinehurst were old and run-down, some of them having been used in the late nineteenth century. Many of them Zeppa saved and refurbished, using for one purpose or another. One building had been part of a commissary in Pinehurst's plantation days, and that was made into a house for employees.

"Mr. Zeppa liked to take old buildings and remodel them and fix them up for his employees to live in," said Winston. This parallel practice was followed by Delta, using second-hand drilling equipment that had been fixed up and put into use. The "big house" at Pinehurst, where the Zeppas repaired for the weekends, traced back to the late 1860s, soon after the end of the Civil War.

Winston remembers when the house was being remodeled. It was in a shambles and had a rickety old-fashioned outdoor toilet behind it. The grounds hadn't been landscaped yet, and some barnyard animals roamed freely about.

"It was just a crackerbox setting up there at first," said Winston. "He went out one day. They had an outhouse and some of us were sitting out on the back porch and he had to go to the outhouse, and he came running out of there and his britches were still down and he said, 'There's a damn chicken in there!' He went and sat down on that hole and there was an old hen started clucking, you know, and here he come. Funniest thing I ever saw. Mrs. Zeppa, I can still see her laughing."

The agricultural operation at Pinehurst was basically the same as that at Oakhurst. "A lot of row commercial crops—corn, cotton, peas, sweet potatoes," said Winston in 1980. "We kinda got away from that and then got into the silage business, corn sorghums, for our own cattle. Then that got so expensive we got away from that, and we buy all of our feed now. It's cheaper to buy it than it is to produce it. We're strictly in the cow and calf business now, for 12 years.

"Part of that time, though, we used to be in the hog business. We had 100 brood sows, which is a lot of hogs. We quit the hog business about '74, about a year before Mr. Zeppa died. The wolves put us out of business. The wolves came out of the [Sabine] river bottom—they just eat 'em up. And the hog business and cow business was a cash income."

Though occasionally a cow would lose a calf to the wolves, they usually kept the raiders off pretty well. And while a sow, riled-up, is normally a formidable foe to man or wolf, the wolves had a system to their predatory behavior.

"They'd get 'em while the sow was pigging in the bed, and they'd slip in there and get one before the old sow knew anything about it, and after we weaned the pigs—put them in the weaning pen, weighed 30 pounds— they'd jump a fence six feet high and scratch under, get a 30-pound pig and leave with it. It surprised me."

Looking at one of his fattening lots which were once used for hogs, Winston said, "I've had 2,000 hogs in there sometimes. And we'd put the shed on it where we could keep the hogs cool. We had them a swimming pool there so they could get in the water."

Just as Joe Zeppa and Delta Drilling Company became inseparable in the minds of those associated with either one, Zeppa's attachment to Pinehurst was as strong as that to his company. "If he didn't have but half a day—if he just got in from New York and had to leave the next day to go to Italy—he'd come out here," said Winston. "He liked to spend his money on the farm. If he was out here, I don't care who was here, he'd buzz me: 'Milton, come on up.' I'd get up there and he'd say, 'Let's go.' We'd get out and ride over the farm. He'd tell them [his guests], 'You all make yourself at home.' But he liked to be out *riding* on the farm. I wore out several station wagons driving him around on the farm. And he just loved it. I've seen him in *hot weather*—he'd *sweat*. He liked to sweat. He liked to *talk* about farming, about which way we were going to run this fence. 'Now let's clear this piece of land.' 'Put a terrace here, Milton.' 'Let's don't go down this way, 'cause your water will start washing [the soil away].' He knew what he was talking about. He just liked farm life. He liked to plan.''

Zeppa gained his knowledge of agricultural practices mainly from observing, reading, and introducing his own ideas, said Winston. He subscribed to a number of farming magazines and also bought books.[3]

Marguerite Marshall was at Pinehurst with some other employees one day and received a tour of the house and gardens. "I remember him saying he kept planting. He said, 'I won't live to see these trees bear, but we plant some every year.' He said, 'Someone will get some benefit from them.' He always was planning and working toward a future even though he wouldn't live to see it."[4]

The big house at Pinehurst where the Zeppas stayed originally was the headquarters house of the plantation. Zeppa added wings to it, underpinned it, and boxed in the hallway to make it more to his liking. Then he had trees set out around the house, to accompany a few that were already there, such as one aging post oak. Gertrude Zeppa maintained Oriental rugs on the floors.

Zeppa saw to it that the house had a library and a bar, both of which symbolize his two dominant activities—the pursuit of learning and socializing with his friends.

"And he liked a rip-roaring fire in the wintertime in the fireplace," said Winston. "He had a table set right here [looking out on the back of the house]. I don't care how hot it got in July, he sat right here and had his highball. I'd have the water [fountain] on and he'd sit here and read his paper. He liked the sun. He really did."

Flowers and a variety of trees graced the homesite and, in fact, most of the farm. Around the house there were cannas, marigolds, roses, salvias, and other flowers. In the vicinity were trees as varied as Chinese oak, magnolia, mimosa, hackberry (the large kind), cedar, and sycamore, in addition to the pines for which the farm was named. For food, the farm had pecan trees, producing the large papershells grown on Stuart and Success trees. Rows of watermelon-red crepe myrtles provided windbreaks to combat wind erosion, and running roses provided not only windbreaks and beauty but wildlife cover as well, particularly for quail. A lot of raccoons also called the farm home.[5]

It was not by accident that Zeppa knew so much about the trees and plants on his farm. John Murrell, chief executive officer of DeGolyer and MacNaughton who came to know Joe Zeppa in the 1930s and eventually came to appreciate his broad knowledge of the natural world, said: "Joe Zeppa at heart was a naturalist. There was not a thing that he didn't know about flora and fauna. He knew every tree, he knew every plant—not only in the East Texas area but in Italy, in New Mexico, in Arkansas, and in Louisiana. He was most interested in anything that grew.

"He knew plants by their scientific names and he knew 'em by their local names. And he could talk intelligently with people who were supposed to know a great deal about it. He visited our home, with my mother there. We lived out on the [White Rock] lake [in Dallas] about two doors from H. L. Hunt, and we had a lot of plants she didn't know. He'd go around and he'd tell her each and every one of the names, and she'd be scribbling it down. She'd ask, 'When's the best time to replant? When's the best time to change them from one area to another?' As far as I know, he was knowledgeable of everything that grew in Texas."[6]

Zeppa was especially proud of his fruit trees and the vegetables that were grown on the farm. Fresh fruit in season was a normal part of the weekend fare when the Zeppas were at the farm.

"He loved it," said Winston. "He had to have farm fruit and vegetables when he'd come to the farm, and he expected it. If he didn't have any, he found out why. His favorite was peaches, I believe.

"And we had chickens—he wanted everything. He was raised on the farm, and when they'd go in to Tyler on Monday, they'd have the maid fill up this big basket of stuff they'd cooked and vegetables and eggs and fruit and carry it to town with them to eat that week.

"And I'd charge him at the end of the year for everything—fruit, vegetables, firewood, labor for maid service, and all that."[7]

IV

In 1975, Joe Zeppa discussed his Pinehurst property. "Well, I've got a big farm. I've got a 3500-acre farm, just here about 15 miles north of town, which I put together and developed. It's a nice place. It's a place that you'd like to go, that'd you'd like to visit. It has several lakes on it. I have about 900 head of cows. I used to have Herefords, but I've got 'em mixed now. I've got a mixture of Brahma and mixed Angus, Herefords also."[8]

Hereford cattle were Zeppa's favorites. In the late 1960s he ran polled Hereford and did some crossbreeding of Brahma bulls with Hereford cows.

From beginning to end, Zeppa was as positive in what he wanted, or did not want, on his farm as he was in his drilling company. And when he wanted something done, he usually had a time limit in mind.

"One day Mr. Zeppa and Mrs. Zeppa were riding around in the car, and I had just fixed the road," said Milton Winston. "Had a little old short stump in it. He crossed this little old draw and he hit high center in his car. It was in July and it was hot and they had to walk back to the house. It was *hot*. He said, 'Milton, my car's down yonder, and I don't want that stump there in the morning.' I brought the tractor back down here, tied a chain to the stump, and pulled it out.

"He was a good feller. I only had one cross word with him all the years I worked for him. Hadn't been working for him but about a year. He'd come to Lindale, Oakhurst Farm. He'd been gone somewhere, I forget where now. I'd sold a bunch of cows. He got back in the office, and he noticed the check where I'd sold some cows, and he didn't want to sell a cow. He wanted to keep all the females, regardless of how good they were or anything.

"He sent word for me. 'Milton, you're selling these cows.' It made me mad, it really made me mad, and finally we had a few words and I told him, 'Mr. Zeppa, you got me hired to run that farm, and I think when a cow is a cull I know it and she ought to be sold. If you're in Europe or somewhere I can't wait till you get back,' and I said, 'You hired me to run it and I'm going to run it!' I said, 'Now under those conditions, you either keep me or fire me.' From that day on, he never did cross me. And all females after that, if I thought they would make a good milk cow I kept them. It was a good lesson for me."

The fact that Zeppa had definite ideas of what should be done did not mean that he always got his way at the farm. Gertrude Zeppa also had her own ideas of things which should or should not be done.

"Mr. Zeppa had his way of wanting things done in the yard," said farm manager Winston. "He'd say, 'Do this and this,' then when he went in to town on Monday morning Mrs. Zeppa would get me and say, 'Now, Milton, do it this way.' So I had to use my own judgment a lot of times. But I handled it pretty good, if I do say so. I'd kinda give and take. I've had to lie a little bit.

"I forget the name of some kind of tree—there were two of them he bought and gave a lot of money for, and put them up there and she didn't like them in the yard. She had them cut down. He [Zeppa] asked, 'What happened to them?' and I had to think. 'Well, a storm hit a few days ago when you were gone to Europe or somewhere and it tore them things all to pieces.' He knew I was lying, but he knew what I had to do.

"And one of the last things she wanted done, he didn't want to do: There's an awning up there on that house. She had it put there. He told her not to put it up there, but anyway she went on and had it put up there. He didn't like it. I finally got him out in the yard and I said, 'Look at that, Mr. Zeppa. It looks good, don't it?' He never did say. He looked at me and kinda grinned, you know. But they were wonderful people."[9]

V

Zeppa intended Pinehurst to be a showplace and a demonstration farm for other farmers in East Texas to use, he hoped, as a model.

Winston said, "Well, he had the money to do it and he wanted to show East Texas farmers what could be done through right, good management and soil and water conservation, and it did rub off on lots of them.

"He terraced land, not only for water but for wind. These hedge rows with these roses growing up and down them, he planted that for wildlife, wind erosion, and windbreaks. And there were some gullies on this place so big you could put this farm in. We put concrete dams across them. Then the rains, flow of water, eventually filled them up and some of them now you can drive across them and you'd never know it was there. Gullies and ditches. He cleared land and left the timber. Remember, it takes 100 years to grow a big tree.

"We cleared out the underbrush, and he liked trees and he planted grasses and legumes—clovers, vetches. And new grasses he would send us. Farm and ranch group or some group would go to Alabama or some place to see new clovers or new grasses, and he was always willing for me to go. He wanted to keep up."

Though he liked to fix up old buildings and used a lot of second-hand equipment in drilling projects, Zeppa frequently spent whatever he thought was necessary to achieve what he wanted.

"He liked to go the expensive route lots of times," said Winston. "We have four deep water wells on the place, and we irrigate when we have to. We have six-inch underground lines down the fence rows and down the road, and about every 100 yards we have a rise and hook in that with aluminum pipe and go out to the fields. These wells, they're about 800 feet deep. We're pumping at 350 feet. About ten years ago we had to lower them. We were pumping at 300 feet and run out of water.

"We use one [of the deep wells] for our central water supply. All the houses and corrals and gardens. All the employees lived on the farm, and he furnished them a good house. At one time he had nine families, and now we don't have but six. It don't take labor like we used to when we had row crops and had to do so much of it by hand; now we have machines. He liked to plan a garden for his employees. We paid one man to take care of the garden. He'd plant it and work it, cultivate it and water it when it got dry. All the employees would have to do is go and pick it.

"Same way with the orchard. We have an orchard with all kinds of fruits in it: pears, grapes, figs, peaches, apples, plum, berries. And then if they wanted it we gave them a milk cow to milk, buy the food for the cow—all they had to do was take care of her and milk her. But now that's a thing of the past. I don't have one family now that has a milk cow. They're making a little bit more money now and they'd rather buy the milk. We used to give them a pig, a butcher hog, to kill every year. We don't have hogs now, so we give them a calf. Each family, give them a calf a year. Mr. Zeppa liked for them to live off the farm. And they can do it if they set their mind to it." [10]

John Murrell was impressed with the thoroughness with which Zeppa took on his farm program. "He found iron rock, as we call it—has a lot of iron in it. He took that rock and he used it for road material—best road material there is. He cleaned out every creek. He took these trees, and he wouldn't just cut them down. If they were trees that he wanted, he'd re-plant. Transplanting the trees in a manner to make it more beautiful. He wanted to make that the most beautiful place."

One of the events which stood out in Murrell's mind was Zeppa's trans-planting trees.

"And he was the superintendent of this tree-moving thing, tree-moving machinery. He had big machinery to lift 'em up and take 'em over here and then down where he wanted them. And he rearranged all of the drainage, all of the creeks, and he was so proud of it that every Saturday and Sunday he'd have visitors. He'd invite them to come stay the weekend. He was a most gregarious-type person, and I know of no one that didn't think Joe Zeppa was probably the nicest guy that ever lived." [11]

Reading up on corn-raising in the North, Zeppa went in for wider spac-ing one year. He planted half his field his way, half the conventional way, then compared the yields. His patch grew 50 bushels to the acre, a better production. One year he grew 93 bales of cotton on 91 acres.

But whether he was improving the yield of a commercial crop or beauti-fying the farm by cleaning out creeks beds to leave shadowed mirror-like pools, he saw his work on the land as a therapeutic innovation.

"It's like healing wounds on a sick person," he once said. [12]

Agricultural groups from time to time visited both Pinehurst and Oakhurst Farms for guided tours by Milton Winston. In 1968 a class from the Ranch Training Program at Texas Christian University in Fort Worth came out to study the pasture development and livestock (cattle and hogs) program. "It is a good example of utilizing available resources to the best advantage," wrote John L. Merrill, director of the TCU program. "There was many a pencil out figuring hog deals for the next several weeks." [13]

Winston himself received recognition for his work at the two farms. A past president of the East Texas Farm and Ranch Association and president of the East Texas Fair, in 1972 Winston's demonstrations in concentrated grazing (30 cows on 30 acres) and the pen raising of channel catfish won him his extension-service district's first award for outstanding work and cooperation with the service's improved practices.[14]

VI

Not all of the Zeppa farm experiments panned out. One of these was an early fertilizer project. "Funniest thing that ever happened to Joe," said John Murrell, "as far as we were concerned. They started manufacturing—from natural gas, air, and carbon dioxide—little white pellets called urea. Joe was growing some cotton out there, so he decided, by golly, he was going to use it. This urea was a comparatively new thing at that time. Oh, this is a long time ago! So Joe treated that field to urea fertilizer, and he grew cotton. It was 14 feet tall. It never would blossom. Joe's cotton field was a famous thing in those days. Twelve to fourteen feet tall. Just went up."[15]

With cotton that tall, it probably was a blessing that it didn't blossom. Picking from stepladders might have been hazardous.

Roy Burchfield recalled two experiments that fell short of their goals. "He did some amazing things, all right. But he bought a bunch of Angora goats and put them out there. He was going to accomplish two things: clean his farm up and raise good wool. One hitch to this was, those long-haired goats got out there in that underbrush and they'd get their hair tangled up in that underbrush. The first thing you know, you would be walking out through there and you'd find one hanging up in a tree. They just weren't accustomed to that type of environment.

"One time he built a deal to set out sweet potatoes. To put sweet potatoes out, you set out slips. So you puncture the ground, you put the slips in there, and you pour water on it and fertilize it and so forth. That is hand work. That's the way it's always been done. He hooked up a deal behind the tractor that pulled the water and the rollers that rolled and punched the hole. He had a little platform where the workers sat and dropped the slip in the hole. Well, the only trouble with it was, they got it all fixed up and loaded down with the water, with the people on it, and put it behind the tractor and tried to go over fresh plowed ground—they didn't have a tractor big enough to pull it.

"Other people wouldn't have thought of doing it. It was a step forward."[16]

VII

Though the Zeppas did not fish, they kept two lakes well-stocked with bass, crappie, and bream, and one of them with channel catfish. Employees and personal friends frequently came out to fish. A pavillion was built next to one of the pine-fringed lakes for the convenience of visitors. Different organizations have used the area for picnics and parties over the years. Children would go out for Easter egg hunts, sometimes five or six grades at a time. Church and school groups used the site on numerous occasions.

Occasionally, employees spent some of their working time at Pinehurst Farm. Sometimes there were delightful interludes.

"One day during the lunch hour," said John Robert Gunn, "I slipped off down there to the lake and brought back three bass, weighed almost four pounds apiece."[17]

Over at Oakhurst Farm, a huge pecan orchard frequently benefited those who worked for Delta. "After they finished harvesting the pecan crop in the fall," said Avis Boren, "they would tell the employees that we could go out and pick up what was left. So around Thanksgiving we used to go out on Saturday or Sunday morning and pick up pecans. We always had all the pecans we could use."[18]

VIII

Although Zeppa continually wanted to enlarge Pinehurst, there were times when he did offer to sell off acreage to Delta. He had started the practice in the earlier years, with full approval of the stockholders and directors, of drawing upon the company for funds he needed, with his income from the company credited against this open account. As the company expanded, dividends were relatively low. Consequently, his withdrawals tended to run higher than the amounts credited him, ultimately leaving him substantially in debt.

Instead of asking for a higher salary, he chose to pay off the resulting indebtedness by selling Delta various properties from time to time. In 1962, for instance, he offered to sell up to 650 acres of Pinehurst Farm, along with mineral rights, at $125 per acre to be credited against his indebtedness. When Sam Dorfman, Jr., suggested Delta should get out of its farming operations rather than increasing them, Zeppa agreed with him but pointed out this land would not be for farming but could be leased back to him or sold. Zeppa, as usually happened in such situations, won his point.[19]

In late 1970 Zeppa took Delta out of the farming business entirely when he accepted an offer by John B. Stephens of Mount Pleasant, Texas, to buy Oakhurst Farm for $400,000, plus $160,000 for the cattle there. This event, coming at a time when the drilling industry also was going through a slowdown, must have been welcomed as a godsend by the minority directors and shareholders who had not been joyful about continuing the agricultural venture. The deal met with the unanimous approval of the shareholders and became final in early 1971. The proceeds of the sale were redistributed among the stockholders. Delta was no longer a farmer.[20]

IX

Though Pinehurst may have represented Zeppa's retreat from the hurly-burly of ordinary business, he never viewed it as an escape from people.

"He had company, a lot of company," said Winston, who was near at hand virtually all of the time Zeppa was at the farm. "Mr. Zeppa liked people and he liked for people to be around him. Even when he was at his hideaway, he wanted company. I can hear him now, coming in from the office, saying, 'Honey, who are we having for dinner tonight? Gertie, who are we having tonight?' He liked people around."

As much as the farm was a showplace, the steady stream of visitors to it was just as remarkable. Trent Zeppa recalled an intriguing parade of visitors to see Joe Zeppa, many of them staying at the farm. "All sorts of people. The Archduke of Austria would come to see him and stay with him and talk to him. Hapsburgs—Otto and Charles, the two Hapsburg brothers. They were living in Europe and they came to stay with Mr. and Mrs. Zeppa, to talk to him. All sorts and varieties of people."

Milton Winston said some Italians made a movie about Joe Zeppa and filmed part of it at Pinehurst. "He's had a lot of people here from Italy that I didn't know, from Venezuela, Mexico," Keating Zeppa said. "I never ceased to be amazed at hearing who had been at the farm last weekend, this sort of thing. Really quite a procession of people from all over the world. We've had the Archbishop of Canterbury, various political luminaries. I think probably the Apostolic Delegate from the Holy See to the United States visited us on several occasions. A lot of the time I was on the road or in Latin America, so I wouldn't be a bit surprised that there was film crew there."

How had Joe Zeppa come to know so many persons from such different backgrounds? Keating Zeppa thinks his father probably met them in his travels, probably because he stood out, himself, as an Italian-American in the oil business at a time when this was a rarity.

"The industry he was in would lead to the contact," said his son, "and he himself was rather unique, being an Italian immigrant. So that uniqueness probably led to curiosity."

And once they got to know him, they weren't likely to forget him.

Sylvester Dayson was probably the most frequent visitor to Pinehurst, for he and Joe Zeppa saw each other almost every weekend.

"They were very close," said Winston. "Mr. Zeppa liked for him to come out. They were in business together and had a lot of holdings together in the early days, and they used to get drunk together. I've heard them talking about it. But they had a lot of good times together and they had a lot of hard times together.

"[Dayson] cussed every breath; he drank an awful lot. He was an atheist, and I've seen him and Mr. Zeppa fight, cuss each other out nearly every weekend. We'd ride through the farm and we'd go up over at Dayson's, or Dayson'd be over here. He had a farm about six miles from here, and he lived in Dallas, and he'd come down from Dallas."

What would they argue about?

"Politics," said Winston. "Lyndon Johnson was in Dayson's home when he was running for President, one weekend, I guess, and Mr. Zeppa was raising hell with him about that. Really, Mr. Zeppa was a Republican, I guess, but he wasn't a die-hard Republican. I think Dayson was a Democrat."

Gertrude Zeppa lost her patience with Dayson practically upon sight of him. "Mrs. Zeppa didn't like Dayson. He was a Frenchman and they had a lot of fights, Mrs. Zeppa and Dayson. She didn't care what she told him. She didn't like him."

One visitor to the farm found Zeppa particularly animated and inspired to such an extent that it showed on his features. Eric Schroeder drove over from Dallas one weekend when he was writing a profile of Zeppa for the *Dallas Morning News* and received a grand tour of Pinehurst.

"I was amazed at Joe's dexterity with his hands," said Schroeder. "He could lay brick. He could do many things on that farm. He put us in a jeep that night and drove us all around the farm. I will never forget that because he drove under a tree with a little overhanging limb, and the windshield protected them in front; that thing came back there and hit me in the head. I will never forget that.

"We had dinner with them, and later on the women went to bed and Joe and I talked, and then we went outside and talked. It seemed to me on that summer night that his face had sort of a glow to it. I could see it there in the dark. He talked and talked. It was about two o'clock in the morning when I finally had to call it quits and went on to bed."[21]

X

That Pinehurst always remained prominent in Joe Zeppa's mind can be documented from his personal correspondence. Scattered through his letters are frequent allusions to the farm. Some examples from his correspondence with his old friend, George T. Keating, will illustrate this. In fact, the birth of Pinehurst is documented in a 1944 letter to Keating.

> I recently purchased a 1500 acre farm about 12 miles from Tyler. Trying to get this place going, together with the hectic conditions prevailing in the oil business, particularly the drilling end, has had me standing on my head most of the time. A lot of

fencing and buildings were required in order to operate the place for cattle, and you now what it is to get anything done at the present time. I think the price was low, about $20.00 per acre, and I have been offered a profit of $10,000.00. If I see that it requires more of my time than I can spare for it, I may let it go although I very much dislike doing so. There are just too few hours in the day and the week is not long enough.[22]

Several years later, in a letter touching upon a wide range of topics, Zeppa told Keating of life at Pinehurst.

We spent the weekend at the farm, and I came to the office from there. Pinny was trying out your Christmas present to him, a .22 high-velocity rifle, with a telescopic sight, which Gertrude contributed for his birthday in the offing. He seems to enjoy hunting in the woods and fields and his bag generally consists of rabbits, possums, crows and such; no foxes yet, though I hope he will kill a few as they are death to quail and other bird life.[23]

Another time, Zeppa declined to discuss some economic and political issues Keating had raised in a letter. Instead, he promised to share his own thoughts in person if George and Hattie Keating would visit the Zeppas at the farm, where they would have "plenty of time to talk and lots of space to holler."[24] And yet another time, Zeppa chose to describe the physical features of Pinehurst and some of its animal population.

The country is particularly beautiful and lush at this time of the year. The farm is a picture, surrounded by the various shades of green and a large part of it carpeted solid with bright crimson clover in full bloom. So far, we have had 96 calves, white of face and red bodies splotched with white. One old cow, No. 13, decided to show us that youth is not everything by giving birth to twins.[25]

If the accountant's hand shows in his head count of his livestock, the artist's eye also appears in his attention to color. Even late in life, he featured the farm and its events when he wrote his sister Paola, by then back in Fubine. But, of course, unlike George Keating who had been reared in New York City, Paola was an old farm girl herself. So it is not surprising that brother Joe provided her with details of East Texas' frequent rains that year and the "excellent farming season" that had produced "lots of grass" for cattle and hay for the barn. He had 700 cows which would be bred during the next few months and, already, 100 calves, with a total of 650 calves expected.[26] Now, 700 cows, 650 calves—in Fubine, that was *huge* news.

XI

There are photographs of Joe Zeppa on a visit to Fubine, as a successful oilman and drilling contractor, in which he is posed behind a plow pulled by oxen. He had plowed the small fields behind oxen as a small boy. It must have been a nostalgic moment as he followed the oxen, but like a person who owns and operates an antique 1930 Model A Ford, he must have been thankful his life had not been frozen to that plowman's role.

"They had a few oxen [in Italy]," said Milton Winston. "I've heard him tell about putting them up in a pen, and they didn't have a place to graze them. The main thing I remember hearing him talking about his home was they didn't waste no land. They used every little corner. If they had room to put a stalk of something, they'd put it. On a mountainside and everything."[27]

Certainly his wishing to be photographed behind the oxen indicated his intention to record this symbolic identification with the land. At Pinehurst, he did everything he could to move away from oxen power, as he improved and modernized every last operation at that farm. Therefore, the photograph of his guiding an oxen-pulled plow also symbolized how far he had gone from the life he had led as a poor little boy in northern Italy.

12
AT THE CENTER
OF AN EMPIRE

I

Like most independent oilmen and contractors, Joe Zeppa was not only the dominant man in Delta Drilling Company, but the single decisive force. Because he was the majority stockholder, he held the trump card if he wanted to use it; but he was more likely to use his powers of persuasion to bring minority directors and stockholders to his point of view. Because he was the majority stockholder, the president, and the chairman of the board, and because he was Joe Zeppa, he made all the decisions on matters he deemed important, and he made a great many decisions on relatively minor transactions.

In addition to the titles just cited, he was, in fact, Delta Drilling Company's lawyer, geologist, and chief accountant. Even if others held these positions or the titles, he treated that as a mere formality when it came to the important, and some small, moves. Both with his family and his business, Joe Zeppa was the firm European father. Many of his employees perceived Delta as a family, partially for this reason. He was Father to a generation or more of employees who shared his goals, identified with Delta, and believed in the work ethic. To them he was both symbol and reality, and the longer they were employed with Delta, the stronger their adherence to this identification was likely to be.

For more than four decades, Joe Zeppa *was* Delta Drilling Company, and everybody knew it. While it may be intriguing to speculate upon what Zeppa might have become had he gone into another career, perhaps not even business related, it is just as intriguing to delve into what he did become and why, and how it happened.

Certainly his high intelligence must be given much of the credit for his accomplishments, but intelligence alone, improperly tapped, may yield little or nothing of note. And Zeppa, also, believed almost religiously in hard work. Then there were intangible qualities that made up his personality, which drew people to him and made them trust him. These personal qualities, when it is all sorted out, may take one further down the road of success than all the intelligence and hard work in the world. And then there is the fact, which Joe Zeppa acknowledged, that luck helps, and he sometimes enjoyed a streak of luck. But luck can go either way, while one's character, however it may be described, tends to follow a steady path.

Zeppa, in 1975, a few months before he died, grappled with the matter of his success and summed it up by saying, "I haven't tried to be successful. I've tried to run my business like I would any other business. If I were in the grocery business, or any other business, I would be doing just like I've been doing. Try to be honest. Try to run the thing like it ought to be run. Try to always have some regard for the other fellow. You know, there's too damned many people that, if they get well, don't give a damn about the other fellow. They don't care whether he swims or sinks. I'm not that way."[1]

He left out a lot in that summation, but he hit some high spots too, and probably put the emphasis where he thought it belonged. The key to his statement is, "I've tried to run my business like I would any other business. If I were in the grocery business, or any other business, I would be doing just like I've been doing." There is little doubt of the truth of that statement. But what does it mean? What was Joe Zeppa like and what were the factors in his life and character that made him what he was?

II

It was hardly a precedent for a youngster like Joe Zeppa to enter a foreign land, lacking knowledge of the new country's language, and become a success. After all, Joseph Conrad left his native Poland as a mature man, went to sea, and later became one of the great English novelists. Napoleon Bonaparte, a native of Corsica, with Italian his mother tongue, became the Emperor of the French. The annals of American business history are filled with immigrants to this land who succeeded in a grand fashion. But the fact is that most emigrant Poles didn't become English stylists, and most Corsicans didn't attain the stature of great generals. By the same token, most Italians who came to this country did not become the controlling heads of their own companies and make millions. It not only took a special kind of person to achieve each of these goals, it probably also required a certain set of circumstances to make it possible. For Napoleon, that opportunity was the French Revolution. For Zeppa, it probably was the Great Depression and the East Texas oil boom. Though it is difficult to believe he would not have succeeded in whatever he undertook, it does not necessarily follow that he would ever have gone into business for himself, had not these two events, with the contacts and the experience he had already acquired, provided him the vehicle for forming his own business.

Joe Zeppa, to use the words of his son Keating, had "the mind of an accountant and the heart of a poet." Probably it took these two dominant features to make up the complex man he was. He was by no means a flawless man and he was frequently difficult to work for, but, as all who knew him attest, he was memorable.

"Joe was a creator of stories," said Red Magner. "People always remembered what he said to them. He always said the right thing." Very few persons who met him failed to carry off an anecdote or a quotation from the encounter.

As Delta grew, the air of northern Italy permeated this domain, suggested Trent Zeppa, who as a boy and young man observed the company politics at close range. "It was like growing up in an Italian Renaissance court. You watched the people come and go, the situations come and go. Who's getting close to the throne and who's getting scorched when they get there. And an awful lot of people'd come consult him. Quietly, you see. Nothing was ever in newspapers or this sort of thing. They would come in [and slip out].

"People'd come from all over the world to see him. A lot of people knew about him and would use him in an advisory sense. This sort of thing went on for years. In the early days they'd come to the house. The Mellons [came]. I forget which of the Mellons. And various people in the industry.

"He never had that stature [of fame and wealth] in the industry, from a financial standpoint or being known as a wildcatter. He was always known as a very fine financial analyst. Or knew people who knew things. And I never saw him write a letter or put down on any piece of paper [his relationship with all these people], and I don't think anyone else did either. He just took it with him to his grave.

"If he'd ever sold the company, he would have done it in a hotel room in Caracas or New York or Italy, or just anywhere but right in his own backyard.

"At one time they said that they wanted him to be head of Standard of New Jersey, and this was 1938, 1939. I just heard it. From my mother. Well, he never discussed business with her, either. He was a past master of the Cartesian method of breaking things up into the smallest common denominators, smallest parts—and hiding all the parts. He would do it to a fare-thee-well in everything, as far as I know.

"If you talked to the old chief accountants at Peat, Marwick, Mitchell & Company on Wall Street, they would always refer to Mr. Zeppa as the accountant that went West and made good. Because in New York he was known as an accountant, by virtue of being a cashier of the law firm, before that part of the law business was put into a bank. When he first came to El Dorado for Captain Holland, his commanding officer in the first World War, it was to straighten out a tangled business situation. And so he did it for Arkansas Fuel, he did it for Cities Service, he did it for Lion Oil. So as for being trained or capable, he was more than able to do that when the Hawkins business was presented to him."[2]

Just as a Renaissance court would have reflected the character of its prince, Delta was an extension of Joe Zeppa's personality.

"It was his life, really," said John Justice. "You read books and articles to tell you about it not being healthy to spend all you time with your work, but he was an exception to this. I won't say he didn't have any diversions, because he did have—his farm. But that was a business, too, really. He had these two businesses, one being Delta and one being his farm. And he thrived on work. It was his vocation and avocation. He thrived on problems."[3]

Although the observation probably also applies to other ethnic groups, Gambino's analysis of the Italian-American may provide some insight into why Joe Zeppa so persistently placed the stamp of his own personality upon Delta Drilling Company almost from the beginning.

> The sense of pride for something done by oneself and for one's family, whether building a brick wall, a small business, or making a fine meal, is essential to the Italian-American psychology . . . . The Italian-American seeks to do something the result of which he can demonstrate to his family. Herein lies another important component of his pride. "With these hands I built *that* wall." "*This* is my restaurant."[4]

There are other clues that substantiate, at least partially, Trent Zeppa's insights. In the culture of Italy, the father used "true Machiavellian principles" to pursue the goals he deemed proper for his family. The responsibility was totally his—*fatti suoi,* his business—and he need not discuss his problems with his wife.[5] Although one must be cautious in slavishly extrapolating from this observation, the methods followed by Zeppa in running Delta seem generally to parallel this old Italian tradition.

Because he did approximate these norms in running his company, Joe Zeppa was not the typical organization man, primarily because he *was* the organization. All of the flow charts pointed toward him. He ran the business as a patriarchal European father of the day ran his family. This emerged in innumerable interactions between Zeppa and his staff. On one occasion Zeppa was discussing a well problem out in West Texas with a vice president, Mark Gardner. After they had worked it over back and forth for a while, Gardner remarked that a qualified man—mentioning his name—was out there. "That's his problem," said Gardner. "I'm going to let him worry about it."

Zeppa drew back a little, as if that reaction were strange to him.

"Mark," he said, "*I'm* going to worry about it."

On another occasion, when Zeppa and Gardner were discussing the pros and cons of a situation, Gardner concluded, "Joe, I think we ought to have a meeting about that, get the pertinent people together."

Zeppa looked back at Gardner with a puzzled expression.

"Mark, we are *having* a meeting.'"[6]

Gardner appreciated organizational structure and knew how to use it. Delta's organization, unlike that of large companies of the time, resided with Zeppa, who could never see it in any other way.

"Mr. Z primarily ran his own show," said Leonard Phillips. "He did it with the acquiescence of Mr. Dorfman and Mr. Sklar, who had absolute, total confidence, not only in his ability but in his integrity. So when Mr. Z wanted to do something, he did it, and that was it. I used to go to all the board meetings with Mr. Sklar, but it was the general acknowledgment that Mr. Z ran the company, knew what was best for it. With the respect the gentlemen had for Mr. Zeppa, if he thought it was good, they thought it was good.

"It was the decision, based entirely upon self-interest. Both Mr. Dorfman and Mr. Sklar were very bright, and they recognized that they had a gentleman in Mr. Zeppa who, obviously, if he did not control his own situation would control another one. And it was *a propos,* good sense, to say, 'Mr. Z, you go ahead and do what the hell you want to do, and we'll back you.' And that's what they did. It was run *entirely* by Joe Zeppa."[7]

If Trent Zeppa's view of his father is any indication, even those persons who found Joe Zeppa difficult also found him to be highly skillful in his dealings and extremely intelligent. How did Trent characterize Joe Zeppa?

"A brilliant salesman—just fantastic," he said. Persuasive? "Oh, very much so! And a brilliant man. Had an immensely high I.Q. And all his confreres in New York and everywhere realized it. As far as his mental acuity in terms of the oilmen he was around, his was much, much higher than almost all of them I ever met. And I met most of them.

"And everyone that he ever met, he didn't forget them, particularly if they had any remote sense of being useful later on.

"He was an extremely brilliant person. I always wanted to know the story of how Lord Cowdray and the rest of them got him to sell Shell Oil in Italy, but I was never the one to ask, and the only people that would, would never do so. [Joe Zeppa] wasn't working for a company, just a group. [Lord Cowdray was involved] way in the background. You get into the murky area of international oil. It is extraordinarily murky! And the people who knew, all died and would never say a thing.

"Mr. Zeppa went to see one of those old contacts in 1958—Shell Transport and Trading House, took my wife with him. She doesn't remember the man's name. Mrs. Zeppa never heard of him, and nobody else ever did. Mr. Zeppa hadn't seen him for 35 years."

Joe Zeppa made an art of business and often transformed a transaction into a thing of beauty.

"I only saw one deal from conception to ending," said Trent Zeppa, "where he came out with all his thoughts from the beginning to the end of it. While he was describing it to someone else. How, and why, and every-thing—it was marvelous to listen to. And I only heard him once. I don't even remember what sort of a deal it was. Something not too great, because he would have never discussed anything with me. It was something rela-tively minor, and it was the most beautiful mental exercise you ever saw. It was like watching a graceful bird fly or some graceful animal going through the water. It was just absolutely perfect, just as smooth, and quick and fast. Just the absolute capability with which it was done."[8]

III

"Joe had an appetite all his life to improve his position, no matter what it was, whether it was vocabulary or whether it was travel," said Watson Wise. "It intrigued him always. We'd be in Rome together, and he'd say, 'Well, Gulf wants to take me to Sicily tomorrow. Let's go down to Sicily.' We'd go down and drive around and fly around Sicily and look at it and get down there and look at the ten-gravity oil—it'd look like tar, you know—and he said, 'Might be a place to put a rig.' Always thinking of a place. An eternal optimist about things of this sort. I've seen him depressed very few times."

This desire to improve his position in life traced way back. Trent Zeppa remembered that as a boy he had to keep accounts of his five-cents allow-ance, just as John D. Rockefeller, Sr., required of his children.

But these observations do not answer the question, Why did Joe Zeppa succeed?

"Well," said C. L. Vickers, now retired from Continental-Emsco, "he succeeded [because] he started in this at a very opportune time. Conditions were very severe and everbody was broke. But that provided the means of getting hold of properties at a very low figure, and if an individual had enough ability to raise a little money, he could make a deal; and I think his continuous work in following it over a period of 40 years with every year stacking up, this amounted to complete success."

At a point in the late 1930s, said Vickers, Delta had production income of around $200,000 a month, an extremely helpful asset to an up-and-com-ing drilling outfit.

Then there were the factors that have always been readily acknowledged in the oil industry.

"Some of it is a lot of luck," said Vickers. "And a lot of it is a willing-ness to take a chance to subject themselves to absolute poverty by failure of a business. In other words, if you haven't got a ship out, you won't have one in. And those who are bold enough to put one out, then you reach a point that you've got enough money that you can buy your way into some-thing that is already proved. Because that is really what the major oil com-panies do. They let the little fellow prove up something, and then when it gets twisted, why, they're able to come, and they're buying a cinch, is what they're doing.''

Although ''being at the right place at the right time'' is an unexcelled form of luck that takes one far, the ability to perceive an opportunity and take advantage of it may be a form of making one's luck.

''A person can work hard, but he's got to use his head and some scheming and be willing to take a chance,'' said Vickers.

Homer Lee Terry recalled how Zeppa had replied to the question about his success.

''They asked him how it was that he made such a success. He said, 'Well, I don't think I made such a success. The only thing I can tell you is that I just work like hell'—that was the way he put it—'and work my credit real good, but don't abuse it, and then just have a lot of intestinal fortitude to step out and do something—you know, make a deal.' And that was his philosophy.'' It is very similar, as a matter of fact, to what Vickers observed.

Joe Bevill considers Joe Zeppa's long-term relationships a factor in his success.

''A friendship with him, once established, was always a friendship. And I've known many, many people in my lifetime, but I've never known another man as well as I knew Mr. Zeppa who, at some time or another, did not make some remark about another person that was belittling. I never heard Joe Zeppa make a belittling statement about anybody. Never. I was with him many hours every day, and he just didn't do it.

''I think it had a lot to do with Delta's growth and success. People respected Joe Zeppa. Everybody who knew him respected him. Now maybe there would be people, I'm sure, who were jealous, perhaps, and didn't like him, simply because he might be more successful than they were, but he was never selfish or hoggish and he never mistreated anyone that I know anything about. He had a very high regard for any human being in any situation in life, from the most meager situation to the ultimate.''

There also was a range of other attributes at work.

''He was very aggressive,'' said Warren L. Baker, who knew him through the American Association of Oilwell Drilling Contractors. ''He was going to pursue that matter. He had a dogged determination to persevere, and he wanted to get it done. He made quick decisions.''

One constant feature of Zeppa may have been one of the most important factors in his career: his concentration on the matter at hand. ''He didn't have a divided attention on anything. He concentrated on what he was doing,'' said C. L. Vickers. Then he added: ''Minds are not made, minds are born; and if they've got that ability natively it will come out. It's been my experience.''

''His ability to read people and to draw from them the maximum amount of performance was phenomenal to me,'' said Tommy J. Blackwell, formerly of Delta's land department. "It was that rare talent that enabled him to look at you and evaluate you within a few sentences of your conversation and know how to utilize that talent in drawing from you the most. I don't want that to imply that he used people. It is good for you to draw from you your best. He was remarkable in that area. After all, this is a people business.

"He also had this rare talent to let people bump around in this box awhile until they found a slot that they fit into quite well. He knew what was going on and he ran the ship with a pretty good hand on the wheel. But you were not a robot. You had the room to move out. You probably knew your job assignment only by experience, and not by explanation. He didn't tell you what your limits were; you found it out by experience, and if you were making good decisions, there really weren't any limits.

"His integrity was absolutely at the top. I've seen the man weigh a problem and get off on the short end of it, to say, 'I want to be more than fair on this.' Very, very fair man.

"He was very successful in the business world. He made a lot of money. But if he were living and this afternoon he made ten million dollars, free and clear of taxes, it would not change one particle of his life. He would take the same route home; he'd wear the same clothes. He would do the same things tonight and eat the same food. It would not change anything in his life. He'd just put it to work tomorrow in new projects.

"It wasn't necessary for him to have the lights flashing on the scoreboard. He knew what the score was, and it wasn't necessary for him to flash it up there for somebody else to see. He was satisfied with himself. He set his own values. He didn't let somebody else set his values. Obviously, he had to get along with people in the business world. But he set his own values as he did."

Coupled with his integrity, said Blackwell, were toughness and shrewdness. Some people are tough and shrewd, he observed, but not honest.

"He knew the end result, but he moved toward it in such a subtle way that you didn't know that always. I've seen people come in and he'd start trading with them. It appeared that he was taking the short end of it, by his ignorance. He really wasn't. He was taking the very route that he had worked out. And while in all his trading, he didn't get more than he was due; he got exactly what he wanted by very careful trading, as he went along. He was a genius at that. An absolute genius. This is a T. J. Blackwell statement: Mr Zeppa wasn't in the oil business; he was in the money business. Oil was his vehicle. If he had been in the automobile business, he would have been just as successful."[9]

IV

No doubt about it, as Trent Zeppa reported, Joe Zeppa was a consummate negotiator—quick to press his advantage, cool when he needed to be, always objective enough to see what was going on, never letting a heated moment blind him to an opportunity that might come forth. Mark Gardner remembered the time Zeppa was in a conference with Lowell Glasco over a well in Val Verde County, Texas, that had been lost, due primarily, said Gardner, to the fact that the promoters had insisted on using junk casing to put in the well.

"I even pointed out at the time whose responsibility it was," said Gardner, "but anyway it turned out to be one hell of a fiasco. They were threatening to sue Delta for losing a jillion dollars worth of leases and all of this, none of which worried Joe very much—he had been there before. There wasn't any question about his liability, but anyway we had a big meeting in Delta's attorney's conference room one Sunday afternoon, in Ed Kliewer's office.

"Glasco and his attorney were there and the other promoter and his attorney, and they were threatening and they were shouting. Joe could get just as rough as the rest of them, but after a couple of hours, Glasco jumped up and started staggering around, grabbing his heart and one thing and another—he was 'having a heart attack.' Nobody paid too much attention to him, and Joe said to Ed, 'Well, why don't we call a recess and have a drink?'

"Well, Lowell Glasco's heart attack eased off when he started drinking some scotch. He and a company called West Coast Pipeline Company envisioned building a line from the Permian Basin to the West Coast. [It later was done after Glasco's death.] But Lowell Glasco was having hell getting it promoted. So while tempers had cooled down and we were having this drink and Glasco was about over his heart attack, Joe said to him, 'How's the West Coast Pipeline Company coming along?' Here we were across the table, bearing fangs at each other. [Glasco] told him a lot about what was going on and Joe said, 'I will take $150,000 worth of the stock.' That was something! Then we went back to arguing and giving them hell about this well in West Texas."

The Glasco contention was finally resolved when Zeppa, though Delta had no financial commitments, offered to take Glasco's place in the deal, since another well had to be drilled. Though the venture failed, the effort was made, to Glasco's advantage rather than Delta's.

"That was Joe's way of working things out," said Gardner. "He felt like there wasn't anything he couldn't work out. Oh, well, there were some lawsuits, but he worked most of them out some way or another. He would bend over backwards."[10]

That Joe Zeppa's personality held the key to the way Delta Drilling Company was run is a fact that few would deny. Mark Gardner put it plainly:

"I thought the direction [of the day's business] was dependent pretty much on what his morning bowel movement was, because I swear to goodness you could take the best deal in the world one day and he would look at it and say, 'No, that isn't worth a damn; I wouldn't have anything to do with this,' and the next day you'd take some sorry deal in there that he would just fall in love with.

"It was absolutely unbelievable sometimes, not to say a bit frustrating too. You would logically conclude that it was just his evaluation of the deal, but the fact of the matter is that on occasion he wouldn't even listen or look long enough to make an evaluation of the deal. He would just condemn it offhand. Just for no discernible reason. This particular day it wasn't worth a damn."

It seemed as if Zeppa had made up his mind before anyone walked in the door. As puzzling as the behavior was to Gardner, he knew he could not attribute it to a lack of perspicacity in Zeppa.

"Many times, while he had the capacity [to figure in his head exactly the cost of his interest], he obviously wouldn't even take the time to do that much," said Gardner. "He would just say, 'I don't like it.' A lot of time it involved personalities. He would make a deal with somebody he liked. He would turn down a bigger deal and a better deal with somebody he didn't like."[11]

Steve Schneider, an independent Dallas oil promoter, first met Joe Zeppa in 1963 when he approached him with a prospect in Kaufman County, Texas. Schneider had gone to Bob Waddell for a bid on drilling the well and, that done, Waddell had suggested young Schneider see Zeppa about taking an interest.

Waddell introduced Schneider to Zeppa, and Schneider went through his pitch.

Zeppa turned to Waddell and said, "Bob, you like this deal?"

"Yes, I think it is a good deal," said Waddell.

"We gonna get to do the drilling on it?" said Zeppa.

Waddell assured him Delta could drill it.

"How much you want for it?" Zeppa asked Schneider.

The promoter told him.

Zeppa said, "Oh, you want too much money for it. It is way too much money for it."

Zeppa pored over the maps again. "Bob, we gonna make any money on this deal?"

"Yeah, we can make some money, Mr. Zeppa."

Zeppa kept fretting to Schneider. "You want too much money for this."

He kept complaining about the high cost.

Finally he said, "Well, I will take it. We will take an eighth. But you're asking too much money for it. Bob, you sure we can make money on the deal?"

This was virtually the same process which repeated itself each time Zeppa took an interest in one of his deals, Schneider said. That particular deal produced a dry hole, but in 1965 Schneider went back to Zeppa with a drilling prospect in northern Hopkins County, in the Birthright area. Zeppa went through the "same song and dance," said Schneider, then announced, "We will take it. I don't think much of it, but we will take it." This one worked, and a number of Smackover wells were drilled at Birthright.

"We never had anything with him that was more than an eighth," said Schneider. "Sometimes he would talk to you about, 'Well, if you have to sell some more of it, let me know. I might be interested.' You knew if he said, 'I might be interested,' that meant if you got in a jam and couldn't quite finish it up, he might take it, but he would probably give less money for it. One thing about him, he never argued or traded price with you. If you came in on a deal and said, 'I want $20,000 for an eighth interest,' even though he might be turnkeying the whole well for $80,000, he never argued the price. He would say you were asking too much money for it, but if he didn't want it, I never had him come back to me and say, 'Well, now would you take this or maybe we can make a trade on such and such a basis.' That may have been unique. I always felt if I came up with a price that I felt was reasonable, then I just stayed with it, even though he would say, 'You are making too much money.' We never really were. Through the years that was pretty much the way we would trade. He was always so cordial to you in his Italian kind of way."

Schneider also discovered that Zeppa had a characteristic way of turning a prospect down, as far as taking an interest in it. "If he didn't like it, he said, 'I don't understand this. It is too complicated. But we will drill the well for you. I don't understand it.' You could talk some more if you wanted to, but it didn't do any good. He would say, 'There is a dry hole over here. Why do you want to drill here?' You would have to know what you were talking about."

Joe Zeppa had been around the oil patch long enough to know that a drilling contractor had to be assured he would get his money, but he seems to have based that assurance upon his assessment of character in the persons he was dealing with, and upon their track records. Schneider recalled the first time he made a drilling deal with Zeppa.

"How you going to pay me?" Zeppa wanted to know.

"Well," said Schneider, "I am selling this interest."

"Well, that's good. When you get enough letters together, trade letters, take them over to Ed Kliewer and he will fix up an assignment of that money."

When Schneider remonstrated that some of the partners might resist that way of doing it, Zeppa said, "Don't worry about it. It's all right. They will sign them."

Schneider did as they had agreed, and Zeppa never asked him for a penny thereafter. Two years later when he sold the deal at Birthright, he asked Zeppa, "Mr. Zeppa, what about paying?" and Zeppa replied, "You're all right."

But another promoter dealing with a Delta staff member ran into a snag. When he went in to make a contract, he grew very upset when he was told he would have to produce the cash for the drilling. He insisted he had known Zeppa for years and there was, therefore, no reason for such a demand.

The Delta employee went to Joe Zeppa and said, "Mr. Zeppa, he said he really knows you well and he says he doesn't understand this. He said you are a very good friend."

"Make him put up the money."[12]

Nick Andretta, who saw deals come and go through Zeppa, saw his soft side more than once.

"He had good judgment," Andretta said. "His only bad judgment was that he had a stack of notes this high, people that he was soft-hearted with, deadbeats, one thing and another. Sounds like fiction."

One man visited Zeppa's office in the 1930s. Zeppa called Andretta in to meet the man, who had been a salesman for a supply company.

"Joe said, 'Look at this. What do you think of this he's got here?' He had this sack on his desk, and he said, 'Smell of that.' I looked in there and there were pieces of oil sand just rich as hell. This fellow told Joe Zeppa that he had cored this in South Texas, down near the border. But this fellow had a reputation of being a hard drinker and the deal was that Joe Zeppa would give him $1000. He had a plat showing the location where his well was; if Joe Zeppa would give him $1000 he would get the casing, complete the well, and give the Delta Drilling Company a half-interest in it.

"So he was staying at the Westbrook Hotel in Fort Worth, and I think he had been fired from [the supply company]. Well, anyway, time went on. Couldn't get him at the Westbrook Hotel. He was registered there, but hadn't been there for two weeks, so Joe Zeppa said, 'What about this deal?' and I took the plat and went down in South Texas below, way out of San Antonio. And I checked the courthouse and talked to some people down there and got oriented, and went out in the country to try to find this location from the description of the acreage.

"Number one: he didn't have one acre of leases. Number two: there was no well that had been drilled—[it was] shallow out there, and this plug, it would be about 1800 feet deep. I hunted all over that country, went to San Antonio and checked; they said he'd hung around the courthouse there, asking questions. So, anyway, I come back, I go to the hotel over there—the Westbrook Hotel—they said he hadn't checked in, hadn't been there in over a month. They had an old suitcase with some dirty old underwear in it, and [we] never heard another word from the guy.

"Now there's a case of Joe Zeppa wanted to be nice—he took his word for it, he had gotten some sand somewhere, might have come out of East Texas field—and we did a lot of things like that where we made a mistake. That was one of those things. Well, it was one of those kinda cute things people will con you out of in the oil business."[13]

"I don't think [Zeppa] even looked at deals geologically," said geologist Jim Ewbank. "I think he considered the economic situation. And he could do the economics of a prospect in his mind as fast as I could do them on a calculator, and I trusted his figures better than I trusted mine. If he thought we could make money on it and could put a rig to work, that was good enough for him.

"He trusted his people and knew that we wouldn't bring a doggie deal to him. Anything we didn't believe in, we never showed him. He had confidence that his people were looking out after his interest.

"He was fair and tough. He demanded a lot of his people. I learned early in the game not to go up there without a definite recommendation, because he wanted you to tell him what you wanted to do. If he said No, he would give you the opportunity to rebut, and if your argument was good enough he would change his mind.

"He very seldom ever took a deal on the first presentation. He would authorize us to participate in a deal, but he always made us go back and try to renegotiate and get a better deal with them. It isn't easily done. It's tough as hell, but it was his philosophy. I think it was just his love of battle. Years of 'Don't ever take the first offer. Always renegotiate. Try to get a better deal.'

"If the other party wouldn't negotiate, we would go back with him and say we tried, and he then would usually say, 'Okay, we will go on the original deal.' But I don't ever remember him allowing us to commit on the first proposal that was made. I think he enjoyed competition.

"Working for Mr. Joe Zeppa, everything was done on the back of an envelope or a piece of scratch paper. There was very little paper work. If you had a prospect you wanted to drill or [if you wanted to] look at a deal, you went up and made your presentation to Mr. Zeppa and he either said Yes or No. If he said Yes, an AFE [Authority for Expenditure] was constructed and we got after it. Ninety percent of our deals were made with a handshake. We dealt primarily with people we knew, people that Delta had been doing business with for 40 years.

"Prospect participation depended a lot on rig activity. If we had rigs stacked, either the land people or myself would get out and start looking for prospects that other people had and needed a partner. If we had had a rig stacked for quite some time, we would commit on almost any kind of deal just with the understanding that we would drill it. And we would take an eighth, fourth, or half of a deal just to put the rig to work.

"Believe it or not, we found a hell of a lot of oil and gas that way, even though the prospects usually were not really up to the quality that you would like."[14]

V

Despite being recognized as a financial genius and a brilliant person on many scores, Joe Zeppa was not what one would call a flashy intellectual. His brain was a storehouse of vast and impressive knowledge, yet he was not the sort of person found on a television quiz show where the mind's instantaneous recall of esoteric detail and even trivia is held at a premium.

"Dad was a very deliberate person," said his son Keating Zeppa. "You could not rush him. I would not characterize Dad as being a fast thinker. His mind didn't work like that. But very methodical, sort of like a bulldozer—once it starts going, it doesn't go like a race car, but don't get in front of it.

"Dad was the same way with his calligraphy. You could never hurry him. When he was making written notes or when he was writing a letter, he would *draw* each letter, almost. That's why it's so beautiful, and even in his latter years [when] it got shaky, it was still very deliberate. You'd never see him jotting notes, that sort of thing. If it wasn't written well—the script itself—if it wasn't up to his standards . . . he just took the time to do it."

These qualities suggest the perfectionist. Was he one?

"He was," said Keating Zeppa. "Very much so. It was the same way with his woodworking that he had as a hobby. When he built the house over on South Broadway—or remodeled and added on to the house that was there—in the back he built a garage with servants' quarters over it, and on the side of it, as part of the frame building, he had a little woodworking shop.

"Dad enjoyed his woodworking very much. Now, he didn't keep it as a lifetime hobby, but he did it for, probably off and on, ten years or so. And you'd see the same perfection coming through in his woodworking. Beautiful scroll work. He built furniture, mainly. The same perfectionist tendencies you'd see there: very deliberate, very careful. He did his own designs for his own furniture and scroll-work in his woodworking shop. These weren't something out of *Popular Mechanics*. He would draw and produce very delicate filigree work, building a little cabinet, this sort of thing.

"He was a consummate doodler, as I am. Dad could no more talk on the telephone without a pen or a pencil in his hand, piece of paper, than I can.

So I come by my doodles honestly. His were generally in flowing lines, and repeated, over and over. Overlining many, many times, this sort of thing. Filigree."[15]

It is worth noting that Joe Zeppa's careful drawing of his name and attention to detail in his woodworking indicate close concentration probably more than anything else, as was his approach in everything he did, whether in work or conversation.

Most of his associates saw these traits carried over into his business practices.

"I always got the feeling that he was a very careful, slow, methodical man, insofar as going into ventures, even though he would take off on a tangent occasionally, like sending two rigs all the way to Australia," said John D. Hall. "It could have turned out to be wonderful, because he got into certain deals here in the States that looked bad, bad, and some way, somehow, he came out smelling like a rose. I think it's just the way things developed. Luck can go either way. But as far as expansion and growth of the company, I would call him conservative, even though he'd plunge out and go in just head over heals [occasionally]."[16]

"He was not a promoter, gambler, nor a fast talker," said Joe Bevill. "He was very honest and forthright, and he'd tell all he knew about [a deal]. He'd put all the cards on the table. Very deliberate. He never got in a hurry. He was usually late to any important meeting because he *didn't get in a hurry.* He just didn't allow it, and if someone was sitting there telling him something, he didn't hurry through with that to get to another one. He finished that one. And that's the way he dealt with those people over there in the East Texas area, and he made a lot of good friends over there. Like he did elsewhere. Wherever he was, once they were his friend they were always his friend. And he never turned his back on a person in need. If he could help them, he did. They just don't make them like that, they couldn't make them like that. Finest man I ever knew.

"Oh, he was a brilliant man. My first impression was that he was so bright, I was a little bit afraid of him. He was just so far ahead of most people when he was talking to them, and you could see it. Not that he tried to show you. It was just the fact that he was way on out there. Really kept you on your toes to be in his presence, carrying on a discussion with him.

"You could just see those gears turning. He was a sharpie. And the move from small things to larger things, Mr. Zeppa was very cautious, careful. He was not prone to make big gambles."[17]

Leonard Phillips first met Zeppa in 1942 and never changed his opinion. "He was highly intelligent, with a tremendously inquisitive mind. It was a pleasure to fly with him from Point A to Point B, because he would historically give you the story of what you had passed over. He was one of the finest human beings I have met."[18]

His mental achievements stand out in most people's reminiscences of him, particularly his memory, though the facts that he remembered were not always to others' advantage. "He would reach back to dry holes," said Steve Schneider. " 'What about that prospect at Dalby Springs? It was a dry hole, wasn't it?' He did it in a pleasant sort of way. It wasn't vindictive at all. It was just sort of to let you know." [19]

Nothing, it seems, could disrupt the organization of his mind or shake his concentration.

"When Mr. Zeppa did start dictating," said Marguerite Marshall, his secretary, "I don't think he ever stopped. I have really taken a whole book full of dictation at one sitting. Both sides, back and front. He had an ability that I haven't seen in very many people. He could be dictating and be right in the middle of a letter and someone would call him on the phone and perhaps he would talk to them for 30 minutes, and I would have to sit there. When he hung up that phone, he picked up right where he left off and would go right on. His office was on the corner, and May Jones and I and Tommie Smart had the office next door to him, and when I walked through the door he started dictating. By the time I got to my chair, he had a letter written."

His forte, of course, was handling the financial aspects of his business.

"He had an ability to deal in different monies in different countries and be able to keep in his head the exchange rates," said Marguerite Marshall. "It was fascinating to listen to." [20]

"He had a fantastic memory for dollar figures," said geologist Jim Ewbank. "I have never seen anybody with the mind that old gentleman had, for remembering dollars and cents. I can remember one example I will never forget. Harry Phillips, an independent oil operator who has been a good friend of Delta's for a good many years, made a geological presentation. But Harry said, 'Here is the deal. I will let you buy into the deal, use your rig, for, say, $200,000, and this will buy you an eighth interest in all of my acreage and this and that.' So I made the presentation to Mr. Zeppa. I said I thought we ought to go, and Mr. Zeppa said, 'You can get into the deal, but you go back and tell Harry that you are not going to give him $200,000; you will give him $155,260.13.' And Harry said, 'No. The deal is this.' That was one time that I made a decision on my own and I said, 'Okay, Harry, we will take it at your price.' Mr. Zeppa was out of town for about two weeks is the reason I went on ahead and committed. Just as soon as he got back into town I went up to him. I said, 'We agreed to Harry Phillips's deal for $200,000.' And he said, 'I told you, you could only spend $155,260.13!' And this was when he was over 80 years old. It was his last year of life. It had been two weeks since I had seen him and he had remembered it to the penny."

How did Zeppa respond to Ewbank's decision?

"I can't repeat it," said Ewbank, "but it was a lot of four-letter words. But he said, 'If you made the deal we will stand behind you and we will take it. Whatever you told him we would do, we will do.' Whatever his people committed to, whether it was right or whether they had his permission or not, if you said, 'We would do it,' well, we did it.

"It was fortunate, though, that that particular well ended up as the discovery well for Grapeland field."[21]

Zeppa amazed veterans of the oil and drilling business. Mark Gardner said, "This isn't only my observation, but an observation of everybody who knew him well: Joe Zeppa could figure faster in his head than most people could with a calculator. He could snap that stuff out, and it didn't take long to learn, where figures were concerned, you'd better not try to vie with Joe Zeppa and think he was going to slip up."

In 1980, Gardner compared the position he was in with that of Zeppa as head of Delta. Gardner, head of Quest Exploration, enumerated his responsibilities.

"Here I am trying to run all the business except the exploration part of this venture, and I am no landman and I am no great financial man. All this falls on me, and I have for the first time in my business career had to think about fragmented working interests and revenue interest and reversionary interests and things like that and oil properties. Joe Zeppa could figure that stuff in his head so fast, while other people had a pencil and piece of paper, and he would say, 'That would result in about a so-and-so interest.'"[22]

Joe Zeppa's command of the English language was one of his most valued skills. His artistry with words rivaled his skill in woodworking. Undoubtedly it was a large factor in his success as a businessman and, indirectly, part of the reason people were drawn to him, for through reading and conversation he gained valuable information and friendships.

"I suspect that of Dad's physical disabilities in his later years," said Keating Zeppa, "the one he resented the most was the loss of vision. Other physical disabilities didn't *molest* him as much. But the degeneration of his eyesight bothered him, because Dad was *always* an avid reader. When he wasn't asleep or talking, he was reading."

What did he read?

"Everything! He had an enormous curiosity, whether it was geography or history or people or business or whatever it was. For an Italian immigrant, Dad *probably* had a better command of the English language than 90 percent of his contemporaries. His business letters—although according to modern style they tended to be wordy, but that was the style he grew up in—were generally a model. They were excellent, a beautiful command of the language. In his later years, as his mental faculties tended to decline, his memory would fail him and what not, but he spoke the language very well, and he wrote it better than he spoke it. And if you go back to some of his early diaries, now there's an Italian immigrant kid, exposed to English for relatively few years, and his written English was beautiful."[23]

The elder Zeppa made a lifelong habit of adding new words to his vocabulary. "He still did that, the last few years," said Milton Winston. "I'd go up there sometimes and he'd be reading. I don't know if it was Latin, French, or what. Mr. Chris would be there. 'Come here, Chris; what's this here? What does that mean?' He was still doing that occasionally. He liked to read, too. I've got some old *Holiday* magazines that he'd had up there. He read the *Holiday* a lot."[24]

"He used to say that if you used a word three times, it was yours," said Marguerite Marshall. "I never will forget the first time he used a word, 'eleemosynary.' I had never heard it at that time. I thought, 'Oh, my word.' I was writing shorthand just as fast as I could go—I wrote *elee*. I thought, 'I'll look it up later' and, sure enough, it was used frequently after that. But that was the first time I had ever heard that word."[25]

Joe Zeppa wrote very little in Italian after he came to this country, and certainly in his later years wrote nothing in his native tongue. Some have suggested that, by then, he was unable to write a letter in Italian. This may be true, but not for the seemingly obvious reason that he simply forgot it through disuse. More relevant, he came to this country at age 12, when he had probably not written letters in Italian, and had the vocabulary of a 12-year-old Italian, albeit a bright 12-year-old. And he probably never added to that Italian vocabulary once he arrived in New York, for he was putting Italy behind him fast as he sought to master the English language.[26]

In 1971 Joe Zeppa confided to his sister Paola that "I cannot accustom myself to writing in Italian very well."[27]

The process of Americanization had been as thorough as anyone could have had reason to expect.

VI

Good credit is essential to a successful businessman. One key to Joe Zeppa's success and his being held in high repute by supply houses, banks, and drilling customers was that he maintained good credit by ensuring that Delta's bills were always paid. In an industry at a time when many independent drilling outfits ended up defaulting on their debts, and in the process going broke, this was no small accomplishment. Frederick Mayer, retired president of Continental-Emsco, was for many years in a position to attest to Delta's superb credit rating. The fact that Zeppa's business matters never came to Mayer's attention documents the fact that Delta's credit never was in jeopardy, since that would be the only occasion for the head man's learning of a customer's problems. Delta was one of Continental-Emsco's larger drilling accounts.

"Joe and I were personal friends," said Mayer; "we talked general business conditions and whenever possible had lunch together. He would drop out to our apartment which was on the way to the airport [in Dallas]. But his finances were always handled through the credit manager, and while there had been some extensive credits extended, they were always on such a basis and handled in such a fine businesslike manner that it was to a large degree routine. Whatever Delta wanted, they would get, and they never made any demands that weren't perfectly in line. Because Joe was a fine businessman.

"Joe never asked for anything out of the way in credit, and I will say this for him, which is very rare: he was one of the few who never failed to meet an obligation on its due date or better. The average independent contractor at that time was an honorable man—he might go broke, but he would come back and take care of his obligations. But he just wasn't meticulous about meeting its exact terms. Well, Joe—I wouldn't say he was an exception, but he was absolutely meticulous in meeting any obligation."[28]

Zeppa's financial skill and excellent credit rating at times proved useful to his friends—at least once without the friend's knowing it. The result might be characterized as a benevolent Machiavellian maneuver. Steve Schneider remembered the time Reagan Carraway, an old-time promoter, went to Zeppa with a deal. Carraway was known as a "real plunger," tending to be about three wells behind in his financing. A piano player originally, Carraway had started out in East Texas during the boom, selling pianos.

"There is a story about how he owed Zeppa some money, and he came over there and wanted to sell a deal," said Schneider. "Zeppa knew he was coming and knew what he wanted, so he arranged down at Citizens Bank for a loan for Reagan, and there were some properties that, I guess, maybe Delta and Carraway had jointly.

"So Reagan came into town and said, 'I want to sell this deal to you, and I got to have the money.'

"Mr. Zeppa said, 'No, I can't pay this. Why don't you go down to the bank and talk to them?'

"He said, 'I can't go to the bank.'

"Mr. Zeppa said, 'Oh, they will do you some good. Just go down there and tell them I sent you.'

"He had the whole thing set up. He said, 'You can borrow $50,000 down there. You have got those properties.'

"So he goes down there and Reagan knew he couldn't make that deal. He goes down and they lend him the money. He came back and Joe said, 'Did you get the money, Reagan?' and Reagan said, 'Yeah.'

"Mr. Zeppa said, 'Oh, you borrowed the $50,000?'

"Reagan said, 'Yeah, if I had known it was that easy, I would have borrowed a hundred!' "[29]

VII

As Red Magner said, Joe Zeppa was a creator of stories, and this seemed to occur even when he was vacationing.

Fred Mayer acquired additional Zeppa anecdotes in 1951 when Mayer, then vice president of Continental Supply Company, and his wife Mildred joined Joe and Gertrude on a tour of Europe. At that time, Mayer observed firsthand Zeppa's persistence that so characterized his pursuit of business. After having split up earlier, they met in Venice.

"Mildred always thought very much of Gertrude and Joe, and I think sometimes she used Joe to work on me. After having a very delightful lunch in Venice, we were crossing the plaza and Joe said, 'Now we're going to see how a present-day Italian family lives.' And I told him, 'I'm not interested. I was reared in Youngstown, Ohio, where about a third of the population was Italian and'—I said—'I went to grade school and high school with children of Italians and some of my dearest friends are Italian and they came into my house and I went into their house daily.' Joe said, 'Well, we're going to see how it is up in Venice.'

"So after climbing about four steps into a nice apartment, it develops [that] these people also sold Venetian laces. Joe talked a little bit then and [the resident] brought out a tablecloth, and Joe and Mildred and Gertrude admired it. And Joe began to dicker in price.

"Now, the least interest I had was in a Venetian tablecloth at that time. Finally Joe said, 'Now, Fred, this'—as I remember—'is about $700 and it would cost you at least two or three times that much in New York.' 'But,' I said, 'Joe, I don't want a tablecloth in New York, and certainly not in Venice.' Well, he told me what a buy it was. And I said, 'I don't know anything about tablecloths. I think it is very lovely and I know Mildred does, but, Joe, I'm not interested.'

"Finally Joe said, 'Fred, if you don't buy it, I'm going to buy it.'

" 'Well,' I told him, 'don't let me detain you, Joe,' and I told Gertie, 'Gertie, I think you've just got yourself a new tablecloth.'

"She said, 'Well, I have several. It's up to Joe.'

"Joe then said, 'Fred, I haven't got my traveler's checks with me. Have you got yours?'

" 'Yes, ' I said, 'I have.'

"He said, 'Well, we'll fix it up.' So I turned it over to him, whereupon Joe said, 'Now, Fred, the deal is this. If, when we get back to New York City, you decide you don't want that, I'll return the money.'

"I said, 'I don't want it in Venice.'

"Needless to say, the Mayer family had a very beautiful tablecloth, which I find Joe was correct in saying that it would cost at least twice as much in the States."

From Venice the two couples drove to Fubine, the ancestral homeplace of the Zeppas. The homestead had been fixed up, and Paola, or Pauline, Zeppa lived in it. Hometown-boy-made-good Joe Zeppa was royally welcomed in the little town. Not only had Zeppa been responsible for refurbishing the old homeplace, he also had been a generous contributor to the church in Fubine. To these friends and distant relatives he was a returning celebrity.

"One day I wanted some razor blades," said Mayer, "and we walked down to a little tiny center of town, and it turned out that these people were distant relatives. So there was nothing to do but for myself and Joe to have a little wine. Well, it ended up they just closed the stores, and about three or four hours later Gertrude and Mildred, wondering where their husbands were, came and joined the party.

"That was a very delightful period. From there we then went to Rome where Joe had arranged a private audience with Pope Pius XII."

The audience with the Pope had been arranged by Monsignor Domenico Tardini, the assistant secretary of state at the Vatican. Zeppa had come to know Msgr. Tardini as a result of his contributions to a boy's institution which was administered by the priest. Tardini, who had been in the Vatican Curia since 1921, was to become pro-secretary of state for extraordinary church affairs in 1952. In 1958, Pope John XXIII made Tardini a cardinal, and he continued as secretary of state until his death at 73 in 1961.

" I could see Joe was a very charitable individual," said Mayer, "but I would say this for him—he never allowed it to be shouted from the housetops. He was very quiet. I picked that up in Fubine, what he had done there; in Rome I don't think it was generally known that Joe made gifts to the institution that Tardini had sponsored among the derelicts."

The trip shared by the two couples had its pluses and minuses.

"The women probably had a hard time on this trip," said Mayer, "because we'd end up each day with Joe and myself having a few drinks and Joe never wanted to eat until about midnight. This wasn't limited to trips; it was as long as I knew him, with the result that I think it got a little boresome to the women. And, of course, Joe and myself would have our arguments at all times, on most subjects. Oh, it would be everything, politics and what have you. The result was, when we got back to the States we felt we had a great trip, but the two women said they would both be glad to travel on a trip to Europe with Joe, or they would be glad to go on a trip with me, but never would they travel with the two of us again!"[30]

In a letter afterward to George Keating, Zeppa constructed a coda to the European trip: "From the social standpoint, I suffered almost as much as I did while traveling with you and Hattie. Gertrude was not in a drinking mood, Mrs. Mayer is a 'one drink' person, and Fred was taking care of his ulcers, which not only made him worthless as a drinking companion but also interfered with the irregularity of my evening meal. They are, however, ideal traveling companions and added greatly to the enjoyment of the trip."[31]

VIII

The work ethic was so strong in Joe Zeppa that he can be said to have personified it.

"He believed in work," said Homer Lee Terry. "That was one criterion that made the company so successful. Everybody worked. No goofing off."

"Joe Zeppa, in a way, was hard," said Milton Winston. "He wanted everybody to work and work hard. That's what he did and he wanted everybody else to do the same thing. But he was a good man. He'd treat you right if you worked hard and showed him that you didn't mind working.

"I've seen him—you know, a crosstie is heavy—when he bought Oakhurst Farm at Lindale and hadn't had it too long; I was up there one day and we're working, had some Negroes out there working, building fences, corral fences. He'd come out there on a Saturday evening [afternoon] and just come from the office and he'd pull his coat off and roll up his shirt sleeves and he'd pick those crossties up and put them on his shoulders and go right on out with them. He was a stout little devil, but he could pick them ties up and throw them over his shoulder and walk off with them. That's a load."

Winston had his love of work in common with his boss, which is probably why Zeppa hired him in the first place. "I had my second vacation last week, that I've had since I've been working for him," said Winston in 1980. "I hate the word 'vacation.' " (After Joe Zeppa died, Keating Zeppa and his mother gave the loyal and hard-working Winston 200 acres of Pinehurst of his selection. Even then, Keating Zeppa asked him if that was all he wanted. "Sure," said Winston. "If they're good enough to give me 200 acres, I'm not going to be greedy.")

Many of Zeppa's men were workhorses like himself—men like Joe Blasingame and Ark Carter. N. L. Webster, heading the land department for many years, was out of this same mold. "He was a dedicated company man," said Avis Boren, who lived next door to Webster before she went to work for Delta. "He spent as much time up here as Mr. Zeppa did."

Zeppa liked to see his employees purposefully busy. One of his most quoted expressions was, "What the hell are *you* doing?" directed at just about everyone in the company at one time or another.

"He said it all the time," said Bob Waddell. " 'What the hell are you doing, Bob? You're not doing anything.' I said, 'Naw, I'm trying to get by.' I remember when I moved into the executive floor when they made me vice president in charge of domestic operations. I'd been on another floor. No sitting back and relaxing and propping my feet on the desk, you know. They kidded me about that.

"After I'd been up there a little while, Joe Zeppa walked in my office one day and I had my feet propped on the desk and he said, 'I'll be. You're getting up in the world. You've got to where you can prop your feet on the desk now.'

"I said, 'You know, that's the thing to do. You ought to do that every once in a while. I just read in a magazine not long ago that a man sitting at a desk needs to ever' so often during the day prop his feet up on the desk, lean back and take off the strain on your heart and blood vessels, and you can relax a little while.'

"He said, 'You know, that might be right.'

"From then on, I propped my feet up on the desk, and never heard him say anything about it."

Zeppa himself worked late by choice probably more than through necessity. In the early days when the company was headquartered at Longview, C. L. Vickers, the Continental Supply Company executive, remembered that Zeppa would combine business with a social visit.

"Mr. Zeppa loved company, and when I was in Longview he would make my appointment to discuss our business at five or five-thirty. We would go through with the business end of it and then we would go upstairs, and after a few scotches and sodas, we would have dinner about 9:00 p.m.

"He was a very gracious gentleman and he met people real well and he liked a good scotch and soda, but he was tough on anybody who couldn't handle their liquor. He was really tough. I know by hearing him talk about something like that. He'd say, 'He's just foolish. He let his whiskey get the best of him.' "

Working for Zeppa could be challenging. Some felt it was worth the effort. Others were less certain.

"He was the type of man whose word was his bond," said his long-time secretary Marguerite Marshall. "If Joe Zeppa said that things were going to be a certain way, they were that way. Everybody recognized that.

"He was not the easiest person in the world to work for. We felt that when you did something for him, you had really done something. It was worth it in the long run.

"I didn't know until just recently that my youngest son felt a great resentment for Delta Drilling Company all those years. I didn't really know that. He made a remark one day about Delta that just wasn't to me the nicest thing to say. I said, 'Why, son . . .' He said, 'Mama, it always came first.' I guess it did really . . . I seldom ever got home before 8:30."

For a long time Marguerite Marshall arrived at the office by 7 a.m.—by which time Chris Zeppa would be there—and left around 8:30 p.m. because Joe Zeppa, coming in at 9:30 a.m., would elect to work late.

There were times when Joe Zeppa was out of town that Mrs. Marshall would get off at a normal time. "I would leave," she said, "and Joe Zeppa would get there maybe after I had left. May Jones was not above putting her head out the window and calling me back where I was standing there waiting for the bus to go home. Sometimes, if I was really desperate to get home, I would go down and catch the elevator on the fifth floor instead of the sixth floor so in case he came in I wouldn't meet him. It didn't do any good. May would stick her head out the window.''[32]

IX

We all have role models, not only our parents and heroes, but other significant individuals we encounter in life. Joe Zeppa was a role model to more persons than we will ever know.

Eddie Durrett said, "I learned hundreds of lessons from that gentleman that you can't learn out of books. Probably the number one lesson that I ever learned from Mr. Zeppa in this business is that you never get greedy. He had the philosophy that he would rather have a quarter interest in four drilling deals than [all of] one. I couldn't agree more with that.

"The integrity that he exhibited toward his partners was to me the extra mile that he didn't have to go.

"I also feel quite strongly about Chris. Chris Zeppa is a man of high integrity, moral fiber, ingenuity, and infinite knowledge. In Delta there was a saying going around that Joe would promise you the moon because he knew Chris wouldn't let you have it. I can tell you this: Never once did I ever tell Chris Zeppa that I needed something [and be] denied that something that I needed. 'Cause Chris knew me well enough to know that when I told him we needed something, we genuinely needed it—it wasn't feathers.''

Another philosophy Joe Zeppa applied to his business was, "Don't worry about those contracts you don't get." Durrett said, "We developed that philosophy back in the early days of Delta Drilling Company that those little bids that beat us out, we wouldn't worry about them. We were worried about the job we got.''[33]

Chris Zeppa expressed what Joe Zeppa also put into practice in his company, when he said, "Above all, what I consider one of the things to be *most* proud of in Delta Drilling Company is the ability to deal well, fairly, and honestly, and deliver the goods the way we said we would do it. It comes from everybody in Delta Drilling Company." Chris Zeppa acknowledged that his personal motivations were instilled in him as a boy in Italy, and it seems safe to say the same for Joe Zeppa: "The honesty, integrity, and the will to work, and the will to treat your fellowman properly and correctly.''

No word has been received yet to the effect that Joe Zeppa was not impeccably honest. He was a superb salesman who could talk people into what he wanted in a business deal. Honest, yes, but also exceedingly shrewd. He drove a hard bargain, which didn't always set well with those he traded with.

"His word was his bond" is a phase that echoes over and over. Homer Lee Terry said: "If he told you he'd do something, he'd do it. And he believed in treating everyone that he did business with fairly and squarely. Now he'd try to skin you on a deal if he could out-trade you, I'd put it that way. But, for example, if we had a jointly operated lease, say one or two partners or three, and when it came time for the division of the properties or if there was anything to be done with regard to it, he was aboveboard. He didn't try to do anything underhanded."

G. I. "Red" Nixon saw Joe Zeppa's character and brains as the foremost factors in his success.

"He got along with everybody, and Joe Zeppa wouldn't cheat you; he just absolutely would not. He would out-trade you, but on joint-interest stuff, he always wanted you to give the partner the benefit of the doubt.

"Joe Zeppa was as near a genius, when it came to money matters, that I have ever seen. Not only in money matters, but in everything he done. Just like the little old airplane they had. He didn't trust any pilot. He had his instrument panel right there in his cabin, just like Jimmy [Rider]. Yes, sir."[34]

Zeppa's letters, if they were any length, usually conveyed nuggets of his outlook on life, and give us some insight into his thoughts. "You were never like the dog who sat on a hot iron, under a hot Mississippi sun, who merely stayed there and howled, instead of getting up and changing places," he wrote an old friend, then retired from Esso.[35] The statement could have been applied to Zeppa himself. On another occasion he advised a young friend, "It is better to be slow and cautious than to be impulsive and sorry."[36] There was a lifetime of experience behind the words.

Although most such nuggets were fragmentary evidences of his outlook, we do have, from his personal correspondence, one detailed reflection of his approach to business ventures.

"A young man starting in business must learn to make money out of other people's capital unless he has a large amount of his own idle," he wrote once. Thereupon he laid out guidelines that, while keyed to his correspondent's situation, might apply generally to a great many endeavors.

> (1) One should not go into a venture which will not return a net income considerably in excess of the interest rates at which he can borrow money. Such being the case, he should make a profit on the money borrowed.
>
> (2) A proposition, to be good, should return the investment in a reasonable time . . . from the profits (including depreciation) over and above a fair allowance for interest on the investment . . .

(3) In financing a project with borrowed money, one must be certain that it is sound, so that the equity money he puts in will not be endangered in case of miscarriage of plans. The security of his equity can always be protected if he has some liquid reserve available. This can be in the form of cash, Government bonds, gilt-edge securities which have a ready market and are available in the form of stocks and bonds which yield from four to six percent.

(4) Improved real estate with a large part of its value financed on a favorable long-term loan is more saleable than one in which the purchaser will have to pay a large amount of cash, particularly as the market for such loans is not always favorable. It is always best to start with the largest loan possible for the longest period of time, provided such loan can be prepaid in whole or in part at any time.[37]

X

Initial impressions of Joe Zeppa were almost always lasting. His dynamic personality thundered forth, showering sparks and frequently generating respect mingled with fear in the hearts of observers.

"The first time I ever spoke on the telephone to Joe Zeppa, he swore at me," Marguerite Marshall said, laughing. She was working for Tyler oilman Ike Rudman at the time, when Zeppa called, asking for Rudman. "It was just that I wasn't allowed to connect anyone to Mr. Rudman without first knowing who it was. Mr. Zeppa was not like that. He was more informal. I asked who it was and he said, 'What the hell difference does it make to you?' Anyway, at that time I found out that was Joe Zeppa."

Later, seeking employment nearer downtown Tyler she took a job at Delta for $150 a month, $30 less than she had been making.

"I'll never forget," said independent oilman Ralph Spence. "I was just a young fellow back from the war. I'd gone out for myself in '49 and I was trying to put some deals together. In the early '50s I'd gone to see Mr. Zeppa and I was showing him something. Someone who worked for him came in and interrupted for some kind of problem, and Mr. Zeppa—he worked him over. He got his attention. In such a way that I don't think I went back to Mr. Zeppa's office for the next two years. I thought he was the roughest fellow I'd ever heard. Thought it might be me, next time. But he was fair. Hard, fair, generous. I miss him."

When Lynna Jo Edwards went to Tyler for training after she was promoted to division clerk in Odessa's West Texas-New Mexico division, Red Magner took her in to see Joe Zeppa.

"And he asked a specific question about a well in New Mexico," said Edwards, "and of course I did not know the answer, because it was on the production end of the business, and I never will forget, he said to me,

'Don't you care?' He looked me right in the eye. I said, 'Yes, sir, but I don't have the opportunity to know.' "

This was approximately a year before Zeppa died.

Bob Waddell's first contact with Delta was through Chris Zeppa, his friend from El Dorado days. It was later, after some time at Delta, before he came to know Joe Zeppa well.

"At first I kinda stood back and looked at him with a little awe, you know, and I didn't understand him too well. He could be a little rough at times, with some people. I have heard him get pretty rough with people. He could when he wanted to. He just told them off in no uncertain terms. Some of the language he used wasn't used in polite society, either. He said pretty much what he thought; he was pretty plain spoken. With anybody. I do have to say, though, that he never, never did anything to cause me to feel anything against him, because I understand he had a big chore to keep this stuff going and finance it, and that was a masterpiece of work that he did."

Later Waddell came to know Joe Zeppa well.

"I think we understood one another and we fussed and argued, but it was all in good mutual respect, and I think he enjoyed a good argument. Nearly anything that I wanted to do that cost money, we had to argue about it a while before he approved it. I think a lot of his resistance to new innovations or new anything was primarily to see how strong you felt: to see how strong you felt about your ideas or what you wanted to do, and that's natural. If you went in and wanted to spend $100,000 for some equipment, and he said, No, we're not going to spend that money, and you walked out, you lost. But if you convinced him that it was worthwhile, then you made your point. I used to say I had a bigger selling job selling him than I did selling the customer who I was trying to get some work from. You had to sell him on everything, which is good business.

"He was good to me and I respected him and I respected his age, and I realized that he was entitled to a lot of eccentricities. He built that company, and I tried to put that across to some of the younger people fussing about him, because he had earned that right. I stood back a little in awe of him for a few years, until I got to know him better and understand him, and then I was wise to him. His outward appearance wasn't really what he felt inside. A lot of it he felt like maybe he had to do those things outwardly; his outward appearance would be a little off at times, but he didn't mean half of it. He didn't mean it."[38]

XI

"One thing I noticed about Joe," said Watson Wise. "When we would talk in the evenings about matters, I noticed Joe would take the opposite side. I knew pretty well what his beliefs were, but he'd take the opposite side in order to see if he was able to make arguments to counteract your belief in something, and he did that with so many people. Lots of times he was gruff, and he'd say things that disturbed people, you know. He was individualistic.''

If Zeppa had remained in Italy and become a priest, the Vatican would have treasured him as a devil's advocate in canonization proceedings. Fred Mayer participated in a great many vigorous discussions with Zeppa.

"He could be argumentative as the devil," said Mayer. "No matter what a set-up was, he had a diametrically opposite viewpoint on it. But then you'd either come around to his viewpoint or he'd come around to yours. He wasn't absolutely inflexible by a long way.

"I was going to take a trip one time to South America, and I was over at his home and explained to one of his men, Nick Andretta, how I'd planned to go through South America, and he laughed when I said, 'I'm going to show it to Joe because Joe knows South America quite well.' Nick said, 'Well, you're planning to start out from Caracas, north, and then around the west coast. Now you tell that to Joe and he'll have you starting from Caracas, south.' The first thing he said was, 'Now, Fred, you don't want to go that way. You want to start here and go south.' ''

Henry Bell, Jr., recalled the time he and his wife Nell were in New Orleans looking at antiques for the Blackstone Hotel. "He was quite a character," said Bell. "I remember when we came back that night, after several drinks, he and my wife got to discussing religion, and they went tooth and toenail at each other over it." What was Zeppa's stand? "I never knew what Joe was," said Bell. Did Zeppa take a devil's advocate approach. "On anything!" said Bell.

Zeppa's argumentation persisted just as strongly in his business dealings, at least with certain persons; and if our samples are representative, he tended to be most argumentative with those on the same social level with him. "Ben Clower (Jerry Hawkins's lawyer) and Joe Zeppa would get into some of the worst arguments you ever saw!" said Red Nixon. Yet probably no debate was ever more fiercely expounded than those between Zeppa and Sylvester Dayson, as will be discussed later.

In the earlier days Zeppa and his two partners presented a fascinating tableau at the board meetings.

"The three of them would get together and cuss and fight, you know— Sklar, Dorfman, Zeppa—but they knew what they wanted," said Seymour Florsheim. "They'd always have a meeting of the minds. Well, when any two of them got together they couldn't talk." He laughed. "That was a scream! They couldn't trade any other way. I mean, they'd have to fuss and fight and raise the devil with each other. They were different, yet alike. They were as sharp as a tack, as far as business was concerned."

"Mr. Sklar would come to Tyler, and I was up on the sixth floor of Citizens Bank at that time, and there was a big fuss," said Homer Lee Terry.

"Marguerite [Marshall] was out there at the desk where the telephone was. I said, 'What is that fuss going on?' 'Oh, that is Joe Zeppa and Sam Sklar,' she said; 'they got into a fight and they are fussing.' "

"They used to have board meetings," said Marguerite Marshall, "and we never could figure out how they knew what they decided, because everyone, the three of them, would get into Joe Zeppa's office and you never heard such noise in your life. Each one was talking louder than the other one. At the same time."

Despite such scenes, Seymour Florsheim doesn't believe there were ever deep conflicts between Sklar and Dorfman with Zeppa. "Course, Sam Dorfman and Sam Sklar would have a meeting and say, 'Well, we're gonna go over to Tyler to this meeting with Joe Zeppa; we're gonna do this and that.' They'd come home, you know, hanging their heads"—he laughed—"and it'd be just like Joe Zeppa wanted it."

As Elizabeth Fischer, Sam Dorfman's second wife, pointed out, "Zeppa was *very* shrewd. He could just twist anything the way he wanted it."

"Oh, it would just be fights on how something should be done," said Florsheim. "Whether to go into a foreign deal, and so forth. But it always came out how Joe wanted it."

Milton Winston summed up this salient characteristic that many business associates of Joe Zeppa's learned, one way or another:

"He knew how to have the last word."[39]

XII

A corollary of Zeppa's argumentative bent was his penchant for providing—usually unsolicited—advice.

"It didn't make any difference what you started to do, he was going to tell you how to do it," said Bob Waddell. "One of the things he did with me that I always remembered: We had an old steam drilling rig running in the city limits in a little town over in North Louisiana, Haynesville, and back at that time the rigs didn't have outdoor johns like they have now—OSHA made them put them on, which is good—but somebody in Haynesville had called him, complaining about the crews' conduct around the rig there, without a john, and he called me in and simply told me what had happened and says, 'We're going to have to build an outdoor john to put on that rig over there. People are complaining.' And then he got him a scratch pad and a piece of paper and sat down and started drawing it off and sketching it off for me to show me how to build an outdoor john, and he'd draw another and I guess after about 30 minutes trying to sketch it off, finally I told him, 'Well, listen, I think I can get one built. I've sat on more of those than you ever saw. You've wasted 30 or 40 minutes here. Now let me alone, I'll get one built.' But that was the type of guy he was; anything like that, he'd start showing you. If you go build a chicken coop he'd tell you how to build it. He liked to do things like that.

"Chris was a lot the same way. He would give you full instructions telling you how to do it."

The advice-giving habit wasn't restricted to matters affecting the company. One day Roy Burchfield was fishing at Joe Zeppa's farm.

"Well, he came down there and walked by where we were fishing and started to give me a lecture on how to catch fish," said Burchfield. "He probably never wet a hook in his life. But this was it: he knew something about it."

His targets did not always respond gratefully to the instructions. Once when some cabinets had to be built on Delta's sixth-floor offices at the Citizens National Bank, Joe Zeppa drew the design himself and a cabinet maker was presented with the drawing and started to work.

"So Mr. Zeppa walked in and began criticizing," said Homer Lee Terry. "He didn't like this and he didn't like that. The old cabinet maker was pretty ornery himself and he said, 'I don't know who you are and what part you play around here, but I'm going according to this drawing right here. If you don't like it, you go tell the man who drew it.'

"Mr. Zeppa said, 'I'm the man who drew it.'

"The cabinet maker said, 'You'd better not change it any more.' Something to that effect. Mr. Zeppa didn't like it. The cabinet man walked off and left the job."[40]

XIII

Once aware of the force of Zeppa's personality, one might conclude that no matter what a native-born American thought of Italians in general, he was unlikely to discriminate against Joe Zeppa in particular. One wonders if Zeppa would have permitted such a thing to happen. He was asked in 1975 if he had experienced any discrimination in East Texas.

"I couldn't tell any," he said. "You know, I'm one of these people that never notices those things. I try to behave like I should, most of the time, and it never occurs to me that people might have some animosity toward me. I am sure there are people that do. But they're never indicated it to me, face to face.

"And if they did, I wouldn't resent it at all, because, you know, the Italians weren't very good for quite a while. They were all right, but they were just working people. They were people that worked with their hands. They wouldn't come to work with their minds. But you can't say that the Italians don't have any brains. Because I know that they have had for centuries.

"But the people that did most of the thinking in Italy were comparatively few, compared to the total population, because there weren't many people that went to school! And one would work in the fields all day and keep on thinking about the hereafter, unless you were a priest.

"But when I was a youngster, I went to school. It disturbed me quite a bit, you know, but I realized that we got the reputation because we behaved that way. Actually, it wasn't that we behaved badly. It was the fact that we just didn't have the *brains* to get in, like the Jew did, and if they couldn't speak English, they still knew how to make *money*. See?

"The Italians just worked for a living. They believed in working their shoulders and their arms and legs. And, of course, I was about that way myself at the time."[41]

XIV

What significance should we attribute to some of the salient symbols which Joe Zeppa attached to Delta Drilling Company? The name, Delta, he once said, came from trying to find a name that would symbolize three people—three divisions of shareholders in the early company, Zeppa, Stacy, and Dorfman *et al.* Delta is the Greek letter *D*. "Well, we just picked the *D*," he said.

But there is doubt as to whether it was as simple as that. "I don't know if he really believed that himself," said Keating Zeppa. "A triangle is probably more appropriate in a logo than just a Roman *D*. So I would think it probably derived from more appropriate graphics than just the Greek letter *D*. I don't know that it really stood for the three original partners."

The triangle, or Δ, also is akin to a pyramid, and for years, until Joe Zeppa's death, there was one person at the apex of the pyramid, which is the traditional organizational hierarchy.

"I would also point out that the circles around the triangle might also represent Delta's organizational structure," said Keating Zeppa. "You know, the wagon-wheel structure, where you have the hub, one person, and all the spokes radiate from or come in to the hub. That was probably an appropriate organizational structure for Delta for many, many years."[42]

This in turn brings in a parallel between Delta's organizational pattern— the wagon wheel, of which Joe Zeppa was the hub, with all the spokes focusing in on him—and the physical layout of Tyler, the city that houses Delta.

In the wagon-wheel analogy, the spokes to the outer rim were the lines of authority, which always flowed directly from this center. The physical layout of Tyler can be compared roughly to a wagon wheel. Today there is a traffic loop around the city which makes the symbolism more apparent, and if one studies a map he will see the many streets which originally were old highways that reached to the heart of the city, or the old town square, just as spokes on a wagon wheel converge upon the hub. Today, the traffic loop serves as the rim. Tyler itself in its early history was an old trade and political center for the region, as the city supplanted Nacogdoches in that role. This bit of history has been incorporated into geography, for the old roads are still represented in the paved streets which replaced them in the growing city: Old Bullard Road, Troup Highway, Kilgore Highway, Old Omen Road, Chandler Highway, Old Jacksonville Road, Dallas Highway, Van Highway, Gladewater Highway, Old Overton Road. The names, each representing a line of communication to the world outside, linger to remind us of Tyler's central position in the region. One wonders if Joe Zeppa in the 1930s, seeing the geographical mapping of this old trade center and "capital of East Texas," recognized, perhaps unconsciously, that his internalization of a business organization modeled on an European patriarchal tradition, had been replicated in the East Texas landscape. There were any number of reasons why one might not want to live in Kilgore or Longview during the boom years, so Tyler was the most logical place to live if one did not move to Dallas. But in addition to his other personal and business reasons, was he drawn there also by his attraction to Tyler's map pattern?

"Joe was a benevolent dictator," said Red Magner, "and in each division he had a benevolent dictator manager. So that when he called he could get answered any question, from one guy, and if the guy didn't have it he went out and got it. So you had a series of triangles, or pyramids; that's just about the way the organization was structured."

This leads to the probability that, if Magner perceived the structure as pyramidal or triangular, Zeppa did, also. Certainly there were actual symbols from the past that fitted in and probably had some effect on his choosing the name Delta with the triangular symbol, for Cities Service, operating from the Shreveport area, used a triangle in its logo. There also had been a Triangle Refinery in the area, which could have served as a reminder.[43] But, of course, none of it would have had any significance to Zeppa if he had not found the symbol appealing.

When Delta started out, Tyler was a perfect headquarters, for the center of Delta's operations was in East Texas. The maintenance yard was in Kilgore. Everything was centrally, and logically, located. But when Delta expanded, a larger city such as Dallas or Houston would have been a logical place for the main office. The main reason the company didn't move, of course, was because Joe Zeppa didn't want it moved. He wanted to stay in Tyler, and that was the only opinion that mattered. Why did he insist on remaining in Tyler? Well, there may be many factors or many facets of one point. His reluctance to move far from Pinehurst Farm was at least a partial explanation. The one that most frequently surfaces is the analogy of the big fish in a small pond. And though he had grown to maturity in one of the largest cities in the world, New York, and was cosmopolitan enough to feel at home in large cities the world over, he appears to have preferred small cities. After all, he had spent his earliest years as a peasant boy on a farm in Italy, an experience totally foreign to the concrete canyons of Manhattan. He had learned all that he needed to know about how to benefit from these large cities—how to gain financing when he needed it—and he could use that knowledge without living there. With his personal relationships and his reputation, he could live in Tyler and have everything he wanted, which was control over his life and his company. He needed Dallas and Houston on occasion, but he did not need to live there. His remaining in Tyler left his personal world unfragmented, in the center of his corporate wagon wheel, at the apex of his pyramid or triangle.

On occasions, Joe Zeppa manipulated material symbols more obviously. He collected relics, many of which conveyed a deep significance to him. One may recall that, after waiting on the steps of the Exchange Bank and Trust in El Dorado for a loan in 1931, Zeppa swore the doors of the bank would never close on him again.

"When they remodeled that bank," said Marguerite Marshall, "I was in his office. He put his feet up on his desk and he called Joe Blasingame over in Kilgore and told him 'to take a truck up there to El Dorado and get those bank doors,' and he didn't 'want to see a damn scratch on them anywhere when they got back.' He bought those bronze doors."[44]

The doors, transported to Tyler, became symbolic of his assertion that the bank's doors would never close on him again. They were in storage, like a trophy one never got around to displaying, when he died.[45]

Probably the most noticeable symbol of Delta over the years has been its color—green. Most people will probably accept that the colors one chooses for a logo, or, for that matter, an automobile or a suit, are meaningful to the chooser. The choice may not be made consciously, exactly, but merely on the basis of, "I just like this color, and that color, better. I don't know why." To make such a choice implies the color has a personal significance that a competing color does not. Colors, like other preferences we express every day of our lives, convey symbolic meaning to us, whether we are aware of it or not.

Why, then, did Joe Zeppa choose green as the basic color for Delta Drilling Company? "Delta green" is the term used by the old-time hands who spent a fair portion of their time painting various pieces of equipment. "Green was his color," said Milton Winston. " 'It stays longer without having to repaint it.' That's the answer he'd give me."[46] But that statement is subject to criticism, for the chemistry of paint is more relevant to longevity than is color. More relevantly, "Delta green" is also the same as "Continental green," the color that Continental Supply Company used for years on its machinery. It is easy to speculate that Joe Zeppa saw the green which Continental used, liked it, and decided it was what he wanted when he founded Delta. But why did the Continental green so appeal to him? There we get into the fundamental attraction of the color.

One thing which green signifies is growth. Grass is green, and it grows; it is an optimistic symbol, for grass comes with spring and sustains animal life and itself is a product of the earth. A farm boy would find the color green a pleasing one for these reasons, if for no other. And since Delta did grow, and since Joe Zeppa obviously had growth as a major goal of his life in business, it is difficult to argue that Delta green did not mean growth to him. Green also is the color of money, but money itself is a symbol of wealth, and wealth is a symbol of having grown economically.

Though green is the dominant color used in Delta's logo, it also shares its place in the logo with red and white. "As long as I've known anything about the company," said Keating Zeppa, "the logo has always had the red and green on a white field. Two red circles, the green triangle, the green lettering, on a white field. Of course, we don't intend to change any time soon, I might add."[47]

It might also be noted, as Keating Zeppa has pointed out, that red, white, and green are the same colors on the Italian flag. The fundamental colors—green for growth, red representing blood for life, white for purity—internalized in the Italian native's mind were then transferred intact into the Delta logo with, perhaps, special meaning, if we are to judge from the subsequent events.

This brings us to the central symbolism—Joe Zeppa himself. As a boy in Italy, his brilliance had set him apart from his peers and, in fact, had assigned him a special place in the family constellation. When his brother Charles left for the United States, Joseph, or Giuseppe, became the eldest son present in the family; and when the father, Vincenzo, broke his leg and became incapacitated, young Joseph, barely 12, became the eldest active male in the family. This did not necessitate his assuming his father's responsibilities, but it undoubtedly must have provided him some additional feeling of maturity, perhaps seeing himself as one capable of being the dominant personality in a group. In New York, he rose fast despite disadvantages. Though his employment for many years had him in less than the top spot, he consistently worked his way up the ladder to highly responsible positions. He knew how to be a corporate father. When the opportunity came to found Delta Drilling Company, he became its president and virtually ran it, despite having only one-third of the stock. When he gained control by buying out Bob Stacy, he, in Freudian terms, had become the "father" of his corporate "family," much as an Italian father—or for that matter, many a native-born American father of that generation—ruled his own biological family.

Delta was his family. Its employees were like members of that extended family. He was undisputed head, or father, of his company, and ran it as a European father of his generation ran his family. He was not necessarily always stern, but he required his "children" to work hard, as he had. In his own family, his competition as the most successful one of his generation, certainly in a material way of measuring such things, had left him Number One. He dispensed fatherly—or benign brotherly—assistance and advice to those who might be termed sibling substitutes, which is to say, those of his generation or younger. Some accepted his advice, others resisted it to the point of loud rebuttals, while still others good-naturedly endured it in acceptance of his eccentricity.

There is nothing to indicate that Zeppa did not look upon Delta in this fashion, and there is a great deal which argues in favor of this view. He ran the company himself. He was the only one who ran it. It was "his business." And like the Italian father, he did not relinquish his position as long as he was capable of asserting himself, which he was till the hour he died.

Images of the father figure arise spontaneously as employees reminisce about Zeppa.

"Everybody loved him," said Jim Ewbank. "He was a mean old man, but I loved him like he was my father."

Whatever disadvantages Joe Zeppa's wheel-like organization might have had, for many employees it also was a comforting symbol, for one could go to Father with a problem. Though all of us seek out our father figures in one way or another, there is a possibility that many, because of Delta's structure and the man at the top, may have preferred, and stayed with, Delta because they were looking for a father figure. And, as Flo Bonham said, if they were looking for a father figure, "they certainly came to the right place."[48]

"Well, if you had a problem, you didn't have to go through channels," said Homer Lee Terry. "If you had a problem that was big enough and you wanted to go talk to Mr. Joe or Mr. Chris, either one, you were welcome."

A personnel clash among some men in East Texas production came to Joe Zeppa's attention when an employee complained that his foreman was trying to force everybody under him to attend the foreman's church.

"So this man gets in the car and goes to Tyler," said Terry, "and he don't go to Mr. Chris, he goes to Mr. Joe and tells him what this man is doing. So Joe and Chris call [production superintendent] Gus Jones and tell him to go up there and get that man straightened out; said, 'We live in a free country where we can do what we want to as far as religion is concerned, and he is not going to force that man to do anything.' That was one instance where this man went over everybody's head."[49]

Many perceived an aura of the strong, protective father. "There was a feeling of closeness between anybody who worked around him," said Marguerite Marshall. "Even though he wouldn't express it to them, you knew that as long as he was there and as long as you did your job, nothing bad was going to happen to you."

In a time of economic uncertainty, this could be comforting indeed.

Because she perceived her boss in these terms, Mrs. Marshall, unlike May Jones, hesitated to even think of talking back to Joe Zeppa, much less point out to him her opinion of his choice of language. Consequently, she had a subtle means of bringing her ideas to his attention.

"Sometimes in dictating he'd say things—a little harsh, something like that—that I'd think, 'I wish he wouldn't say that,' " she said. "The only way I could get around it—and I'm sure he saw through it, but it was never mentioned—when I'd take the letter in to him I'd make a gross typographical error right next to where it was. Then he'd chew me out for not reading my letters, but in changing the mistake, he would change what I'd [wanted him to] see. Now I couldn't have told him, 'I don't think you ought to say that.' He was the same age as my father, and I could no more criticize him than I could my father."[50]

The fact that the Zeppa family was involved in the company—Joe Zeppa and his brother, then the two sons—helped many employees perceive the company as a family business. "You had that family feeling," said Grace Baker. "You felt that they really cared about you, not just as a machine, but as a person. That feeling of closeness and of really caring." Mrs. Baker cited an instance when her son's daughter required the services of a specialist in Oklahoma City. "Mr. [Joe] Zeppa put the company plane at their disposal, and they flew to Oklahoma City. That's the way he was. I remember we had a young man who worked for us, and he suddenly developed a rare blood disease. He died that year. Mr. Zeppa just did everything that he could to see that the man had accurate medical help. He was just a caring person. Although he was a person that it was not easy to chit-chat with. There wasn't a lot of small talk. Mr. Chris did better at that."[51]

"What stands out in my mind was the closeness of everybody," said Avis Boren. "We were just like one nice big family. Everybody was friends with everybody else, we knew everyone, and it was about that time that Lake Tyler was built. We acquired those three lots out there at Lake Tyler and built an employee recreation club, and the men would go fishing and catch fish and then we would all have a fish fry out there. We used to have treasure hunts and have lots of fun.

"At Christmas time we would rent Smith Memorial Building and have a Christmas party for our children. One of those earlier ones was when Trent Zeppa was working here, and he talked his father into supplying silver dollars for each of the children. So Santa Claus passed out silver dollars."[52]

One impression that shored up Zeppa's paternalistic image was his apparent reluctance to dismiss an employee. "He had the ability to get rough and raunchy," said John D. Hall, "and even in spite of that, everybody loved him. One thing about him: he never would fire anybody. I don't remember anybody he ever fired."

Bob Waddell agreed. "He'd put up with a lot out of a man before he'd fire him. I don't know if he ever fired anybody, really. I don't know that he ever did." But, Bob Waddell added, others would have to do the firing for Zeppa.

"I caught some of that," said Waddell. "At one time I had the reputation, probably, of being a little rough, because I had to let some people go or ask them to go, but I didn't mind. And, too, we had problems overseas at times, and I never did do foreign work other than trips for troubleshooting and looking out, and as a result of my trips we had to let some people go; and that got to the point [that] when I had to go somewhere people would say, 'Well, somebody's going to get the ax because Bob's gone.' It wasn't too pleasant. The company felt like I was enjoying some of these foreign trips, and, really, I didn't enjoy it because most of the time when I went on foreign trips it was an unpleasant situation."[53]

Part of the reason Zeppa shunned the act may have been that since he felt Delta was an extension of his family, firing someone was like having to get rid of a member of his family. But Mark Gardner thinks Zeppa's ego was involved in his putting up with some people he might otherwise have fired. "I think it is a characteristic of people, if they fire somebody it reflects on their judgment for having them there in the first place. 'But I don't want to admit it.' "

Gardner said Zeppa told him one time that he guessed he had never fired anybody. On occasions when Chris Zeppa or other executives complained sufficiently, Joe Zeppa might put pressure on others to let a man go, but, said Gardner, "the people they would decide to let go were the best people we had." On the other hand, he cited one instance of one employee whom Gardner wanted to fire. "Every time I would get into it with Joe about firing him, he would give him a raise." This, concluded Gardner, was consistent with Zeppa's permeating desire to run everything.[54]

XV

Because the overwhelming majority of Delta employees have remembered Joe Zeppa with dizzying admiration despite his rough edges, it is well to bear in mind, in order to gain a balanced view of him, that he inspired a variety of reactions.

While many persons loved Zeppa, many more respected him, and some disliked him to varying degrees or at least felt ambivalent toward him. Mark Gardner, his vice president for a number of years, fitted into the latter category.

"I guess I never hated a man any more and loved a man any more," said Gardner. "Sometimes I would have gladly killed him." Though Gardner recognized and respected Zeppa's genius, he frequently found the president's personal quirks and dominance too frustrating to function pleasantly. This was one reason Gardner resisted moving to Tyler for most of the time he was with Delta; instead he lived and operated out of Dallas, where he had been when he started with Delta Gulf and remained after the process of digesting that company had begun. In fact, Gardner stayed in Dallas until five months before he left Delta. How did he manage to do that?

"I did it just by merely agreeing with Joe that I ought to move to Tyler every time he would bring it up. He would say, 'That is something we have got to do,' and finally I had to succumb to the pressure.

"I couldn't stand the regimentation of a big company headquarters in a small town. Joe would sit and piddle around all day long, then about six o'clock he would want to dictate to Marguerite [Marshall]. He had all day to do it. I used to say, 'I wonder why she didn't tell him to go to hell.' She was loyal.

"I would go over there [from Dallas] in the morning to get something straightened out, and I couldn't see him; then [in late afternoon] he would want to dictate to Marguerite, and then he would want to get the scotch out and then we could talk, and he would want all of these people sitting around—Joe Bevill and John Hall—and they couldn't go home for dinner. He wanted them all sitting around there, drinking scotch. Every night, and I just didn't want to be there. I finally did [move to Tyler] but by that time the doctor had told him to slow down on the drinking a little bit."

Gardner, like many another who knew Zeppa well, saw him as a unique person. "He was a great, great man," said Gardner, "and he had a recipe for being very successful, to build a company in his own unique way, starting with junk. He always stayed with the junk the whole time of his life. That was the reason Delta Marine didn't ever amount to anything. He wanted to try to get in the offshore business the same way everything else was done.

"On the other hand, Joe had a tremendous mind. Probably no other person ever could have done what Joe Zeppa did. The way he did it, it took a unique individual."[55]

XVI

Despite what many Delta employees from earlier days seem to think, Delta's pay scale was not attractive compared with other companies in the petroleum industry. Mark Gardner, who both admired and criticized Zeppa, attributes such policies to the head man.

"When I talked to him a second time about going to work, and we talked about salary consideration, he said, 'Well, Mark, that is more than I pay Chris!' And I said, 'Well, maybe I can be helpful to Chris. This is what I want.' And he paid me that and he paid me well, and I hope that I did a fairly decent job with Delta Drilling Company. But the general salary levels, both field and office, were always a little below the industry. In fact, he had a very peculiar view of office personnel. He would espouse the theory that, regarding clerical help and accounting people, secretarial help, when they got up to a certain level you couldn't raise them any more. If they wanted to quit and go somewhere else, just let them quit and go someplace else. This was a theory of his. He didn't put any particular value on experience gained by years of association. You would just get somebody else."

Nonetheless, toolpushers, drillers, and roughnecks have frequently stated that Delta was a better employer in days gone by, and some have even insisted that Delta paid better than other independent contractors around East Texas.

"They may have paid better in East Texas," said Gardner, "but I would say generally Delta paid in the low range. In most places I think they paid in the low range. There was a problem to get it up."

Yet many of the drillers or roughnecks would work for another outfit when their Delta rig was down, then go back to Delta when work became available. How does one explain such loyalty when the pay scale didn't warrant it?

"The old-timers with Delta Drilling Company were extremely loyal and their regard for Joe Zeppa was good," said Gardner. "I think it was the force of Joe Zeppa's personality. Certainly it wasn't the company, pay scales, or any benefit plan or anything else. Of course, the natural human tendency is to stay with someone you are familiar with. Ninety-five percent of all of the people in the country are reluctant to change, so they knew Delta, they knew the people, and when the rig would start up, they would go back."[56]

Also, Delta never missed a payroll at a time when many independent contractors made their crews wait for their money, sometimes even going broke before they paid off their employees.

Grace Baker encountered an experience that suggests an insight into Zeppa's pattern of granting raises. "There was a time when we got a $25 a month raise once a year," she said, "That was a good raise. This was in about '57; the oil industry sort of went down, and it was in this recession for about five years. I had gone without a raise for some time. After a few years, Joe Bevill thought maybe I should have a $50 a month raise instead of $25. Mr. Zeppa would go over all the raises and when he came to that, Joe Bevill was trying to explain to him he thought I deserved that. He said, 'Hell, Joe, that's a lot of money and she's got a husband who's got a good job and she doesn't need it.' I didn't get it. I got $25. It was not your merit, but your need, which would not go at all now."[57]

Zeppa's conservative management of wages and salaries permeated all levels of the organization. Even his own salary lagged behind those of most executives in other companies who held the responsibility that he did. In 1948, his annual compensation from Delta was only $25,000 and in 1972 it was raised from $40,000 to $50,000, retroactive to January 1, 1971.[58] Considering that he also built an indebtedness to Delta through his open account, it can even be argued that he paid for the privilege of running the outfit.

As if to make up for any economic deficiency, sometimes one received a title, ostensibly a promotion, instead of a raise in salary, according to Bob Waddell.

"I remember when they had a board meeting and finally elected me vice president in charge of the East Texas division. I was in Joe Zeppa's office and I had already heard it, but he said, 'Well, we've decided to make a vice president out of you, Bob. What do you think about that?' He liked to joke. He was like a kid. Most people didn't know that. I said, 'Well, I'll tell you. I think it sounds great. But, you know, you can't eat titles.' He got the message, and when my next paycheck came out it had a pretty good raise. Some people wouldn't have said that to him at all, because they were afraid of him. But he enjoyed your talking back to him. At times."[59]

XVII

Though Joe Zeppa was the patriarch in every way, his contact with his two sons was relatively slight. Childhood in the Zeppa home, for both Trent and Keating, meant growing up with a minimum of give-and-take between father and sons. His *authority* was always there, but his actual presence was another thing. At this time he was busily building Delta, working late, and on the road a lot.

"Dad was not a friend to his children," said Keating Zeppa. "Either to his stepson Trent or to myself. He was not antagonistic, but he never saw his role as being a friend. He was Father, and that was sort of like being Zeus. I mean, he sent down from on high, and it was supposed to be done. I think he secretly harbored some disappointment in the fact that I would never buck him, even when I was 40 years old, but I would somehow or other find some way to avoid doing those things which he told me to do, with which I was not in agreement. I would try to do it quietly and gently . . .

"But Dad was gone quite a bit and, as I reflect back, as I have many times on the relationship and our family, my father really was not present very much, even when he was home. Because when he was home it was late in the evening. He was tired; we would have dinner. You know, it was probably late anyway, by East Texas standards—7:40, something like that. He worked late, and then it was just sort of the style, too, even when he wasn't working, when it was a holiday or something.

"Really, Dad's only child was Delta Drilling Company. We were his biological family, but the real family that took most of his time and attention is Delta Drilling Company. I don't *resent* it. But, really, that's about the way it was. He showed up at home when he was hungry and when he had to sleep, and when he was in town, because many times he wasn't. In the early days he probably was out on the rig, or he was in Dallas arranging financing, somewhere else getting contracts, or whatever. Dad, really, just wasn't home that much."

Being the "crown prince" and living in the shadow of a patriarch like Joe Zeppa made times difficult for Keating Zeppa. The situation grew even more frustrating during Joe Zeppa's declining years when Keating had become executive vice president following Chris Zeppa's retirement. But, as has been observed repeatedly, Joe Zeppa continued to run the company till the day he died.

"If I had been a professional executive vice president, with no family ties to the president," said Keating Zeppa, "I probably would have quit six months after I got the job. I would have realized that I was essentially ineffectual. You know, I had the title, I had the authority and the responsibilities, but I really didn't.

"Occasionally I could put some things over, but you remember that when I chose to do battle on a given project with Dad, it was a battle, a few times. You've got to remember that I was having to reconcile my filial respect for a very strong father with professional regard for business considerations. And I promise you it was not easy. Because Joe Zeppa was a very patriarchal father image, and I probably lived in as much fear of my father as any kid did. You just didn't buck him! What he said was law. And to this day I probably carry some of that baggage around with me, psychologically. I was just never brought up to, say, pound the desk at your father and [tell] him that, by God, you were going to do this!

"It was an unhappy time, frankly. I did not like it and I can assure you that there were times, in the late '60s and early '70s, when I would happily have gone to work somewhere other than Delta. There were a couple of times when, if somebody had walked in and said, 'Hey, do you want to go to work for So-and-So?,' I would have said, 'You bet!' "[60]

Not only did the elder virtually run the company and overrule his executive vice president, but the low salary he paid his son also became a sore point. (Trent Zeppa also complained of his low salary, the ostensible reason he left Delta.)

"Dad always believed, among other things, that I should never be concerned about my salary, for example," said Keating Zeppa. "Of course, I didn't participate in any dividends. Furthermore, he always felt that, I think, my attitude should be one of extreme gratitude that I was put into a position where I could learn about business and this sort of thing and not worry about tomorrow. Well, it never really occurred to him—I should live within my means, and that's fine, with four kids and a wife, very nice clothes and this sort of thing—but my means were substantially less than that of a toolpusher. He didn't believe that I couldn't reconcile the two positions on my salary. And certainly I've never had the reputation for being a high liver. We live very quietly. But that sort of added to the irritation from time to time."

Joe Zeppa's salary probably holds a key to the matter. Through the years, he had supplemented his salary with bonuses and dividends as well as building up an indebtedness to the company as the result of a system of withdrawals that handled his expenses and living costs. From time to time he sold to the company various properties he had acquired personally, in order to reduce his indebtedness. Other presidents, no doubt, would have managed this in another way, probably by increasing their salaries commensurate with their needs.

"Dad was essentially hypersensitive about the salary he drew from Delta Drilling Company," said Keating Zeppa. "He drew a very low salary compared to the rest of industry, for what was a normal salary for a person in his position. His theory being that he shouldn't draw a big salary because that came out of the other stockholders' pockets and the company was paying dividends and that made up for his low salary.

"But what his low salary did was cap everybody else's. So when you tried to go out and hire somebody, on our salary structure you couldn't do it. I'm talking about executives, engineers, whatever, because once you cap the pyramid, then everything flows down from that. How do you hire an engineer that everybody else is paying $1000 a month and your structure allows you to pay him only $700?"

Occasionally, a few employees resorted to unorthodox methods to "equalize" their restricted salaries.

"They accepted the lower salary and just stole the difference," said Keating Zeppa. "That was not uncommon in the oil patch, I hate to say, and this wasn't the only company it went on in. You know, you'd see company equipment, trucks off on somebody's private job. People didn't talk about it, but . . . I don't know that it was anything of major damage to the company, but it was sort of like cockroaches—it's not so much what they eat, it's what they mess up."

Based on oil-patch gossip, he estimated that this pattern existed from the middle 1930s to possibly as late as the 1960s.[61]

XVIII

"He was not a demonstrative type of person," Marguerite Marshall said of her boss. "The one thing that bothered me when he died was the fact that I never knew if he really liked me or not. I'm sure he did, or I wouldn't have been there all those years. He didn't have to keep me around. He would have worked it around some way. But it did bother me. It really did.

"But about six months after he died, his good friend, Mr. George Keating, called long distance to talk to Keating [Zeppa], and it was during the noon hour, and I answered the phone. Of course, he recognized my voice, and he said, 'Marguerite, I am so glad to talk to you. I have been wanting to talk to you ever since we lost Joe. I just wanted to tell you how much he thought of you, because I'm sure he never did it himself.' "[62]

"That was against his religion," said Milton Winston, his farm manager. "He'd never say, 'I'm sorry,' apologize for nothing—never—or

'Thank you.' He'd gotten where in the last few years, when he'd go home to Italy he'd bring me something. He brought me a beautiful silk robe from Italy one summer. Mrs. Zeppa said, 'Milton, he never did his own children that way.' "[63]

Keating Zeppa reflected upon his impressions of his parents. "I have thought about it a lot and I can recall few instances—and I have to go way back into my earliest memory—of really any demonstration of real caring for each other in the presently accepted sense of touching, of caressing, this sort of thing. I think there was a very strong bond between them, but at times it was based more upon mutual respect than it was the romantic notions of love. Oh, I'm sure that existed in the early years. I just don't recall great demonstrations—neither one of them touching people. They could care very deeply about things, but they were reserved people. You didn't carry your emotions on your sleeve, didn't demonstrate it."[64]

Parsimonious though he may have been in expressing appreciation, he did find appropriate ways of reminding friends that he thought of them. Each Christmas he would send Helen Adams, from the Shearman and Sterling days, azaleas. He never forgot what a strong and positive force she had been in his life.

Sometimes his remembrances took unexpected turns.

"When my Chris was born," said Marguerite Marshall, "I was in the hospital for ten days. There was a girl who shared the room with me, and she asked me if my boss had sent me anything for the baby, and I really didn't think he would, but [I said], 'I wish that he would be practical and send me something like a war bond.' I had been there several days when here came someone carrying this huge bouquet of red carnations. This girl says, 'There goes your war bond.' But then I was home and had not come back to work yet, when May [Jones] came out to the house and brought me a note from Joe Zeppa and inside the note was my war bond for my baby. It was a war bond for $500. When he grew up and married and needed some money to make a down payment on a house, I gave him his war bond and that was what he used it for.

"Mr. Chris also sent a bond when he was a year old for $25. Now that's the difference between the two men. Mr. Chris gave of himself; Mr. Joe gave of his money. When they [employees] would have a crisis in the family, or a death in their family, Mr. Joe would send money; Mr. Chris would go."[65]

XIV

Joe Zeppa loved to socialize. In a way, it was his means of relaxing from his long hours of work. In these leisure moments, he enjoyed a good story and banter with others, and he cared not at all about the social standing of the persons in his presence.

And, observed Mark Gardner, he had a fine sense of humor, even when the joke was on him. "No matter who you were, if Joe Zeppa was taking the president of Exxon out to dinner and you were some damn roughneck that just happened to be with him, he would take you along, and all of us would go, no matter who Joe had. One night Joe was with a group in a hotel suite—and Joe had a marvelous mind and he knew a lot about the world—and for some reason or other there were some women around and Joe was telling them about the differences between the Italian alphabet and the American alphabet, and he said, 'They are pretty much the same except there are four letters that don't exist in the Italian alphabet that do in the American.' He named them and one of them was *W*. And Jimmy Rider [the company pilot] who never would say much, said, 'Mr. Zeppa, if there isn't a *W* in the Italian alphabet, how do they spell *wop*?' And Joe used to tell that story on himself. He loved to tell it."[66]

In his everyday activities, Zeppa bolstered his world with small rituals that indicated he was a creature of habit.

Bob Waddell remembered when he, along with others at Delta, would eat lunch with the boss in the old Georgian Room at the Blackstone Hotel. The big dining room was about the best place in Tyler to eat, at that time. Joe Zeppa lived in the Blackstone. The waitresses had been there a long time.

"They knew him and it was always a routine with him to fuss with the waitresses and they knew it, and if the food came out, it never was just right. The plate was cold, the food was cold, or the plate was too hot, and he'd fuss at them. There'd be three, four, or five of us sitting around the table, and we'd argue about who was going to pay for it and it would wind up with him signing the check every time. Once in a while he'd force me into paying for it.

"After we got through he wanted to know if anybody wanted a dessert. It was the same thing every time you went in there. And he decided whether anybody wanted one, and he picked on me about the dessert and I said, 'I don't want any more of that food,' and he'd say, 'Well, won't you eat a half a piece of pie—half a piece?'

"He always wanted lemon icebox pie. I didn't care much for it, but he had to work us every time and he'd say, 'Now go get us a piece of lemon icebox pie,' He'd tell her, 'Cut the piece in half and put one piece on one plate and one piece on another and give Bob half of it and me half of it. We just want a half.' I had to eat that half piece of pie. Once in a while I'd say, 'I don't want lemon pie. I want chocolate,' and occasionally I could talk him into getting a piece of chocolate icebox pie instead of lemon.

"That got to be a joke. One day I was in there with my family, and one of the waitresses that always waited on us said, 'Mr. Waddell, you can have a whole piece of pie today.' "[67]

Away from Tyler, visiting was a habit with Zeppa. Warren L. Baker, former executive vice president of the American Association of Oilwell Drilling Contractors, came to know Zeppa well while Zeppa was president of the organization. The president was expected to visit all of the chapters which were spread over the nation.

"So our first crack out of the box, the meeting was scheduled in Los Angeles, the following night to be in Bakersville," said Baker. "The thing about this trip that I remember is that we went to the meeting that night and then we got back to the hotel at 9:30 or 9:45, and Joe said, 'Well, let's visit a little while.' So we sat down in the lobby, and finally at 2:30 in the morning I just had to say to him, 'Look, Joe, its 4:30 back home and I am just going to sleep.' I found out right then that Joe would forget time. He just absolutely forgot it. It didn't mean anything to him if he would get interested in something."[68]

No matter where he was or whom he was with at the moment, Zeppa never forgot his origins. Once he and Mark Gardner were in Rockefeller Center in New York, around the forty-fifth floor, to say hello to the export manager of Continental Supply Company.

"Joe was great about that," said Gardner. "He loved to just go around and say hello to people and visit, even though he didn't have any business with them. This was one reason everybody liked him. We went up there to see this guy, and he was busy right at the moment, so some young fellow said, 'Well, he is busy just at the moment, and he is very anxious to see you.'

"So Joe and I wandered over to the window and this young boy came over and he said, 'You see right down there? That is what they call Hell's Kitchen. You know, immigrants come over that can't even talk English and they end up down there in that area.' And Joe said, 'Yeah, I know something firsthand about that.' He started telling this kid about that, and it embarrassed the hell out of this kid, and I told him he didn't need to be embarrassed: 'Who in the hell would ever figure that was where Joe Zeppa started his career?' "[69]

XX

Joe Zeppa was conservative in politics, and on the national level habitually voted Republican, though he supported conservative Democrats on the local and state levels. Some of the leading Texas political figures made a point of seeking him out in Tyler: Lyndon B. Johnson (though Zeppa was not an admirer of his), Texas Governors Beauford Jester and Allan Shivers, and governors from Oklahoma and other states. Grace Baker, in charge of Delta's central files, recalled Zeppa had received correspondence from Presidents Lyndon Johnson and Richard Nixon. Marguerite Marshall recalled that George Bush, elected Vice President of the United

States five years after Zeppa's death, was a friend who sought counsel from time to time. "Before George *ever* ran for anything, he came and spent an awful lot of time with Mr. Zeppa," she said.

To Marguerite Marshall, one of the bonuses of her job was meeting the people through her office.

"I remember waiting for the bus on the corner by the office one day," she said, "and [Texas Governor] Allan Shivers walked by and called me by name, and my son just nearly dropped his teeth and he said, 'Mother, that's the Governor!' "

Zeppa's conservatism, as we have seen in some of his business transactions, extended beyond his politics. Perhaps one aspect of his sense of convention could be traced to his formative years at Shearman and Sterling, regarding what one wore. Though the term was never used, a "dress code" was in effect at Delta's Tyler office as soon as one walked through the door. Whether weekend or not, Zeppa wore a coat, white shirt, and tie, and expected the other men to do likewise. Early F. Smith, like everyone else, knew this, but thought when he went to work one Saturday morning that, since hardly anyone would be in that day, it wouldn't matter much what he wore, as long as it was neat.

"I had some good clothes on," said Smith, "but they weren't very formal. They were tweed slacks, nice cordurory jacket, and a green plaid hand-tailored shirt. I was going down to get a cup of coffee, and Mr. Zeppa came in the door and said, 'Going hunting, Smith?' That's all he said, then just walked up the hall. I knew what he meant though. I didn't ever go quite that informal any more."[70]

XXI

In 1961 when John Bainbridge wrote *The Super-Americans* about Texas' wealthy, he categorized Joe Zeppa in "the rabbit's-foot group" of oilmen, describing him as "an Italian immigrant who arrived in this country in steerage at 12 and is now, at 67, reported to be worth more than $50 million . . . ."[71] Bainbridge's figures were somewhat inflated, and he probably, like virtually all outsiders in and out of Texas, was unaware that the Sklar and Dorfman families also held close to half of the Delta stock.

Trent Zeppa believed that Bainbridge, because of the elder Zeppa's Italian ancestry, suspected a connection with the Mafia. Trent chuckled at the absurdity of such a possibility. "Mr. Zeppa and Uncle Chris wouldn't have known the difference between a button man and a buttonhole," he said. Trent, then working for Delta, reported that he kept Bainbridge at arm's length which probably left the author wondering what, indeed, was going on.[72]

Bainbridge's categorization of Zeppa among the lucky oilmen clearly is contradictory to the burden of evidence we have already seen. Zeppa's policy was one of steady growth, whenever possible leaving little or nothing to chance. He progressed steadily and surely up the rungs of recognition in his industry, over the years participating in its organizations and, eventually, receiving its honors.

His honors and activities were many. Zeppa was elected president of the American Association of Oilwell Drilling Contractors (AAODC) in the fall of 1958 and took office on January 1, 1959. Delta, and Joe Zeppa, had been one of the original members when the organization was formed in 1940. Zeppa was a director of the association from its founding in September, 1940, until his death in 1975.

In effect, he was elected by his competitors—by 200 members representing around 175 different companies.

While Zeppa was its president, the organization began, for the first time, to take an active role in studying and, on occasion, protesting motor tariff rates, which affected the special rates charged by the trucking companies that moved drilling rigs.

"We organized a transportation committee, and when the carriers would ask for increases, we joined with a number of producers that operated under Oil Field Shippers Association," said Warren L. Baker, AAODC's executive vice president. "Usually we would join with them if we thought unreasonable rates were being sought. We would be protesting before the Interstate Commerce Commission on federal rates or interstate rates, and then in individual states like Texas, Oklahoma, Colorado, and so forth. Well, that became a very important part in the Association's work over the next 15 years or so. That started in '59 while Joe was there.

"Also, the Association had temporarily abandoned an annual safety clinic, and one of the early steps taken during Joe's time, and with him advocating it, we reestablished the clinic, which was a good thing to do." [73]

Zeppa was also active in the Independent Petroleum Association of America (IPAA), as a director; the Texas Independent Producers and Royalty Owners Association (TIPRO); and Texas Mid-Continent Oil and Gas Association. In 1968 he was named "Wildcatter of the Year."

In 1971 he received his highest accolade, the Grand Old Man of Drilling Award—a gold medal—from the International Petroleum Exposition in Tulsa. The award was originated in 1927 by John D. Rockefeller, which probably made it all the more appreciated by the Texan.

In August of that year, Zeppa was honored for his business and civic efforts in a Joint Texas House-Senate resolution of the Sixty-Second Legislature. Representative Fred Head of Henderson, Texas, made the presentation.

Though he obviously was proud of his industry awards, he modestly took little credit for having earned them.

"You know," he said, "after you get so old, they figure, 'Now we've got to get somebody to give something to! Well, let's see who we've got.' And they look around"—he laughed—"and see I'm still living, so they find something to give me!"[74]

The IPE's gold medal represented a difficult act to follow, but it did have a competitor of sorts. Dallas oilman M. B. "Bill" Rudman founded an organization called the All-American Wildcatters, with Joe Zeppa a charter member. Steve Schneider, another charter member, put together an annual fun type of prize called the BFU Award—Big Foul-Up. Schneider set up the committee. He was, as a matter of fact, the lone member of the committee. Among the winners of the dubious award have been Rudman, Robert Mosbacher, and Bunker Hunt. Rudman earned his for starting the organization. Mosbacher was recognized for scheduling the group's second meeting in Los Angeles, where there had been an earthquake about a week earlier. Selecting the epicenter of an earthquake for a meeting site had been no small feat. Hunt's "honor" came after he had scored his big strike in Libya and had been shut down by the new revolutionary Libyan government. His BFU Award cited him for going to Libya and always referring to it as "the little country with the big heart."

"When I give the BFU Award, truths do not have to enter into it," chuckled Schneider. The motto of the BFU Award as designed by its creator: 'To laugh at onself is a subtle expression of inner strength, of self-confidence, and a true understanding of one's place in the ever ruling matrix of mankind. Whereas, to play with oneself is socially not acceptable."

In 1973 the BFU Award went to Joe Zeppa for hiring a female roughneck.[75]

Zeppa delighted in attending industry-related conventions, but he insisted on a logical *raison d'être* for every organization he joined. To one man who urged Joe Zeppa to become a charter member of an association of millionaires, Zeppa declined, noting that he saw no reason for such an organization. "First of all," he wrote, "there are too many millionaires in the country and, secondly, many of those who are millionaires today can be poor tomorrow without exciting the attention of any of the rest of the population. Futhermore, I think it is a poor policy to class yourself on the basis of money earned or saved, under any condition." He urged the correspondent to drop the idea.[76]

As tends to be true with most highly successful businessmen, Zeppa served on the boards of a number of enterprises and business-related associations. A long-term association was with the Citizens First National Bank in Tyler. He was elected to its board on September 4, 1951, and elected to the executive committee on October 23, 1956. He went off the board the third Tuesday of June, 1975, at his own request—approximately two weeks before he died. He also was a director of the Texas Research

League, a business-oriented organization. In the 1950s he was a large stockholder in Pan American Sulphur Company. He also was a substantial stockholder in Lone Star Steel, as well as a director and member of its executive committee.[77]

Local honors also accompanied his other successes. He received the T. B. Butler Award as Tyler's outstanding citizen, was chosen East Texas "Man of the Month," and received the Fred D. Patterson Leadership Award.

Zeppa said, "You know, I've lived in Tyler now since 1933 and I don't go around as much as I used to, but one time I was very active in the Chamber of Commerce and the Community Chest, and active in the Tyler Symphony. I've known the damned thing [the symphony] when it didn't get money enough to pay its expenses. We got it in good shape and it's a beautiful symphony now."[78]

XXII

Joe Zeppa's last "rescue mission" came in the 1960s in behalf of the creditors of a Longview firm. When Trice Production Company went into the red to the extent of millions of dollars, its creditors eventually went to Joe Zeppa, requesting that he serve as referee in bankruptcy and straighten out the financial quagmire, as he had solved other such problems in the past.

Though Joe Zeppa may have considered buying Trice Production Company and operating it, the complications were "too big a can of worms," said Keating Zeppa.

When Zeppa first began to examine the complexities of Trice Production, Maxine Fant, working temporarily in its offices, glanced up one day, startled, to hear Zeppa, standing in the hallway, boom out, "I'd write a goddamned book about this place, but nobody would believe it!"[79] It was strong testimony to how snarled the company's affairs had become.

With many of the participants in the firm in the East, Trice Production, according to Roy Burchfield, had promised to fill out these participants' tax returns.

"Wound up they were keeping the equivalent of 15,000 sets [of records] a month," said Burchfield. "This was a tremendous undertaking. They were running the computer 'round the clock. Besides, in all cases they had made all the pipeline companies—the purchasers of the oil—pay them 100 percent direct and they were distributing. This was a tremendous thing.

"They went ahead and issued the checks on everything that was payable, and then they signed them as they could get the money. So to tell what their liabilities were at any one time was almost impossible. They had a five-drawer file cabinet full of royalty checks that had been processed and signed but never mailed out.

"[Joe Zeppa] got into this and he managed to get it all straightened out. He got the company reorganized. He liked meeting these problems. Mr. Zeppa was this way: he liked to make all the decisions. If he had problems coming to him, he got real itchy. He would assign you something to do and at one time tell you exactly how to do it. If you did it the way he said it, although you knew it was wrong [and] it backfired, he would say, 'We didn't go about that right, did we?' He always remembered he told you to do it and took responsibility for it."

The Trice creditors first secured Zeppa's services as a consultant. Then they reorganized the company and elected him president. He served as head of Trice Production Company then, with the creditors taking stock and the oil being used to liquidate debts. [80]

His efforts resulted in the unsecured creditors recovering most, if not all, of their potential losses. Trice Production was reorganized into Oleum Incorporated, with O. Wayne Crisman assuming the presidency; subsequently, in 1970, Oleum merged with Falcon Seaboard. [81]

XXIII

H. J. "Red" Magner had roughnecked for Delta in the summers of the late 1930s while working on a degree in business administration at The University of Texas. But as the Depression ground on, he suddenly found his education in jeopardy.

"I was broke," said Magner, "and I wrote [Joe Zeppa] a letter and told him that I needed $138 to be able to finish and get out of there. He sent the money."

After Magner graduated in 1940, he went to work permanently for Delta and paid Zeppa back.

"He was very fond of education, and if you'd stick with it, he'd stick with you." [82]

Perhaps Magner, because Dutch Meinert was his uncle and Meinert and Zeppa had been boon bachelor companions back in the El Dorado days, had more of an "in" with Zeppa than a stranger would, but there was a pattern throughout Zeppa's life of helping those who needed assistance in finishing school.

This is not to say that Zeppa invariably acceded to every request for financial aid to continue education. In turning down one applicant he noted: "Frankly, your spelling, composition, and handwriting seem to indicate that you are not prepared for college and that you probably have not made as good use of your schooling, thus far, as you might have." Furthermore, he said, he felt the youth should seek assistance closer to home, which was Dallas, because Zeppa felt his first duty lay with boys and girls in the Tyler area. [83]

Keating Zeppa estimated that his father spent half or more of his "extra-curricular" activities on matters related to education. He not only worked with all-black Jarvis Christian College, but also Texas College, a Methodist-sponsored black school in Tyler, both institutions which, in the days of segregation, were separate but never equal to the white ones. The elder Zeppa carried his zest for education even into seemingly unrelated events. As his son recalled, there were many instances when Joe Zeppa was in a position to express his paternalism outside the family by helping someone in need, and he would tell them, in effect, "I'll help you if you continue your education, but if you don't, I won't."

Zeppa held positions on two boards which recognized his ability and enabled him to exert his influence on education. One was as a regent of East Texas State University at Commerce, Texas, on which board he was still serving when he died. Another was the Board for Fundamental Education.

The Board for Fundamental Education, headquartered in Indianapolis, was dedicated to improving education for minority and underprivileged children. Primarily, this meant black people. Its members felt that education and training could solve the problems facing these underprivileged citizens and raise their standard of living. Not only did Zeppa serve on the group's governing board, but Delta contributed substantially to its objectives.

In the 1950s and 1960s, he served in a number of governing and advisory positions related to education: advisory board, Texas College; board of trustees, St. Stephen's Episcopal School in Austin; advisory board, University of Dallas; advisory council, Graduate Research Center of the Southwest; board of regents, East Texas State University; and board, Jarvis Christian College, at Hawkins, Texas. Gertrude Zeppa subsequently served on the board of trustees of St. Stephen's after he stepped down.

He made a practice of contributing to a variety of educational institutions.

"Well, we always do," said Joe Zeppa. "Somebody writes me a letter and it strikes me the man's being worthy of it. So I send him a check. And generally when I do that, they remind me once a year, you know. So, you see, they won't let you forget." He laughed. "Naturally, we give things to a great many institutions—SMU, Dallas University, Baylor, several of them."[84]

He could easily have broadened his list to include St. Stephen's Episcopal School, McCallie Patrons Association in Chattanooga, and St. Paul Industrial Training School at Malakoff, Texas, among others.

St. Paul School was a black institution. His support, stemming from oilman Bill Rudman's bringing its needs to his attention, is typical of Zeppa's concern for black people. Even in the early 1930s, Trent Zeppa recalled, Gertrude Zeppa established a day-care nursery for blacks, which prompted local racists to bombard the Zeppa home with eggs, which, Trent remembered, he had to clean off.[85]

"[Joe Zeppa] had a high regard for colored people as he would anyone else," said Joe Bevill. "But his regard for those people was probably two-fold: one, I'm sure he regretted that they did not have the same opportunities here that other people had, but he understood the reasons; and those who were capable and who had spent their life attempting to improve themselves received his whole-hearted support, which was very unusual in this part of the country."[86]

Keating Zeppa said, "He was very catholic—and I use this in a general sense—in his acceptance of different religions, different philosophies. Dad was not a judgmental person. He felt very strongly about certain moral and ethical rules—if you will, the Judeo-Christian heritage of ethics—but he didn't care whether a man was a Mormon or a Jew or a Roman Catholic or whatever he was. It just didn't enter into things. If he was a moral person, it didn't make any difference.

"By the same token, Dad was one of those people who was ahead of his time; he was as close to being color-blind as you can be. Dad simply did not accept that a difference in color implied a difference in ability. That's one reason he took so much time and effort to work with Jarvis Christian College."

Jarvis Christian College was a black college at Big Sandy, Texas, not far from Tyler. Zeppa contributed to it and served on its board of directors. At one time he permitted the college students and faculty the use of land he owned near it for agriculture. A number of leaders in black organizations solicited Zeppa's assistance from time to time, and undoubtedly his sympathy became known through word of mouth.

His color-blind attitude toward blacks was exhibited in other ways more striking for the times.

As Keating Zeppa remembered, "Among these people parading through Pinehurst Farm on weekends and what not, you'd have black people—in the early '50s, coming as house guests. In the early '50s in the South, that was unheard of. Granted they were educated, cultured people, but, you know, that didn't make any difference in this area."

Why did he do this?

"Basically because, as far as he was concerned," continued Keating Zeppa, "his life represented the triumph of what became his philosophy. Now whether he had the philosophy early on, I can't say, but I doubt it. His philosophy basically was: hard work plus education would get you anywhere you wanted to go. But the absence of either one or both of them was condemnation to poverty or to whatever.

"He never knew anything but hard work. He grew up that way, on a farm in Italy, so that was second nature. But he recognized that without the education, the hard work meant you stayed digging ditches or farming. Simply, your horizon was very limited without it."[87]

Joe Zeppa was one of the purchasers of charter stock in the National Security Bank in Tyler, the first and only minority-run bank in East Texas, according to Ralph Spence. "Everyone of us agreed to buy charter stock on condition that they [the black businessmen] ran the bank," said Spence. "It was their bank. None of us would serve in an official capacity. But Mr. Zeppa would even go further. He picked up the last part [of the stock] that they needed so they could get started. All of us had taken what we promised. Had a little bit left over. Mr. Zeppa took it. And he was just that kind of a man."[88]

Though he was an active philanthropist, many of Joe Zeppa's good deeds never came to public light. The stories dribble in of how he was in Mexico and the village priest told him about a local boy's brilliance, resulting in Zeppa's helping the boy get an education in the States. One reflects back to Zeppa's own childhood in Italy, and how an opportunity to learn and improve himself had changed his life.

"I expect you'd be surprise to know how many educations he paid for, not only in Mexico but other places too," said Milton Winston. "Italy, Venezuela, a lot of them right here in Smith County. He and Mrs. Zeppa too. They did a lot of things, going to pay somebody's hospital bill. I don't say it boastingly. I expect I know about as many of those kind of things as anybody, or more than anybody. He'd tell me lots of times he did such and such a thing—'Don't say nothing about it.' Mrs. Zeppa would also do the same thing. She would go to the hospital and pay somebody's bill and she'd tell me what she did and she'd tell me, 'Don't tell anybody.' The hospital didn't know. She paid cash or gave me a check.

He recalled one such such case.

"It was a black boy that was in a car wreck. Broke his back, was in the hospital I don't know how long. Mrs. Zeppa, this particular time, she give me some cash and said go give that to the hospital for him. That was $1,000, and $1,000 was a lot of money then. That was 20 years ago. That was a lot of money.

"They did lots of things for these black churches over in here. They always liked to come around and ask him for money anyhow. But he'd give to them real freely."[89]

As an example of such gifts, the Zeppas would donate television sets, as many as ten at a time, to the tubercular hospital near Tyler, which is now the University of Texas Health Science Center. "With no names on them," said Henry Bell, Jr.[90]

Some requests came from far off.

"He used to get a letter periodically from a Mr. Mpom Ulukule," said Marguerite Marshall. "He wanted an accordian. He was in Africa. A priest somewhere told him. He wrote to Joe Zeppa, and Joe Zeppa sent him money to buy an accordian. Periodically after that, whenever he wanted something, he would write to Joe Zeppa. It didn't always work."[91]

Often recipients of his generosity were Delta hands. Warren Hirsch said, "When we had a severe injury [to an employee]—and we did have one once in a while; men were permanently disabled and compensation was very meager and the family was in need—I personally went to Mr. Joe and got quite a few hundred dollars on the side, over and above and aside from compensation, to give to these people. He was quite a large man at heart, and that's how I came to know about other things that he had done, in exchanging ideas with the old-timers."[92]

Soliciting funds from him was not an open-and-shut case, however. Zeppa insisted on a sound use of any money he contributed. Henry Bell, Jr., recalled the time his wife and another women went to see Zeppa to get help in building a museum in Tyler.

"He had a list of questions for them," said Bell, "which they couldn't answer. He said, 'Now, you girls go back and do your homework and then come back and see me.' They did. And they went back to see him. He'd given them a hard time about how they would operate it, you know, expenses and where the income was coming from and all this. Sound business. And they went back and did their homework and came back and told him about it and he gave them $25,000. But, as my wife said, it was the best education that the two of them could have had. 'Cause it told them how to raise money. How a businessman would look at it. She said she was scared to death that first time, but second time she wasn't scared. She was ready to go.

"I saw him do the same thing when we were building the YMCA. He said, 'I'll support it, but I want to see how it is going to be maintained, where the income's coming from, and what the expenses are going to be.' And he made them budget. Which is sound and helpful, really, to the fund drive. He just didn't believe you ought to build something when you didn't know how you were going to take care of it."[93]

Zeppa's philanthropies were linked infrangibly to his personal belief system. Grace Baker recalled, "There was this young woman who had five children and her husband was a sergeant in the Army and he was stationed overseas. She wanted to go and be with him. She wrote Mr. Zeppa and asked him if he would be interested in buying an acre of ground that she had, so that she could get back to her husband. He set up an appointment and said, 'Come and see me.' She went in, and the outcome was that he said, 'No, I don't want to buy your acre of land, but I will advance you money, and you can pay me back.' He did not know this woman. She meant nothing to him, no relationship to the company. She was just someone that had heard of him. He did give her the money. There were a lot of people that said, 'You'll never see that money again.' She did pay him back the money.

"Sometime later she wrote to him again: 'I want to take a secretarial course so that I can go to work and help the family income, and would you loan me the money.' Financially, there was no reason why he shouldn't have, but he wrote her and said, 'No, I don't think that you should do that. You have all these children, and you need to be home, and therefore I am not going to have a part in this.'

"I thought that was quite an insight into his character. It wasn't the money, because she had paid him back. He sort of reminded me of someone that thought he should take care of all of us. He should make the decisions."[94]

XXIV

Of the several persons outside Delta and the family with whom Zeppa maintained close ties over a period of many decades, the names of three tend to pop out—Sylvester Dayson, Anthony Gibbon, and George Keating.

Sylvester Dayson was a "wonderful mystery," even to his family. As his only daughter, Suzette Dayson Shelmire, explained, "His past was never fully understood by any of us, my mother included. She never asked questions; she just loved him. And I asked a lot of questions and a lot of them were evaded, and a few of them came through."

He was born in France, from the best accounts, in 1892, which made him about a year older than Joe Zeppa. His father and uncles were in the oil-refining business, and when World War I broke out in 1914, young Sylvester Dayson was in Romania working in a refinery. Those in charge burned the refineries so the Germans wouldn't get the oil. Young Dayson returned to France. He served in the French army during the war, flying a single-wing Bleriot as a bombardier in reconnaissance missions.

A central event in his life occurred after the Armistice when troops of several nations, including France, occupied the city of Odessa, in Russia. The object was to protect the oil there for the White Russians during the Russian civil war. But the town rose in arms to support the Reds, and the foreigners, to keep from being annihilated, fled however they could from the Black Sea port. Dayson scurried to the air field, but no planes could be moved. He ran to a ship.

The unceremonious escape under such circumstances may have been acceptable in other nations, but in the French army, the episode was treated as traitorous. In combat, when an outfit ran in battle, men sometimes were drawn from the ranks and shot as examples. This retreat was viewed no differently. He told his daughter that after reaching Marseilles, some of the men were executed and, fearing his fate, he sought out a friend who worked for the Royal Dutch Shell Oil Company. The friend slipped him off the ship and onto another headed for Holland. The episode was the beginning of a new life for Sylvester Dayson.

He went to work for the Royal Dutch Shell in Borneo. From Borneo he went to Mexico, at a time when that country was still unstable after years of revolution. In 1921 he came to the United States on a Standard Oil tanker and built a refinery at Everett, Massachusetts. At the time he was 29 years old. From Everett he moved about the North American landscape, everywhere involved in refineries—Bowling Green, Kentucky; Rawlins, Wyoming; Port Arthur, Texas; Sand Springs, Oklahoma. When the El Dorado, Arkansas, boom came in, he went there. There his life crossed Joe Zeppa's for the first time.

"His life seemed to have started in El Dorado, Arkansas," said Suzette Shelmire. "He found my mother in the area and married her, and from that point on, that was his life. His wandering suddenly stopped."

His bride had been Colonel T. H. Barton's secretary. To escape her persistent suitor she had fled to Chicago to get a job far away from him. She thought he was "too volatile." But he followed her and won her over. By the late 1920s Dayson was a vice president of Lion Oil in El Dorado, dealing with refineries. Trained as an engineer, he became known in the States for his contributions to the cracking process whereby petroleum fractions are broken down to manufacture the commercial elements. He had been used to working with heavy crude oil in Romania, so the experience proved to be helpful in changing refineries to process heavy crude into gasoline. His experimental work in Port Arthur enabled him to work out a cracking process, according to his long-time friend, Joe Massie.

Dayson lost his money in the 1929 crash. In debt, he struck out on his own to build a fortune. Working out a deal with J. R. Parten, whom he had met in El Dorado, he set up a refinery at Baird, in West Texas. Then after the East Texas oil field was opened up, Dayson, Parten, and Zeppa built the refinery near Longview which became the Premier Oil Refining Company, with the stock divided three ways between them. At first it was a duplication of the Baird refinery. As we have seen, Parten became chairman of the board, Dayson president, with Zeppa joining them on the executive committee. The Daysons lived on the premises of Premier refinery until the company was sold in 1948 to Farmer's Co-op in Minnesota.

In 1939 Dayson bought a 660-acre farm, Holly Tree, near Winona, north of Tyler, for the recreation of Premier employees. When the refining company was sold, however, he retained this as his personal property and spent weekends there.

Dayson was flamboyant and, in many other ways, the antithesis of Zeppa. Dressing like a dude, he was almost never without a coat. Slender, at five-feet-seven, and sporting a mustache all his life, he smoked cigars and wore Texas-style hats and, frequently, black boots. "People never forgot him," said his daughter. "People never knew what to make of him. He loved power better than anything. Money got him what he wanted at the time, but doing things was much more important to him. He made a lot of huge mistakes by overdoing in a lot of ways. But that was his way."

Dayson's language was sprinkled liberally with "goddamn" and "son of a bitch," to the dismay of many around him. "Mama would say, 'How can you?' " said Suzette Shelmire. "But that was the way he talked. Joe Zeppa did too, but Joe cleaned his up as he got older. Mrs. Zeppa cleaned him up pretty good. She would not put up with that. Papa never did change."

Dayson's language was a major reason Gertrude Zeppa did not like him.

"They really did have a vendetta going," said Dayson's daughter. "Toward the end they began to tolerate each other, but before that, they just couldn't stand each other. Mama and Gertrude Zeppa liked each other, but Gertrude thought Papa was a rude, crude, unattractive man. Yet toward the end, after Mama died in 1959 and everything, they began to laugh at each other and their ways and all, and they got a kick out of the way they disliked each other. It was fun to be around them. Because they were being totally nice to one another."

Suzette Dayson and Keating Zeppa grew up together as friends, a relationship that both have described as "good friends" or a "brother-sister relationship." After Suzette became engaged to Overton Shelmire, an architect, Dayson laughed as he told Zeppa, "I tried the best I could to get Suzette interested in your son. I wanted so desperately to get into the Joe Zeppa millions."

Whereupon Zeppa said, "Sylvester, we can't do a thing with our children these days!"

Dayson's politics were those of an independent liberal, something of a rarity in the oil industry. He identified strongly with the New Deal and voted for Franklin D. Roosevelt each time he ran for President, except 1940 when, opposing a third term, he opted for Republican Wendell Wilkie. However, he returned to the Democratic fold in 1944 and remained there. Politics, then, frequently provided the fuel for a great many Zeppa-Dayson verbal conflagrations. Whenever they got together, they had ready-made differences on which to base their next argument.

Suzette Shelmire recalled some of the arguments.

"Joe would usually come over to the farm about 4:30 or 5:00 and they would have a scotch, and Papa would start the arguments. They argued over the Vietnam war. Papa would say, 'Goddamnit, why don't we buy the country? We don't need to spend all this money with Brown and Root, sending them over there to build all these air bases. We could buy the country.'

"Then Joe Z would say, 'You're nothing but a pinko. You're just a Red. How can you talk about our country? You're just not patriotic.'

"Then Papa would say, 'I chose this country, and you came over here, you were too young to choose. I came here because I wanted this country.'

"It was usually like that—the news, just anything he could stick a knife in Joe about.

"Papa and Joe Zeppa went to the World Oil Congress together in Moscow [in June, 1971]. That must have been some trip. They got left by the buses way out in the boondocks. Somehow they wandered around and couldn't find their bus and there they were stranded. I can just imagine. They were about 80 then."

Perhaps one reason Zeppa and Dayson clashed so frequently is that they were both highly intelligent persons competing for the top rung in life. Each one wanted to be Number One, which Zeppa was in his own company; while Dayson was president of Premier until it was sold, he never had the control over it that Zeppa had over Delta. Yet they also complemented each other: Zeppa was a genius of finance, and Dayson's engineering talents combined with his knack for putting ideas together—the skills of a promoter—made him quite valuable in his field. In business, these skills could mesh together with considerable profit, as they did from time to time. As Trent Zeppa said, "Well, one recognized the other as technologically sound." Furthermore, Zeppa usually had the advantage of having more money or better access to it. But this apparently was a point of irritation to Dayson.

Dayson went into a few drilling ventures that were dry holes. Probably the thing that nettled Dayson most was that Zeppa never invited him into any of Zeppa's profitable deals. Zeppa did bring a lot of people into deals, though as Mark Gardner has pointed out, they already had enough money to satisfy them. Did Dayson represent a rival in certain ways to Zeppa, perhaps a kind of substitute sibling whom he was afraid to see get a chance to surpass him? Mark Gardner offered another explanation.

"One reason might have been that Joe didn't want to have to fight Gertrude any more over his relationship with Sylvester Dayson. She hated Sylvester's guts."

"Strongly in the back of Papa's mind he thought that every venture that Joe went into with him, made money," said Suzette Shelmire, "but Joe never asked him into anything that he was going into to drill. He never asked Papa in to anything. And Joe got very much ahead of him. That ate Papa up inside.

"Papa looked like he had as much money as Joe Zeppa when he had Holly Tree Farm. But, probably, Joe had millions. Papa may have had a million, in comparison, but somehow or another he had to look as good as Joe, and act like he was. He lived way beyond his means, and as Papa always said, 'Shrouds have no pockets.' But he didn't leave me owing anybody anything. He left me Holly Tree Farm."

Zeppa and Dayson carried their predilection for argument wherever they went. Mark Gardner witnessed this at a hotel in Chicago during an American Petroleum Institute meeting. "I heard them arguing about a dam Sylvester Dayson was going to build on his farm, and Joe got to telling him how to build the dam, and Sylvester was shouting about how he was going to build the dam, and Joe kept shouting, and I sat there and listened to them and they got so loud I was afraid the hotel management would throw them out. And finally when they stopped to listen to each other, they were talking about doing it the same say!"

Almost everyone who knew the two men eventually witnessed an explosive encounter.

"At least Joe and myself would try to hold it on a more temperate basis," said Fred Mayer, "but between Joe and Sylvester—that was an uproar. I don't think you could have ever taken a tape of that. A tape couldn't have stood it. I know on one occasion I was over spending the weekend with Sylvester at Holly Tree, and he said, 'Now Joe will be over.' Pretty soon Joe drove up. He had driven over from his farm.

"A little while later the argument started. Now Joe and myself fundamentally agreed on many things, which was not always the case between Sylvester and Joe—especially Sylvester and Joe. Sylvester had a tendency to be a little left in speech. Sylvester had just returned from a trip to Russia, and was speaking rather laudatory about it. That started an uproar and we nearly got personal. I never heard hardly anything as bitter as [those] two men [going] after each other in that, and pretty soon Joe said, 'Aw, to hell with you,' and got in his automobile and drove home.

"Sylvester was there having a drink. I said, 'Well, Sylvester, with this great friendship through all the years, to let this explode . . . ' He said, 'Oh, this is no explosion. This happens every week. Now, next Sunday morning my friend Joe will be back over.' Which was true. I did mention it later to Joe and he said, 'Oh, no, just one of those things.'

"Those two had so many explosions and were so diverse in a lot of their viewpoints, but they still respected each other. I think Joe and Sylvester loved those fights."[95]

XXV

Anthony Gibbon's father had been a consul in Mexico, where Tony was born in 1900. Gibbon's uncle, Francisco León de la Barra, served as interim president of Mexico in 1911. Gibbon himself was an editor with *World Oil* for many years. He was a chubby, kind of round-faced man who reminded Grace Baker of a leprechaun. "He was a very friendly, outgoing person," she said. "He would stop by and talk to us when he was visiting

Joe Zeppa."[96] In Zeppa's personal correspondence files, a thick stack of Zeppa's carbons and Gibbon's originals dominate the G section. For the last several decades of his life, only Zeppa's correspondence to George Keating eclipsed his to Tony Gibbon in quantity.

Joe Zeppa and Tony Gibbon met in Louisiana in 1926 when Zeppa was secretary/treasurer of the Natural Gas and Fuel Company, in El Dorado, Arkansas, and Gibbon was managing director of the Urania Petroleum Company, from London. The Urania company had discovered the Urania oil field, the first commercial Eocene Wilcox sand production in this country. As a result of the sale of Urania's properties to Natural Gas and Fuel, Gibbon and Zeppa forged a friendship that was to persist until death parted them.

Gibbon subsequently became an independent producer and drilling contractor. He entered the newspaper business as oil editor for the Houston *Press*, and in 1944 went to Tulsa as Mid-continent editor for Gulf Publishing and *World Oil* until his retirement. In 1952 during a visit to Tyler as Zeppa's house guest, Gibbon told him of a story he had heard in 1919 about pitchblende deposits supposedly on the Island of Tiburón, off the coast of Sonora in western Mexico. Gibbon was well acquainted with Mexico from childhood on, and he had traveled widely over the country. According to the story told Gibbon, a colorful prospector named "Arizona Charlie" Meadows from Yuma, Arizona, had found "a massive outcrop of what he took to be coal" on Tiburón in 1902. When he had his samples assayed they were said to be pitchblende, which then was worth $10,000 a ton for its ore. But Meadows was never able to raise enough capital to mine the ore and died in Arizona in 1932.

Gibbon suggested that Zeppa finance an expedition to Tiburón to search for the site Meadows had found, even though the old prospector had left behind no clear evidence of the pitchblende's location.

"It's an interesting yarn," said Zeppa, "but it sounds crazy to me. However, if you want to go ahead, prepare an outline of your plans. I'll put up the money and see what comes of it."

The only clue Gibbon had to the ore's location was from his memory of the earlier informant's report that it was on the "south end of the island." Gibbon organized a small party for the search and they drove to Hermosillo, Sonora, and from there to the coast where they boarded a boat for the island. After zealously collecting mineral samples, they returned to the States for a petrographic analysis. The reaction was negative. The rock samples were obsidian or volcanic glass, from an old lava flow.

One positive result of the expedition was Gibbon's becoming friends with Roberto Thomson of Hermosillo, who for a number of years had been the Mexican government's agent to the Seri Indians, a small, diminishing group that once had numbered 10,000 on the Island of Tiburón and had shrunk to about 200 on the Sonora mainland. Through Thomson, Zeppa became acquainted with the plight of the Seris and, as a result, periodically made contributions through Thomson to the poverty-stricken tribe.

During the expedition to Tiburón, Gibbon and his associates also met Francisco "Chico" Romero, the chief of the Seris. Romero, as it turned out, saved the life of a member of the expedition, and in return Gibbon informed Zeppa of the event and the fact that Romero had been saving for three years to acquire enough goods to pay the family of his intended bride, Candelaria, for her hand, as was the custom of the tribe. As the party left the Seris, Gibbon slipped some bills into Romero's hand for the expense. His entry in the journal of expenditures read: "Towards the purchase of a wife for Chief Chico Romero $50.00 (U.S.)." It was a huge sum of money for them. That Christmas Chico and Candelaria were married.

The following year Gibbon decided to try once again to locate the elusive mineral on the Island of Tiburón and, though he hoped it would not be sending good money after bad, Zeppa agreed to provide the funds, while declining Gibbon's invitation to join the adventure. A week of work, during which the party diligently and intensively searched the island, failed to produce anything of economic value, and finally Gibbon abandoned the dream.[97]

Then in 1961, Chief Romero lost his wife. Tony Gibbon reported the sad news to Zeppa, with a reminder that the aging Romero now would need another wife. Zeppa sent a check for $50 toward the new expense and commented that "it appears that the customs in the Tiburón area are just the opposite of those here in the United States. There, it is the cost of the acquisition of the wife that is high, whereas, here, the original cost is low and the upkeep high. However, considering Chico's age, it seems to me that $75.00 or $80.00 for a companion and housekeeper (such housekeeping as they may do) is not too much."[98]

Chief Romero, disconsolate over the loss of his beloved Candelaria, however, was unable to find one he wished to marry. The North American contribution found other uses, not difficult to find among the Seris.

The Zeppa-Gibbon friendship continued until Zeppa's death, primarily through correspondence and telephone conversations as the men monitored each other's declining health. Though Gibbon retired while Zeppa worked on, Zeppa periodically proffered advice on how to keep busy. "Continue to work, as that keeps your mind occupied and keeps your body out of trouble," wrote Zeppa in 1974. "It is all right for you to complain about having too much work, but you would complain much more if you did not have any."[99]

XXVI

Zeppa's most voluminous personal correspondence was with George Keating. The two men saw little of each other the last few decades of their lives (Keating died within years of Zeppa), but they regularly exchanged messages by mail and telephone. Keating invested in Zeppa's drilling projects in the 1930s and 1940s, and Zeppa performed a series of professional services for Keating—maintaining trust records, distributing income from partnership wells, investing for Keating (very successfully), and repeatedly advising Keating on a broad spectrum of financial and other matters. Zeppa's cutting Keating into various oil deals made the New Yorker huge sums of money.

Once Keating retired and moved to California, he devoted much of his time to his two great loves—travel and book collecting. One gains the impression, in reviewing the correspondence, that Keating deferred to Zeppa's judgment in all financial decisions. Reading between the lines, one may feel that, as the years stretched on, Zeppa occasionally grew impatient with Keating's seemingly disorganized handling of his resources and, particularly, with Keating's habitual gloom-and-doom expatiations upon political events. Though both men were political and economic conservatives, Keating seemed to perceive a downward slide to the world's political and economic institutions that failed to lather up Zeppa's emotions. From time to time Zeppa would chide his old friend about his pessimistic ways.

"If I were you," he wrote once, "I would read my newspaper with the thought that everything which occurs today will be past tomorrow, and the worries which we entertain daily generally do not amount to very much, especially as we get older. The trouble with you is that you take everything too damned seriously, in the belief that, for some reason, you can rectify all the ills of the country, some real and some imaginary, even though they have always existed." [100]

Surprisingly, in person Keating does not seem to have projected this anger and dissatisfaction toward the modern world that had changed so greatly in his lifetime. In his latter years, he was a roundish, jolly man of five feet eight with silver-grey hair. "Oh, he could discuss any subject under the sun," said Marguerite Marshall. "He would find, in talking to you, what you were interested in, and that is what he would talk about. He was nice to be with and he thought the world revolved around Joe Zeppa." [101]

Taking a perfectionist, almost obsessive, approach to his hobby, Keating acquired several notable collections over the years, which he presented, finally, to public repositories. In 1938 he presented his assemblage of the works of novelist Joseph Conrad to Yale University. The collection contained not only rare first editions inscribed by Conrad but valuable manuscript material, as well. Keating's friend, the writer Ford Madox Ford, termed it a "monumental achievement." In 1948 Keating's musical collection, consisting of 1250 original autographed musical scores and manuscripts, went to Stanford University. In time, he also accumulated rare operatic recordings, 17,000 originals, in which he included, as he called it, "All the Great Names of the Golden Age of Opera." He shared the rarities with friends by playing the records in annual concerts at New York, San Francisco, and Los Altos. This collection went to Yale.

His interest in the history of warfare resulted, first, in putting together a relatively small but valuable collection, "Battles and Leaders of the Civil War," which boasted documents, autographed letters, and autographed photographs. There were the original autographed diary of Admiral Farragut, four autographed letters of Lincoln's, and many important letters of Jeb Stuart, Grant, Lee, Jackson, and others. These treasures became a part of the Huntington Library.[102]

Keating's final large collection was composed of "All the books I can locate in autographed state dealing with two world wars." The volumes represented the great names of both World War I and World War II. "Every volume is a first edition," he said. "Every volume is autographed."[103] He had gone to great effort and expense to have many of the books inscribed. He had been working on the library since the 1920s. In some respects he considered it his most interesting one. The volumes were bound or cased in morocco, with the World War I materials in red, the others in dark blue. In 1954 he had 650 volumes in all; by 1971, additions had pushed the total past 1,000.

By early 1964, Keating had decided it was time to find the appropriate library where the books could be readily available to scholars, and for this purpose he consulted "my oldest and closest friend, Joseph Zeppa, of Tyler, Texas." Keating thought $36,000 would be a fair price for the collection, considering he had spent about $27,000 or more himself. Bindings alone had run from $20 to $50 each, and "ten years from now [it] will be impossible to form, even tho one spent a million or so," he wrote Zeppa. "I am old, I am tired, they are a hazard to fire, water damage, and theft, altho insured could never again be assembled by H. L. Hunt!"[104]

By the end of the year Zeppa had suggested Southern Methodist University as a proper home for the library, and Keating had agreed. Zeppa discussed purchasing the collection and presenting it as a gift to the Dallas school; he broached the matter to Robert M. Trent, director of libraries at SMU, offering to arrange for Trent to go to California and examine the books.[105] With SMU interested in receiving the collection, the two old friends discussed whose name or names should be part of the collection's formal title. Earlier Keating had indicated he would not be adverse to having his name associated, and Zeppa suggested that Keating's name accompany his, Zeppa's. At this point Keating demurred. "It would be an honor," he told Zeppa, "but I feel that as a Texan, a friend of the school, and as deserving some recognition for your many good deeds, it should be called the Joseph Zeppa Collection. Remember I have my name associated at Yale with the Conrad collection, and that is quite enough . . . My career has been uneventful, at least not of general interest, and I really prefer it so. You, too, outside of your oil interests, are not sufficiently known for your good works, and while I know you feel about that much [as] I do myself, I hope, in this case, you yield to my decision."[106]

By 1967, when the Joseph Zeppa Collection of War, Diplomacy and Peace was announced by SMU's Fondren Library, it contained more than 800 volumes, each enclosed in a morocco slipcase, all first editions and inscribed by the authors or important persons. SMU announced that the collection "includes almost every name from President Woodrow Wilson to General [Curtis] LeMay," and "is as fine as anything in private hands."[107]

XXVII

Joe Zeppa's positive traits—fierce determination, powers of intellect, persuasion, unwavering concentration, compulsive honesty, loyalty to friends, and a knack for penetrating analysis—stand out to those who witnessed him in action. Probably these facets of his personality will be remembered long after his observed inconsistancies and sometimes exasperating arbitrariness are forgotten. Many whose lives he touched have even internalized some of those qualities in their own personalities, thus both memorializing the man and lending a touch of immortality to him.

"Very often," said Tommy Blackwell, "I have caught myself weighing a problem, deciding how would I handle this problem. Even unthinkingly, I say to myself, 'How would Joe Zeppa have handled this?' and go through the motions of how I would have sat down and talked to him about it, and he would say, 'Tommy, why don't you do such and such? This will, I believe, handle this problem.' I go through the motions as if it were an interview with him, to see how he would handle this problem.

"He was a great teacher, a father figure, a friend along the way."[108]

In 1976, Delta Drilling Company built the largest land-based drilling rig in the world. Rig #76 went on location that year near Brookhaven, Mississippi.

"Big iron" became familiar to Delta hands throughout the company as 21 new rigs were added to the rig fleet between 1976 and 1980.

In 1956, Delta Marine Drilling Company was formed to provide offshore operations to customers. That same year, the Joseph Zeppa, *a drill tender, was launched.*

The Festival of American Folklife, sponsored by the Smithsonian Institution in October 1978, had an energy theme. Delta sent the Northeast division's Rig #72 to the exposition to give the public an opportunity to view a drilling rig and drilling procedures firsthand.

The Maintenance & Equipment Division underwent a major facelift at its Kilgore, Texas, facilities in 1979. The "zero downtime" maintenance control program which M & E oversees is one of the most innovative maintenance programs in the industry.

Joseph Zeppa, far left, was honored with a party on the occasion of his 80th birthday. Also in attendance were (from left) his son Keating, his brother Chris, and J. M. "Joe" Bevill, corporate secretary/treasurer.

Operations have expanded over the years, and by the end of 1980 Delta had drilling, exploration, land, and production divisions in East Texas, West Texas, South Texas, the Gulf Coast, and the Northeast.

Keating Zeppa became president and chairman of the board of Delta Drilling Company on July 9, 1975, assuming positions his father had held for more than 40 years.

On March 17, 1981, Delta Drilling Company offered its common stock to the public, ending 50 years as a privately-held corporation.

13
CHANGING TIMES

I

Like most changes in our own lives, changes in the petroleum industry came from several directions at roughly the same time. The period from the late 1950s until the watershed Arab oil embargo of 1973-1974 can be viewed as a transition period during which many of these changes began to make their presence felt.

Following the peak drilling year 1957, domestic drilling fell off, leading to long slack periods. The major oil companies went overseas in search of cheaper crude oil. With inexpensive imported oil competing with increasingly more expensive domestic crude, many independent operators and drilling contractors were driven out of business. By then most of the "easy oil" had been found in this country—in the shallow, accessible, and readily drilled fields. One had to go deeper, into rougher terrain, and face severe weather conditions.

How did Delta respond to these early symptoms, largely unrecognized, of later crises? How did Delta approach some of the other changing demands of the times, such as safety and employment practices?

The fact that Delta Drilling Company was a privately owned company, run virtually by one man, might be expected to add some unique twists to a general situation faced by the entire industry.

II

Although Joe Zeppa was usually reluctant to move fast in making changes in his operations, there is no doubt that he was aware of economic trends and had anticipated certain problem areas in his own business. As early as 1962, he summarized, at a convention of landmen, some of the salient trends related to the cost of drilling and completing wells. By then the deeper wells, though constituting the smallest percentage of the total wells drilled, already represented the larger percentage of expense. He noted that while the cost per foot for actual drilling had decreased over the years, other factors were raising costs. One of these was the growing inaccessibility of drilling sites in certain terrain, and he specifically mentioned two areas where Delta was active—the Rocky Mountains and the Louisiana swamps.

"The cost of these operations is not controllable unless one can prevail on the geologists to make their locations close to a highway," he stated. "My experience is that most of them feel that oil or gas can only be found in a very particular point on a block of acreage, and that, generally, the most inaccessible."[1] Although completion methods for many wells would feature increased costs, the majority of wells, he said, continued to require simple completion procedures.

He then addressed himself to the nondrilling costs of gasoline—primarily marketing. At that time, he stated, regular gasoline cost 12 cents a gallon at the refinery in the Mid-continental region. The 12-cent price tag included leasing, geology, drilling, producing, pipelines, refining, and all of the taxes up to that point. Zeppa left in the minds of his audience that at least half of the cost of a gallon of gasoline was added on by advertising, various marketing steps, and service stations before it was delivered to the consumer. He felt the public and political leaders should be more concerned with what took place between the refinery and the service-station purchase than in the earlier steps, for, as he argued, drilling contractors were going to have to increase their charges if they were to continue in business and attract capable personnel.

The remarkable thing about his analysis in 1962 is that, in spite of the drilling industry's being in the doldrums, there was an air of optimism amidst the statistics and facts which, just as easily, could have been interpreted gloomily. One gathers, not from what he says but from the general tone of his paper, that Zeppa believed the malaise of the drilling business at the time could be cured. This view was not much different from the way Zeppa generally approached difficult problems.

III

"Years ago," said Flo Bonham, the veteran division secretary of Delta's Gulf Coast region, "a strong back and no mind and an ability just to 'Yes sir' and tug the forelock and lift that barge and tote that bale—that was all that was required. It is no longer the case, and something that is not peculiar to Delta Drilling Company."

Technology has made oilwell drilling infinitely more complex, but there have been some significant improvements in the lot of labor. Although it still helps to be big, strong, and tough, the importance of brute strength on a drilling rig is not as it once was. "You don't exert yourself as much as you used to," said Bert Gauntt. "I've seen guys years ago get out under a joint of drill pipe and just get in the middle of it and pick it up on their shoulders—500 pounds. That was a pretty good lift. I've seen them do that."

Louis A. "Gus" Mayton, a veteran toolpusher, compared his working conditions in 1980 with those when he started pushing tools in 1956.

"I had a little shack. It was about 8 feet wide and about 18 feet long, with the sides and the ends raised up, and had some screens so the winds would circulate through it. But nowadays you got color TVs, good beds to lay down on, stove and everything—man, the works. We got it all. Everything is electric. Wish they had it then; we'd been in fine shape."

In those earlier days a crisis at the well might keep the toolpusher virtually unable to leave, even to eat.

"When you could have gone, you didn't get to go. You'd better have a can of sardines or something. You might not get to go. Yet we thought we had it pretty good."

Other things have changed little or not at all. One of these is the difficulty outsiders experience in understanding the problems and terminology of the oil patch.

"I've always held to the idea that I work 24 hours a day, and if they need me, I'm available," said Bill Goolsby. "And my wife says, 'You're quite unusual. You never complain. When they call at two in the morning, you answer just as happy as four in the evening, and if you've got to go, you start packing right then.' Probably one of the hardest things: My wife was married to a dentist before, and he never traveled. We'd been married about a month and a half, and they had a fishing job, and they called and they had a problem, wanted me to come out.

"She didn't say anything about that till I got back. Then she said, 'I've got one question to ask you: How come they fish at night and everybody else fishes in the daytime?' That's hard to explain to somebody who doesn't know anything about the oil field!"[2]

IV

While one generation grew older, another generation was coming on, training for and awaiting its mature role. But though Joe Zeppa clung tenaciously onto all his titles and power within the company, his brother Chris decided to retire in 1968. He was old enough to collect his retirement and Social Security, for one thing. "Second," Chris Zeppa said in 1980, "when I retired I felt like, well, I'm in pretty good shape now and besides I want to rest a while. And I retired to make room for the heir to the throne. Incidentally, that played a large part, for there was a youngster coming up: he needed responsibility, he needed to be in a place where he could speak his own word and answer for his own deeds and otherwise come up the ladder to be ready—and he was." He chuckled. "Keat—that's the heir to the throne."[3]

When Chris retired, Keating Zeppa succeeded to his position as executive vice president, at least in theory the manager of the company. By that time Joe Zeppa was 74, ordinarily a ripe retirement age. Whether anybody else suggested to Joe Zeppa that he, too, should step down and take a less active part in business or retire altogether, we have no record, but there is evidence that his old friend George Keating indulged in an uncharacteristic bit of counsel in 1967, not long before Chris retired. Why shorten one's life by working on and on? essentially is what George Keating asked. And then he said:

> . . . You have a good man in Keating; he seems to me to have developed much faster than I thought possible, at least in knowledge of your complicated structure. Give him plenty of his head and less and less think you must personally do everything, you can't, you know by now. I would you take this advise [sic] as much as conditions and your disposition allows, as it meant [sic], I think you admit, from not only only [sic] your oldest friend but your best one too.[4]

During the 1960s, as Keating Zeppa has said, there were several occasions on which he felt frustrated enough in his role with Delta to have gone elsewhere to work, had the right opportunity presented itself at the right time.

"Because Dad was getting old and he was getting, not just set in his ways, but times were changing—it was a difficult time, and it became evident that there had been some judgments made earlier about expansion of the company and this sort of thing, and the results of these decisions were not turning out well. They had turned out fine in the earlier years, but it was time to change again, and I can appreciate his feelings; he was just getting too old and tired to change. He didn't want to have to make changes. I don't blame him a bit. But the natural conflict between the, if you will, inertia of exhausted older age and, you know, the young buck that thinks he knows all the answers."[5]

But as the patriarch declined in vitality, it was impossible for him to maintain as tight a grip on the company as he had earlier.

"In the last few years of Joe Zeppa's life," Bob Waddell said, "he thought he was running things, but there were a lot of things he didn't know about. A lot more going on, and he couldn't keep up with all of that, but he didn't have anything else to do. He couldn't get out and go. He still wanted to know what was going on. He'd wander in my office and sit down. He'd get lonesome. Or he'd have his secretary get hold of me—'Joe Zeppa wants to see you'—and I'd go in there and he just wanted to talk, you know, and he was just lonesome."[6]

"Really before his death, the changes in the guard were beginning," said lawyer Ed Kliewer, "and looking back on it, it was a rather trying time and trying most of all, I'm sure, for Keating.

"As I reflect on it, Keating handled himself well and he ran the company, and he did run it beautifully during those last few years. Mr. Zeppa had had a stroke. He had lost a great deal of the vitality that he had had and yet, unlike the average person, he was still determined to be the master of his own destiny, and that included his company—especially his company—of which he was very proud, and rightly so.

"So that posed a great problem for Keating, and he would be the last person in the world to classify it as that, of course, but the problem's inherent in that situation. As I reflect on it now, he handled it beautifully.

"In his latter days Joe would become somewhat short and, by reason of this illness he had, through no fault of his own, had somewhat a change of personality. But that transition period was a very difficult one, primarily for Keating, and in a secondary way, for everybody else around. But Keating managed that and bridged the gap, and then when the changing of the guard occurred, he had fairly well in place a concept of what he wanted to do with this company, and there wasn't any fumbling around."[7]

Tommy Blackwell, head of the land department, was one who perceived the generations' differences at close range.

"I've told Keating many times, 'Working for your father is the hardest job in the world.' He said, 'Tell me about it.' But he did a good job of it.

"One day Keating said to me, 'Deacon, I think you've got Dad's ear better than anybody else in the company. Would you, if you agree on this project, suggest to him that we do thus and thus?' Then Keating and I went over the project, and I said, 'I agree on that,' and I said, 'Let's do it.' I was highly complimented by that, and I did spend a couple of hours a day with Joe Zeppa in my last few years at Delta and enjoyed it immensely because it was the greatest school anybody could have gone to."

Blackwell interpreted the give-and-take between father and son as generating less friction than others have seen.

"Keating had to have a rare talent to walk this line between performance and checking with his father on things you knew he wanted to have a voice in. At least handling it with me, they did not put themselves into a position to contradict each other. If Keating gave me instructions, Joe Zeppa might have felt differently about it, but when I told him those instructions, he immediately endorsed them. He didn't reverse the things. He respected his son's judgment.

"Keating and I were on Delta's plane one day. I believe we were going to New York. It was just the two of us. It was one of those rare times when you get to talking about these things. I asked him what he had in mind for Delta Drilling Company, and his philosophy was just like Mr. Zeppa's: 'Well, we would like the world headquarters to be in Tyler; we would like to continue what we're doing the same way; and while we may not have the same corporate structure to do so—that is, it may take more people to do the same thing as we grow—I want to be basically honest, have a good reputation, continue the drilling end, and continue to find production.' It sounded like his father had done a good job of instilling in him a set of values. I think it shows today."[8]

V

Delta's retirement program began in 1964. Until that time, the company's Christmas bonus policy was based upon longevity. When the retirement policy was instituted, the money that had been going into the annual bonuses was then diverted into the retirement program.

It was not, however, a policy that came about overnight.

"We'd talked about it for many, many years," retired corporate secretary and vice president Joe Bevill said. "Mr. Zeppa was well aware that other companies the size of this one had long-established retirement programs that were helpful to their older employees, some of whom became disabled. He was always grateful to the employees for their service, and it just broke his heart to see some of them become unable to work, physically.

"A lot of field people, you know, would get hurt and could no longer work, and it just left them with nothing, and he wanted to do something about it. But in all those years we were so busy expanding and developing the company's foreign operations, we never could seem to find the finances available, or the time. We didn't have many people that he could put on it, like we do now, to study the program and do anything about it. But in 1964 it became so acute—I had talked with him so many times about it—that we just sat down one day and he said, 'Well, as a starter let's give up this annual bonus thing'—that was costing $200,000 a year or so—'and put that in the retirement fund.' "[9]

If the retirement program was a landmark fringe benefit for the employees of Delta Drilling Company, the participation plan was right behind it. Some steps in this direction had been taken by permitting employees to purchase stock in Delta Marine and in the Venezuelan company. Both of these operations had lagged considerably behind expectations, but nonetheless Zeppa saw to it that those employee stockholders did not lose any money and, in fact, gained. But these were hardly giant steps in the road toward partial employee ownership. The participation plan, set forth in March, 1974, was a larger, firmer step.

Recognizing that "A company is not only capital but is also people," the plan offered long-standing, loyal employees "an opportunity to participate in the earnings and equity of the company." Eligible would be employees with 15 or more years of service. Not only would they receive dividends as declared by the directors, but in the event of the company's sale, merger, or going public, the employees would receive their accumulated dividends and the value of one share of stock for each unit of participation awarded to each employee.[10]

VI

As a privately-owned company grows larger, with the majority stock in one man's hands, the question inevitably arises of what will happen when that man dies. While there is an expectation that in certain ways management of that company is likely to change, of equal interest is the question of what happens to that dominant stockholder's estate. Because his stock would have increased so much in value since the company's fledgling days, the very serious problem becomes one of finding a way to enable that man's heirs to survive both his death economically and the accompanying tax bill for the estate.

Joe Zeppa's estate and the probable impact of his death upon Delta Drilling Company, his heirs, and the minority owners as well, were matters of concern at least 25 years before he died. In 1950 he was looking into a legal structure that would permit the company to carry on without him. At the crux of the problem was the fact that, even then, estate taxes might reach several million dollars—enough to consume the bulk of his estate and, in the process, wreck Delta Drilling Company. If nothing were done, the estate taxes would climb so high that in order to pay them the heirs would be forced to sell Delta, and under adverse conditions. The problem, then, was liquidity of assets. The heirs would need cash with which to pay the death taxes without jeopardizing the continued existence of Delta.

The book value of Delta in 1950 was around $5 million, but its actual value—which would be determined by appraisers at the time of Zeppa's death—was closer to $20 million. At that time, Zeppa held 58 percent of the stock, probably valued at $12 million. His half of the community property in Delta would be $6 million. Estate taxes and other expenses likely would reach $3.5 million, a very large sum for 1950 or any other time—and an exceedingly large amount of money for his heirs to produce on relatively short notice. This evoked the question: How could some of his holdings be arranged so that they could be converted into cash without harming Delta's operations? Boiled down, the problem involved liquidity of assets.

An estate planner in 1950 recommended that Zeppa increase his company-owned life insurance to $550,000, and his personal insurance to at least $500,000. In addition, a system of trusts would be established which would allow for assets to be passed to the next-generation heirs, with both the elder Zeppas' wills being revised to better fit into the planning. Furthermore, it was recommended that Delta Gulf, which was then a new entity, take out a life-insurance policy of from $500,000 to $1 million on Zeppa's life, because the company's fate depended so integrally upon his personal role. A final suggestion was that Delta be reorganized to divide its assets into two corporations, one for drilling and the other for production. At death the production company, containing the bulk of Delta's oil properties, could be liquidated.[11]

Other possible solutions would include a carefully structured merger with another company or a public offering of stock, so that cash could be readily made available upon Zeppa's death. Certainly, the necessity of approaching the eventuality of Zeppa's death was of concern not only to his friends and associates but to Zeppa himself, as Mark Gardner, a Delta executive in the 1950s, recalled.

"Even Joe had made the statement from time to time," said Gardner, "and Fred Mayer [of Continental-Emsco] would beg me with tears in his eyes to get Joe to do something about that company, in the way of making it public, or something before he died and left all of these problems.

"And I would say, 'Fred, how the hell do you think I can do it when you and Philip Russell [vice president of Metropolitan Life Insurance Company and a close personal friend of Zeppa] and some of his old, old friends can't get him to listen at all? Why would I be able to do it?' "

Gardner then left Delta and went to work for George Bush's Zapata Drilling Company in Houston. As he would meet other executives in the industry, from time to time the question of Joe Zeppa's situation at Delta would come up. Later, Gardner went to work for High Seas, a development company in Houston (which also had contacts with a number of financial institutions, including Drexel, Harriman). One day Kenneth Montague, head of General Crude Oil Corporation, and Gardner discussed the possibility of a deal whereby General Crude could utilize its substantial cash reserves. General Crude was a stockholder in High Seas. It became apparent that a merger, bringing in the cash reserves to go with Delta's production and rigs, could be favorable to all concerned, and at the same time be a pleasant solution to Zeppa's estate problem.

"So I went up and I talked to Joe about it," said Gardner, "and I said, 'Joe, now I don't want to be presumptuous, up here talking to somebody that knows the financial business and has a hell of a company, but here is what I know exists as a possibility.'

"We sat and talked for two hours or more about it, and he said, 'Why don't you just get those Drexel, Harriman people and Ken Montague and we will set a date?'

"We talked a little while longer, and after we talked about all this big deal, all of a sudden he said, 'Mark, you know now, we have got this building.' They had just moved into the Delta Building down there north of the hotel. 'Now we have got this building here and I don't know just what would be done with this building. I just think the headquarters would have to stay right here.' And from then on, the conversation just deteriorated and there was never anything concrete that came about.

"He almost got too far along and made a commitment to at least talk about it, but then he caught himself up and he started throwing road blocks. Of course, he would have been the controlling stockholder, and I think he knew that he ought to, but [he didn't want to share the ownership with any more people]. And you know, very few people in the industry today would have ever known that there were other stockholders in Delta Drilling Company. Because as long as Sam Sklar and Sam Dorfman were there, there was not even any question about what Joe Zeppa did.

"But, anyway, this was a problem that Joe recognized, but he could not bring himself to do anything about it.

"I think Joe knew that I really was sincere when I said, 'I don't want you to think I am presumptuous, Joe; I am a hired hand and have learned a lot from you. My knowledge of this field would be minuscule, compared to yours, and I don't propose to come up here and advise you what to do.' But he knew Fred Mayer had talked to me. It was ripe with possibilities, and he knew it, and we had a fine conversation until he thought, 'Well, now, wait a minute. I have gone too far. I really don't want to talk to these people.' Then he had to back off from it.

"He thought it was reasonable enough. There were all kinds of people making passes at him. Joe Zeppa could have sold out to any number of huge companies. But, you know, Joe wasn't going to buy that. He wasn't going to go that route because then he would be the little fish, but here [the merger deal Gardner talked with him about] was a situation where he is going to end up the big wheel and yet have his business in shape when he died."[12]

Before Joe Zeppa died, at least two other companies expressed interest in buying, or at least buying into, Delta. This came in the 1970s, a few years before the patriarch died. Two of these were Texas Oil and Gas, and W. R. Grace whose TRG is probably one of the top five drilling contractors.[13]

By placing his Delta stock in a trust, however, Zeppa considerably simplified the situation his heirs would face upon his death.

VII

Safety is a topic that everyone acknowledges to be important, particularly in the oil patch where every work day is fraught with physical hazards. A drilling rig's crew may be exposed to fire, falls, poison gas, and heavy iron that can maim or kill. In the early days of the oil industry, preventing injury was generally up to the individual. In the Spindletop boom that opened up at Beaumont, Texas, in 1901, a man who wanted to work had to sign what was known as a "death warrant." He swore that he was white, 21 years old, knew the work was dangerous, and assumed all risks.

This absolved the employer from all responsibility for whatever harm might befall an employee.

Times have changed substantially from those Spindletop days eight decades ago. Yet safety as a special area of concern has had a slow evolution in Delta Drilling Company, as in the industry. As with most other independent contractors, no special steps were taken to ensure safety until after the 1930s. There was no special department or person who supervised and trained personnel in safety procedures. This is not to say that nobody was concerned with safety, for nearly everyone was.

The first officially designated "safety man" for Delta was Robert W. "Bob" Waddell, who had that assignment added to his work as a mud engineer.

"Since I was going to the rigs so much, they just said, 'Well, you might as well be a safety man, too,' " said Waddell. "So, in conjunction with my mud engineering work, I started trying to improve our safety conditions on the rigs, and took part of that."

Safety breaks down into two factors: equipment and personnel. At that time Delta still had mostly steam engines on its rigs and was beginning to acquire power rigs. But Waddell's work as a safety engineer was restricted by the fact that it was a part-time responsibility, necessarily sharing his time with his drilling-fluid duties.[14]

Over the years, safety has improved. "They go out for safety now," said driller J. W. "Turk" Hardin. "It used to be every man had to look out for himself. We never had a safety meeting or safety equipment. A lot of things that were real dangerous, you would just have to walk around them. Pump belts, they never had guards on them, and chains wouldn't have guards on them. There would be holes in the floor that you could step through and they never did put up handrails around the rig floor. A lot of steps didn't even have rails. Derrick men rode blocks up, and when you got up to the board, you would just jump off. They gradually got to changing it."[15]

What were some of the kinds of accidents that occurred in the earlier days?

"Things falling out of a derrick and hitting a man, or you'd be under the structure and bumping your head on the structure, which would be your own fault," said Fax C. Marshall, a drilling superintendent.

When did hard hats come into general use?

"They were here when I came to work," said Marshall. "We had felt hats back in the '30's or '40's. Of course, when I came to work for Delta in the '40's it wasn't a must at that time that you had to have that hat; about everybody was going to it, so my brother-in-law gave me one that he had in the shipyards. I started out with that one. Best I remember, the first outfit I worked for I was wearing an old straw hat. Then when I went to work for Delta the guy that got me the job gave me an old shipyard hat.

"Back then, we could get hurt and that old boy, if he didn't want to lay off, if he wasn't nearly dead, he'd go ahead and we'd bring him out there and we'd do his work for him, what he couldn't do, and he'd bring us water or make out a drilling report or something, and we just kept him and we'd all work together to take care or him. Everybody was like that; of course, we show more lost-time accidents than we did then, because we'd go on out there and we wouldn't ask that doctor if we could go to work; we'd just go get sewed up and go on back to work."[16]

At one point when safety standards were often unspecific, Gordon Burke in South Louisiana instituted a kind of a rudimentary safety belt for his derrick men. This came about after a crewman's fall that, fortunately, was aborted. The man fell from the derrick but managed to catch his grip to prevent a fall to the floor. "That gave me the idea of giving the derrick hand some security up there," said Burke. "So I put the rope on him, tied, which wasn't too good an idea to me, because you were tied stationary. Then later you got the safety belt."[17]

Because life on a drilling rig is an intimate work situation in which one man's life may depend on what another does or does not do, a new man goes through a period of testing.

"It's real dangerous, and you have to bring everybody together," said driller Charlie Butler. "You depend on each other. Everybody is holding each other up. You make a crew. You get a new man, you don't know much about him, and you can't put no real responsibility on him until you give him all these little ol' details. After a couple of weeks, that is when you start putting him on these little details. We call it the shit detail. You'll put him on this job—well, you are not running him off; you got to find out if he will make a hand or not. You have to watch him real close, 'cause he can get killed, get wiped out, just like that, and it's all over. You can tell pretty much whether he will work out or not."[18]

Hugh McKenzie compared the close cooperation on a drilling rig to that of an athletic group. "It's just like a basketball team," he said. "Prairie basketball team. Everybody knows where the other man's supposed to be. Just like you see a lot of pro basketball players throw the ball away. Well, he didn't actually throw it away. A play maker told him what he was going to do and that other fellow just didn't read it, and he wasn't there and the ball went. And that's the same way with making a trip or anything." The analogy is an apt one, down to the number of men on a "team." In basketball, it is five, and also on a rig, though sometimes six men are used there. At one time McKenzie worked 3,300 days without a lost-time accident on his crew. "I'll knock on wood," he said. "I never had a man break a bone and never got a man killed while I was roughnecking and running a rig."[19]

Crew members who are familiar with one another and who have been working together for some time are apt to be safer. As Willis S. Howard said, "They are acquainted and they probably talk to each other more. They discuss safety and they watch after each other drilling. Boys working together as a team. The longer they work together, the less chances they are going to have an accident."[20]

All other things being equal, the longer a crew works together, the safer they are likely to be. South Louisiana drilling superintendent Doris Darbonne cited a personal example to illustrate this. "It seemed like at the time, way back, the older your hands were, the less accidents we had," he said. "For instance, when I first was drilling for Delta, the best I can remember, I had one crew. And I made 900, almost 1,000 working days without a lost-time acident, but I had the same crew. They learned how to work together and watch. One take care of the other. They had more experience. We got one or two rigs that's got old hands. Very seldom we have accidents on them. But when you first start like new rigs, that's where we have a lotta lost-time accidents, hiring new hands. Usually they have mashed fingers, the main one."[21]

The first full-time safety director was Warren A. Hirsch, who joined Delta in September 1951, by which time Delta was completing its twentieth year of existence. He was to remain with the company until his retirement in 1969. Chris Zeppa was the one who had added safety to Waddell's job description, and he also initiated hiring Hirsch.

Hirsch first came into contact with Chris Zeppa in South Arkansas when Hirsch provided safety service for Arkansas Natural Gas Company as the representative of the insurance company carrying Arkansas Natural Gas's liability insurance. At that time, Chris was a plant manager of the compressor station at Louann, Arkansas. At the monthly safety meetings, the field hands and supervisors were called in, usually to hear a lecture from Hirsch.

"Chris was always a very attentive fellow," said Hirsch. "Since it had to be, I guess Chris just decided to get the most out of it. Our meetings in those early days, in the mid-30s, consisted mostly of lectures. And Chris asked questions and showed a lot of interest in safety work."

Hirsch went from that job to another one. In the late 1940s when he heard that Delta Drilling Company might be looking for a safety director, Hirsch's ears perked up.

"The man who had succeeded me in the insurance company came to me one day and he said, 'Warren, did you know that Delta is having problems with their safety program? Bad, bad. And I understand that Chris is thinking about hiring a man.' Then I called Chris, made an appointment and went over to Tyler and interviewed him. What I was trying to do was sell Chris on my services, which he was already partly familiar with from the '30s when he attended safety meetings that I conducted. That helped me.

At least it gave me an entrée to him. I didn't have to come in and show all my medals."

Chris discussed it with Mark Gardner, who approved the hiring, and Hirsch was then with Delta. Chris Zeppa and Gardner wanted Hirsch to live in Houston so that he would be residing closer to the rigs along the Lousiana gulf coast. That division's office was in Houston then. The Hawkins acquisition still operated under the Delta Gulf name, and Hirsch's primary concern at first was the Louisiana Gulf Coast and West Texas, working out of Odessa. At first, with Waddell doubling as a safety man in the East Texas area, Hirsch was serving the regions where Delta Gulf operated. Later, as his duties expanded, Hirsch took over the safety work in the Rocky Mountains division, then the East Texas division and Pennsylvania. Thus he was safety director for all five districts. He added the East Texas district after Waddell became division manager, and he moved to Tyler.

At that time there was no unified safety program for all of the company. "There was some interim field inspection going on, but there wasn't any systematic procedure being followed that I knew anything about," said Hirsch. "So Chris asked me to survey all districts and come back with what I thought might be a workable program.

"We [meaning only Hirsch] were what you might call manager of a one-chair barber shop in those days. We didn't have any assistants. We did all the work by ourselves and utilized the superintendents, managers, and the pushers in every district to carry out or follow the program of safety activity that we tried to set up.

"Safety revolves about two factors: the physical equipment—keeping good, safe physical equipment—and then the maintenance of safety consciousness, an awareness of accident-prevention work on the part of all people. And so, naturally, being alone I had to call on the help of all the district managers to follow the general program that we outlined for safety inspection—routine meetings every month and so on. And I tried to make each district every month. It kept me on the road pretty regularly.

"In other words, I was trying to be an educator to the people in the field and to furnish them with material that they could use in their daily safety work, pretty much as it's still done today—to keep the men on the rigs or in a factory aware of the fact that since they're doing the work, they can, if they will, accept the responsibility for preventing a man from being injured.

"And it was a combination of an educational program plus a routine physical inspection of the equipment to see that we weren't working with beat-up, worn-out, and practically useless, unsafe equipment and material. So what we had to do, since we only made each rig once every month—and we were lucky to do that; really, it was closer to once every two months—we had to depend upon the local management in each district to see that his key people, mostly the pushers, were made aware of what should be done by a physical inspection, what turned up on needed repairs and maintenance, to do it and don't wait for somebody to get severely injured or killed by it."

Delta's poor safety record had become a matter of concern to its management, and Hirsch found it easy to set his goals.

"Back in those days they used figures promulgated by insurance companies called 'frequency and severity.' Frequency is, as the word implies, a great number of them [accidents] clustered together. Severity is the cost factor involved. So, as I understood it when I went to interview with Chris, insurance rates were pretty high as a result of high frequency and severity. In other words, nothing had been done of a systematic nature to more or less control the thinking of the men in the field.

"It was not that the people in the oil patch didn't know as much and more than I did about the physical equipment, but they didn't coordinate the idea of safe working practices with what they knew. In other words, they took chances. Most anybody that's working will never stop to think, unless he's had some training, that 'This is rather an unsafe procedure, and if I don't know what I'm doing, then I'd better go ask my boss.' He'll usually take a chance and very often get hurt as a result of it.

"So, it meant setting up a procedure with the key people, with the superintendents and the pushers, that was the safest practice to follow in rigging up, going in the hole, pulling out, mixing mud, rigging down, and all the rest of the procedure that goes with drilling, so that the people on the rig would have a fair chance of doing what they had to do in a safe manner.

"And, as a matter of fact, what contributed, I guess, to the high frequency in the days before me and some frequency during my period was the fact that there was a rather large labor turnover, and the men didn't have time to learn all the intricacies of their work. And drilling-rig practices—that's an operation where a man can get hurt, very often killed, very quickly if he is a little bit unaware, not necessarily stupid, but if he hasn't been trained to handle his work safely. Because there are big pieces of equipment to handle, there are high pressures to be dealt with, and in later years gases were a big problem that usually meant death if they weren't properly taken care of."

Did the systematic approach to instilling safety in the men work? Hirsch reported that the results justified the effort. Improvement was discernible within two years.

"I didn't keep any records of what the actual drop was in percentage figures or frequency and severity," he said, "but I know it was well enough to gain the favor of the company, Chris and Joe and everybody else, and the accident rate did come down quite nicely."

The safety program, most of all, required coordination of efforts among all who were involved.

"There weren't any hard and fast rules to follow: do this or do that and don't do this and don't do that. It was a gradual steady educational program in the proper method of doing their job, and we leaned heavily on the pushers who were the older men that had come up through the ranks, and the drillers, who taught their crews. We had to do that, because we couldn't simply spread ourselves thin enough to cover, in those days, approximately 40 to 45 rigs scattered through five districts. I was on the road most of the time and spent about ten days or two weeks in every district."

At first the focus of safety meetings was upon the drilling rigs. "They were the hot spots of the industry," said Hirsch. "And then we later spread our wings a little further to include all the other people on the payroll because they were exposed to pumping wells and driving trucks and repair and maintenance and welding, pipelining, and so forth."[22]

Many employees remember that Hirsch was also like a news messenger, passing along information from other divisions while making his rounds through the divisions and rigs, inspecting and making his safety talks.

"Ol' Warren would help you any way in the world he could," said Willis S. Howard, a toolpusher. "He was a wonderful man. The crews, they'd look forward to his visits. He was a good speaker. He could make the hands listen. He'd usually get you, if you possibly could, to shut down a few minutes and let him go over some of the points of safety and this and that and so forth."[23]

Handling poison gas, particularly in the Smackover formation, was a challenge, as far as safety was concerned, that assumed importance in the 1950s. Ed Gandy, former manager of the East Texas Division, remembered his first experience with this gas—before the company had a systematic approach to it.

He had been transferred from North Louisiana in 1949 to push tools on Delta's Rig No. 6 in Camp County, in East Texas.

"We drilled the Chance No. 1 there," said Gandy. "Had we made a well there, I don't know—might not none of us been here, because we didn't know nothing about the Smackover, which is poison gas. We did drill it into the Smackover and we got the smell, as we called it in the Smackover lime, but did not make a well."

By the 1950s Delta was busily drilling Smackover wells and handling the hazards by using gas masks which Safety Director Warren Hirsch had provided and explained.

"In '57 I came back to Texas and drilled a Smackover well up here in Hopkins County," said Gandy. "Sulphur Bluff. There we made a well, but we knew a little more about the Smackover at that time. And we did have gas masks. Warren Hirsch was the safety engineer. He was up there with me on that job, several times. We knew what was in the formation and better how to handle it. On that particular well, of course, we made a well out of it. We did run a drill-stem test, but we wore masks. The boys wore masks on the [rotary] floor, and I had my mask on, of course.

"I was off the floor and on the floor, just trying to keep the [observing] people back. We had quite a crowd there, making a discovery well. Trying to keep people out of there was something else. Landowners—they wanted to see what they had and I kept trying to tell them how dangerous it was around there. If something happened, gas got loose . . . But it's hard to convince them of that. 'If gas gets loose, it will kill every one of you.' But we had it roped off and managed to keep them behind the ropes."[24]

Donald G. Peters, now retired, had a memorable personal encounter with poison, or sour, gas.

"One [experience] that kinda decided me that I didn't need any more oil fielding—that I was almost old enough to get out—was that gas from a Smackover well right here between Tyler and Canton. We had trouble down in the cellar. I had to go down in there and put some connections back together, and I got too much of that gas and fell out down there one night, and that like to got me. You get enough of a load of that stuff in your lungs, it will knock you out. They made you wear a gas mask when you were pulling a test and, you know, you get all of that mud, salt water, and stuff on your mask, you couldn't see, and I couldn't breathe with one of those things hardly.

"That particular time I didn't have anything on; I was down there in that cellar when the connection blew off. It all popped up in my face, you know, and I had to run out from under the cellar. I got out to where it wasn't blowing oil on me, and I come to all right. Boy, it liked to got me. I just sat and got to thinking. My God, if I'd been there just a little longer they'd probably had to drag me out of there."[25]

Though in small quantities hydrogen sulphide gas, or H_2S, has the odor of rotten eggs, the sense of smell cannot be depended upon for detection, because the gas may cause paralysis of the olfactory nerves. This loss of sense of smell can occur within 60 seconds or less at high concentrations, and within 2-15 minutes of exposure at low concentrations. Colorless, heavier than air, the gas is almost as toxic as hydrogen cynanide, and five or six times as toxic as carbon monoxide. Severity of symptoms depends upon the concentration, with coughing, eye burning, and throat irritation coming from low exposure, and death occurring swiftly from high concentrations. The gas can paralyze the respiratory and heart functions. As the company manual warns. "One deep sniff of high concentration can cause death."[26]

Although general awareness of safe procedures might not involve a list of "do-this's" and "do-thats," when it came to sour, or poison, gas in wells, there were a number of specifics to consider. As Delta began drilling Smackover formation wells in the 1950s, sour gas became a very serious problem. Out of this experience and considerable research, Hirsch wrote Delta's first poison gas, or hydrogen sulphide, manual. A pioneer effort, it also was the first such manual in the industry. The revised edition, put out by Hirsch's successor, Bill Goolsby, remains in use to this day.

"In the early days we had certain physical equipment like the Chemox gas mask," said Hirsch. "It had to be demonstrated and shown to the people in the field how to use it. We had to test for the presence of hydrogen sulphide, which is sour gas, and it was usually from what they called the Smackover zone where the sour gases come from.

"Well, when you got into that area, you had to use quite a bit of testing for the presence and concentration of sour gases, so then it became necessary, according to my thinking, to train the people that were working in the sour gas condition how to use the protective equipment, and that was the reason for the creation of the sour gas manual. The field instructions and the equipment to be used with it, we purchased, with the company's blessings, and trained all men in East Texas, South Louisiana, and in Pennsylvania who were exposed to that type of thing."

Delta soon was sharing information on sour gas with the rest of the industry.

"We gave copies to all the major companies that operated," said Hirsch. "In fact, [before this manual] there wasn't very much assembled between one set of covers, in one book. We hounded the Bureau of Mines and all the chemists and everybody we could get information from, on the characteristics of this sour gas."

Sour-gas problems kept Hirsch busy up to the day he retired, January 1, 1969.

"In late December [1968] we had a gas well blow out in Hopkins County, Texas, and it was salty enough to bring Red Adair's [firefighting] crew up from Houston. And he had two key men who went to different jobs when he was busy. One of them was 'Coots' Matthews; one was 'Boots' Hanson. Well, Coots Matthews, during an effort to cap the well, dropped at my feet after a couple of charges of hydrogen sulphide gas. Well, I was close enough to him that we got the inhalator and attached the oxygen mask to it and pulled him out of it, and before the next hour had passed, we had four of our men overcome."

Bill Goolsby, who had been hired about three days before as Hirsch's replacement, had joined Hirsch at the well, to gain some impressions of a problem he was likely to face himself in the future. The gas had caught some of the men by surprise as it escaped from the drill hole and spread out like smog over the area. Though Hirsch had a gas mask, the men working on the blowout had not expected to need masks.

"Evidently they weren't aware of the quantity of gas at the time and the concentration. But we had the equipment to resuscitate them with. I'll always remember that as a sort of going-away party or gift or what-not—the fact that Delta as a company did something specifically to save lives in working with sour gases."

Over the years since Hirsch retired, improved equipment has made the sour-gas procedures even safer.

"The manufacturers—mine-safety and other people making these things—devised better and more comfortable-fitting face pieces and more readily usable packs to go on the back and things like that."[27]

Goolsby's "baptism by fire" as incoming safety director came with the blowout at Birthright, which he observed along with Hirsch. Red Adair by then had looked it over, then left. Coots Matthews, one of Adair's men, was deciding what needed to be done, and Joe Blasingame, Delta's maintenance supervisor, was on the scene to supply the firefighters what they needed.

"Joe was tight as skin on a drum," said Goolsby, "and so Joe was following Coots around, and every time Coots would name something he needed—'We need two 'dozers out here'—well, he'd write that down. 'We need some soft line out here, and we need two or three catlines and we need some half-inch cable.' Well, the next morning at daylight Joe had all that out there. But Coots wanted brand new stuff. New cable, new catline and all, but old Joe went over there and got all used equipment. No new stuff at all. Used boards, used catline. But it was real funny: When ol' Coots got over there, the doggone fire had gone out.

"And Russ Hudeck and Cleo Lowery had to go out there and re-light it. It was night, and they went out there with Roman candles—just real Roman candles, fireworks—to shoot at it and light it, but they didn't know where the gas was. They didn't know which way the wind was from and they just let it blow till daylight till they could see which way the wind was from. They finally got it lit, so that meant we had to put it out. They had to light it because we didn't know how much poison gas was coming out. You want to leave the gas burning so you can see where it is—if it's in a low place. That's one of the things you do, you don't put it out. You just hold the fire away from you and walk in there and close the valve. Then the fire goes out on its own.

"Coots and a bunch of them, they wouldn't wear gas masks. We tried to get them to, but then our men wouldn't wear them. Well, Red Adair and them are kinda macho, and if they didn't want to wear them our boys thought they shouldn't wear them either. Gas overcome ol' Coots Matthews, and he fell down in behind a bulldozer in the track. And Warren had a resuscitator out there, and they grabbed ol' Coots and drug him over there and put it on him, and in the meantime two other boys went down. Two big boys. Delta boys. I saw them go down and I knew they were in that flume, and I went in the edge of the flume and pulled them out.

"I pulled one out, and nobody would help me. So I brought him out where they could get to him with the resuscitator. I knew there was another man in there, but nobody could see him from out there. No fire then. So I went in and drug the second man out, and I'd been in there about half a minute and these people had been in there for 20 minutes before they were overcome, so I knew I could go in there and get them out. So I drug them out, and one of these boys weighed at least 225, and I weighed about 210 then, and I grabbed him by the overalls and I drug him out far enough where they could give him oxygen.

"So we got one of those boys up and he helped us load the other one in the car. We got Coots Matthews up on the first breath, and we took that other boy and he was 'heaving,' and we took him to the hospital, and it was kinda funny—he couldn't burp and he couldn't pass no gas, and if we had known then what we know now, we would have pumped his stomach, but they didn't know to do that then. So he layed there about a day and a half and his stomach swelled with that gas, and finally he passed some gas, he was all right. He had terrific headaches from diesel-fuel fumes.

"After that, Warren retired and we sat on that job 53 days fishing, just to drill another 100 foot for this operator."

Hirsch had already produced his manual on handling sour gases, which Goolsby revised, and it is now used in the drilling industry by a great many companies.

"Now Warren had published a book like that for in-house, and this was an in-house book also, but it grabbed hold. Every time we would take one to Exxon or someone, and we'd give them a copy of it while we were drilling that well, they would ask for additional copies. Initially we gave it to our people and to the operators, and then they started asking for them.

"Then they ran a little article in the trade magazines. Said we had this manual and if they would write to us, we would send them a copy. So, the first printing of like a thousand, it didn't last no time. We're in our fourth or fifth printing. We revised it, and we printed it twice."

What were the major contributions of the sour-gas manual?

"I think the major contribution at that particular time was making the people aware of the hazards. When we first started this thing, people thought you could put a bandanna around your nose and keep the hydrogen sulphide out of your nose, but they didn't realize just how dangerous it was and how deadly it is.

"The next thing was in the 'plan.' How to plan your well and take advantage of the breezes to help dissipate the hydrogen sulphide gas. The gas disperses quickly in the wind and distance, and so you could have a high concentration right here at the wellbore coming out, and ten feet away that concentration fell way down, and 20 feet away, and 50 feet, and 100 feet. We probably drilled as many hydrogen-sulphide wells as anybody, but it was the idea of knowing how to cope with it. It's a heavy gas. It lays out on the ground.

"One thing about it, you do have to watch on still days. It goes downhill and lays in low places. That's the reason in West Texas, every now and then, you hear of a man walking off down behind a rig somewhere and they find him dead later on. He has walked into one of those hydrogen-sulphide gas pockets. So this is the thing we wanted to educate our people and train them in: the use of gas equipment and gas testing equipment. We wanted them to be sure they were aware of the hazards and how to use the equipment."

The Smackover formation tends to run about 11,000 feet deep, though it will vary at different locales.

"Really," said Goolsby, "the control of hydrogen-sulphide drilling is good drilling practice. Men well-trained and good drilling practice. Good supervision, good drilling practices, and training. If you do those things, you won't need these other things. Won't have all these problems."

Labor turnover may have a significant effect on accidents. This was most evident following slack drilling periods. When the work would slow down, men would be laid off; the older experienced ones would be moved down a notch or so, such as a toolpusher becoming a driller, a driller going to roughneck. These changes didn't affect safety much. But when drilling would pick up and these men would be promoted back up again, and with additional jobs opening, new problems would occur.

"When we'd go back to work," said Goolsby, "we'd bring in an influx of new people. Going down didn't hurt; it was going back up [that] our accidents would start. Bringing new people in. Today we've got some smarter people than we've ever had before. Sharp young drillers and toolpushers. But they don't have the experience. Last week we had some drillers in the [training] class that was 22 years old. They are doing a good job, but they don't have the past experience to rely on when something comes up. They can't train people. With maturity comes some leadership ability. The younger man is running into a situation of having to lead people older than he is."

Communication is vital to safety, and the ages of crewmen are crucial factors in that communication, Goolsby believes.

"You take an old boy, he's got a crew that's 22, 23, 24, and we'll say he's about 24 himself. Okay. Then over here on another crew you've got a driller that's 40—if you'll notice, he's got a crew that's 43, 30, 28, and then one at 19. Then you look at another crew, he's 52 and got a motor man that's 45, 19, 27, 28. Okay, we've got a funny situation. The 52-year-old can communicate with his crew 'cause 45, 28, 27 are in the same generation. The 40-year-old, he's got a balanced crew. You move the 24 over here, and you're going to have a turnover. If you move the older driller over to the younger hands, there is a good chance some of them will leave. Just from lack of communication. Generation gap.

"I've asked every driller that comes in, What is the age of your crew? If he's 50, he'll have an older crew. If he's 22, he'll have a younger crew. You surround yourself with people you can communicate with."

Age seems to be the most significant variable in communication, and, therefore, safety problems. Race seems to have no impact. "Really don't see any problems. We don't see any more accidents in blacks than we do in whites." How about mixed crews, or all-white, or all-black? "We don't see that it matters at all."

Of most significance is age, particularly when a young, relatively inexperienced man is promoted to his next position. He will not have the experience required to handle any emergency, because he hasn't had time to learn everything he needs to know in certain situations.

And age affects safety in other ways. As a man reaches his late forties, he faces a possibility of back problems in doing heavy work, oftentimes in cold weather. "He gets to the point where he can't bounce back. Falls are more likely to occur, perhaps with broken bones. Whether he had been with us all his life, or just come with us, he doesn't bounce back from these injuries," said Goolsby.

The most frequent kinds of accidents, said Goolsby, are hand injuries—fingers and the hand. The susceptibility of fingers to the gears and heavy iron around a drilling rig seems to be one constant in the oil patch. A relatively high percentage of the long-time crewmen seem to have a finger or a portion of one missing.

One approach to accident analysis which Goolsby instituted was to use the computer to study the various factors involved.

"What we do is take the lost-time accidents versus the no lost-time accidents. Then we go back and find the injury group that's having the accidents. Then we have it broken down—head, eye, neck, and by parts. Certain time of the year, torso. Insect bites, overheating would come in here. Weather conditions. We put it into a report and send it to the roughneck right out on the rig where he can see it."

Goolsby served as safety director for ten years following his employment in 1968. In 1978 he was assigned as instructor at Delta's training center in Tyler. A result of the reorganization at that time was that Goolsby was not replaced: There was no longer a safety director overseeing it all from Tyler; instead there were ten staff assistants for safety and personnel working for the various division managers. In 1980, however, Allan Collins was named safety director for the Drilling Group.

By 1980 Bill Goolsby had been on the National Safety Council executive committee for about a decade. The committee is made up of safety directors of three of the larger drilling companies and all of the major oil companies, and the 32 members meet three times annually.[28]

VIII

Bill Goolsby considers communication an important part of safety. A test of communication is to follow oral directions to a drilling well, which often is apt to be in the boondocks. Goolsby recalled the time he and several others were trying to find one of the company's rigs.

"Delta was thinking about doing some kind of documentary film some years ago, and we were down in Livingston, Texas, and we talked to a farmer, and he was telling us how to get to the rig. He told us just how to go, and told every curve, every bridge, how far it was between each, and so forth. When he got through and we started to drive off, he said, 'But they moved day before yesterday.' He said, 'I don't know where the hell they moved to!'

"I've had a lot of fun trying to locate rigs," continued Goolsby. "So I came up with a deal where we would have a rig locator that would tell you how to get to a rig, all the people's phone numbers and addresses, and what rig they are on. To me that was a must. A form of communications. Other companies are the same way. I never saw one written until I got over here.

"Between myself and Joe Blasingame, we finally got the thing going. He needed it for his mechanics. The Northeast and the other divisions were not using it, and that is where we really had to work on it. Wyoming was real bad. What would happen out there, you had a fork in the road, and say it was 11 miles to the rig; so you fire a roughneck, he goes up there and takes the rig sign and turns it and points it down the other fork down there. You drive 20 miles; you look at your gas gauge; you're lost; but you still got to go back. Lose a lot of time. Say you had a flat on that road, or it snowed on you."

Yet initially there was resistance to the rig locator system.

"It was another report to write," said Goolsby. "They knew where they were, but nobody else did. They would tell you how to get to it, but you couldn't find it."

The rig locators are keyed to particular landmarks or places, with specific measurements.

"There is one thing you learn in the oil field," added Goolsby. "It don't hurt to ask one more question. I learned that from Red Magner. Red said, 'Always ask one more question.' It pays to know.

"I've got a theory about being lost that I think is real important. You take when an ol' boy gets lost, he's going to the rig, most people start driving slower and slower. Well, if you're lost you might as well drive as fast as you can, 'cause the quicker you find the road mark, the faster you're going to know when to turn around and come back."[29]

IX

A revolutionary change in employment practices occurred in the middle 1960s, when the first blacks were hired on the drilling rigs. Traditionally, black men in the oil patch had been practically nonexistent. Even in the early days, back at Spindletop in the first decade of this century, the only Negroes were those who worked as teamsters or dug earthen pits, and they were soon driven off by whites wanting their jobs.

When Esset Ates went to work for Delta as a janitor in the Kilgore yard right after World War II, he, Sylvester Shelton, and a few others, were the only black employees in the company, and they were restricted to similar low-level service jobs.

"There weren't any colored people on the rigs," said Ates. "All of them that worked out on the rigs were white men."[30]

"I can remember when we didn't have any blacks," said retired assistant East Texas assistant drilling superintendent Z. L. "Nig" Spraggins. "I never did have any problems [when it changed] because I had worked with blacks for quite a while at the mills and so forth, and I didn't see why they wouldn't make good hands. A lot of the boys kinda resented them. Around 1976 I had one [white] driller for a while who had a solid black crew. Then I had a black driller, Charlie Butler, who had a solid white crew. You might think that there would be some friction there, but I never did know of any."[31]

Delta hired its first black roughneck when federal antidiscrimination policies in hiring began to apply pressure to industries, including the oil business.

Hugh McKenzie, now a retired toolpusher, hired the first black man to work on a drilling rig for Delta. McKenzie was a Democrat, something of a rarity in the company, and he suspects that is why he was given the assignment to hire a black roughneck. He didn't mind arguing politics.

"They knew I was a Democrat," he said, "and everybody just gouging you from one end to the other, you know. And I'd fight, boy. I'd have them so mad in that office down there. Man, it just done them good to see me come in there early some morning and set down when there wasn't no rigs running. They'd all begin to gather, and we'd tie up."

But though he backed the Democrats, and a Democratic administration was pushing equal job opportunities for black Americans, McKenzie had never been around black people much at all. No Negro had ever lived in Alba, his little hometown in East Texas.

"I wasn't raised around them. If you know how people was raised up, you can handle that situation pretty good. But, never around a colored. It was pretty rough on me, too, to get used to them."

At that time McKenzie was a driller on a rig near Winkler, Texas, in the Corsicana area, when he was told to hire a black man for his crew. At first he said he wasn't going to go looking for a black man to hire. "Y'all send him out here, I'll work him. I'm not going to hire him myself. First place, I don't know how he's going to get to work."

McKenzie and others were then driving 60 miles to the rig every day, and he didn't expect the other crewmen would want a black man riding with them.

McKenzie kept looking for ways out. He said, "Wait a minute. Y'all got a company car. Let the toolpusher hire him. Safety engineer. They're always around."

But, as it turned out, he was only delaying the inevitable. To recount it in McKenzie's own inimitable words:

"One morning [drilling superintendent] Ed Gandy told me on the radio, he said, 'You go to the telephone and call me.' I was running days and I said, 'Yeah.'

"I got on that telephone and he said, 'My God, you hire a nigger. I don't care if it's a nigger woman!'

"I said, 'Well, I don't know how he's going to get to work.'

"He said, 'Don't worry about that. One up in Emory, wanting to go to work.' "

McKenzie didn't know but one black man in the community where the prospective employee lived, and that was the bootlegger. He mentioned the name to the bootlegger, who advised him to wait a few minutes, as the man would soon be there. When the man drove up, McKenzie was amazed.

"He drove up in one of them Corvettes, four on the floor. Better car than any of us roughnecks had. I was worrying about how he was going to get to work. I said, 'You still want to work for Delta?' And he said, 'No, McKenzie, I hired out day before yesterday at Chance-Vought [an aircraft plant] over there in Greenville.'

"Well, that kind of relieved my mind. I come back in. I radioed in next morning that I tried to hire him, and ol' Bob Waddell said, 'You hire one. Don't you be all day about it!' "

McKenzie headed for a country store in the area and started inquiring about good black workers who might be available to work for Delta around there. The groceryman knew one—"if he'll go to work for you"—though the man had a herd of Brahma cattle in the Trinity River bottoms that kept him busy. McKenzie sent word for the fellow to see him at the rig.

"Well," said McKenzie, "here he drove up in an old pickup. His name was Blue. I said, 'Blue, you want to work?'

" 'Well, I don't know nothing about it.'

"I says, 'We'll teach you.' And I said, 'We want you to be one of us, now. Don't think we're getting you out here to do the dirty work. You won't be asked to do none.'

"Well, sure enough, he came out there. I noticed that day he didn't drink no water. And the next day I said, 'Blue, you bring your own water?'

"He said, 'No, sir, Mr. McKenzie. I didn't bring no water.'

"I said, 'That water can's there and there's a paper cup. That's your water can, same as it is ours. I want you to feel just exactly like one of us boys out here.' And I said, 'We ain't going to ask you to do something that we haven't done, or bully. You just haul in there with the boys, and they'll treat you right. Them's all good boys out there.'

"And he sauntered down there around the light plant and he come back up on the floor and he said, 'That light plant's missing, ain't it?'

"I said, 'Yeah.'

"He said, 'I'll fix that for you.'

"Well, he looked at it and the points was burned up. I said, 'Well, I'll just have to get some points out of Kilgore in the morning. I'll call in and the supply boys will bring us the points.'

"He said, 'I got a set out there in the pickup that'll just fit it.'

"I said, 'Well, if you don't mind using them, put them in there and I'll pay them back.'

"Well, he went and put them in that old thing just as tight. And I bought him six sets of points and give them to him."

On another occasion when the valves stuck on another piece of machinery—the desander—Blue analyzed it, too.

Finally the drilling job at that site was finished. It was time to move on to the next location.

"And, you know, when we tore down, he said, 'Well, I'm sorry I can't go with you boys, but I've got those cattle. Can't nobody see after them but me.' And he just set down and went to crying.

"I said, 'Blue, has somebody mistreated you?'

"He said, 'No, ain't nobody mistreated me.' He said, 'I didn't know white men were this good.'

"And every time one of our rigs went over there for ten years, he's worked for them. And he'd tell them. Oh, he was good enough to tell them not to depend on him if they could get a regular man."

Later, when McKenzie was promoted to toolpusher, he made a policy of keeping one or two black hands on the rig all the time. But when he was drilling in the Como field, "I run out of blacks," as he put it.

"Bob Waddell drove up there at Como one day. By then the Republicans had taken over [the White House]. So ol' Bob, he said, 'Where's your nigger?' I said, 'I ain't got none.' 'Well, hire some.' I said, 'The hell with you, Bob.' I said, 'You Republicans hire them. I hired them when the Democrats was in.' Next morning there was four there."[32]

As toolpusher, McKenzie had "put out" the first black driller for Delta, Finnis Cooper, Jr. Cooper had been roughnecking for one of McKenzie's drillers.

Cooper had worked at a canning factory and a clay company at Lindale, Texas, before he joined Delta in December, 1967. He worked with whites at the clay company, but just as a laborer. A friend from Lindale who had worked on a Delta rig told him about the job.

"At the time we weren't working too much, just a couple of days a week," said Cooper, "and I began to need a little more money. My family was getting a little larger, and my income just wasn't tallying out to my needs. My friend was telling me about how I could make some money, so I thought I'd try. I started out from scratch. I hadn't never seen a drilling rig before.

"It was rough there, though. It was sleeting and snowing the second day I was out. I stayed with it. I worked 15, 16 months, and I got to thinking I was too far from home. I came back and stayed awhile and then I went back and asked for my job back and they give it back to me, and I promised them I'd stay as long as I could."

He started out as a floorman, working with tongs, and later was promoted to motors. He worked motors till he started drilling. His co-workers all were white.

"They treated me all right. Never had a minute's trouble," he said.

He became a driller in 1976 on Rig 22. "They kept after me to drill, so I tried it and I liked it all right. I told 'em I'd try and if I didn't like it I'd like to have my motors job back. I always liked to work motors."

His initial experience as a driller was memorable.

"The first day I run it, I bumped the crown," he said. "I got excited then, and other than that I haven't been too excited. I thought I'd kicked it out of gear, but I must evidently have kicked it out and kicked it back in. I seed what was fixing to happen and I started stopping, but I still had it in before I realized what I was doing."

He attributed the near mishap to his being nervous on his first day.

"Well, I give it up right then. They told me, Naw. One reason, they hadn't showed me how. I didn't hurt anything, but it was good experience, right there. When it happens, the traveling block goes up. I can knock the top out, but I just bumped it. I could have hurt myself and everybody else."

He started out as evening-tour driller; later when Delbert Miller was promoted to toolpusher, Cooper moved into his daylight drilling job. At first he had an all-white crew, and at various times has had all whites on his rig. He reported no serious difficulties.

He drilled in Arkansas, then in East Texas, around Marshall, Trinity, Moscow.

"Well, course, the oil field's been a great pleasure of life to me, and I don't think I'd want anything else."[33]

Since Finnis Cooper, Jr. "broke out" under Hugh McKenzie's tutelage, other black drillers have joined the ranks, and there have been several combinations of mixed crews. White driller L. D. Hurt, for instance, once had an all-black crew at LaSalle, Texas, with no problems; and Charlie Lee Butler, a black driller, much of the time has had an all-white crew.

Butler was 32 years old when he first went to the oil patch after he was laid off his previous job when the business closed. Son of a sharecropper who lived on a white man's farm, Butler left the farm at 18 to go to work in a Tyler foundry as a laborer. He worked his way through the plant in three and a half years, landing a top job there. Of his 14½ years at the foundry he had about ten years' supervisory experience, with 29 people working under him. At a given time he would have around ten whites working for him, which gave him a wealth of experience in working with different types of people.

"I can work with people," said Butler. "I can get along with people. You learn how to handle people. I learned how out at the foundry. It was good experience."

But the work he had been doing in the foundry was nothing like what lay ahead of him in the oil field. Nor did he even suspect he would ever go to work on a rig. However, in August 1974, on his birthday, the foundry closed and he was without a job.

"My wife said, 'What are you going to do now?' I said, 'Well, I'm going to call all my bill collectors now and tell them I don't have a job.'"

Later that year, on the basis of information from a neighbor who was working for Delta, Butler went to see Delta's personnel manager in Tyler. "We need you," he was told, but was cautioned that the crews were working seven days a week and he would be sent to the Beaumont area. In January 1975, Butler struck out to his rig and new job.

"I had seen the oil rigs, but I didn't know nothing about it," said Butler. "I was on Rig 21. I was the only black person. I worked a couple of days and I guess you could say I went to work at the wrong time, because we were tripping pipe, and in and out of the hole. I said, 'Oh, Lord, do we do this all the time?' It was out there, no break or nothing, just grab a sandwich or something and get back to it. I [had] wondered why they called it roughnecking, and I called my wife and I told here, 'Now I know why they call it roughnecking. It's rough down here.'

"Turk Hardin was my driller and he told me, 'Just stay with me, Charlie, and it will get better; when we get all this out of the hole, everything going to be all right.' In a couple of days we got it all out of the hole and everything got all right. We started scrubbing. I said, 'Well, I can call my wife back, 'cause I think I'm going to make it.'"

He not only became a roughneck, he also acquired an enduring appreciation for his driller.

"Turk Hardin—I don't believe there is a prejudiced bone in his body," said Butler. "You can tell it. There is this little feel. But Turk—no way. He would talk to me and tell me just like he did the rest of them. I told my wife, if I ever go back to roughnecking, I want to go back to work for him.''

About 15 months after he had gone to work for Delta, Butler was promoted to a drilling job on Rig 74.

"They say that is the workhorse rig," said Butler. " 'Cause on that little ol' rig you are going full blast all the time. It is a good rig. It makes a hole real fast. You are moving it fast. I worked on that rig for three years. And then [they] sent me to Rig 22 to drill. I told Mr. [Z. L. "Nig"] Spraggins that I wanted off the rig, that I wanted another rig. A bigger rig. I wanted more experience. I called the drilling superintendent, Fax Marshall. He said, 'Well, we don't have an opening right now.' It looked slow for a while, and I was going to quit, and go to hauling logs. My wife didn't want me to quit. And one morning I was laying there in the bed and they called me, said come up to the office. They transferred me down to Rig 90, so this makes the fourth rig I've worked on for Delta. I'm on Rig 90 out of Beaumont. It's a diesel electric; it's beautiful.''

At the time he broke into the oil patch, Butler said, there were few blacks on drilling rigs. "It was kindly tough, color-wise." But since then he has found his racial background to be no barrier. "I don't care what color you are, or nothing. Well, they tell you in the office: Delta has one color, and it's green.''

When he first started drilling, Butler had an all-white crew, and since then, though he had some blacks working under him, he has never had an all-black crew. His problems in dealing with his men have been relatively few.

"We were down near Louisiana. One white man was kinda a problem. He didn't tell me, but he told my hands. One of them came and told me this guy said he didn't like me, said he didn't like working for a black guy. But I knew those weren't the words he said. I said, 'He got to do what I say.' About three months, he was gone.

"And this black guy, he was a problem, a real bad problem. He thought people owed [him] something. I had two blacks, three whites, and he wanted me to work the white guys and let him sit down in the doghouse with me. I said, 'No, you got to work with the other guys.' He said, 'Man, they owe us something.' I said, 'No, they don't owe us nothing. We are out here to do a job. Delta Drilling Company did most of us a favor when they hired us.' And he said, 'You being black and you are a brother and I'm a brother, say, we got this thing together.' I said, 'We got a good job and all we got to do is do this work.' ''

Butler made his point, that the man had to measure up. The man left. [34]

A major reason the hiring policy worked so smoothly is that top management pushed it. Joe Zeppa had long been noted for his interest, albeit paternalistic, in blacks, and Keating Zeppa staunchly supported employing black roughnecks.

Keating Zeppa said, "As I made a point of explaining to our brethren in the drilling contractors association, at a director's meeting of the association, probably back in about 1967 or '68—they were all grumbling about problems of equal employment opportunity and all this sort of thing the government was getting in; I got tired of hearing it—I pointed out to the assembled throng, maybe a little too impassioned, that we had four rigs running down in Brazil and there wasn't a white face in sight, and they could drill rings around these crews up here. So don't tell me that black people can't run a drilling rig. You know, were it a matter of putting them on a plane and bringing them up here and everybody could understand everybody else and there were no immigration problems and all that, I guarantee you we'd have a lot more black people up here. Just because a fellow's black doesn't mean he can't run a drilling rig. Now the fact that he's never seen one before might influence him."[35]

In 1980 Delta had three black drillers in the East Texas division: Charlie Butler, Finnis Cooper, Jr., and Van Turner.[36]

X

An even more revolutionary social change in the oil field has been the hiring of women on the drilling rigs, which also was a result of federal pressure upon the industry. While blacks had to contend with long decades of a whites-only policy on drilling rigs that was industry-wide, nobody ever argued that they weren't strong enough to do the work. On the other hand, most people didn't think women *were* strong enough to handle the heavy work—and most women probably aren't, and wouldn't want to try. Of course, there were other entrenched prejudices against women at the drilling sites. In early days, some workers used to say a woman at the location brought bad luck. Most of all, a woman's presence cramped the men's language, which on a drilling job tended to be decidedly sexually-oriented. Work on an oil well separated the men from the boys—and what would happen to that image if a woman were permitted to join the crew? But no matter what the reasons given, no matter what the real reasons for wanting an all-male crew, the argument inevitably returned to the simple refrain: Women aren't strong enough to do all the work required.

One cannot say with certainty who was the first female roughneck and when she first worked on a rig. H. H. Howell's "Egg Shell Drilling Company" had women working on it.

As T. A. Everett remembered it, "Mr. Howell had a sister. Her husband got killed in a wreck. He was one of the drillers. She had four or five kids. So she went to roughnecking. That was in 1935. She didn't ask nobody for any help. She reached and got every piece of iron just like the men did. I worked with her on three or four wells. Then they drilled a wildcat well up close to Wills Point in Van Zandt County. I went up there and she was a toolpusher on that rig."[37]

A few women had done jobs previously reserved for men, but not on the rigs. In the late 1970s, Delta hired its first geologists other than Jim Ewbank, and one was its first woman geologist, Georgiana Ashford; at one point she served as acting division manager.[38] Sarah Clark, administrative assistant in the East Texas division, doubled as a scout for a year in the late 1950s. Though she did not go into the field, she called all her contacts first thing in the morning and typed up a daily report to Chris Zeppa, explaining what was happening to the different wells being drilled in the region.

However, Claudine Shapley—now Claudine Rogers—received national, perhaps international, attention as the "first lady roughneck in the Western world" when she went to work for Delta in December, 1972. She roughnecked until June, 1973, when she married and quit roughnecking at her husband's request.

Obviously one cannot say that she was, indeed, the first woman in history to roughneck on a well, for certainly during the 1930s, as we have seen, some women did work on drilling rigs. But if we jump to the post World War II era, it is likely that Claudine Shapley is deserving of the distinction bestowed upon her.

At the time Shapley first considered roughnecking, she was a legal secretary for a law firm in Tyler where she had been for eight years. A divorcee, she had a grown daughter.

She first learned such a job was even possible because she knew a Delta toolpusher and a driller and their wives. One evening she was visiting driller Lelon Griggs and his wife Faye while toolpusher Glen Koonce and his wife were there.

"I heard Glen say, 'Hell, I heard we are going to have to start hiring women.'

"I walked in there and said, 'Did I hear you right?'

"He said, 'You sure did.'

"I said, 'Well, I want to be the first lady you hire.' I was just kidding, really kidding.

"So anyway, he went out to the car that night and got me an application blank. Course, nothing else was said. So a few weeks passed by, and I was over at their home again one night and Lelon said, 'Did you ever fill out that application blank?'

"I said, 'No.'

"He said, 'Well, go get it now and fill it out. Let's fill it out and see what happens.' "

She filled it out and Griggs took it to Personnel. Not long afterward she received word that an interview was set up at Delta with Bob Waddell and Homer Stokes, the personnel manager. She had to explain her absence to her present employer, the attorney, and he said, "You have got to be kidding!"

At Delta the interview seemed, to her, to go on and on.

"I told them, 'Now, if I go on a rig and I can't do it, I'll let you know, but if you hire me I'll give it all I've got.' So I guess I convinced them. Anyway, Mr. Waddell said, 'What are you going to do with those long fingernails?' I said, 'Cut them off.' He said, 'What are you going to do with rings on your hand? What are you going to do with long hair?' So then he said, 'What are you going to do when you get your hands mashed and you pull your hand out of a glove and the fingers stays in?' I mashed my hands so many times, when I would pull my hand out of my glove I would think, 'What if my finger is still in there?' I'll never forget it."

That weekend, before she had quit her office job, she went to the drilling site in north Louisiana for a try-out.

"They were rigging up. It was cold, and ice everywhere. I said, 'Well, I'm going to work this weekend and then come back Monday and give notice, and it will be two weeks before I can go.' Mr. [Alex] Beall [her lawyer employer] said, 'I hope you freeze to death down there.' His wife said, 'Alex just can't get over this. He just can't believe it.' So, anyway, I went down there and worked that weekend and I said, 'I'm going to take the job.' I came back and gave them notice, two weeks, and on Friday afternoon I got off and went to work, and worked that night. It was the ninth day of December, and icicles were this long. And I stayed seven months, and I made three wells. I was going to stay a lot longer than that, but I got married."

She worked the 11 p.m. till 7 a.m. shift on Rig 15, near Arcadia, Louisiana. The work was hard, but she took it in stride.

"It was a big challenge to me," she said. "A big challenge. And I thought, 'I'm going to go out there and do it or else.' It was a big challenge. And I loved to work outside, and I like to work at night. I'm a night person."

As hard as the work was, the publicity that resulted from her employment was the toughest part of it for her. "I know I wouldn't have taken the job if I'd known there would have been that much publicity," she said. "It made me sick. I had to take a nerve pill." At the outset, she was featured on a newscast on the Tyler television station, KLTV, utilizing an East Texas site.

"They put me in some men's boots and coveralls, coat, and wrapped my head up and put that ol' hard had on, and Roy Young was toolpusher on that rig. And they told me to take those tongs. Well, I didn't even know what they were. And they were making a trip at the time [at the East Texas well where the interview was filmed]. Never been on an oil well. I didn't even know what I was doing. I didn't even know what an oil derrick was."

When the news got out of "the first lady roughneck," she received letters from people all over the country, attention from magazines, and visits from curious observers.

"The guys would come out on the rig, and I'd hear Lelon say, 'She's around here' or 'She's changing oil in the motors' or this or that. 'There she is' and I would look up and there would be somebody in the doghouse, 'cause they would want to see me work. They would come all the time. They couldn't believe it.

"It was real, real hard work, but I did it. And carrying that pipe is the hardest thing I've ever done. Where you carry it on your shoulders. Course, I had those pads on my shoulders. Carry it up the ravines and woods, laying pipe. Everybody was just carrying one [length of pipe]. They are real long, and one place where they laid that line, it was a mile or two you had to walk."

The most pleasant part of her work was that first payday.

"My first paycheck was so big. You worked seven days and it also had Christmas and New Year's on there. Double time. And I thought, 'Man, this is really living!' All this money. I was going to stay on there. I was going to work and buy me a Cougar and a diamond ring. Pay for it, that was what my joke was."

Because she knew Lelon Griggs and Glen Koonce before she started working, in many ways she had her way paved for her, possibly precluding some of the adverse reactions she might otherwise have experienced in her pioneer role.

"I had it over them, 'cause Lelon told them all that I was a lady and I didn't want to hear any cursing around here, and it was just like a broken record. We just had fun. We just got along great except one man, this first well. This man was kinda kooky. He would not keep his hands off me when we would sit down, and when he would sit down by me he would always slap his hand on me. It would make me so mad. I told Lelon I was going to pick up something and knock him down. I told him to keep his hands to himself. Lelon fired him. Other than that, I had no problems.

"Glen said the rig was cleaner than it ever had been, 'cause I scrubbed that thing out. He said he had the cleanest rig he'd ever had since he went to work for Delta. I said, 'Well, you ought to, you know. That was one of my jobs.' I kept those runs all spotless. He said that doghouse [had] never been that clean."

As a floor hand, she learned, first hand, of the hazards of working on a drilling rig.

"You can get killed real easy. It is dangerous. I got my right hand little finger mashed. I think we were running tubing. It was on a Saturday night, and it was pouring down rain. I remember one of those guys, we would look at each other and start singing, 'In the Good Ol' Summertime.' It was cold. We were running casing, and I got it smashed.

"You always had on gloves. When we would make a trip, sometimes you would just be solid black. I've got some pictures of me where you can't even tell who it was. I wore scarves most of the time. But I always had the hard hat on, and this breast protector thing on, and I wore boots. Steel-toe boots."

The breast protector is specially made from hard plastic, and it fits over a brassiere. "They told me not to go on the [rig] floor without it on." She got a second one when she wore the first one out. The main reason for the breast protector was that it shielded her body where the pipe would hit her while she was making a connection.

"I wore blue jeans, and I wore coveralls, long shirts. Another problem was changing clothes. Glen always changed clothes in his doghouse. That was where I would go to change, and nobody would bother me. See, this was another thing: everybody used the same privy. I never went inside of that. See, it was nighttime. I went down the road. I was always in the dark, so I had no problem with that bathroom. Several asked me about that and I said I just went down the road. Lelon and Ray Amos always told me, said, 'Always let me know when you are going to leave to go to the bathroom.' 'Cause if anything happened to me up there, nobody would know I was missing. I'd just tell one of them. No problem."

Some of the lifting was more than she could handle.

"The mud hopper, those 50- and 100-pound sacks. I could lift the 50, but I could not pick up those 100-pound sacks. They didn't want me to, anyway. Two of us would throw those. Or I would stand at the hopper all night. We really took turns. I'd help one guy lift, and you'd come out just white. All night long. But they didn't fuss about that. They would just work right along with me. We would work like on an assembly line. Then I would take turns on the hopper. And I could do that alone."

There were rewards other than monetary ones, at the end of a shift.

"Daylight was always pretty on the oil rig. I loved that. The sunrise was beautiful."

When she got off work, she'd usually stay up two or three hours before going to bed. Occasionally she'd even go fishing till noon with her elderly landlord, but most of the time she would go home, put on a Tom Jones tape, write letters, take a bath, and fix her hair. Then she'd sleep till around 6 p.m. She wore out a number of Tom Jones tapes. One time she had a week off between locations; another time, ten days, during which she returned to East Texas to visit her parents and friends.

One morning, though, her time was not her own.

"I had just gotten in bed and [the landlady] Mrs. Rich came in there and said, 'I hate to bother you, but Glen and Lelon are out here. You've got to get up. The people from the TV station are going out to the rig and you got to go back there.' I said, 'No, I'm not going back. I'm not going through that again.' I had to go through all that again. Taking my picture and asking all those questions."

But the publicity did have its pleasant side effects.

"No one ever looked down on me. You know, I thought they would. People in Arcadia, why, they accepted me. I hesitated saying anything at the washateria. I know one day I went to this department store, and I was looking around for some clothes. They were quizzing me about where I was from, and the ladies got to talking, and I said, 'Well, I'm a lady roughneck.' She got everybody in the store and told them, 'This is that lady roughneck.' They treated me royally. They didn't look down on me at all. Nobody did. Everybody was so nice . . . well, there was just one lady. She wrote me an awful letter. Just terrible. I told Mother it was no different than working for a lawyer, 'cause I worked with men all the time. Mother said, 'That is right.' No different, just working harder and less time off. But I made more money.

"It was a great experience. It was an experience of a lifetime."

In the summer of 1972, several months before Shapley joined Delta, Trent Zeppa had attended a party where she was. As a result of that gathering, Zeppa was able to quip later, when her change of jobs was publicized, "She was the only roughneck I ever danced with."[39]

The "first lady roughneck" did not go unnoticed in the industry. When the one-man committee of the "coveted and highly sought-after" All-American Wildcatters BFU, or Big Foul-Up, Award was announced for 1973, Joe Zeppa received it. Tongue in cheek, Stephen W. Schneider, the chairman of the one-man committee, noted, regarding Zeppa, in the presentation:

"Actually his past BFUs have been few. A native of Italy, he grew up in New York, fought with the AEF in World War I, and if he had really been smart, he would have invented the pizza pie. But no, he drifted to the oil patch of Arkansas in 1921 and got into the oil game. This was his first BFU. Ten years later, in 1931, he topped his earlier BFU by pulling an even bigger BFU: he went into the drilling business where he was told 'easy money' was to be made. Not knowing any better, he hung in there and built one of the largest and finest drilling companies in the business. To do this he virtually eliminated BFUs—*until* December, 1972, when he unquestionably committed the top BFU of the year and clinched this year's award: He hired the first woman roughneck. This has required that the jargon and verbal communication used on the derrick floor in the past be extensively revised.

"On the plus side, the supply and service companies have no trouble spotting Delta's Rig 15. It is the one with the flower garden around the toolpusher's trailer."[40]

Though Claudine Shapley Rogers was the first female roughneck, she was not the last one for Delta. Early in 1973 Jacqueline Wilson LaCouture went to work on Rig 31 at Montegut, Louisiana, in the Gulf Coast division. She had read a story in her local newspaper about Shapley's roughnecking and decided she would apply at the Delta office in Lafayette.[41] The next year, Ester Owens, a petite 39-year-old black mother of seven, signed on as an operator trainee at the Etexas Producers Gas Plant east of Tyler. A widow, she had never seen a gas plant before, but was soon handling a three-foot wrench and shinnying up the towers like a veteran. Because of her size, her fellow workers followed an old oil-patch custom and gave her a nickname—Half Pint.[42]

One of the early women roughnecks labored in South Louisiana while the late Marcus Black was division manager there.

"She was a rather large young woman, not that tall, but very, very sturdily built," said Flo Bonham, "and [she] was quite at home with the roughnecks in that she was not a shy or retiring person, and you could call her speech very picturesque.

"And what really did get the whole thing off the ground with a bang is that there were some federal regulations that any woman working in such a hazardous occupation as this must wear this chest protector, and so Mr. [Walter L.] McElroy from our purchasing department called down to get the young lady's chest size, which was substantial, and this created quite a little conversation. The nature of the atmosphere is relaxed, to say the least.

"I had to call her for Mr. Black one time and I finally ran her down at her home. She was on days off, and a kid brother answered the phone. The rig was down temporarily and she was very anxious to get back to work, and I was calling her to give her this news, and so the kid brother went screaming through the house, yelling for her, and she finally got to the phone. I was really fond of her. She just absolutely destroyed me, because she was so funny. She came to the phone and I said, 'Babe, did I take you away from something important?' 'Oh,' she said, 'I was out trying to cut the goddamned grass, and I reached over to turn the machine off and grabbed the spark plug and the son-of-a-bitch shocked the shit out of me!' I said, 'Well, that is a shame.' It didn't upset me. I thought it was funny."[43]

XI

One of the most significant moves for Delta during this transitional period was the changeover in the early 1970s, just before the Arab Oil embargo, to diesel-powered rigs. The prime force behind this action was Keating Zeppa, the executive vice president. By moving from gas to diesel rigs, Delta found itself in a strong competitive position when domestic drilling finally began to flourish again in the aftermath of the embargo.

"He was trying to modernize us because we were slipping a little bit behind, I'd say," said Red Magner. "Getting the pension plan, updating the insurance programs, getting in computers, and I'm sure a whole lot of that was his [doing].

"I know one of the first things I was involved [in] with Keating was, he went out and stuck his neck out and said, 'Well, we've got to convert to diesel engines.' And we made a deal, after exhaustive studies, to buy $2.5 million worth of Caterpillar engines—and $2.5 million was a *lot* of money to Delta Drilling Company, more than we'd make in whatever. But we were at a point where gas was getting more expensive and the gas lines were getting more difficult to lay across people's land, all those things. Gasoline engines were not as efficient, took time to lay lines, and all that, so it just looked like that was the way to go. But how do you come up with $2.5 million when you don't even come close to making half that much in a year?

"Anyway, he pushed it and pushed it. Just so happened that it hit just right, the embargo thing, in '73. And the demand for rigs and demand for engines all was created overnight.

"So that was in the early part of '72 or last part, right in there. That was the lowest ebb. We could have actually sold our position with the Caterpillar engines and doubled our money. Which is nice, to have new engines and all that, and so since then we've plowed back into the rigs and equipment a whole lot more than we've made, I'm sure."[44]

Bernie Wolford looked at the new engines from his vantage point as head of maintenance.

"The single most beneficial thing that Delta ever did as a drilling company was when they decided to buy all diesel engines, about '73 or '74. I guess it must have been about '72, this decision was made. They looked and decided to go Caterpillar from past experience, and we made a deal with the Caterpillar factory to buy the engines through their dealers. One rig per month of engines at a discount of ten percent of the list, less two percent for cash, and this contract was signed. Anyway it was all supposed to transpire over a period of about 18 months or two years. Just a few months after that, this oil embargo came along. And when that came along you couldn't buy a Caterpillar engine at any price.

"Well, when this '73 embargo came along, the rig activities got 100 percent, when it had been about 60 percent or 70 percent. They had new equipment coming in, and diesel equipment, and the saving they had on the fuel was actually paying their engine cost. In my opinion it did more for Delta's reputation, besides making money, than any single thing. Everyone goes diesel now. A diesel engine will wear twice as long as a gas engine. You don't have the safety hazards your butane or natural gas has, you know, fire hazard. That diesel costs more, but if you compared the amount to the usage it takes, it would still be more economical."[45]

"They had really started trying to upgrade their rigs and stuff before Joe Zeppa died," said Ted T. Ferguson, retired production superintendent of the West Texas division. "I'll buy that a 110 percent. They were trying to."[46]

When Red Magner returned to Tyler from a stint in the Southeastern Division, Delta was beginning a production and exploration department, which it had never had before.

"We had never had a production and exploration department *per se,*" said Magner, "just had a production department and had a geologist and so forth. So that was the beginning of that type organization. So Keating was structuring it, as difficult as it was. It was difficult to do it and defy his dad. His dad was, I'll say, reluctant to change, and it took some convincing.

"Then all of a sudden the prices changed. Mr. Zeppa couldn't ever envision that oil could be that high priced or gas could be that high priced [as it is now]. I came in one time—I made a contract, just signed it, and he said, 'What did you get for that gas?' I said, '$1.10.' He studied a minute and he said, 'It's just not worth it.' He'd never heard of gas any more than 25 cents. And he just couldn't envision that something could be worth four times that in that short a period of time."[47]

The Arab embargo represented for Delta, then, possibly the most dramatic event that had occurred to the company. The impact, of course, was industry wide. As J. M. "Joe" Bevill put it, the oil crisis of 1973 changed one's viewpoint from "wondering where the next activity would come from, to the broader view of 'Well, this country is going to have to get busy and develop its own energy resources.' Changed our entire way of thinking."

Although there was confusion in the wake of this watershed, the resulting price of oil generated expectations that swelled into a drilling boom in the late 1970s. And overseas, Delta saw an upsurge.

"In Mexico," continued Bevill, "the operation became more profitable, and I would assume that everywhere the company operated, except Venezuela and Argentina, became more active."[48]

XII

In November, 1971, Delta marked its fortieth anniversary, and *Drilling-DCW* noted the event with a feature article on Joe Zeppa, then 78, and his company. By then, the magazine stated, Delta had a total of 77 rigs and 1300 employees. Earlier in the year Zeppa had been honored by the International Petroleum Exposition as the "Grand Old Man of Drilling," and the Texas legislature had similarly honored him. Photographs depicted Zeppa, vice president Keating Zeppa, some operational scenes in Italy and in the States, and "mug" shots of some of the long-time employees who had been with Delta 25 years or more: Dallas Bryant, Auburn Bryant, Miss Tommie Smart, M. W. O'Dell, Hugh McKenzie, D. H. Martin, Robert W. "Bob" Waddell, J. M. "Joe" Bevill, E. M. "Ed" Gandy, Z. L. "Nig" Spraggins, W. O. "Shorty" Meyer, H. J. "Red" Magner, Nathon Bacle, Fax Marshall, Bert Gauntt, Homer Lee Terry, T. A. Everett, Gus Jones, Roy Bentley, Joe Little, Roy Burchfield, W. B. "Jack" Kennedy, Milton Winston, Marguerite Marshall, and Joe F. White. Except for Joe Zeppa, Nathon Bacle had been with the company the longest—38 years at the time. The total rig count came to 42 in the United States (39 land rigs and three marine rigs) and 35 in foreign countries (22 land rigs, three marine, and ten workover rigs). By then, someone had computed Delta had drilled around 7500 wells and 50 million feet of hole.[49]

14

THE DEATH OF A PATRIARCH

I

When the long-time president/founder of a company retires, or dies while still running the business, that event inevitably becomes a historical watershed for the enterprise. This is particularly true when that business happens to be almost totally dominated by that single personality, as Delta Drilling Company was by Joe Zeppa.

By 1975, Delta was approaching this watershed, which was coincident with one of the great watersheds of petroleum history—the Arab oil embargo of 1973-1974. The jolting shock of recognition that the United States was dependent upon others for some of its energy production had sent stunned, often irate, motorists into long gasoline lines where they found not only scarcity but sharply rising prices. Perhaps the majority of Americans did not totally accept the petroleum shortage as a reality, but the crisis held their attention as nothing else, short of war, could do. The times were changing; they were changing faster than anyone had anticipated.

By then the veterans of the Arkansas and East Texas booms were fading away, their numbers dwindling rapidly. H. L. Hunt died in 1974 at age 85. By that time many others had also died, and their contemporaries were thinning out. An entire generation was going, leaving in its wake younger generations to face the new problems of a new era characterized by rapidly depleting resources, steeply rising prices, and confusing economic and political signals.

By 1975, few members of the generation that made the Arkansas and East Texas booms were still active. Those who had not died had, with very few exceptions, retired.

Joe Zeppa was a member of that select minority. At the age of 81 he was still carrying on as the active head of Delta despite the burdens of his age.

II

When Joe Zeppa asked an interviewer in 1975 if he knew a certain old-timer, the interviewer replied in the negative: "That was before my time." Zeppa paused a moment and said, suppressing a sigh, "Well, I guess it would be. You know, the trouble with me, I've outlived everybody I know!"[1]

By 1971, all of Zeppa's original partners in Delta Drilling Company were gone. Sam Gold, of course, had died back in 1935, but the others had survived Gold by decades. Sam Dorfman was killed in an automobile accident in Longview on February 17, 1957; Bob Stacy died on July 2, 1964, in Shreveport.

Sam Sklar died in his sleep on December 16, 1971, at age 81. He had suffered badly from depression in his later years, a condition apparently triggered by his son Fred's death in the Battle of the Bulge in 1944. The elder Sklar never really recovered from the loss.

With Sam Sklar's death, Joe Zeppa was the only Delta founder left.

The old-time employees, representative of those who had provided the labor that got Delta started and who then stuck with the company as long as they could work, were also fading away.

Annie May Jones died on July 3, 1964, at 68. She had been plagued by hardening of the arteries for several years. She had suffered some light strokes and had undergone a series of operations on her arteries and legs. After a final stroke, she lapsed into unconsciousness for weeks until her death.

Her death was almost as much of a landmark event as the demise of one of the original partners. She had been with the company from the first days and was the only employee who ever regularly matched wills with Joe Zeppa. Loyal always, and as hard-working as her boss, she became the living ideal of the dedicated Delta worker.

Joe and Gertrude Zeppa had returned to Tyler from South America the day before Annie May Jones died. A few days later, Joe Zeppa devoted the first section of a letter to George Keating to her death.

> While we all hated to see her go, her physical condition had reached such a state of deterioration that there was no hope of recovery and all she could look forward to was a few months of bed-ridden semi-consciousness. It is a mercy that she did not have to go through prolonged pain.[3]

Other old-timers were disappearing. L. M. "Ark" Carter fell dead of an apparent heart attack while with his family at his cabin on a lake near El Dorado, Arkansas, on August 31, 1971. L. V. "Frenchy" Portier, retired in Tyler, was in poor health. He had stopped attending the annual company party because he was so nervous around crowds. His wife Shorty would always take home a plate of food for him. In 1972, Zeppa sent the old driller a get-well note while he was in intensive care at Mother Frances Hospital in Tyler. The bell soon tolled for Portier, who died at 77 in August, 1974, following a long illness with an enlarged heart.[4]

The 1960s were especially filled with news of deaths of many of Zeppa's close friends. Domenico Cardinal Tardini, secretary of state of the Roman Catholic Church, died at 73 in 1961. Pope John XXIII called him his "strongest and nearest aide." Cardinal Tardini suffered from a heart ailment as well as hardening of the arteries. Tardini had been made pro-secretary of state for extraordinary church affairs by Pope Pius XII in 1952. Pope John named then-Monsignor Tardini pro-secretary of state the day after he was elected pope in 1958, and a few weeks later made Tardini a cardinal.[5] Zeppa had been closely interested in Tardini's home for boys in Rome and had contributed regularly to it. Cardinal Tardini, while still a monsignor, had arranged for the audience with Pope Pius XII for the Zeppas and other couples from Texas.

Miss Helen C. Adams, Zeppa's old friend from his Shearman and Sterling days on Wall Street, had broken a hip in 1960 and become ill and infirm in 1962. Consequently, she had gone to live at the Miriam Osborn Memorial Home for aging women in Rye, New York, an institution she had maintained a vital interest in from 1904 until she became ill. She had held a variety of positions on its board of trustees until 1947, including secretary, treasurer, and executive superintendent. She had retired in 1947. During her years with the Osborn Home she had developed a reputation for great acumen by the way she helped handle the investments and other financial decisions at the Home.

Through the years Joe Zeppa had maintained contact with Miss Adams, sending her flowers at Christmas and on other occasions, and corresponding from time to time.

When she died on March 30, 1967, at age 87, her passing was noted with a story and photograph in *The New York Times*.[6]

Though he dutifully took notice of these accumulating evidences of human fragility, Zeppa watched his generation die off with a philosophical outlook—he kept doing what he had always done, as well as he could. And he advised others, as he did Marvin Williams, president of Spencer Harris Machine Company. Marvin Williams and Joe Blasingame had been close friends. Williams was with Blasingame when the Delta maintenance chief had his heart attack in October, 1970, which proved to be fatal; Williams had rushed Blasingame to the hospital.

"After Joe [Blasingame] passed away, I was talking to Mr. Joe Zeppa, and he told me, he said, 'Marvin, as you continue to get older and older, you are going to lose friends as long as you are still living.' He told me that. That did pacify me a lot."[7]

III

Over the years Joe Zeppa had survived a variety of medical problems. He had his gall bladder removed about 1939, and several years later underwent surgery for kidney stones. Neither of the surgical maneuvers seemed to slow him down or dampen his spirits.

In 1965 he underwent a cataract operation on his right eye. When old friend Eugene Constantin, Jr., sent him a basket of flowers, Zeppa admitted he was "immensely pleased," for he "was not aware that news of my slight indisposition had traveled to the top of the Mercantile Bank Building [in Dallas]." Zeppa rated his progress as satisfactory. "In fact," he concluded, "I am not sick at all, but merely confined to my residence until, probably, the end of next week when the stitch will be removed and I shall be able to use my eyesight for reading and not merely for television."[8]

A powerful precursor of declining health came that same year when Zeppa was 72.

"Dad had a slight stroke in September of 1965, at the farm," said Keating Zeppa. "It was on a Saturday. Ellie and I were in Houston. We got a call from Mother about seven o'clock in the evening that Dad had a slight stroke, that he was home and all right. We came on back early the next morning—couldn't get back any sooner anyway, we had flown there; we didn't have a car we could drive back."[9]

"He had a stroke walking from his house to my house," said farm manager Milton Winston. "They were walking, Mr. and Mrs. Zeppa, up the road right there, and my wife looked out the window and Pat said, 'Milton, come here and look at Mr. Zeppa.' Mrs. Zeppa was holding onto him and he was just kinda fighting her and I went and got him then and brought him up to the house and he didn't know hardly what he was doing. He couldn't say a word. Couldn't talk. He was kinda staggering, coming up the road.

"So I ran and jumped in my car and came on back around in front of my house and put him in and took off to the hospital."[10]

"He had stumbled and fallen and had become somewhat incoherent," said Keating Zeppa, "and for several days did not have full communication capability, couldn't speak too well. Recovered quite well from it, but as a result of that it was determined that the left side of his brain, essentially, was getting no blood. The right carotid artery had become almost completely occluded. Apparently, over a period of many, many years his left ones were completely occluded and inoperable. They said he had been working mainly on the left side of his brain, so shortly thereafter he went to Houston for treatment.

"Jimmy Howell was the surgeon. He cleaned out these arteries and patched them up and what-not. He was in Methodist Hospital, Baylor School of Medicine. Howell is one of Dr. [Michael E.] DeBakey's protégés. Did a good job on him.

"Frankly, at that time—this would have been about December '65—there was some risk associated with this, because the circulatory system was in poor shape. And I think at that time he and Mother did some sort of widespread estate planning and this sort of thing, and I think some papers were probably executed just before he went into surgery.

"Even at that time he was slightly impaired on his right side, as I recall. His signature was very scrawly. As it turns out, of course, the surgery was very successful and he became very active again and probably looked better than he had in a number of years, and he was not left with any permanent damage of any sort.

"But certainly that was the first sign that that was what was gonna carry him off, if you will."[11]

Slightly over a year following his carotid-artery surgery, a checkup revealed marked improvement in most ways, but also some impairment of circulation in an artery going to the back side of the heart. This problem was causing angina pectoris, and he had reported the chest pains to his doctors. He was placed on anticoagulants to help keep his blood from clotting in the heart's coronary arteries.[12]

IV

Though Zeppa had his share of health problems through the years, particularly in the last decade of his life, he not only never dwelt upon them, he tended to deny them.

As his long-time secretary, Marguerite Marshall, expressed it, "He hated the thought of sickness. He would be sitting in his office, just snorting and blowing and his nose would be twice its size and his eyes would be running, but, no, he didn't have a cold! Absolutely refused to acknowledge it, and if anybody in the family was sick you didn't dare ask him about it. You could be concerned if we'd *know* that someone was in the hospital, but he would cut you off very, very short."

"He would!" agreed Grace Baker. "One very cute thing—Marguerite remembers this, too. He had written this letter and this was toward the end of his life. And he said,'If, and when, I should die . . .' "

"He wasn't at all sure he was going to!' said Mrs. Marshall. " 'If and when I die.' "

Grace Baker said, "He *refused* sickness, and I think that was apparent, even to the end of his life."[13]

If his tentative approach to prospects of his demise sounded like the preamble to an "if and when" oil and gas contract, it can be said that he had read a lot of them in his business career, and no doubt found the language appealing enough to apply to his own life.

V

As Zeppa's health began to decline, he and Gertrude sold their home on South Broadway and took a suite at the Blackstone Hotel in downtown Tyler, next door to the Delta Building. A number of elderly people lived in the

Blackstone at the time, and though it was a losing proposition, the hotel was still being operated. There were no Delta offices in the Blackstone at the time, though a few other outfits, such as the East Texas Fair Association, did headquarter there.

And he found more leisure time for one of his chief loves, as he continued to spend his weekends at Pinehurst.

"He was out here [at the farm] more in the last five years before he died than he used to be, because he rested a little bit more," said Milton Winston. "He realized that he wasn't feeling good." [14]

Though age had slowed him down, Zeppa retained his penchant for conversation. One day in the 1970s, Homer Lee Terry came in from the Kilgore yard on business at the corporate offices.

"He met me on the elevator," said Terry, "and he said, 'Come up with me.' I was busy and I had things to do, but still he was president of the company and asked me to come up to his office, so to be courteous I go. So I went in and sat down and talked to him, and he just wanted to visit with me. Well, after he got through talking for a few minutes, he just didn't have anything else to say. I said, 'Mr. Zeppa, have you got anything else to say?' He said, 'No, what have you got to say?' "—Terry laughed—"I said, 'Well, I've said all I'm gonna say.' I felt kinda bad to get up and leave, but he had talked himself out of what he wanted to say. I went out and told Marguerite, 'Well, he got through with me in there. I don't know what he wanted.' She said, 'Homer Lee, he's just lonesome and he wants to talk to somebody.' " [15]

For the last few years of Zeppa's life, his close employees feared the worst, and did what they could to prepare for it. "We had expected it," said Marguerite Marshall. "For years, under all of the telephones, I had a little tape of his doctors' numbers, because he stayed long hours in the office. For a long time I wouldn't go home. I was afraid to leave him there. I knew that some of us would be very nervous and upset if we found him at his desk and something had happened. I pasted his doctors' numbers under all of the telephones so it would be really handy for everyone." [16]

Bob Waddell entertained similar trepidations. "He was old and I had fears all along that probably we might walk in his office and find him dead at his desk, because he had lapses. I'd go in his office sometime and he'd be there with his head down on his arms, you know, and he'd get tired and he'd get sleepy and he'd doze off. I wouldn't have been a bit surprised to find him dead in his office. But even though he'd had those lapses, he'd raise up and go right on with it, just as alert. He didn't miss much. He was very alert, as far as mentally alert, at all times. You'd think that he was missing a lot, but he wasn't missing much." [17]

"I have made presentations to him and he would go to sleep while I was going through a geological explanation," said geologist Jim Ewbank. "He would be sitting there with a cigarette in his hand when he would go to sleep. I would just stop. Then he would wake up in a few seconds and I would just pick up where I had stopped."[18]

His memory sometimes was impaired.

"In his later years," said Marguerite Marshall, "he might forget who lived next door to him and he might forget who came to see him from Houston, and I was supposed to remember all those things, but you put a dollar mark in front of him and he never forgot anything about the company."[19]

His loyal aides usually came to his rescue when his powers failed him.

"The last two years or so, Marguerite's memory was absolutely fantastic in helping him," said Joann Blalock, "because he could remember things 20 years ago just very, very clearly, but he would have his days when he could not remember very well, and people would come in to see him and he would not know them until Marguerite would let him know who they were and what. A lot of these old-timers from other companies just dropped by. They thought Joe would live forever."[20]

VI

In 1971, for the first time since he was a boy fresh from Italy in 1906, Joe Zeppa visited South Glastonbury, Connecticut, where he had picked fruit that long-ago summer. Gertrude's nephew, Larry Barnard, had driven him through parts of New England which he had first seen as an impressionable lad. They purchased some apples and a pumpkin on the excursion. When he returned to Tyler he wrote to thank Jan and Larry Barnard and said, "The country really has changed, except the main highway which was graveled then but now is blacktopped."

The visit to a pleasant childhood scene held sentimental value to him, as was indicated by the fact that he dwelled at some length on it in his 1975 interview.

"So, I was visiting some friends in Boston and I said, 'Let's drive down to Hartford, Connecticut. Let's take a look at the territory that I spent my first summer in. We did. And I recognized one of the old houses, the place where Hale lived—Senator Hale had peach orchards in Georgia, I think, and he got the idea of having peach orchards in Connecticut! And he took a lot of rundown farms—mostly they had grown up in second-growth oak and so forth—and cleared them out and planted peaches. Anyhow, it was amazing. The town hadn't changed at all. Of course, they had a new highway going through there, but the old highway is still there. And the old Victorian house is right on the main drag, [a] great big home. But the place is being developed as a suburban place, you know. It's kind of hilly and they've got beautiful streets and homes laid out."[52]

The vacation apparently whetted his desire to return to New England. In June, 1975, he wrote his old New York friend, Philip A. Russell, who faced an active golf schedule in retirement, "I envy you old people who are retired and have nothing to do except look for a place to amuse yourself." He eschewed any interest in spending any time that summer in New York City, "with its terrific amount of unpaid debts and its difficulties in obtaining refinancing," but he did indicate his travel plans for later that year: "We plan to go up to Boston and then trek through Maine, New Hampshire, and Vermont in the fall when the leaves change color." And if he and Gertrude did that, he advised Russell, he would make a point of visiting with his old friend.[23]

VII

In his latter days Zeppa seemed to be easing his way, almost tentatively, toward a limited retirement. In 1975, for instance, he indicated he was about to step down as regent at East Texas State University.

"Don't think I'll be there after this term," he said, "because, after all, I need to quit working for somebody [else]. I need to take care of myself a little bit."[24]

There was no hint, however, that he was ready yet to phase out his responsibilities at Delta.

On June 13, 1975, Delta's board of directors met in Tyler for a special meeting that, while also reviewing estimated earnings of more than $2 million for the first three months of the year, primarily was spent in discussing the latest chapter in the Delta Marine story. The drilling tender, Delta Marine 8, was under contract to go to Brazil for three years, but the hull had so deteriorated that it would be uneconomical to repair it. It was recommended that another vessel be purchased and converted into a drilling tender. This project would cost $2.5 million or more. Consequently, the directors agreed for Delta to guarantee Delta Marine's $2.5 million loan from Republic National Bank of Dallas.

Also coming before the board was Zeppa's negotiations for another loan with The First National Bank in Dallas for $1.7 million to pay the remaining half of 1974 estimated income taxes, plus the second-quarter installment on the 1975 estimated tax. And as an indication that Delta was continuing to grow, the board authorized the facsimile signature to be used on checks issued for $5,000 or less. The previous limitation had been $1,000.

The only clue that might tip one off to the fact that this board meeting was at all different from any of the hundreds that had been held over the years was Joe Zeppa's signature. It was shakier than in the past. Though readily recognizable as his, it seemed to lack the force that had always characterized his signature.[25]

It was the last Delta board meeting he would attend.

A few days later, another hint came that Zeppa might be gradually relinquishing some of his active business roles: On June 17 he turned in his resignation as an advisory director of the Citizens First National Bank in Tyler. It was not a sudden move, as he himself had suggested an age limitation several years before. Bank president Henry Bell, Jr., read the letter of resignation. Upon a motion by Ralph Spence, seconded by Bob Lake—the son of Pete Lake who had bankrolled H. L. Hunt in 1930 when the East Texas field came in—the resignation was accepted with regret.[26] Zeppa was then 81.

VIII

Whatever clues he may have dropped about reducing some of his outside responsibilities, Joe Zeppa ran Delta Drilling Company to the last moments.

Some shook their heads in doubt, while others defended him. Tommy Blackwell recalled: "A fellow said to me in Dallas, 'Are you sure that a 79-year-old man should be holding the wheel on this ship called Delta Drilling Company and directing it? Don't you think that that's a little too much age to do that?' I said, 'If you think he's weak, why don't you go down there and take something away from him?' I said, 'You'll find that he's more than capable of tackling it.' The man was most capable, at all ages, when he was there."[27]

To the end, Zeppa cherished the work ethic and strove to maintain control over virtually all company policies. Though Keating Zeppa was attempting to exert his influence and management philosophy to the extent he could, he kept seeing his orders countermanded or, at least, resisted. This would occur even on relatively small things, as when Keating told his father, late in the afternoon of July 3, 1975, of a memo he had distributed throughout the company that day.

The Fourth of July that year came on Friday. Keating Zeppa had put out a memo saying the office would be closed for the holiday and Saturday morning as well, for those that might affect.

"You know," said Keating Zeppa, "who's going to come back for one day? They're looking for a long weekend, so we'll just close the office. Well, he just raised hell. 'Goddamn rigs are running out there. What makes these people think they can't work? Well, what do you mean we're going to take the day off? We've got work to do!'

" 'But, Dad, I've already put out the memo. Everybody's going to be gone.'

"This was the day prior to the Fourth, that I put out the memo. He really went home disgruntled on that note. I hadn't given him time to countermand it or anything else before I put it out. He was really upset that we were going to take that sandwich day off. It would have been a three-day weekend. He really didn't like that."[28]

IX

The Fourth of July—despite the contretemps late the afternoon before over Keating's giving the employees Saturday morning off—was a classic, almost idyllic, holiday in the Zeppa fashion at Pinehurst. Trent Zeppa's children were at the farm during the day. Keating, Ellie, and their children came out for the family dinner.

Joe Zeppa's old arguing buddy Sylvester Dayson was at Holly Tree Farm that day. As Suzette Shelmire remembered it, "Joe came over to see us, and I was so glad. I was off on a raft and had to swim and come up on a hill. And I saw Joe every weekend, but there was just something inside of me that made me make that extra effort that day. He looked fine to me. He made his usual comment. 'What do you have on under there?' He would say, 'You girls just run around naked all the time.' And I was so glad I had seen him."[29]

The elder Zeppas had houseguests—Mr. and Mrs. Paul Schultz, he being a landman in Tyler. It was a typical weekend dinner. In addition to the Schultzes and Keating and Ellie and their children, Milton and Pat Winston joined them.

"We all had dinner there," said Milton Winston. "We had been up there [at the Zeppa house] and we had already gone home. Then Pinny—Keating, I call him Pinny—went home. Everyone left but the Schultzes. And I saw Pinny when he passed my house, waved at them, and he hadn't been gone just five minutes by my house when Mrs. Zeppa buzzed me and said, 'Milton, Pops is sick. Can you come up here?'

"And I went up there and he was sitting up in his chair, done pulled his shirt off. Mrs. Schultz was standing behind him, holding his head. And I felt of his pulse and couldn't get any and I slapped him a little bit on the face and talked to him. I said, 'Let's put him on the bed.' And then we put him on the bed and I couldn't get nothing out of him.

"And I gave him mouth-to-mouth resuscitation. When I first started he moved his hand about like that [slightly]. That's all he moved. He went real easy, no struggle.

"It was not long after dinner, because they [Keating Zeppa's family] left pretty soon after dinner.

"Well, after he died we put him on the bed and I told Mrs. Zeppa she might as well call for them to come get him. Mrs. Zeppa called Keating while I was working with him."[30]

Keating Zeppa recalled the sequence of events that evening.

"We'd headed back into Tyler, and I remember we commented, about the time we got into town, at the store on the highway—an ambulance passed us with lights going and everything, headed *out* the highway—and either Ellie or myself, you know: 'Must have been a wreck out there or something.' As it turns out, it was the ambulance to pick him up. But he was dead, of course, when it got there.

"We got home. The phone was ringing, and it was Mother calling, saying that Dad had died. And there was no reason to go back out to the farm for anything, but I did meet her at the hospital.

"And it was just that simple. When we had left, he had walked upstairs, climbed the stairs to go to bed, had undressed and, I think, had started to get in bed, and suffered a heart attack and died. It didn't take long.

"So the cause of his death on July 4, 1975, was basically congestive heart failure, which says the heart just got tired and wouldn't pump. But it had been pumping against some pretty good back pressures and resistance in the circulatory system."

Consistent with his life and mode of living, he hadn't been stricken so that he would linger on in disability and inactivity, as many others before him had suffered.

"For him," said Keating Zeppa, "it would have been unacceptable."[31]

Many saw an appropriateness that blended in with long-established patterns of his life.

"He and the Lord got together and picked the time and the place, didn't they?" said Marguerite Marshall. "After he had had a lovely day and had worked the whole day the day before."[32]

One also might note that, for a man whose patriotism was so deeply ingrained, he could not have selected a more fitting time to die than on his adopted country's birthday.

X

The necessity of notifying all interested persons of the death required compiling a press release and messages the following Saturday morning, ironically countermanding, at least for some of the staff, the day off which Keating Zeppa had ordered over his father's objections.

The press release outlined the dead patriarch's life and achievements following his humble beginnings in Fubine back in 1893. And it noted that at his death, Delta had grown to become "the second-largest drilling company in the world, with more than 1,300 employees operating 83 rigs in six domestic divisions, five foreign countries, an offshore company, and two gas processing plants."

Telegrams were dispatched to friends, business associates, industry organizations, and subsidiaries. Perforadora Central in Mexico City, Delta Overseas Drilling Company in Rome, and the others were notified, along with industry organizations such as the Texas Independent Producers and Royalty Owners Associations (TIPRO) and American Petroleum Institute. The president and chairman of the board of Jarvis Christian College, as well as all those institutions at which Zeppa held an official position, received mailgrams beginning, "Regret to inform you of the death of your friend and co-worker Joseph Zeppa who passed away peacefully at his farm home about 10:30 p.m. July 4 . . . ."

Chris Zeppa informed the family and friends in Italy, simply: "JOE E MORTO VENERDI IM PACE SENZE PENE. LETTERA DOPO FUNERALE." It was signed: STOFU.

Few employees were surprised to hear of his death, for they had monitored his declining health. But invariably they recognized that a personal era had ended.

Marguerite Marshall was talking to her brother, who was visiting her from California, when the phone rang. Joann Blalock told her that Joe Zeppa had died. "I was not surprised," said Mrs. Marshall. "I just wondered what our life would be like from then on, because he had been the center of it for a long time."[33]

XI

By Sunday, the day of the funeral, newspapers were reporting the death, and stories, usually with a photograph, appeared in the *Tyler Morning Telegraph,* Fort Worth *Star-Telegram,* Shreveport *Times,* Dallas *Morning News,* and Houston *Post.* The weekly Tyler *Leader* subsequently published a two-part profile of Zeppa, noting he had climbed up from Hell's Kitchen in New York to build the second-largest drilling company in the world. Industry publications such as *Drilling-DCW* and *Petroleum Independent* followed suit in due time.

The hometown Tyler *Morning Telegraph* reviewed Zeppa's life in a long editorial and pronounced it an uncommon success story.

> Joseph (Joe) Zeppa was one of those rare men who in his lifetime saw his dream—the American dream—fulfilled.
>
> From a rural background in northern Italy, Mr. Zeppa came to the United States at the age of 12 and, during the remaining 69 years of his lifetime, lived an American success story that no fiction writer could pen.
>
> It was a success story composed day-by-day through his dedication, initiative, and ability, plus unlimited hard work and determination.
>
> Every phase of his career obviously was a credit to the individual he was at the time. He fulfilled his U.S. citizenship with the unusual pride and responsibility that is felt by some special non-natives.
>
> By any measure, Joe Zeppa was a man of success. Perhaps the thing he considered most vital to success in anything he undertook was a lot of plain old hard work, and he was still pouring that out up to a matter of hours before he died. [34]

In thanking Calvin Clyde, Jr., at the newspaper for the editorial, Keating Zeppa commented, "The closing sentence indicates that you knew him well, since, in fact, Joe Zeppa never believed that there was any substitute for plain hard work. Fortunately, he also believed that his sons should have the benefit of a good education, since he felt that the combination of the two was unbeatable."[35]

XII

Joe Zeppa had left his widow, Gertrude; his two sons, Trent and Keating; seven grandchildren; and his brother Chris.

The funeral was held the following Sunday afternoon, July 6, at Rose Hill Cemetery. The family requested that no flowers be sent but that memorials, if one wished, might be made to one's favorite charity or to Boys Towns of Italy, Inc.

"And that was it," said Keating Zeppa. "We didn't want a big funeral service. It was a private service at the graveyard. Plain wooden casket. Very close friends. Apart from family, there probably were 15 other people there."[36]

XIII

Zeppa's death had a devastating impact upon his old friend Sylvester Dayson, who was ailing with arteriosclerosis. He grew despondent.

"When he heard that Joe had died, he never got up any more," said Suzette Shelmire. "He cried and cried. He was miserable. Joe was, to him, in better health than he was. I could feel his misery. He felt there was nothing else to live for. His arguments were over."

Dayson remained at Holly Tree Farm, an invalid from hardening of the arteries, with nurses attending him and Suzette driving over from Dallas every weekend. He continued smoking cigarettes, as had Joe Zeppa. "Toward the end," said his daughter, "we were just sure the place was going to burn up. We had a smoke detector over his bed, because it was one cigarette after another." He stayed at Holly Tree until he was taken to Dallas a week before he died of heart disease, three years after Joe Zeppa had.

One day Watson Wise was reminiscing about Joe Zeppa, and he concluded with, "Bless his heart, he was wonderful." He smiled. "He may be listening right now. If he and Papa [Dayson] are up there together right now, there's a lot going on, I'll tell you that!"[37]

XIV

Keating Zeppa, as executive vice president, called a special meeting of the Delta board on the following Wednesday after his father's death. Resolutions of condolence on Zeppa's death were read and passed. A gift of $30,000 was voted to Mrs. Zeppa.

Keating Zeppa reported that his parents had placed all of his father's Delta stock, both Class A Common and Class B Common, in a revocable trust that, upon Joe Zeppa's death, became irrevocable and would be administered as two separate entities, the decedent's trust and the survivor's trust.

To fill the vacancy on the board, Keating nominated his mother, at which event Dr. Sam Y. Dorfman, Jr., expressed his pleasure. She was elected. The election of a new president then came before the board, and Joe Bevill nominated Keating Zeppa, which was seconded by Albert Sklar. Ed Kliewer moved that nominations cease, which Dorfman seconded, and Keating Zeppa was elected president unanimously.

The new president expressed his feeling of responsibility to the board of directors and the stockholders, and asked the board to fix a salary which they believed to be "commensurate with the duties and obligations of the office, permitting the President to pursue corporate activities without undue concern about personal financial matters." They fixed it at $75,000, whereupon the President reported that his father's longstanding personal account with the company would be liquidated "in an orderly manner." There was a brief discussion of specific company business, and the meeting was adjourned.[38]

Keating Zeppa filled in the details of the session that escaped the minute book.

"As executor of the estate I had control of no stock, because the Delta stock that both he and Mother owned was put into a trust some years prior. The co-trustees are Citizens First National Bank, Edward Kliewer, and myself. So the Delta stock was not in his estate. Now for estate-tax purposes, it was included in his estate. But, in fact, as executor I didn't have to worry about the distribution of that stock or the handling of it, as it was already in a trust.

"Upon the death of either Mother or Dad, whichever occurred first, the trust became irrevocable, with respect to all the stock. Mother exercised no fiduciary powers with respect to that trust. She was a beneficiary in the trust for her lifetime.

"So I called a meeting of the board to discuss with them the events. Certainly the minutes of the meeting will not necessarily reflect all of the conversations. At that point the other stockholders really had little choice other than to accede to my recommendations because, number one, they had never been allowed to have any say, really, in what direction the company took. For Dad was a very strong majority of one. You know, they didn't have enough knowledge of the company's operations to say, 'We ought to hire a manager' or 'I want to do it.' There really wasn't any room for any squabbling because of their ignorance in the matter. And I suggested to them that I was prepared to carry on, with their cooperation and their help.

"I didn't make them any guarantees when I took over. I guess as long as you don't have much choice, you might as well enjoy it.

"Really, it was a fairly short meeting. I was elected president. I briefly outlined for them the direction I would like to go, which is, number one, greater board participation. I pointed out to them that the company had a life-insurance policy on Dad's life, but what difference does that make when you're trying to run a company? It was fine for a partnership, with a small organization. I suggested to them, probably at that time, that I felt the best insurance the stockholders had was the bringing in of competent people, management development, organizational development, so that in the event of my untimely demise—my demise will always be untimely, from my point of view!—their best guarantee was that they have some good strong management on board: 'And let's don't go out and buy insurance policies for a million dollars or whatever. Hell, let's spend the money and start building an organization.'

"Whether they believed it then or not, I don't know. But that was my pitch to them at the time. Simply, my pitch was, I wanted greater input and participation by the board. I needed their help. I would bring decisions to the board, not as a matter of form but as a matter of substance. I felt I was working for *all* the stockholders, not just the Zeppa family, and I felt that the company was well-positioned to take advantage of what appeared to be boom times for many years ahead.

"The meeting really was very short. I don't know that we did a lot business. They did not seem to be reluctant in granting to me the powers of the presidency of the company. But it also meant they didn't have to go look for somebody."

Though a logical decision, it was not necessarily a foregone conclusion.

"I didn't feel that it was," said Keating Zeppa. "They didn't know me, really, and they didn't know what was going on in the company, really. I had no track record."

Minority board members at that time were Albert Sklar, Leonard Phillips, Louis Dorfman, and Sam Dorfman, Jr. "Zeppa" places were held by Keating Zeppa, Joe Bevill, Chris Zeppa, Ed Kliewer, and Gertrude Zeppa, replacing her husband.[39]

A new era had begun.

XV

In the next issue of *The Delta Digger,* Keating Zeppa, now filling his father's shoes in both name and in fact, wrote a page in memory.

Mr. Joe Zeppa lived 82 years.
He was Delta Drilling Company for 44 years.

He was my father for 42 years.

Numbers have a certain coldness and finality about them. But there is no thought of coldness nor of finality when considering the life of Joe Zeppa. He was a tough, practical businessman, who believed in hard work and perseverance. He very humanly enjoyed the fruits of his efforts, but was happiest when able to help someone less fortunate than he, and most of those kind and generous acts will be known only to those who benefitted from them.

Although we will no longer hear the familiar "What the hell are YOU doing?" his presence will continue throughout the company, the industry, and his world of friends and acquaintances. From those people have come an outpouring of cards, letters, and tributes to his memory. Thus, there can be no note of finality to the life of Joe Zeppa until the last of those who knew and loved him pass from this life.

His memory will be best honored by all of us in Delta by maintaining the high ideals that he set of strict honesty, complete integrity and plenty of plain hard work.

He may have been in error . . . but he was never in doubt.[40]

For one who was a stylist of the English language, it was a fitting and accurate epitaph.

XVI

Gertrude Zeppa followed her husband in death on February 15, 1979, following declining years that had afflicted her with a variety of medical complications associated with old age—a heart condition, hardening of the arteries, cataracts, and hearing difficulties. In her terminal illness she refused any artificial life support. She insisted on dying with dignity, and at home, and she did.

15

A NEW GENERATION

I

Nations and companies are run by generations. Each generation almost always has its distinctive characteristics as it impresses its goals and style upon society or the business in which it operates. The flux of life being as it is, there is scarcely any way one generation, reared in a different time, with different circumstances, can become a virtual carbon copy of its predecessor. Change inspires, if not demands, individuality. One generation falters, retires, or dies; the next supplants it, sometimes seizing leadership before the old has bowed out.

The new generation, though, is doing more than merely replacing its predecessor. The men and women of each generation, to a certain extent have been shaped by common experiences and mores. The generation that fought World War II is a world apart from those veterans of World War I, and those of the Vietnam era are unlike either of those. Even when the younger cohorts share goals similar to those held by their parents and wish to perpetuate the ideals of their common ancestors, they are living in a different world than the one that spawned their elders and respond in their own ways to the different stimuli.

Delta, like the nation, had been changing in the years before Joe Zeppa died. Although Joe Zeppa was open to change, his attitude was not that of a man a generation younger than he. We tend to live emotionally in the world that first nurtured us, and our perceptions of the present day are often colored by that earlier experience. Those who grew up in the Great Depression have different fears and goals than those who came to maturity during a period of unparalleled affluence.

So these factors which influenced Joe Zeppa had, in turn, a profound effect upon Delta Drilling Company. Delta was changing, but at a slower pace than many of the younger generation, such as Zeppa's son Keating, would have wished.

Joe Zeppa was, after all, nearly 82 when death snatched him from the helm of his company. Very few, if any, companies could say that its president was over 80. This may be a large reason why the changes·which Delta underwent following the patriarch's death would seem all the more striking to those who had been with Delta for decades.

The early signs of far-reaching global ferment that would affect not only the petroleum industry and Delta but the nation as well had all appeared by the time Joe Zeppa died. His death marked the end of an era for Delta, coinciding almost perfectly with the end of an era in petroleum history and the coming of the Age of Energy.

II

Trent Zeppa, from his retirement vantage point in Fort Townsend, Washington, summed up, in 1980, his view of the decisions his brother had made in Delta.

"Keating, you see, has done away with most of the corporate nonsense, and has just been making money like mad."[1]

The Delta minute book for the period from July 1975 through 1980, shows a major clean-up and simplification of the operation. While contract drilling operations continued busily and on a large scale, there was an increased emphasis on exploration and acquiring additional production. As for foreign operations, the tendency has been toward reducing them, with a stated policy of eventual withdrawal from South America. Almost immediately following his accession to the offices of president and chairman of the board, Keating Zeppa took steps to liquidate the company's interests in the Tyler Hotel Corporation and eliminate the drain from Delta Marine, with a long-range goal of liquidating it.

A restructuring of the organization, with new departments and chains of command with designated responsibilities, left the company with three groups—Drilling, Finance, and Exploration and Production. An influx of new personnel swelled the payroll, and new rigs and equipment replaced or were added to existing ones. While the company's debt soared to previously unheard-of levels, so did its net earnings and value.

Much of the marked success of the company during the second half of the '70s could be attributed to the increased prices for petroleum products in the radically changed world market, due not only to the accelerated depletion of these nonrenewable resources but also to the arbitrariness of the Organization of Petroleum Exporting Countries (OPEC) in setting prices. But preparing the company for the new conditions meant investments: in new equipment that could compete favorably with competitors', in trained personnel by offering attractive salaries and wages and incentives, in personnel training for more effective results on drilling rigs, and, as much as possible, in streamlined operations. For all of this, an organization was needed that had not existed, nor been required, before 1975.

The first step toward simplifying Delta's balance sheet and reducing its losses came in August, 1975, when the board discussed closing of the

Blackstone Hotel at the end of the year. By then the Tyler Hotel Company had already lost $113,950, with the total projected loss to reach $200,000 by year end. Mrs. Gertrude Zeppa, a full-time resident of the hotel, told her fellow directors she intended to move into an apartment within a few weeks, and saw no reason to continue spending money on the hotel. Accordingly, the hotel was closed that year, no doubt to the deep satisfaction of all of the directors.[2]

Three years later the shareholders were informed that the Citizens First National Bank of Tyler had verbally agreed to purchase the property owned by the Tyler Hotel Company, including the buildings, for $750,000 cash. Delta would continue to be a tenant in both the Delta Building and the Blackstone Hotel as long as necessary, paying annual rent of $2.50 per square foot of space occupied.

By then Zeppa had decided to move into a new headquarters in the office tower being built a few blocks away on the town square by the Peoples National Bank.[3]

Delta Marine became one of the first targets of medication following the assumption of power by Keating Zeppa. The only functioning portion of its operation was Delta 9, which had been put into operation with an investment of $8 million, to be recovered in three years from a contract with Petrobras in Brazil. By 1976 Delta Marine had no personnel except the people on Delta 9, and its performance was rated as good; the Brazilian operation, of which Delta 9 was a part, was profitable.

By the middle of the following year, Zeppa could report that Delta Marine held the offshore Brazilian drilling record: The firm had drilled more hole, faster and with fewer fishing jobs, at the cheapest rate per foot drilled. The customer, he reported was "completely satisfied."[4] By the next spring (1978), Petrobras was considering purchasing Delta 9, which would enable Delta Marine to be finally liquidated, but in the fall of that year Delta Marine was poised to sign a two-year contract with Petrobras for Delta 9. However, Delta 9, headed for drydock, would require $1.7 million to be ready to carry out the contract.

At his first meeting as chairman, Zeppa announced his intentions of doing everything he could to dispose of the primary assets of Perforaciones Delta in Venezuela, where conditions continued to be poor. In 1975 Delta agreed to organize a joint Venezuelan corporation, Delta Venezolana Compañía Anónima (DELTAVENCA), which would be owned 20 percent by Delta and 80 percent by Fedepetrol, the petroleum workers union. In turn, Delta's Venezuelan company would sell to the new entity its Drilling Barge and Rig 6—a step toward liquidating the Venezuelan operation. Rigs 9 and 26, now surplus, were shipped to the States and sold at auction in Lafayette, Louisiana.

In November 1976, following a fact-finding trip to Venezuela, the new executive vice president Jim Wible reported that Delta could make more money at home than in South America. The company's operations there, he explained, were too small to bring any worthwhile profit. This prompted Zeppa to point out that the company was committed to withdrawal from South America.

In May 1979, the company reached a verbal agreement to assign its stock in Perforaciones Delta and its accounts receivable—amounting to $3,243,153.06, as of December 31, 1978—for $2,750,000 to Anson Drilling Company of Colombia. The sale was completed for that sum in May, 1980.[5]

Business conditions in Argentina remained unstable during this period, and a resolution for this long-time problem child appeared in 1979 when William A. Triest and associates purchased Con-Tra-Pet for $4,498,500, to be paid off by 1981.

Even the Mexican operation, heretofore one of the bright spots in Delta's foreign ventures, was being sold off by 1980. When Zeppa reported the possibility of selling half of Delta's 49 percent interest in Perforadora Central, some of the directors urged him to see if he could sell all of it. In May, 1980, the board approved the sale of half of Delta's interest to its partner, the Álvarez Morphy family.[6]

While Delta was retrenching in South America and even in Mexico, there was slight expansion in Italy in the late 1970s. Delta purchased two existing Italian companies in Rome—Mediofine Compagnia Finanziaria S.p.A. and Saga Petroleum Italia, S.p.A., the latter to be merged into Delta Overseas Drilling Company, which celebrated its twenty-fifth anniversary in 1980 as two new rigs were dispatched there.

Domestically, a consolidation of effort whenever possible, and liquidation of certain operations and equipment, were part of the move to streamline the company's functions. The Southeastern division was discontinued and its rigs placed under the Gulf Coast division. The drilling in that region would then be supervised by the Lafayette office, thus trimming the overhead. The Rocky Mountain division was eliminated when its four small rigs and equipment were sold, and T. C. Carlton took charge of auctioning the property at Casper, Wyoming, in 1976, grossing about $1.7 million.

In 1975 the company plane was sold for $10,000.

"We decided to shut down the Rocky Mountain division and sell it off," said Keating Zeppa, "because we either had to fish or cut bait. It was a small division. Four rigs. They were old. They were little. The overhead was high. We either had to commit a lot of capital to build up that division, or bail out.

"And the Arab oil embargo had occurred three years earlier. Things were beginning to heat up. We were coming into a position where even that old iron would fetch a pretty good price. At that time—I guess it was June, 1976—probably brought the rig count down to 34, and we scrapped out one or two rigs. I would say in July '75, our rig count was at about 34 domestic land rigs."[7]

In 1978, as Delta was expanding and reorganizing, a new drilling division, the South Texas division, was established at Victoria, Texas, on the coast. To get it started, two rigs were transferred from East Texas, two more from South Louisiana, along with new equipment. Eventually, Zeppa said, the company would have ten rigs in South Texas, most of them drilling from 10,000 to 14,000 feet, one or two to 18,000 feet, plus some small rigs.[8]

III

Sam Dorfman, Jr. said that at board meetings, Joe Zeppa "just sat down and sort of ruled the roost. Then he died, and since then we elected Keating to become president, and Keating has been *very* good. He has been fair with everybody. I think that feeling is gone now, the feeling that 'It's us'n against them'n,' which was there when Joe Zeppa was president. Joe Zeppa felt it should have all been his, that he made the whole company and it was his baby. And I don't think Keating feels that way."[9]

Although recognition of a generational conflict is implicit in Dorfman's impressions, the managerial styles of the two Zeppas have been strikingly different. The way Keating Zeppa has invited participation from the board members must have seemed something of a novelty, compared with his father's manner.

"Just to show 'em I really meant what I was saying, I started calling monthly board meetings," said Keating Zeppa, "and they thought this was great—until they realized that these monthly board meetings were getting to be a drag. Cut into their schedule. So we went to the current schedule of bi-monthly board meetings. And probably in the near future we'll go to quarterly meetings and appoint an executive committee that can work in the interim.

"Our board meetings now generally last three to four hours. They are given any information they ask for, more than they want sometimes. Management recommends. We occasionally argue about things, and I don't always win. And that's probably for the better. If I felt it were a critical matter, I would win, but I think it's wise that I don't always win, as long as it's not critical, in my estimation. And I've had good input from them and I've had excellent cooperation and help."

Since dividends were small, the minority stockholders' only income was virtually restricted to their directors fees. When Delta started, fees amounted to $20 per session, then later were raised to $50, and at the time Joe Zeppa died, they were $100 for each infrequent meeting.

Then when Keating Zeppa took over, fees went to $4,800 a year in 1978, and to $6,000 the following year. "But, of course, I asked more from them than Dad did, and I expected them to meet," he said. "I call meetings of the board, and I consult with them on major decisions, on the telephone, and poll them. So if I'm going to take their time, the company should pay for it."[10]

The improved, two-way communication between chairman and board members met with the overwhelming approval of minority stockholders on the board, as did the man responsible.

Although the radically changed conditions in the oil business were huge factors in Delta's growing success in the 1970s and the beginning of the 1980s, there was also the fact that the company was in a position to take advantage of the drilling boom.

"It's grown also because Keating became president and was able to take advantage of opportunities that of late would have been missed, had the old man been there." said Sam Dorfman, Jr. "Oh, they might not have been, you never know about that, but I really believe Keating has done very well for the company."

One of the younger Zeppa's most meaningful acts, as far as minority stockholders were concerned, was to deal with the $342,000 debt his father had run up through his open account with the company. As independent executor of the estates of Joseph and Gertrude Zeppa, Keating Zeppa paid off the sum in December 1980, after which Delta's board adopted a policy prohibiting any such future loans to directors, officers, or employees.[11]

IV

One of the most significant—perhaps scarcely noted or even known by but a few within the company—changes over the years has been the impact of population growth. Just as the nation's numbers have proliferated over a 50-year period, so has the population of Delta shareholders. From the three original stockholders—Zeppa, Stacy, and Dorfman—the number of present-day stockholders may be said to have multiplied almost geometrically. With Joe Zeppa's death in 1975, the original founders were gone, but the second generation remained, plus the third generation of heirs, with a fourth generation to come within the near future. With each generation, the number of divisions of shares and trusts for the various offspring has increased substantially.

From 1931 until 1959 there were never more than three stockholders of record. In 1959, there suddenly were 15 different dividend checks, as stock was divided into Class A and Class B, and trust funds were established for members of all three families—the Zeppas, Sklars, and Dorfmans—officially bringing in, for the first time, members of both the second and third generations. The pattern was to continue. By 1967 the number had reached 49, and the dividends report was prepared by computer.

V

Keating Zeppa, in a 1980 address to an industry conference in Dallas, stipulated what he considered to be strong points of Delta's contract-drilling services. By extracting the major points of his speech, we gain an overview of what he had seen as his goals upon taking over the leadership in 1975. We will then examine what, in fact, did occur.

He listed these positive features as:

1. Superior equipment.
2. Good maintenance program.
3. Superior training of personnel.
4. Good safety record.
5. Diversity of equipment types, both as to geography and to capacities.
6. Diversity of experience and technical skills.
7. Financial stability and strength. [12]

VI

Instead of the slow, methodical expansion that characterized most of Joe Zeppa's incumbency, the period since his death has produced rapid expansion. By 1981 the company had practically doubled its personnel. Two major features of this era were, as Flo Bonham put it, "the advent of so much new blood into the company and the absence of the powerful dominant male figure." [13]

There was, of course, nothing haphazard about what was to ensue. Keating Zeppa had given considerable thought to his plans for growth.

"My perception of the situation was that, although we had a lot of good people in Delta Drilling Company, they had grown up under one man and they were not accustomed to responding in what I would call an organization manner, or structured manner," he said. "They worked for Joe Zeppa; whatever Joe said was fine; that's what they did. But nobody was ever encouraged to think, and there was nothing like participatory decision making. I knew we needed some structure.

"I'd never worked anywhere else either, but one thing that had always stuck with me was how much the U.S. Navy could accomplish, utilizing people who knew nothing. And the secret in the Navy and the military is that you've got a job description, you train a man, you put him in that slot, and say, 'Here's what you're supposed to do. Here are your authorities and your responsibilities and here's how you do it.' And as long as they have a book to go by, they're fine. And if they do well, you know, experience teaches them things that the book can't.

"So, really my feeling was, the first thing we needed was an organization other than the wagon wheel, and we needed something that was a little more modern, something in which you had managers who could manage and who expected to, and which did not require the president to make operating decisions as to whether we should buy ten nuts and bolts or whether we should buy six of them. That just wasn't my style. And I thought we also needed it from the point of view of eventual financial requirements, because I knew, from my limited experience, that banks are very sensitive to organizational capabilities. In other words, they enjoy dealing with Joe Blow, but what happens if Joe Blow isn't there, all of a sudden? Does the company collapse? So this was a need I felt we had to cover, and in casting about and thinking of people whom I believed would be capable of bringing that expertise to the company, one of them that occurred to me as being very good was Jim Wible, whom I had known from my days in Argentina. He was there with Esso. We had sort of kept up and were friends. Once or twice a year we'd hear from each other.

"So after doing some checking out and what not, I got in touch with Jim, and we visited and struck a deal whereby he would come up, work for Delta Drilling Company for a limited period of time. Jim said it was not his intention to die in Delta; there were other things he wanted to do. And I told Jim what we needed and what we expected to accomplish, and once that was done, then he should feel free to go on his way, if he wanted to. That was the deal.

"He came in September of '76. So that was really the beginning of the restructuring of Delta. The game plan was to essentially bring Delta into a posture of being more and more an exploration and production company. Not doing away with the drilling, but the balance has always been 90 percent drilling and 10 percent production. And from what we thought would be true in the future—well, my point very simply has been: for the foreseeable future, domestic demand for oil and gas will exceed domestic supply. A very simplistic analysis says that's where you need to be—on the supply side of that equation.

"So the idea then was to bring up the E&P [exploration and production] side of the business rapidly, faster than drilling. But that takes time, to lay in people and to get them functioning and to start exploring, so in the interim—and we were earning good money, we had good cash flow, and we had no debt essentially, so we had plenty of borrowing capacity—the place to put money to work was in the drilling rigs. Because we could relatively quickly buy new rigs, get them staffed, and get them out and generating cash flow, and hopefully by the time we had finished that, then their capital requirements would be greatly reduced, and by that time we would have the E&P effort coming along and it would need that capital and be able to use it intelligently.

"And that basically is the way it has worked out. The drilling group—their investment at this point, new rigs, and what not, is leveling off now. I'm not saying we may not add a rig here and there. But the big buildup is over. So now they are a large net contributor to the corporate cash flow. Meanwhile, exploration and production are coming along nicely, and we will be able to use all of that cash flow, plus borrow some more.

"I'm not saying it has just been a bed of roses. It hasn't been. It's diffi-cult. It has been particularly difficult for all of us old-timers around here, because it has been a wrenching change from the way things used to be, and a lot of people haven't been comfortable. Very few have chosen to quit, and probably no greater number have been asked to. You know, there hasn't been a mass bloodletting or anything like that. But it has been diffi-cult for all of us."[14]

James L. Wible, who served as executive vice president until May, 1980, was a former Exxon executive, and attracted a number of former Exxon people to Delta. In February 1977, Wible and a management consulting firm presented an organizational chart designed to reflect Delta's new look. The same month Wible held his first meeting of the executives from the different divisions, at a motel near the Dallas-Fort Worth airport. As Thornton Tarvin put it, "He was there a few months getting his feet on the ground, and when he did get 'em, it started."[15]

One sure sign of the expansion was that within a few years, the 77 names on the confidential payroll in Tyler had grown to 146, just about double the old figure.

During the height of the expansion, as Wible reorganized and added new employees, the effect was sometimes breath-taking. "They started gradually," said Flo Bonham, who viewed it from Lafayette, "and began to pick up momentum and finally it got to the point that it seemed twice a week you were being advised that you had a new department, and here were six new people which you had never head of. Well, you hadn't digested the last one yet, and here you would go. We couldn't even keep the scorecard up-to-date."[16]

There suddenly was a mix of the new and the old, with most employees in 1980 having been there less than five years.

"I guess if you had to characterize Delta's employees," said Merrel Greenwell, himself one of the new executives, "you had three groups: those in the 20- to 30-year range of service; those in the 20- to 30-day service; and there's a few in that 5- to 19-year [service], I guess."

And as new employees began swarming in, many of them from Exxon, Texaco, Hunt Oil, and other large companies, the older Delta employees began to see a pattern. "They're trying to make Exxon out of Delta," one said. "It's a bigger company today than it was five years ago. Maybe everything's for the better. I don't know."

Different "company vocabularies" played a part in some skewed communications during the expansion in the late 1970s. As Merrel Greenwell put it, "You bring in new people that have had their training in some other organization—I don't care whose it is, and when they say something, they mean something, but they're saying something from their own experience, and me, coming from another organization, I hear those words, and it means something else to me. It takes time to develop a company vocabulary; I don't care who it is. If it's an old organization that's been going 30 years, you have this vocabulary; but if you take the organization and double the number of the people, you don't have that vocabulary. And it takes time to do it. Joe Blow has been here for 20 years, and he has a Delta vocabulary. Jack comes in from John Smith's organization with 20 years experience—he's got a different vocabulary, and it takes time." [17]

Some of the responsibility for better communication fell upon the public-relations department, itself an expanding section. Originally a one-person operation headed, for several years, by Doris Kivel, public relations has been run since 1978 by Gloria Stackpole, who replaced Kivel. While continuing to publish *The Delta Digger,* the company magazine, Stackpole's staff by 1980 had more than doubled, parallel to the overall growth, and a separate publications staff had been established within the department. The *Digger* was to become a handy instrument for improving communication within the company.

Among the new programs was one setting up an employee feedback program. "For so many years the employees in the field have felt left out—that they were sort of the stepchildren and everything went on in the office," said Gloria Stackpole. "So we have some programs set up to make them feel they *are* Delta Drilling Company, because everybody who works in this corporate office gets paid because of those guys in the field."

For the magazine, the staff developed on-location stories. With the company's emphasis on planning, efforts were made to let both employees and the various divisions know what these plans were. [18]

Long-time employees once felt that Delta had been like a family. Is it still a family?

"It was," said T. A. Everett. "I'll put it that way. It's no more."

When did it stop being a family?

"When Joe Zeppa died." [19]

Most employees believed Delta's "family" feeling had become a victim of the times and fast growth, not necessarily peculiar to Delta alone. With the large growth especially in the corporate office in Tyler, many of those in the division offices tended to feel less of a personal relationship with many of the new personnel, partly because there were so many, and partly because the division personnel visited Tyler infrequently, if at all.

Promotions were likely to come with dizzying speed, compared with the slow progress a man made up the ladder in pre-World War II days. "You just about had to wait until somebody died before you could be promoted to a driller," said Nig Spraggins. "Now we have expansion, new rigs, and we have to supply our need for drillers. All the time we are working with personnel and trying to teach them how to be operators or drillers.

"I will say for Delta that they have always tried to use their personnel. In other words, promote from within. Then when you get all your qualified drillers and make pushers out of them, then you have to replace all of them; then you get your qualified roughnecks and move them into a driller's job. Well, that don't leave you very many qualified roughnecks."[20]

Paper work—reports, statistical data, invoices—have proliferated over the years, and the burden has fallen upon division and central offices alike. "We are doing things in this office now that used to be handled by Tyler," said Flo Bonham. "Like coding of invoices—the numbers we get are staggering. It is mind-boggling. A few years ago there was a small bit of coding that would be done at this level by a clerk. It would be approved and sent to Tyler. Now it is done in a more detailed fashion, all put on a computer. Tons of material that we used to send to Tyler is being paid from here in this invoicing area."[21]

Growth can be measured by several gauges.

"We checked the mail that we sent out yesterday," said Grace Baker in 1980, "and we had run 1,600 pieces of mail through. When Arthur Clark came in with the mail from the post office, he had three large mailbags full. Not more than three years ago, we would buy $500 worth of postage for our postage meter. I believe that would last for two weeks. We had one day recently where we ran $600 on the meter. So now, instead of buying $500 we buy $2500. If you compare, two days' mail output is as much as we did in two weeks."[22]

"There are so many people now that you don't know all of them," said Avis Boren. "It is amazing how many I don't even know. When I meet somebody in the elevator now, I hesitate to say, 'Well, do you work here or are you just a visitor?' I don't know whether to introduce myself or not."[23]

The rapid growth, coupled with the previously unheard-of accumulation of debt as the result of buying new rigs and equipment, etched deep worry lines into the brows of many of the older Delta employees and retirees. Used to the slow growth that had been characteristic of Joe Zeppa's management, some feared that the company would become overextended.

In 1980, Ed Kliewer, a member of the board and a long-time Delta watcher, assessed such views: "Well, if you lay aside that part of their concern, it really results from the fact that the company isn't being run the way it was. After all, you get used to something over 30 years. And if somebody else comes in and completely reverses the trend, you're not going like it. Any time you reverse that trend and change the course of action, there's reason to be concerned. Because you're going off in essentially uncharted waters. And so to a degree there is merit in the thought that if you're going to change this company, you'd better be damn careful how you do it, because you're playing with a porcupine.

"Now I've watched this and I think that this change—and it is a change, God knows it is—is being handled in a very logical way. I think it's being handled competently. Now there's a lot of people getting stepped on, and a lot of people don't like it, but that's just the nature of the beast; and I think the financial results are indicative of at least the beginning success of this effort."[24]

Though the wagon-wheel structure may be said to have ended with Joe Zeppa's death, in a sense the hub concept is not gone. Each division, to a certain extent, is autonomous but answerable to the central management in Tyler, so the effect is something like that of satellites linked to the Tyler office, perhaps a more sophisticated reconstruction of the wheel organization, with Tyler the hub and the divisions on the rim, the spokes going to them. Though chain-of-command organizations, such as one might find in the Navy as well as in most large businesses today, have supplanted the more directly personal approach of Joe Zeppa, the old system has been modified and streamlined, rather than eradicated and supplanted.

With so many other changes, many wondered if Delta would keep its headquarters in Tyler.

"Why haven't I moved it out of Tyler?" said Keating Zeppa in 1980. "All I have to do is get out the back of an envelope and a pencil and when I'm in Houston, say, ask some of my friends: 'What are you paying for office rent down here? What are you paying for an executive secretary? What are your working hours?' Very frankly, we stay here because it's cheaper to be here. And, as we've gone through this big expansion in the last three years—we've really been recruiting a lot of people at all levels— by and large we have found Tyler to be an effective recruiting tool. People want to move to a smaller town. You can get 'em out of Dallas and Houston and Chicago and Pittsburgh by the droves. You have [in Tyler] about a ten-minute traffic jam, from 5:00 until 5:10 on South Broadway. Both [Dallas and Houston], in my mind, are intolerable. And I think this affects productivity. You get to the office already PO-ed at the world.

"There is a billboard, coming out of Dallas to Tyler. It says 'Rush Hour in Tyler.' Big, sort of standard-sized billboard, and it shows a scene out on Lake Tyler and there are about five sailboats, sort of drifting to the finish of a race; no big charging thing with waves and all this, but tranquil, and ripples on the lake, and here are these five sailboats just drifting down to the finish line.

"We can recruit, particularly among young professionals who have kids now that are five, six, seven, eight years old. They're getting into school-going age. They get the image of a small town, where everybody knows everybody else. They're never more than five minutes away from somewhere.

"I anticipate that we'll probably stay here. If we're successful, we can get back and forth to Dallas with no problem. We use a charter air service here, as needed. For instance, I was in San Diego last weekend, got on the Metro, flew to Dallas, went to San Diego and back—no big deal."[25]

VII

A sharp break with the past was Delta's policy of buying new equipment and rigs, instead of used ones that could be refurbished. In May 1977, for instance, the company sold some used drill pipe at $6 and $7 per foot, bringing in $251,000 on the sale. Keating Zeppa, explaining the transaction to his board, considered it better to replace the pipe with new tubular goods than to risk expensive fishing jobs which could result in lost income. His decision typified the company's approach to equipment by the late 1970s.

Zeppa laid out his position on machinery at an industry meeting in 1980.

"We do not try to be price competitive with everybody," said Keating Zeppa. "So when we bid or when we make our pitch to those companies that have their own drilling staff, who have their own in-house capabilities of evaluating the services being offered to them, we feel that some of the more important elements of our pitch should be the following:

"*Superior equipment.* In other words, if we're going to get the highest price possible, we need to offer, among other things, the best and most modern equipment we can.

"This is of interest to our customers because it tends to reduce the risks of delays, blowouts, fires, all of these nasty things that happen occasionally. And, in general, whether right or wrong, the operator generally considers that the contractor with the newer equipment or the better-maintained equipment is probably going to do a better job. It does not always work out that way, but that's the way it appears.

"We also believe of substantial importance to the large operators is a *good maintenance program.* The operator doesn't want to pay for something he isn't getting, and I can't blame him. Why should he pay for our down-time? If we can't keep our equipment up, why should he pay for it?

"Well, we've taken the approach, in Delta Drilling Company, that he shouldn't have to pay for it and, therefore, we have been bidding on the basis of, and accepting contracts with, no maintenance down-time chargeable to the customer. If our rig won't work, that's our problem.

"Now, you say, 'Well, somebody's got to pay for it.' Well, of course, somebody does. It's in that daywork rate. You bump the price $50.00, $75.00 a day, whatever it is. But we feel that, in the long run, both of us come out better. The operator likes to see that if he's going to be paying so much a day for a rig, he gets it for 24 hours. If he doesn't get it for 24 hours, he doesn't pay for it for 24 hours.

"On the other hand, for our company, it puts upon us the burden of looking after what we should be looking after, and that is our people and our equipment. And if we're not prepared to accept responsibility for that which we are providing, then we just aren't doing our job.

"On a consistent dollar basis—to the extent that Delta is successful in its enterprise, obviously—we will be engaging in an increasing level of investments in equipment, machinery, technology. And I would footnote that by saying we are buying more technology than we ever have in the past, and I guarantee you we're going to be spending more on technology in the future, whether in the form of Simtran trainers, whether in the form of pocket calculators for toolpushers, whether in the form of software for our computer system. Technology is going to be a very saleable item for us."[26]

The implementation of this policy has had a breath-taking effect.

"I never knew of Delta spending money like they've spent in the last year, year and a half," said retired toolpusher Glen Koonce in 1980. "Rigs, rigs, rigs, they've bought brand new rigs—that's why I'm mad at them. I didn't get to work on any of their new rigs."[27]

"We really upgraded the equipment," said Bernie Wolford. "I think we had a budget of something like—the last year I was there—$32 million in new equipment. The prior year, about $26 million; and the prior year, $21 million—a lot of new machinery."[28]

South Louisiana, like the other divisions, was sporting new equipment. "That is a big-g-g improvement," said drilling superintendent Doris Darbonne. In turn, with the word getting around that a company has new equipment, good workers are attracted. This, emphasized Darbonne, creates a strong combination: new equipment and good hands.[29]

Flo Bonham summed up the changes from her vantage point as division secretary in the Gulf Coast region.

"All of the rigs have been upgraded over the past few years, with the replacement of the major components which have to do with power, the pumps, different blowout preventers—systems to make them more efficient, more powerful, to do their job better, more efficiently. They are more safe now. A good bit of the equipment would be removed from these rigs. There would be absolutely nothing wrong with it and not necessarily outdated, but perhaps not at the level that they wanted to achieve."[30]

The expansion in the late 1970s translated into bustling shops as new equipment headed for the fields. "You go around Kilgore where they are building those rigs," said "Nig" Spraggins in 1980, "and it just looks like a rig factory. They are working on rigs just about all the time. They get it together and ready to go and they move it out in the field."[31]

Kenneth Fowler, looking at the results in South Louisiana, assessed the effect upon customers. "Delta has built itself a strong reputation of having excellent men, and even if our iron cratered we have another piece cut there to take its place. Delta enjoys a lot of backup. The customers know that we have a company policy about inspecting pipe and collars, and with Delta's intellegent approach to inspecting these tools, we're not going to have this type of failures."[32]

With new equipment have come pleasant side effects beyond lucrative contracts. "I remember working on some of the equipment day and night [several years ago], and now the night calls have been nearly cut completely out, as far as the West Texas division," said Charlie Martin. "We get emergency calls, but we're able to work during the day and be home at night with Momma. That saves Delta time. That's what the name of the game is."[33]

In 1977, Delta had around 36 rigs operating company-wide. By August 1980, the company had 56 domestic rigs. The cost of rigs by then started at $4 million. "Rig 76, if you were to replace it today, would cost about $7.5 million," said Jack D. Robinson. "So in the past three years we have put out 20 rigs."[34]

Expansion affected all divisions of Delta in the second half of the 1970s. The East Texas division, for instance, was operating six to eight rigs in the 1975-1976 period, and by 1980 there were 11 rigs operating there. Traditionally, the East Texas division has been one with the highest rig count. Ed Gandy said, "I've had as high as 14 rigs operating here in East Texas district."[35]

When Kenneth Fowler joined Delta as its Gulf Coast-South Louisiana manager in 1977, the division had nine rigs. By 1980 it had 14 rigs. Four rigs had been overhauled. This rig count made the division one of the largest ones.

By then Fowler was optimistic about the company's prospects in that region. "This is a very healthly environment down here right now," he said in 1980. "We have jobs for all 14 of our rigs, and work ahead through the end of the year for sure. The people who have our rigs running now not only want to keep these, but would like to have some more of Delta's rigs.

"We have the good rigs now. We have an excellent maintenance department, and we have left the trucking where it belongs—to the trucking contractor.

"We've gone to 'no down-time' contracts. When an operator starts paying upwards to $200 an hour for a land rig, he'd like to get it off his back while you've got something shut down. Mostly, if you had to quote a norm, I'd say four hours per shut-down up to 24 hours in a 30-day month is considered normal. In spite of our offering these no down-time contracts, most of the operators still send us contracts with 4 and 24. We accept them, but we're putting the dollars in there, as though it were a no down-time contract."

"Delta's been here for a long time and has earned the respect of these hands, and that's important," said Fowler, "as much as it is earning the respect of the operators. And the hands know they can trust Delta. We probably have more older men working for us—and when I say 'old,' I mean 40 years up—than any other contractor down here. With our growth in the last three and half years, we've hired from the outside a total of two toolpushers. We have those older men that paid their dues the hard way and they are a stabilizing factor for these young kids coming along. We don't have a driller here with less than five years' oil-field experience. And we have some derrick men and motor men out there who have been working 20 years that know that they don't want a drilling job. And we've tried unsuccessfully to get them to take one."

As Delta approached its fiftieth year, Fowler felt that his company was "probably the best contractor" in the Gulf Coast region of South Louisiana, particularly in quality of service. "I think that covers a lot of sins—our quality of service," he said. "All of the majors are apprised that we are training our people, and they're all apprised of the type of iron we have out there. Good equipment and good men. Delta has always enjoyed good men. I can't say the same for the equipment, but Delta did a complete 180-degrees [turnaround] here in the last three years, and we only have to thank the owners for that.

"I would guess that Delta's probably in a stronger position than any other contractor their size right now, and I include all of them in that. Of course, there are not that many contractors our size. Of the ten contractors that are probably Delta's size, Delta's got to be in a stronger financial position. And they didn't get that way easy. They had to work at it all during the years when they scratched with everybody else."[36]

"There is a motto, if you will, for the company now," said Flo Bonham. "It is: 'Good equipment well-maintained. Good people well-trained. That to me has been the biggest really significant change, and it was a long time coming."[37]

When Thornton Tarvin retired in 1977, Delta had 33 rigs. Three years later, it had 56. "So they have really spent a lot of money and have some fine rigs. They have about as fine a rig as there is on the land. I would say they are one of the outstanding companies today. 'Cause they are very well thought of. There's not much lost time. It makes a lot of difference with the companies if you've got something that you can fix immediately without breaking down and wait, wait, to get something."[38]

Fax C. Marshall summed up the situation in his division and the ideal of all divisions. "We sure got some good equipment. We're better than anybody we're competing with in East Texas. We want zero down-time."[39]

VIII

A number of innovations affected employees personally. Holidays were expanded from five to six. The vacation schedule was enlarged. A policy manual, new to Delta, spelled out what the employees could expect. But none of these has had the impact upon both personnel and the company that formal training sessions have. The key to efficient operations, Keating Zeppa has said, is employee training. "People are the key to efficiency," he said. "That drawworks setting out there is neither efficient nor inefficient. It depends on the person running it. Training of people: that's where your efficiency is. We believe that *superior training of personnel* is of essential interest to our customer, and we are putting a lot of emphasis on training. [Furthermore] we believe a personnel *safety record* is a saleable item, something that should indicate what sort of operation the operator is buying."[40]

Jack Robinson considers Delta outstanding in the way it has arranged for class attendance. "Delta Drilling Company owns about two percent of the domestic drilling rigs," he said, "and I would say [that] out of all the drilling contractors operating in the United States, Delta is about the only one that pays their employees to go to school. We pay the roughnecks, toolpushers, drillers, derrickmen—we pay them full wages to go to school. Other companies do have training centers, and I would say a majority of them. It is mandatory that they do attend school, but they are not paid wages while they are attending. We also have set up a Driller-1 category, which is classified as a salaried employee. These drillers are guaranteed a 40-hour work-week. I would say right now 60 percent of our drillers are salaried employees. So that has helped to reduce our turnover. It has helped build loyalty, dedication."[41]

In 1978 the training center was opened in Tyler, with Ray McClendon its first manager. Joe M. Baker became head of the center in November of that year, and remains in that position as of 1981.

Although Delta had previously sent employees to outside training schools, the establishment of the company's own training center in Tyler led to its first centralized program. Joe M. Baker, center manager, came to Delta with a variety of drilling engineering experience over most of the globe that included ten years with Dresser Industries.

"The main thing," said Baker, "we started tailoring the courses for Delta personnel. I guess the first course that we developed and really put into action was pressure control or blowout prevention. At that time there were several schools throughout the industry that Delta personnel had gone to, but we pretty well developed this thing to not only meet Delta's specifications but also USGS [United States Geological Survey]. At the present time we have four different certification programs from the USGS [which include the full school, land blowout prevention, offshore techniques, as well as refresher training]."

All drilling personnel considered to be potential supervisors or drillers through managers are required to complete the pressure control and drilling practices schools successfully. This includes, especially, engineers, drillers, and toolpushers.

Baker said, "I would say that our field personnel are exposed to more technology than any [other] drilling contractor's personnel in the world." Conceding he was making a large claim, he added, "Well, I can almost guarantee that [is correct]."

Citing the cost of sending a driller to school for five days as being in excess of $1500, Baker pointed out that the procedure pays off in benefits for the employees, the company, and Delta's customers. One major company, he said, had compared the different drilling contractors doing its work and concluded that Delta had the best trained, most knowledgeable crews. Baker believes this accolade was the result of Delta's having the most comprehensive training program in the industry. Furthermore, he said, the Delta schools, unlike many other industry courses, actually fail students.

"If they don't make the grade, we don't give them a driver's license," Baker said. "I look at it like this, that if I put my life in the supervisor's hands out there and we have a threat of blowout, I want to be assured that he knows what he's doing. If he doesn't, we fail him, and he has to go back through."

Delta's pressure control school is one of its most vital training programs.

"I would say 95 + percent of the blowouts could be prevented," said instructor Bill Goolsby. "The big thing is knowing your country, knowing your geology, and good drilling practices and good supervision. So many people think of a well just blowing out, but a well doesn't just blow out—pop! It starts over a long period of time. It starts telling you something is wrong, but you got to know what to look for. If it takes a bubble of gas an hour and a half to come from the bottom to the top of the hole, you been in trouble an hour and a half and didn't know it.

"We can come in here and simulate a blowout, we can make it do and sound like one, or we can cause a bubble to come up. We use simulators. It's like an old flight simulator: You can put the problems in there—you run out of fuel, propeller comes off, a flap don't work right, or the weather don't work right, or your compass quits working."[42]

Pressure-control school not only attracts students from the ranks of Delta but also from other companies such as Mobil Drilling, Hunt Oil, Texas Oil and Gas. The subject matter includes pressure control, how to kill a well, and how to handle blowouts and prevent them. It might be entitled, What to Do Until Red Adair Arrives, or, as training center secretary Claudine Rogers says, more to the point, "How to Save Your Life."[43]

In addition to schools for field personnel, Delta holds management-experience seminars conducted by consultant Ted Willey, with the emphasis placed on the individual's responsibility for his own life.

IX

"I like to use the terms, the old Delta and the new Delta, which is an easy distinction between the two companies," said vice president John H. Justice.

To understand the differences between the new and the old Deltas, it is instructive to examine the financial side of the company. Joe Bevill was the chief financial officer until his retirement in 1978. Until the day he died, Joe Zeppa kept himself a part of all financial operations. As manager of the tax department, Justice, a CPA, reported to Bevill.

"But Mr. Zeppa was such a man that he didn't place too much emphasis on organization," said Justice. "Following lines of authority and responsibility, I guess in later years I reported directly to him. There was no financial group."

When the company was reorganized, the financial group, of which Justice became vice president, was composed of separate departments involving taxes, financial accounting, financial auditing, treasury, and financial analysis. The manager of each of these departments reported to Justice. A similar scheme was followed in other groups.[44]

One of the most dramatic features of the new Delta has been its handling of job pricing. "We do not try to be price competitive with everybody," said Keating Zeppa at an industry conference. "It's very simple: We would like to be the highest-paid drilling contractor on every company's payroll. We don't feel that you get any stars in heaven by being the low bidder all the time."

After laughter from the audience, he continued:

"And I mean that seriously. You know, a lot of people have gone out of this business by being the low bidder, and we intend to stay. But if we're going to be the highest-priced one, we want it to be on the basis of cost effectiveness. And we feel this real corporate commitment to charging the highest price that we can get, but provide the best damn service that anybody can give.

"Now, we don't always succeed at this. We have found ourselves being the low bidder with some consternation, and, at the same time, providing more services than we probably should and accepting more liability and responsibilities than we should. And on the other hand, we have found ourselves in the extremely embarrassing position of having been awarded a job at an almost unconscionably high price, and then falling down flat on our face and doing a lousy job. And if I have to take my druthers, I'll take the first case rather than the last one."[45]

Jack D. Robinson, manager of drilling, pointed out that Delta's rig efficiency has enhanced the company's image in the industry. "Delta enjoys an excellent reputation. We are not the cheapest drilling contractor going, but that is one thing [about which] Keating Zeppa has made several public statements: 'I don't pretend to be the cheapest drilling contractor, but I do feel that we are the best.' We believe him.

"And in the long run that's economical. It can save time or money. There are a lot of ways. Good machinery, good people well-trained. One mistake can cost an operator an arm and a leg."[46]

Charlie Martin, former Odessa-yard maintenance man, probably was accurate as any analyst when he summed up the difference between the old and the new regimes.

"Maybe say this," said Martin, "Mr. Joe Zeppa wanted to be the largest drilling contractor. I think Keating wants to be the best."[47]

X

When the net earnings were in for 1975, the new chairman could report they had reached $10,517,309.13—"the highest in the company's history."[48] Though net earnings were not to rise progressively thereafter but to display ups and downs, they continued to be very satisfactory over the next several years. By 1976, production was expected to be up by ten percent.

By early 1977 the company was obligated to spend $4 million on two years of in-fill drilling in the Ozona gas field in Crockett County in West Texas. This would involve a maximum of 45 wells on a new contract with Northern Natural Gas Company for the gas.[49]

The first five months of 1977 brought $22 million in revenues, compared with $17 million for the same period of 1976. But costs and expenses were also up. Optimistically Zeppa estimated that 1977 revenues could reach $60 million, with $12 million in net income.[50]

By mid-1978, production revenues were up 25 percent, even when income from South Culebra Bluff No. 1 in New Mexico, a highly profitable property, was excluded.

Though earnings in early 1979 were not as good as in 1978, due to a lull in drilling, the directors voted Zeppa a $50,000 bonus for having managed the company so skillfully since taking over.

While profits increased, the company's indebtedness soared as a result of expansion. In a series of revolving credit agreements with the First National Bank in Dallas, Delta, in January 1978, had available $25 million in credit, of which by the end of 1977 it had actually borrowed $13.5 million; in September 1978, it had a debt increase to $35 million; and in July 1979, $60 million. The Dallas bank served as the lead bank in a syndicate composed of it, Citibank of New York, and Peoples National Bank of Tyler.

In May 1980, Keating Zeppa explained:

"Our current debt stands at about $60 million, which for the company is not an untoward amount. In fact, we are concluding today an amendment to the basic credit agreement with the banks that would immediately essentially run us to a borrowing base of about $125 million. So we're only borrowing about 50 percent of what the bankers are willing to let us have."

At one point early in 1980, the company had reached a peak of about $64 million in debts.

"And I know this makes some of our people very uncomfortable," continued Zeppa, "including some of our directors. Albert Sklar is very uncomfortable with debt, and he tells me so. And I say, 'I'm sorry, Albert. This is *not* a one-horse operation; this is a rather large corporate enterprise. If the bankers are comfortable. . . .' "

Keating Zeppa went on the board of Peoples National Bank in the late 1960s, an event which became the cause of a difference of opinion between him and his father, since the elder had been on the board of Citizens National Bank for so long.

Before Joe Zeppa's death, Keating Zeppa swung no business Peoples' way. After his father's death, Keating did gradually steer Delta business toward Peoples, though substantial deposits remained at Citizens. The key move was Peoples' willingness to accede to Keating Zeppa's requests as a lead tenant for the new building office—Peoples Plaza—as to space, price, and eating facilities (he insisted on a cafeteria in the building that everybody could use). In April 1981, Delta's corporate offices moved to Peoples Plaza, a few blocks away from the old location. Blending the old with the new, Keating Zeppa installed, as the doors to the executive offices, the brass doors his father had purchased from the bank in El Dorado, Arkansas.[51]

XI

By early 1976, Keating Zeppa had already made clear that he thought most of Delta's earnings should be plowed back into exploring for new oil and gas reserves, instead of buying more drilling rigs. "Putting profits back into the ground" was to become a recurring refrain in the years to come. The bulk of such funds was to be spent in East and West Texas.[52]

In 1977 at the great expansion, Red Magner, now senior vice president of exploration and production, became vice president of the marketing group. "All of a sudden," he said, "we were taking greater participation in deals. Instead of taking an eighth or sixteenth, we were taking a quarter or a half or even 100 percent. And we were becoming a net operator, using our rigs. We'd instigate the deals, sell the deals down and so forth. That type of endeavor was developing and it would be self-paying, so they created a marketing department."[53]

Delta's directors have viewed favorably the shift in the late 1970s toward exploration and production.

"Delta's primary interest, of course, is drilling, but the secondary, with strong emphasis, is in the acquisition of reserves," said board member Leonard Phillips. "And I think that will ultimately be the most valuable part of Delta Drilling Company. There always was an attempt [to do this], but to utilize only the drilling profit, so that at worst you ended up with a nonprofit situation in which you didn't lose any money. But Keating has gone further than that and has established a large budget for exploration, and the two are now two separate branches of the company, one that looks to make a profit on drilling and one that looks to drilling successfully in geological prospects."

What has been the reaction of the minority interest directors to this shift of emphasis?

"Well," said Phillips, "it has been the desire of the minority stockholders to go the route we presently are going, and we foresaw this many years ago, but again, we relate back to the personalities that were then involved, who could not be changed. So Mr. [Joe] Z and the rest of them saw it that [other] way. Course, nobody could visualize that $2.65 oil would go to $37. Had we been able to see this then, obviously our emphasis would have changed. I think Keating is doing a very admirable and successful thing in what he's doing."[54]

Change has come, most of all, to the process of finding oil and gas. "Just the methods are changing, and depths are changing," said geologist Jim Ewbank. "You are going for deeper horizons now that in the past were uneconomical to explore. Ten years ago, you could not justify drilling a $3 million well because it wouldn't pay out on 27-cent gas. Now the economics have changed, but the exploration techniques are basically the same. Instead of looking for a 100 feet of pay, you are looking for ten feet of pay. It's just a matter of economics." When Jim Ewbank went to Delta as its geologist, he was the third one the company had ever hired, and at that time he was the only one on the staff. By the end of 1980, the company had 23 geologists.[55]

XII

Although the opportunity for profit had never been better than in the 1970s, risks climbed accordingly. A case in point is that of the wild gas well, South Culebra No. 1, in Eddy County, New Mexico, in the Permian Basin. In the beginning, Delta wasn't even drilling the well; it belonged to Amoco. But around 4000 feet in the hole, Amoco for some reason decided to abandon the operation, an unusual decision to make at that stage. Delta had a new employee who had worked for Amoco and was familiar with the well, and because of this the company succeeded in obtaining a 100 percent farmout for Delta in November, 1977. Amoco retained a one-sixteenth overriding royalty, convertible to a 50 percent working interest after payout of the well.

The well blew out on January 3, 1978, and caught fire the next day. The fire was finally put out on January 10. Gas continued flowing, and a new wellhead assembly was installed and the gas flared on January 15.[56]

Delta attorney Ed Kliewer recalled the sequence of events. "So Delta drilled the well on," he said. "Before they had one piece of paper on the damn thing, the well blew out, caught fire, cratered, and burned down the drilling rig and was out of control. On fire. Delta brought back Red Adair from wherever he happened to be at the time putting out a fire, and when he got through there he came back and finally they got the fire out, but they didn't get the well under control. It was flowing at the rate of around 45 million cubic feet a day, just out of the ground. They had enough pipe in the hole. They were able to get some surface connections on it, and they just turned this well, open flow, in El Paso [Natural Gas]'s line. It was out of control. It was producing over 40 million cubic feet a day."[57]

T. C. Carlton was the division manager for West Texas-New Mexico at the time. "They went ahead with the rig. Drilled down and set their protection string and proceeded to drill ahead with water, which is not uncommon out there. All of the history around that well, all the way around where Delta drilled, no problem—you just drill them down and plug 'em and move off. Production department was looking after it. They were located in the same building we were in, but we weren't paying attention to it. That's the normal way they did back them. They looked after their own wells.

"They drilled into that gas pressure, abnormal gas pressure, unexpectedly. Nobody was expecting it to be there. It was about 11,000 feet [10,784 feet].

"Well, when it unloaded all the fluid out of the hole, a valve broke out on the manifold out there, and when it did, then the well was loose and there was nothing you could do about it. It caught on fire. If that valve had not broken, we could have penned that gas until we could have fixed it; but once that valve broke, we had to waste it.

"I was at home about 11 o'clock one night. I got a call from Ed Arnold [in Tyler] and they had assigned that rig to the drilling department that night. Ed told me that I needed to get my big butt there and see if we could get that well under control—and at that point it was not on fire. So they sent these people on up there with the instructions not to go out until they could see and give me an assessment. I had an appointment in Houston the next morning, so I went on to Houston, and I had a call waiting when I got to Houston, that it had caught on fire. So we immediately got hold of Red Adair.

"Nobody knows how it caught fire. No light out there. It just ignited. Anything could have caused it. Static electricity may have started it. But I do know one thing: it kept West Texas [division] busy. Of course, when the firefighters got in there, first, they got tons of water that they squirted on that thing. I think they did that and took all the equipment off and shot the blowout preventers off with dynamite, and the drill pipe. We wound up digging a hole around that well about knee-deep and cutting that pipe off down below there. They finally killed the well and fished out the drill bit. I think it took us about 12 days to get that thing under control."[58]

Merrel Greenwell was manager of production when South Culebra No. 1 blew out.

"It blew out on the third of January," said Greenwell, "on a Monday morning right after we'd been off on the weekend. Came back that morning and they called; I think it blew out something like, if I remember the time, about 6:30 that morning. Great way to start the first day of the week. It started off with a bang. The well was drilled into a high-pressure area and it got away from them, caused it to blow out and burn the rig down. They never did kill the well, but they got it under control. They dug down, cut off the pipe and put new blowout preventors on, all while this thing was continuing to flow. The best estimate that we had, it was flowing something like 45 million a day, which is a lot of gas and a lot of pressure. We couldn't really kill it without damage to that old hole. There were some new zones. We negotiated a contract with El Paso [Natural Gas] that they would take that gas, if we could get it to them.

"It caught on fire a couple hours after it blew out.

"This was not a Delta rig. It was somebody else's rig, drilling the well. Amoco initially spudded the well. It was a wildcat well."

Greenwell recalls that the costs to Delta ran to $1.7 million. This included Red Adair's firefighting bill and damages caused by the escaping gas.[59]

Because of the emergency conditions, New Mexico authorized Delta to sell gas at an unlimited flow rate into El Paso Natural Gas' system, and an emergency sales contract was signed. By the end of January the well was producing 40,000,000 cubic feet a day with a reserve projection estimated at 10 to 20 billion cubic feet.[60]

Because the New Mexico Oil and Gas Commission gave Delta permission to produce the well under those conditions until it could be safely completed, the expenses were soon paid off.

The income was startling. "I think it amounted to something like 30 percent of our income for that one year," said Merrel Greenwell, "off of oil and gas properties. Well, at 40 million [cubic feet] a day, at $1.50 per thousand, that's about 60,000 dollars a day—it adds up fast."[61]

On January 26, 1978, the sale of gas began. As of February 4, Delta had spent $1.4 million on the well. The estimated cost of cleaning out the well and recompleting it was put at $350,000. When Amoco exercised its option to convert its override to a 50 percent working interest, Delta's share of the income would decrease.

The well was producing 31 million cubic feet daily of gas and 100-110 barrels of oil daily. The recoverable reserve estimate had been hiked to 30 billion cubic feet. By then the well seemed to have three different hydrocarbon zones, but still the main objective, the Morrow zone, had not been reached.[62]

According to Richard J. McDonough, accountant in the exploration and production group, the well paid itself out in about a month, an amazing statistic.

"They got the fire out, but it was still too much pressure in there to complete the well, and they produced it until about September of 1978, providing a casing in the bottom of the well.

"It produced four days in January and all of the month of February, and they considered the payout date the first of March. That was close to the biggest well we had ever got.

"We were on computer already, of course, by that time, and they showed me the proceeds from that well. That was one which we were to distribute 100 percent, and it would not fit in our computer. The figures were too large. We had to break it down into halves, and pay it as two separate items."[63]

"That well, as it turned out, has been an excellent asset," said Ed Kliewer. "But if they hadn't succeeded in controlling the well, in the sense of getting the fire out and getting that flow of gas channeled some place until such time as the bottom-hole pressure was decreased to the point they could get in there and control the well, that well could have broken Delta. That's the kind of risks that are involved in these things. There's no way to measure the exposure that Delta had."[64]

XIII

The rate of change has been dizzying in the oil patch in the 1970s: increased rig counts, proliferating costs, and increased demand for petroleum. Around 1971-1972, there were about 900 rigs operating in the United States. By 1980 there were 2600, and the number was still growing.

Veteran Frank Dykstra, a drilling consultant after his retirement as division manager of Delta's Rocky Mountains division, pointed out that wells which were plugged years ago because they were producing only 25-30 barrels a day—a losing proposition at $2 a barrel—are now profitable, and drilling such fields is now attractive. He believes, as do many others in the industry, "There is still plenty of oil, but you have to go deep for it. People don't realize what it costs to drill a well. I just came off a dry hole that had $2 million in it. And that's just gone. Not a real deep well—13,000 feet. Most of them, you can't drill [for] under $1 million."

Inflation and soaring operating costs characterize the new era. "When I left Casper, Wyoming, in 1976, the J-750 rig would go for about $1300 a day, daywork," said Dykstra. "Now the same rig is going for $6000 a day." By the late 1970s, with the drilling rigs so expensive and the day rate for them so high, an operator tended to get the drilling rig off his budget as soon as possible and complete the well himself. Because there was plenty of work, Delta benefitted by such a trimmed-down schedule.

"We just drill a well and move," said Bill Goolsby. "Leave it. Then they come in with a little workover rig and they complete it. So that is working real good for us. 'Cause, really, we make the big majority of our money on footage. Most of our work now is called 'daywork.' They pay us so much a day for the rig and the crew. But on footage, if you get a good run, you can take your money and go on to the house real quick. Kinda like picking cotton. If you get out there early, you get yours before the others do."

Frank Dykstra summed up the frantic search that has come to dominate the petroleum industry's activities: "They are digging anywhere they can now to find anything they can."[65]

XIV

As pointed out earlier, generational differences influence a company's posture. Stockholder-director Albert Sklar agreed. "Keating is from a different generation than his father," Sklar said. "I'm from a different generation from my father. We don't run things the same way. We are educated in different areas and we have different feelings than they had. One of the biggest fights I had here [at Sklar Oil] was trying to modernize the accounting system. Because the old man didn't like it. In fact, he refused to have the accountants keep the books in the way that they should have been kept, in a more organized and, I would call it, a more business-like manner. He told me where to go, but I fought it until we got that [the accounting modernized]. Different strokes for different folks.

"We now have [in Delta] a young man with a philosophy of building a large company, whereas you had an older man that liked to have, I would say, his personal hands on everything, and this is very similar to the way my father was.

"Somebody has to be the guiding force, and at this time it's Mr. Keating Zeppa, who is a very intelligent, fine businessman, and a fine man with other interests, as well."

The late John D. Hall, who had known both Zeppas, agreed in principle. "Keating is different from his dad," he said. "He's not afraid to delegate authority. His dad liked to keep a finger on it. He'd keep it in a position where he'd have a say-so in any major change, purchases, sales, or anything. I'd even make it a 'minor' instead of 'major.' "

Leonard Phillips saw a serendipitous heritage in Delta's chief. "Amazingly, Joe Zeppa's son has inherited the better of two worlds. He inherited his father's intelligence and his mother's niceties, and the combination is excellent."

"He's of a better disposition than all of the rest of the family," said Trent Zeppa. "Keating's disposition is better than mine, his mother's, or his father's, or anyone else's. He's more like his grandfather that no one ever met in this country—Vincenzo, he's named for him. [I know that] because of what the older Italians have told me. Vincenzo Zeppa was a very nice person. The people in Fubine liked him. And people like my brother. He's intelligent enough and he has a pleasant manner about him. He doesn't try to bully or hector people. He's God on wheels when he does get embroiled over something, but other than that he's very nice."

It is hardly surprising that some people emphasize the similarities, others the differences, between father and son. A lot probably depends upon the individual's dealings with the elder Zeppa, which becomes the yardstick by which comparisons are made. Pinehurst Farm manager Milton Winston, for instance, emphasized reliability.

"Pinny is just like his daddy," said Milton Winston. "And when he *tells* you something, it's just that way."

The overall impression one gains from Keating Zeppa's associates and employees eventually adds up to a picture of admiration, unmarred by criticism—a kind of younger Joe Zeppa without the rough edges, perhaps more relaxed, less obviously aggressive.

Joann Blalock, Keating Zeppa's long-time secretary, did not have in mind a comparison of the two men, but without specifying it, her characterization of her boss did contain a mélange of similarities—knowledge and thoroughness—and differences—delegation of authority—of father and son.

"I'm constantly in awe of his [Keating Zeppa's] massive knowledge of so many subjects," she said. "There's just not too many things you can talk to Keating Zeppa about that he doesn't know, that he doesn't have a working knowledge of. Very well-read, remembers everything.

"He is very much a people person. He likes to know about his people, and I say 'his people' because this is what we refer to around here as a family-type situation. He has a very strong feeling for them on a personal basis; even though he may not see them, he recognizes that they have a personal life, that they have families, that they have needs as a person.

"Keating is a very thorough person, and before he goes into anything he makes sure he has all the information needed, and he does delegate; he believes in having people around him that he feels have the intelligence and the ability, and gives them a chance, a right for them to try and do it, and then gives them all opportunities to perform. Then, of course, people have to perform themselves, because they just really are surprised at how much they can do when they really have an opportunity.

"It's very interesting to watch people who have their first dealings with him, because they always come away feeling very, very good, if they're coming to work for us or thinking of working for us."

Lawyer Ed Kliewer classified each man as appropriate for, as well as a product of, his times. "Keating early on proved that he was a damn good manager, and I think that what we're seeing in the company is a very competent son taking over where a very competent father left off. Totally different world and a totally different concept. The fatherly concept, the one-man company, that was Mr. Zeppa's way, and it's a thing of the past. Keating is an organization man."

Leonard Phillips, Sam Sklar's son-in-law and a Delta director, traced some of the salient factors in the two regimes.

"Change occurs by, number one, the difference in all fathers and sons, because of age and because of times and difference in education. Secondly, and most important with me, have been our 'friends,' the Arabs—the ball game is changed. Take, as in baseball: instead of nine innings, they have 37 innings, and instead of three outs, 12 outs. And unless you can shift with the new rules, you're in trouble.

"It hadn't occurred during Joe's lifetime, any more than during Sam Sklar's lifetime. I don't think that Joe responded as well [to the Arab oil domination and its consequences] as Keating has. Only because his experience was rooted in the past, and he viewed things in the past as a measuring stick. Talk to people today and look at a $15 steak and they say, 'I remember when it was $3.50.' Well, that's the way anyone's mind works. And when Joe, having grown up in an era where he had to scratch to make a living, and Keating, having grown up in an era where it was 'How many dollars do you spend?' it's a different ball game. And presently, because of the circumstances that exist in the world, Keating is more adjusted to running a company of this nature than Joe had been. As my son could probably do a better job than I do."[66]

Keating's formal technical education as a petroleum engineer sets him apart from the world of his father, but there is no better symbol of the differences between first- and second- generation directors than Sam Dorfman, Jr., a specialist in internal medicine who first took a degree in petroleum geology. He is now able to characterize himself as the only person in Dallas who can read both a well log and an electrocardiogram.

XV

Among the symbols of continuity which Keating Zeppa's succession to his father's mantle summons up is his election as president of the International Association of Drilling Contractors (IADC) for 1978. By then IADC had changed its name from the American Association of Oilwell Drilling Contractors (AAODC), which was its name in 1958 when Joe Zeppa had served. The two events had come exactly 20 years apart. And just as his father had, in his time, the son seemed on his way to receiving a variety of honors related to the industry. In 1980 he was selected as a distinguished graduate of the College of Engineering at his alma mater, The University of Texas at Austin.

In addition to the IADC, Keating Zeppa was a member of the Society of Petroleum Engineers, the American Institute of Mining and Metallurgical Engineers, and the visiting committee of the School of Petroleum Engineering at The University of Texas at Austin.

Closer to home, in Tyler, he busied himself in community affairs much to the degree his father had, though reflecting different interests. An active member of Christ Episcopal Church in his hometown, he has also served as president of the board for the Tyler Catholic School System, as well as chairman of the East Texas Hospital Foundation and the City Planning Commission. Previously he had served on the board of St. Andrews Day School in Austin, and on the Airport Advisory Commission.

Reflecting an old interest in the sea he had come to love in California as a boy, he has served as commodore of the Tyler Yacht Club.

Quietly cheerful, with a multi-faceted mind, Keating Zeppa appears to be a man who relishes running Delta. Patient but capable of swift action when he needs to assert his authority, he has keyed his policies to a pursuit of quality, which has been reflected in the company's revamped inventory of rigs and equipment, its training programs, and its services. He has been particularly adept at dealing with people, whether employees or business associates. If Joe Zeppa had the mind of an accountant and the heart of an artist, Keating Zeppa has the mind of an engineer, the heart of a scholar, and the style of the diplomat.

His *persona* has evolved through the years, as no doubt his philosophy has. For many years a symbol of his individuality could be found in the heavy black mustache he wore, a heritage of his Argentine days. In 1979 the mustache had grown into a full, salt-and-pepper beard that, of itself, was likely to set him apart from the usual run of drilling contractors. But the unexpectedness of his image, set against that of the traditional concept of the drilling contractor or oilman, does not end there. Joann Blalock was thinking of that when she commented on Zeppa's leaving for Washington a day early for a board meeting of the IADC in order to see the National Gallery of Art and the Smithsonian Institution—"to feed the inner man," as she put it. "Not exactly what you'd expect of the typical drilling contractor."[67]

In 1980, as Keating Zeppa was rounding out his fifth year as president of Delta, he examined both personal feelings and professional goals as the company was on the verge of entering its second 50 years. He looked both backward in memory and forward in planning.

What, he was asked, had been his most difficult decision through the years? "To stay with Delta," he said. "I had to make it, over and over again for a period of time. It was a repetitive decision."

Since 1975? "Again, to stay with Delta. When you go to a board of directors and say, 'Okay, the president's dead and I'm the executive vice president and I want to take this on,' you have do some thinking before you make that statement to a board. At least I gave it quite a bit of thought. Here's a chance to tell them, 'Go find yourself a manager!' I think those are pivotal decisions.

"Hanging over the whole analysis in the decision-making process was that, according to the terms of the trust that Mother and Dad established and into which they put their Delta stock, the trust provided for essentially one-half of the assets of that trust to be available for me and my family's benefit. And so, really, what it boiled down to was, did I really want someone else looking after my rice bowl? It's a pretty big rice bowl, frankly, because of subsequent events—not necessarily because of what I've done, but all of a sudden the domestic oil and gas industry is pretty healthy. There is a lot of activity. So you might say it was almost self-preservation that said, 'You'd better stay with Delta and look after your interests in it.' At least that way you can't blame anybody else if it goes sour.

"I can think of all sorts of things that I'd rather be doing. It's that simple. That sounds selfish, but I feel a little selfish some times because there, again, I never really felt that I had that much of a childhood. And traditionally when you're most selfish is when you're young. And the older you become, the less selfish you become, theoretically. On the other hand, I always felt that I was called upon to be more than I was or older than I was. Why did I come into Delta? I never felt I had a choice. Not that I felt beaten down. I just never realized there was a choice. And now I'd like to make some choices. Not that I want to go work somewhere else. I've got a meal ticket now, but fortunately I inherited a curiosity from both my parents. There are a lot of things I'd like to know more about. Very selfishly. Not because I intend to teach, or pass on to my children, but just for my own personal gratification. So, really, my financial aspirations aren't that great. I don't want to be a Bunker Hunt or this sort of thing.

"So, selfishly, there are things I'd like to do and places I'd like to see. Some of them involve other people, some don't. So I have all sorts of reasons for not wanting to perpetuate or at least try and establish a second dynasty.

"And if I see a big difference between Dad and myself, it is that I don't want necessarily the next stage in Delta history to be: 'Well, Delta is Keating Zeppa.' We're in a different era. I don't think it's any longer appropriate, and I don't think it's healthy for the company, for it to be known as, 'Oh, that's Zeppa's company.' There are too many implications there that are stifling in some respects. I'm simply not the same person Dad was. Can't, and don't want to be. Not that he didn't have many admirable traits which I want to be able to assimilate for my own. In the first place, I don't think being a carbon copy is appropriate, because we don't have carbon-copy times and conditions. And, although I can assess my own weaknesses, I can assess, retrospectively, Dad's shortcomings as charitably as I do my current shortcomings, and learn from both of them.

"We're in an era when young bright people expect to move quickly and are not prepared to have the same job for 20 years. If we're going to attract and keep good people they've got to have a clear line of promotion. And that to me is the vitality of any organization—the new ones coming in, and the experience that can be passed on from the older ones and the enthusiasm that the younger ones bring aboard."

"I choose to believe that Delta, generally speaking, in the industry has stood for integrity. Delta was never the glamorous, the bright star, the flashy kind of company. It was staid, plodding, but paid its bills, maybe late when times were tough, but always paid, met the payroll. Did a good job. Sometimes didn't run the best iron out there, but it always ran, though it was patched. Mainly I would say: honesty, integrity. And this is something I feel very strongly about.

"It has always been the company's position—it was Dad's and it is mine—that no matter how bitter the pill may be to swallow, when an authorized employee makes a commitment on behalf of the company it is a corporate commitment. And no matter what the results may be, the company will keep that commitment. Now, we may have words with the person who made the representation, but that's our business and that's in-house. If the company is committed, it is committed, and that is an obligation of this company, and it will be fulfilled. Now, if it's obviously such a thing that will break the company to keep it, we're going to go visit with the people and say, 'Look, here's the situation. We're prepared to do all we can to salvage, but we can't throw the company out the window!' I don't know if that's ever occurred.

"We want to be known for up-to-date methods, equipment, good training of our people. I'm less interested in being absolutely the most profitable company than I am in having the reputation for just being one of the best companies, whether it's as a place to work or as a supplier of goods and services. A place where people are proud to be part of the team."[68]

XVI

A rare opportunity for communication with the petroleum-consuming public, as well as with the government, came Delta's way in 1978 during the Festival of American Folklife sponsored by the Smithsonian Institution. Featuring energy as a theme, the week-long festival brought together a real drilling rig, a mock coal mine, and a refinery exhibit. The rig was Delta's—No. 72 from the Northeast division, a Spencer-Harris 7000 which drills to about 9,000 feet. The opportunity had come about because Keating Zeppa was president of the IADC; he volunteered to bring a rig over from Pennsylvania for the occasion. No expense or effort was spared to produce a memorable exhibit. "We swallowed, I know, $100,000 on the project," said Gloria Stackpole. "We talked about geology, how much it cost to drill, and took people on the rig floor. Secretary of Energy James Schlesinger and many Congressmen came. We had a lot of people who came over from the Department of Energy who had never seen a drilling rig. They wanted to see what one looked like."[69]

Burly toolpusher Hugh McKenzie, a man of many and apt words, was chosen to describe the drilling procedure to the people who came to view it. Armed with a microphone and his native wit, master of ceremonies McKenzie entertained and informed 10,000-30,000 persons a day. The whole rig was made ready for operation. A doghouse for the crew was nearby, just as if they were on an actual location. They even made 50 feet of hole, and went through the motions during the day, for each new crowd's benefit, as if they were actually drilling. In the crew a black man, a Chicano, and a Cajun symbolized equal employment opportunity in the industry.

"We had a lot of fun up there," said McKenzie. "Then, of course, you had to be a-talking and showing them and telling them all this, and try to keep the crowd [with you]. I told them a joke about when I first went to work. 'I asked the driller, "What do we take off for dinner?" And he said, "One glove." When we got ready to quit that evening, he turned around and said, "Worm? Have your wife work buttonholes in them sandwiches and we'll just hang them up and eat them a-walking tomorrow." ' And they got a big kick out of it. Just old oil-field lies, that's all. There was a nice-looking woman sitting down there. She said, 'I want your autograph.' I said, 'Good God, woman, autograph of a roughneck?' So a year ago last Christmas I got the nicest Christmas card from her.

"After the first show, Keating come around to me and he said, 'You've been lying to me'

"I said, 'I ain't lied to you, Keating.'

"He said, 'You never did tell me that you'd ever worked for a carnival.' "[70]

XVIII

The question of whether Delta might sell out, merge with another company, or make a public offering of stock certainly was not a matter which surfaced suddenly in the 1970s. In 1968 it was discussed with a member of the New York Stock Exchange, apparently at some length, and Keating Zeppa indicated, even then, the direction he preferred to take in later years.

"I, frankly, do not feel as enthusiastic about a merger as I do about the possibility of a public offering of Delta Drilling Company's stock," he wrote, "primarily because I feel Delta has, or will get, the management necessary to properly handle the company under public ownership. Nevertheless, attractive merger proposals would not be rejected out of hand. I hasten to add, however, that we are certainly not shopping around."[71]

A variety of prospective offers and discussions, ranging from sale to merger to "going public," were aimed at Delta in the ensuing years, particularly toward the end of Joe Zeppa's life and in the first few years after his death. A major company, Sun Oil, for one, expressed interest in buying Delta's production, should the company be so inclined. When Delta was approached as "an acquisition candidate" by a firm specializing in mergers and acquisitions in 1977, Keating Zeppa replied that "to my knowledge, neither Delta's management nor shareholders have indicated an interest in selling or merging Delta Drilling Company." He did go on to say, however, that though "in my opinion, the chances are slim to none that the present stockholders of Delta would entertain a sale or merger, I would do them a disservice if I rejected out of hand any genuine expression of interest by one of your clients," and he offered to discuss it further.[72]

In reviewing the correspondence files for approximately a 12-year period, it appears that from the beginning there was an interest on Keating Zeppa's part to consider "going public," an approach which his father apparently did not choose to pursue past a certain point. On the other hand, the possibility of an outright sale or a merger seems to have been quite distant over these years.

By the late 1970s Delta had become a rarity. As Leonard Phillips said in 1980, "There are very few companies of the stature and size of Delta Drilling Company where capital is formulated and created within the company itself. In fact, I don't know of any drilling company, the size of Delta, that is not financed publicly.

"The next question is pretty obvious. To buy drilling rigs today—the rigs of the type that Delta has, that will run from [a total] $10 million up to $50 million—you put that on a 20 percent interest scale, and you find that you can't generate 20 percent off your profits of drilling, so consequently you're running a negative cash balance. And the only way you can generate that type of funds is through the public, and that's why Delta's a rarity and has been a rarity. The future depends on Keating, where he wants to go."[73]

Keating Zeppa had made specific commitments to the stockholders relating to liquidity and the best interest of Delta.

"We now have some third-generation stockholders," he said in 1980. "These third-generation stockholders could care less about Delta Drilling Company as an entity. It's not like Sam Gold and Sam Sklar and Sam Dorfman and these people. Plus the fact, the third generation—they're young, they're in their twenties. So what the hell to them is 1,000 shares of Delta Drilling Company stock when they can't sell it? There's no market in it. They look at that, and they look at their needs on the other hand, their desires, and they say, 'How do I get from here to there?' The answer is, 'I can't.' And that builds up frustrations, because their fathers and mothers and their grandfather may have told them, 'Now you hang on to this. It will be real valuable.' Young people think that's fine; it's going to be real valuable someday, but what's it doing for me now?

"So I feel a very strong obligation to take the company in the direction that will generate liquidity for all the stockholders."

Looking back over the decades, Keating Zeppa saw considerable evidence that the idea of a public offering of stock had not originated with him. "I have to suspect that sometime in the past Dad had probably considered a public offering," he said, "because, after all, he went public with Delta Marine. That was a public offering, essentially. The financing for Delta Marine was set up in a much more complex fashion than the financing for Delta ever was in the beginning or was contemplated to be subsequently. So I'm sure he must have, from time to time, contemplated it.

"On the other hand, until the 1960s, I won't say there was no public market for drilling companies, but there just was no track record. And I would have to suspect that by the middle '60s, Dad, particularly after a stroke, was prepared to deal with a known mortality, as the rest of his life being of a finite determinant, and that he probably may have come to the conclusion that, first of all, it wasn't worth the struggle [to go public], but more importantly that there was a better chance of having his estate valued, for tax purposes, at a lower value as a privately held company than if we'd have been publicly held with quoted stock out there, with no question of valuation. I don't *know* that to be the case. He never discussed it with me, but knowing him I'm sure that was a consideration in not bringing the company public."[74]

One precipitating nudge toward a more general solution came with one stockholder's desire to sell his Delta stock. For several years Louis Dorfman—a son of founder Sam Dorfman, Sr.—had indicated an interest in disposing of his stock in Delta. In 1980, Zeppa had an agreement worked out with Dorfman whereby the corporation purchased his stock and two trusts' stock, of which he was trustee, amounting to a total of 9,387.5 shares, Class A and B shares combined, paying him $945 per share.[75]

Although there were other factors involved in the complex maneuvering to provide liquidity for holders of both Class A (voting) stock and Class B (nonvoting) stock, Louis Dorfman's move helped bring matters to a head. In early January 1980, Dorfman had entered into an agreement with Rotan Mosle, authorizing them to dispose of his block of stock in the company.

Keating Zeppa elaborated upon the subsequent events.

"Knowing that the company's by-laws provided for preemptive purchase rights by shareholders and the company regarding any stock offered for sale by a shareholder, Rotan Mosle kept the company informed of the progress of its negotiations and efforts to sell Louis Dorfman's stock for him. As a result of that firm's efforts, Delta did, in fact, exercise its preemptive right to purchase Louis's stock at a price of $945 per share, which was consummated in late March, 1980.

"Louis was apprised of the possibility that we might go public. At the time I met with him to close the deal and sign the agreement to make the purchase, we were sitting in his office in Dallas and I said, 'Now, Louis, you know it's possible that we might go public.' He said, 'Well, I know; I've heard that before.' 'Well, I just want to make sure that you understand that this stock would probably be worth more in a public company, based on what's happening in the marketplace,' and he said, yes, he understood that, but he was just happy to find a buyer.

"During Louis's attempt to market the stock, I had been thrown into contact with a number of investment bankers from time to time. He had retained Rotan Mosle, and I came to know those people quite well—Jon Mosle in Dallas and John E. Justice with Rotan Mosle in Houston. They were very helpful. In fact, they did a lot of work on evaluating Delta Drilling Company in order to be able to help sell Louis's stock for him. And in doing so they came to the company for information, which we gave them, and in going through this whole drill I became pretty well educated in, really, what the stock market was beginning to pay for drilling companies, particularly in the middle of '80, and toward early fall of '80 it became apparent that stock prices were just soaring, for companies such as Delta. So, really, it came to the point of 'Well, there probably is no better time.'

In fact, our timing was late; we were probably six months late coming to market, in terms of actually getting top dollar. But, of course, the objective was not so much to get top dollar but to generate liquidity and to fund the participation plan.

"Over the period April through September of 1980, the company held sporadic discussions with investment bankers and other parties to assess the probable outcome of a public offering by the company of its common stock. Principal shareholders were contacted on an informal basis to determine their feelings about the matter, and formal action was taken by the board of directors in November, 1980, directing management to proceed with the steps necessary for a public offering of Delta's common stock.

"With the counsel and advice of Rotan Mosle Incorporated, the company interviewed four prospective co-managers for the proposed underwriting. These interviews were held in Dallas, on November 18 and 19. Present during these meetings were company management personnel and principal shareholders. Also present as advisor to the company was Mr. J. Wesley Hickman of Dallas, who had enjoyed a long and close friendship with Mr. Joe Zeppa, and who had been instrumental in handling the financing of Delta Marine Drilling Company.

"On November 19 the decision was made to ask the firms of Morgan Stanley & Company Incorporated and Rotan Mosle Incorporated to act as co-managers in the underwriting of a 2,000,000 share offering by the company, to be effective as soon as possible. The company was immediately informed by the managing underwriters of the teams that would be working together on the Delta offering."

Originally, plans had called for filing the registration statement with the Securities and Exchange Commission at the end of 1980. It was not possible to meet this deadline. However, as one step toward the public offering, Delta's stock was split, 150 for 1, in December 1980, with the par value of its common stock reduced from $10 to $.10 a share. By that time Delta's identifiable assets totalled $233,323,000, having risen steadily from $76,093,000 in 1976. During 1980 Delta produced 723,000 barrels of oil and 8.6 billion cubic feet of gas. As 1981 began, Delta had estimated total proved reserves of 3,465,231 barrels of oil, 2,282,110 barrels of plant liquids, and 53.7 billion cubic feet of gas. The company was operating domestically 56 onshore rigs, including six limited-partnership rigs in which the company held a 20 percent interest, while also owning and operating six rigs in Italy.

On January 14, 1981, the company filed its registration statement with the Securities and Exchange Commission, followed two days later by another registration statement on the exchange offer, related to the participation plan established in 1974. All Delta employees with 15 or more years of credited sevice with the company were entitled to be included in the plan. Upon the public offering of stock by the company, each unit of participation would be convertible into the value of a share of common stock. The exchange offer by the company provided these employees with cash and shares of common stock for their units of participation.

Adding the exchange offer lent an unusual angle to the ordinary features of the public offering. Because both registration statements were very similar, there was always the possibility readers would become confused. Furthermore, the exchange offer would have to be completed and closed before the company could offer its common stock.

During the period from November until March, a number of Delta employees worked feverishly. The accounting department toiled seemingly without stopping, turning out the information required for filing with the SEC. Long hours were devoted in the drilling and exploration groups to compiling statistical data. Keating Zeppa and others in management spent tedious hours reviewing the drafting of the documents. Charles White, manager of operations in the financial group, was responsible for gathering most of the data. Robert L. Hampton, manager of investor relations, served as the principal liaison between the company and the drafting group which included underwriters and attorneys. The accounting firm of Peat, Marwick, Mitchell & Company assigned a large team to the account. For many of these individuals, days off, free weekends, and the eight-hour day became merely memories.

After waiting almost six weeks for response from the SEC, Delta filed the first amendment to its registration statement in late February. A new schedule was set, tentatively calling for bringing the stock public during the week of March 16. Based upon this estimate, meetings with the financial communities of various cities over the nation were scheduled—the "road show" through which members of Delta's senior management team would present the company's features to potential investors. This team consisted of Keating Zeppa, Red Magner, T. C. Carlton, Jim Ewbank, Bob Hampton, and representatives of the underwriters, Morgan Stanley & Company and Rotan Mosle.

Starting approximately one week before the company was to go public, the team appeared in Houston, Dallas, New York, Boston, Chicago, Los Angeles and San Francisco. Keating Zeppa recalled, "At a luncheon or a breakfast or an afternoon meeting we would have a group from the investment community in that city—stockbrokers, analysts, representatives from money-managing firms, banks—and we would go through our little dog and pony show, with slides: What is Delta Drilling Company? Where did it come from? Where are we? Our earnings and sales. Then we'd answer questions. It was very tiring."

Meanwhile, employees holding all of the participation units had accepted the company's exchange offer, and by the middle of March this transaction had been completed. Delta distributed $15,965,000 in cash and issued 2,128,665 shares of common stock to 88 employees who held a total of 3,040,950 participation units. The newly distributed stock had a total value of $37,252,000, based upon what was to be the public offering price. The participation plan was based on salary and tenure, compared to the salary and tenure of everyone else who qualified for the plan.

While a few long-time employees pocketed more than $1 million as a result of the exchange offer, even those who had barely qualified were left with neat sums. "For example," said Keating Zeppa, "one person came into the plan in May of 1980 [upon completing 15 years] and when the date of anticipation [of going public] was set as of November, 1980, that person, who was making maybe $25,000 to $30,000 a year, in that six-month span of time had accrued about ten units, I believe, and each unit worth about $2600."

At least 13 employees became millionaires, and others received handsome windfalls. Elder "Catfish" Daniels, a heavy-equipment operator, was paid $11,600 in cash and $94,000 in stock. Jack Elkins, now a trucking coordinator, became a millionaire, along with Alton Blow, Red Magner, and others. The average benefit came to $612,000.

Keating Zeppa was quoted: "Some of these people have worked in the oil patch for 40 years. They worked hard and long, sometimes under the worst conditions you can imagine. The company didn't give them anything. They damned well earned it. Every penny."

Upon completion of the exchange offer, Delta offered to the public two million shares of common stock at $17.50 per share, the total price coming to $35 million, with proceeds to Delta amounting to $32,560,000 after expenses of selling the stock. After the company had made cash payments to the individuals in the participation plan, the remainder of the income from the public sale was applied toward reducing its debts under the bank revolving credit agreement, which on March 10, 1981, came to about $57 million.

On Monday, March 16, 1981, the board of directors met in New York City to review and approve final pricing of the issue, and by resolution dated March 16, 1981, approved an initial offering price of $17.50 per share, and a "gross spread" of $1.22 per share. The managing underwriters, attorneys, and public accountants worked throughout the night of the 16th to have finally printed the third amendment to the registration statement so it could be filed with the Securities and Exchange Commission early in the morning of the 17th.

All of Delta's directors—except Chris Zeppa, 80, who did not feel up to the trip— and their wives gathered in Morgan Stanley's section of the trading room, a vantage point equipped with computer terminals and cathode ray tubes where they could monitor the activity.

Keating Zeppa recalled the scene.

"We were going to be trading on Over the Counter, or NASDAQ—the National Association of Security Dealers' Automated Quotations—market, which is an electronic market. It's not like the New York Stock Exchange or American Stock Exchange where you have a trading floor and specialists that handle different things. The market is handled in an orderly fashion by a number of traders in your stock, with various securities firms around the country. There were 19 companies handling Delta stock over the country.

"Fifteen minutes before we went public, the underwriters had purchased their stock from the company, and they, in turn, were putting it out through their syndicated network. There were maybe 90 underwriters in the syndicate, and they pretty well had that stock placed. They knew who was interested.

"At about 10:15 on the morning of March 17, trading was authorized by the SEC. They called and said we could be permitted to 'go effective,' that they had reviewed the documentation and that they were satisfied with it. That means literally go effective that moment, and you can start selling your stock to the public.

"Here were all the directors, except Chris, and their wives. A lot of them were grouped around, sitting side by side around these terminals. The news went out on the wire—the Dow Jones news tape and on the NASDAQ system—to all brokerage houses. For about 30 minutes only syndicate members could sell the stock, for only they had it. Then they 'broke syndicate,' so others could buy. It was like a country-wide auction."

Interestingly, Delta had "gone public" on St. Patrick's Day, which has as its symbol the shamrock, the color of which is green—an appropriate coincidence to fit the occasion.

"To my mind," wrote Keating Zeppa in *The Delta Digger,* "New Yorkers showed uncommon support for this Texas-based event by adorning themselves with bright green all day long. I suppose there are some who would say the color was Kelly green, but most of us with Delta in the city on that day chose to believe it was Delta green."[76]

XIX

As Delta Drilling Company approached its golden anniversary on November 17, 1981, it was vigorous and healthy as it moved forcefully into its future as a publicly owned company.

Throughout the year, reminders of Delta's past were frequent as the public relations department coordinated a traveling display to the divisions, published a schedule of articles in *The Delta Digger* keyed to specific decades since 1931, sponsored the making of a historical film on the early 1930s in the East Texas and North Louisiana oil fields, and in November placed anniversary advertising in trade journals and newspapers serving regions where Delta operated. An industry-wide reception at the Petroleum Club of Houston on April 23 and a book party in Tyler on November 17, linked with the publication of this history, were the formal corporate affairs recognizing the fiftieth anniversary.

Ed Kliewer, who from his vantage point as corporate attorney and director had become a veteran Delta watcher over the decades, took cognizance of a clouded economic horizon as he speculated upon the company's future.

"I think that to the extent that one can at all sense what's going to happen, Delta is getting itself in place to take a damn good run at the future," he said. "It's increasing and updating its drilling equipment. There's got to be, to survive, an increase in the amount of drilling activity in this country. And Delta's getting ready for that, and some day is going to find additional reserves of oil and gas in this country."[77]

One could not help but harken back to that modest beginning in another economically clouded time 50 years before, in 1931.

"It is a remarkable success story," said Leonard Phillips, "and it could only exist in America. That a man could start with two junk drilling rigs and build an empire!"[78]

NOTES

Prologue: A Dream That Grew

1. Marcus O. Jones Interview.
2. Nicholas Andretta Interview.

1. Native and European Roots

1. Elizabeth Powell Fullilove Interview.
2. *Ibid.*
3. Robert A. Stacy, Jr., Interview.
4. Fullilove Interview.
5. Stacy Interview.
6. Fullilove Interview.
7. P. D. Connolly Interview.
8. Stacy Interview.
9. Connolly Interview.
10. Stacy Interview.
11. Fullilove Interview.
12. Stacy Interview.
13. Irving Howe, *World of Our Fathers* (New York: Harcourt, Brace, Jovanovich, 1976), p. 7.
14. Bernard Pares, *Russia* (New York: New American Library, 1949). p. 39.
15. Howe, *op. cit.* p.10.
16. Abraham J. Karp, ed., *Golden Door to America: The Jewish Immigrant Experience* (New York: Viking Press, 1976), p.8.
17. Samuel Joseph, *Jewish Immigration to the United States from 1881 to 1910* (New York: AMS Press, 1967), pp. 65-66.
18. *Ibid.* pp. 93, 100-101.
19. *Ibid.* p. 47.
20. *Ibid.* p. 42.
21. Thomas Kessner, *The Golden Door: Italian and Jewish Immigrant Mobility in New York City 1880-1915* (New York: Oxford University Press, 1977), p. 38.
22. Joseph, *op. cit.* p. 46.
23. Howe, *op. cit.* p. 39.
24. *Ibid.* p. 41; Milton Meltzer, *Taking Root: Jewish Immigrants in America* (New York: Farrar, Straus and Giroux, 1976), pp.34-36.
25. Howe, *op. cit.* pp.58-59.
26. Karp, *op. cit.* p. 15
27. Joseph, *op.cit.* p. 128.
28. Kessner, *op. cit.* xii-xiii.
29. *Ibid.* pp. 102-103.
30. Howe, *op. cit.* p. 62.
31. Kessner, *op. cit.* pp. 171-172.
32. Karp, *op. cit.* p. 216.
33. Abe Rozeman Interview.
34. Goldie Rappeport Interview.
35. *Ibid.*
36. J. R. Parten Interview.
37. Betty LaCour Interview, citing legal instruments filed at time of Gold's death.
38. Rappeport Interview.

39. Rose (Mrs. Sam) Bishkin Interview.
40. Dr. Myron H. Dorfman Interview.
41. Rappeport Interview.
42. Samuel J. Pearlman, M. D. to James Presley, October 20, 1980, February 16, 1981
43. Albert Sklar Interview.
44. Leonard Phillips Interview.
45. Sklar Interview.
46. Dr. Myron Dorfman Interview.
47. Sklar Interview.
48. Seymour L. Florsheim Interview.
49. Rappeport Interview.
50. Sklar Interview.
51. Marguerite Marshall Interview.
52. Sklar Interview.
53. Abe Rozeman Interview.
54. Elizabeth Fischer Interview.
55. *Ibid.;* Dr. Myron Dorfman Interview.
56. Rozeman Interview.
57. Fischer Interview; Florsheim Interview.
58. Phillips Interview.
59. Florsheim, Fischer, Sam Y. Dorfman, Jr., M. D., Interviews.
60. Dr. Myron Dorfman Interview.

2. A Boy from Italy

1. Joseph Lopreato, *Italian Americans* (New York: Random House, 1970), pp. 34-35.
2. Richard Gambino, *Blood of My Blood: The Dilemna of the Italian American* (Garden City, N. Y.: Doubleday, 1974), pp. 64-65.
3. Lopreato,*op. cit.* pp. 102-103.
4. Francesco Cordasco and Eugene Bucchioni, eds., *The Italians: Social Backgrounds of an American Group* (Clifton, N. J.: Augustas McNelley, 1974), p. 347. (Quotation from a reprinted article by Francis A. J. Ianni, "The Mafia and the Web of Kinship," *Public Interest* (Winter, 1971), pp. 78-100.)
5. Lopreato, *op. cit.* p. 101.
6. Thomas Kessner, *The Golden Door: Italian and Jewish Immigrant Mobility in New York City 1880-1915* (New York: Oxford University Press, 1977), pp. 28-31.
7. Joseph Zeppa to Mrs. Glen Jeffrey, May 10, 1963.
8. Zeppa Genealogy (Joe Zeppa Personal Files); Chris Zeppa Interview; Trent Zeppa Interview; Eric Schroeder notes of Joe Zeppa Interview.
9. Chris Zeppa Interview.
10. Joe Zeppa to Paola Zeppa, November 9, 1972.
11. Chris Zeppa Interview; Eric Schroeder notes of Joe Zeppa Interview; "Mud and Money," *The Delta Digger* (May-June, 1970), p. 1
12. Chris Zeppa Interview.
13. Eric Schroeder notes; Gambino, *op. cit.* p. 87; Chris Zeppa Interview.
14. Kessner, *op. cit.* p. 10.
15. *Ibid.* pp. 22-23.
16. *Ibid.* pp.13, 16-17; Chris Zeppa Interview.
17. Kessner, *op. cit.* pp. 169-170.
18. Joe Zeppa Interview.
19. Lopreato, *op. cit.* p. 157
20. Schroeder notes.
21. Joe Zeppa to Chris Camp, December 19, 1963.
22. Joe Zeppa Interview.
23. Chris Zeppa, Nick Andretta, Edward M. Keating, Leo Sides, Joe Zeppa Interviews; George T. Keating to Joe Zeppa, November 20, 1956; Schroeder notes.
24. Schroeder notes.
25. *New York Times,* April 1, 1967.
26. Cordasco and Bucchioni, *op. cit.* pp. 206-209 (reprint of an article by F. Aurelio Palmieri, "Italian Protestantism in the United States,"*Catholic World* (May, 1918; Vol. 107), pp. 177-189; Schroeder notes; Trent H. Zeppa Interview.
27. Keating V. Zeppa Interview.
28. For a pro-Republican editiorial from Chicago's *L'Italia* in 1894, "Italians Should Be Republicans," see Wayne Moquin and Charles Van Doren, *A Documentary History of the Italian Americans* (New York: Praeger Publications, 1974), pp. 304-305.

29. Giuseppe Zeppa, "Dichiarazione di arruolamento e di dispensa provvisoria dal servizio alle armi in tempo di pace di un militare nato all' estero od emigrato prima del 16° anno di eta," 26 settiembre 1913.
30. Joseph Zeppa, Declaration of Intention, June 30, 1916.
31. Edward M. Keating Interview.
32. Keating V. Zeppa Interview.
33. _____ Salmon to Joseph Zappa (telegram), May 22, 1917.
34. Joe Zeppa Interview; Joseph Zeppa to Helen C. Adams, July 26, 1917.
35. Zeppa to Adams, August 19, 1917.
36. Zeppa to Adams, September 9, 1917.
37. Zeppa to Adams, January 10, 1918.
38. Zeppa to Adams, March 22, 1918.
39. Zeppa to Adams, March 16, 1918.
40. Zeppa to Adams, April 28, 1918.
41. Zeppa to Adams, May 24, 1918.
42. Zeppa to Adams, July 5, 1918.
43. Zeppa to Adams, July 8, 1918.
44. Zeppa to Adams, August 3, 1918.
45. Zeppa to Adams, August 31, 1918.
46. *Ibid.*
47. Zeppa to Adams, October 1, 1918.
48. Zeppa to Adams, October 12, 1918.
49. Zeppa to Mrs. Sallie T. Ames, October 26, 1918.
50. *Ibid.*
51. Zeppa to Adams, November 15, 1918.
52. Zeppa to Adams, November 25, 1918.
53. Zeppa to Adams, November 30, 1918.
54. Joseph Zeppa, notebooks, Toulouse University Faculty of Law, April 8-May 1, May 22, June 18, April 4-27, June 9-25, 1919.
55. Joseph Zeppa, notebook, entry "GHC," April 27, 1919.
56. Chris Zeppa Interview.
57. Special Orders No. 87, A. E. F., University of Toulouse, Toulouse, France, May 27, 1919.
58. Joseph Zeppa, notebook, July 1, 1919, entry.
59. Zeppa notebook, July 14, 1919, entry.
60. *Ibid.*
61. *Ibid.* July 18, 1919, entry.

3. The Arkansas Connection

1. Deanie Beazley, "El Dorado—Boomtown, 1921 and 1971," *Petroleum Independent* (May-June, 1971), pp. 28-30 ff.
2. Joe Zeppa Interview.
3. Eric Schroeder notes.
4. Joe Zeppa Interview.
5. *Ibid.*
6. Schroeder notes.
7. Joe Zeppa Interview
8. Eula Carter Interview.
9. Joe Zeppa Interview.
10. Schroeder Notes.
11. Joe Zeppa Interview.
12. Beazley, *op.cit.*
13. Elizabeth Powell Fullilove, Robert A.Stacy, Jr., P. D. Connolly Interviews.
14. Goldie Rappeport, Seymour L. Florsheim, Elizabeth Fischer, Albert Sklar Interviews.
15. Joe Zeppa Interview.
16. Suzette Dayson Shelmire Interview.
17. J.R. Parten Interview.
18. Marcus O. Jones Interview.
19. Guy B. White Interview.
20. Ada Viola Portier Interview.
21. Nick Andretta Interview; *The Delta Digger* (November-December, 1976), pp. 10,14-15.
22. Mrs. O. F. Vollmer to James Presley, October 13, 1980.
23. Eula Carter Interview "A Different Breed of Cat,"*The Delta Digger* (May-June, 1977), pp. 6-7.
24. J. M. "Joe" Bevill Interview.
25. "The Lord's Been Good to Me," *The Delta Digger* (November-December, 1972), p. 12.

26. Robert W. Waddell Interview.
27. Chris Zeppa Interview.
28. "The Lord's Been Good to Me," *op. cit.*
29. Joe Zeppa Interview.

4. East Texas—The Beginning

1. Joseph Zeppa, Student Pilot's Permit, February 24, 1930, Personal Files (J.Z.).
2. E. G. "Eddie" Durrett Interview.
3. Trent H. Zeppa Interview.
4. Joe Zeppa Interview.
5. Trent Zeppa Interview.
6. Joe Zeppa Interview.
7. This section is based upon the Elizabeth Powell Fullilove and Robert A. Stacy, Jr., Interviews.
8. Staff, East Texas Salt Water Disposal Company, *Salt Water Disposal, East Texas Oil Field* (Austin: Petroleum Extension Service, The University of Texas, 1958; 2nd ed.), pp. 1-6. For more detailed discussions of the East Texas field, see James A. Clark and Michel T. Halbouty, *The Last Boom* (New York: Random House, 1972) and James Presley, *A Saga of Wealth: The Rise of the Texas Oilmen* (New York: G.P. Putnam's Sons, 1978), pp.111-180.
9. W. S. "Buck" Morris Interview.
10. Fullilove Interview.
11. Joe Zeppa Interview.
12. Trent Zeppa Interview.
13. Marvin Williams Interview.
14. Joe Zeppa Interview.
15. Fullilove Interview.
16. Minutes, Delta Drilling Company, November 17, 1931.
17. Minutes, December 9, 1931. (The word "unanimous" was handwritten into the typed minutes).
18. Joe Zeppa Interview.
19. Nathon Bacle Interview.
20. Joe Zeppa Interview.
21. Leo T. Sides and Edward M. Keating Interviews.
22. Trent Zeppa Interview.
23. Joe Zeppa Interview.
24. Joe Zeppa to Earl N. Wood, April 7, 1972.
25. Joe Zeppa Interview.
26. Eric G. Schroeder Notes; "Mud and Money," *The Delta Driller* (May-June, 1970). pp. 1-3; Ibid. (July-August, 1970), pp.1-3.
27. Fullilove Interview.
28. Nick Andretta Interview.
29. Fullilove Interview.
30. Minutes, Stockholders, December 22, 1931.
31. Minutes, Special Board Meeting, February 18, 1932.
32. Joe Zeppa Interview.
33. *Ibid.*
34. Joe Bevill Interview.
35. Chris Zeppa Interview.
36. Joe Zeppa Interview.
37. Andretta, Bacle Interviews.
38. S. L. Florsheim, Jr., Interview.
39. Minutes, Board, April 3, 1933.
40. *Ibid,* December 19, 1934.
41. *Ibid,* May 25, 1932.
42. Joe Zeppa Interview.
43. Andretta, Stacy, Bill Dailey Interviews.
44. Herman F. Murray Interview.
45. Trent Zeppa Interview.
46. Minutes, Stockholders, April 23, 1937.
47. Joe Zeppa Interview.
48. Floyd Sexton Interview.
49. Andretta Interview.
50. Dewey Strength Interview.
51. Williams Interview.
52. W. L. "Buster" Medlin Interview.
53. Strength Interview.

54. Ted T. Ferguson Interview.
55. Eula Carter, Herman F. Murray Interviews.
56. Medlin Interview.
57. Carter, Strength, Williams Interviews.
58. Medlin Interview.
59. Williams Interview.
60. Bacle Interview.
61. Willie "Bill" Dailey Interview.
62. Mr. and Mrs. Elmer S. Young Interview.
63. Medlin Interview.
64. Williams, Guy B. White, Marcus O. Jones Interviews.
65. Florsheim Interview.
66. Medlin Interview.
67. Trent Zeppa Interview.
68. Trent Zeppa, Andretta Interviews.
69. Keating V. Zeppa Interview.
70. Joe Milton Winston Interview.
71. Keating Zeppa Interview.
72. Marcus O. Jones Interview.
73. Ada Viola Portier Interview.
74. Herman F. Murray Interview.
75. Joe Beasley Interview.
76. Eula Carter Interview.
77. William B. Kennedy Interview; Bacle Interview; Portier Interview; Mrs. O. F. Vollmer to J. P., October 13, 1980.
78. Guy B. White Interview.
79. Hiram S. Cole Interview.
80. Portier Interview.
81. Herman Murray Interview.
82. Portier Interview.
83. H. J. Magner Interview.
84. Eula Carter, Cole, Guy White, Joe White, Herman Murray Interviews.
85. Cole Interview.
86. Magner Interview.
87. Joe White Interview.
88. Cole Interview.
89. Murray Interview.
90. Marcus O. Jones Interview.
91. C. L. Vickers Interview.
92. Joe White Interview.
93. Bacle Interview.
94. Joe Zeppa Interview.
95. Minutes, Board, October 4, 1932.
96. *Ibid.* August 10, 1937.
97. J. M. Bevill Interview.
98. Marcus Jones Interview.
99. Guy White, P.D. Connolly, Nick Andretta Interviews.
100. Marcus Jones Interview.
101. Nick Andretta Interview.
102. Joe White Interview.
103. J. M. Bevill Interview.
104. Robert A. Stacy, Jr., Marcus Jones, Bacle Interviews.
105. Fullilove Interview.
106. Minutes, Board, August 18, 1932.
107. *Ibid.* December 30, 1933.
108. *Ibid.* January 10, 1934.
109. Willie "Bill" Dailey Interview.
110. Joe Zeppa Interview.
111. Betty LaCour Interview, regarding Louisiana Iron and Supply charter and minutes; Rose Bishkin, Goldie Rappeport, Albert Sklar Interviews.
112. Frederick M. Mayer Interview.
113. J. R. Parten Interview; J. R. Parten to James Presley, January 29, 1981.
114. Joe Zeppa Interview; W. S. "Buck" Morris Interview; staff, East Texas Salt Water Disposal Company, *Salt Water Disposal, East Texas Oil Field* (Austin: Petroleum Extension Service, 1958; 2nd ed.), pp. 1-21.
115. Minutes, Board, January 8, 1935.

116. *Ibid.* January 14, 1936.
117. *Ibid.* September 14, 1938.
118. *Ibid.* December 14, 1936.
119. Minutes, Stockholders, December 9, 1937; Minutes, Board, December 9, 1937.
120. Dividend Record, Minute Book, Minutes, Board, September 17, 1937.
121. Murray Interview.
122. Bevill Interview.

5. The Stacy Interest

1. Minutes, Board, August 12, 1937.
2. Nicholas A. Andretta Interview.
3. Joe Zeppa Interview.
4. Elizabeth Powell Fullilove Interview.
5. R. A. Stacy to Board of Directors, Delta Drilling Company, November 21, 1940.
6. Minutes, Board, November 30, 1940.
7. Sklar to Delta Drilling Company, undated.
8. J. Zeppa to Delta Drilling Company, undated.
9. R. A. Stacy to Board of Directors, Delta Drilling Company, December 1, 1940.
10. J. Zeppa to R. A. Stacy, December 9, 1940, with endorsements.
11. Sam Sklar to J. Zeppa, President, Delta Drilling Company, December 15, 1940; Sam Y. Dorfman to J. Zeppa, President, Delta Drilling Company, December 15, 1940.
12. Minutes, Stockholders Annual Meeting, January 14, 1941.
13. Minutes, Board, January 14, 1941.
14. Joe Zeppa Interview.
15. Minutes, Board, January 29, 1941.
16. P.D. Connolly Interview.
17. Robert A. Stacy, Jr., Interview.
18. *Ibid.*

6. The Hawkins Deal and Delta Gulf

1. Edward Kliewer, G. I. "Red" Nixon, William O. "Shorty" Meyer Interviews.
2. G. I. "Red" Nixon Interview.
3. *Ibid.*
4. Ted T. Ferguson Interview.
5. Nixon Interview.
6. *Ibid.*
7. *Ibid.*
8. Nick Andretta Interview.
9. J. M. Bevill Interview.
10. Nixon Interview.
11. John D. Hall Interview.
12. Nixon Interview.
13. Chris Zeppa Interview.
14. Nixon Interview.
15. Joe Zeppa Interview.
16. C. L. Vickers Interview.
17. Nixon Interview.
18. Edward Kliewer Interview.
19. Chris Zeppa Interview.
20. J. M. "Mark" Gardner Interview.
21. Hall Interview.
22. Bevill Interview.
23. Agreement as recorded in Minute Book, Delta Drilling Company.
24. Minutes, Board, July 7, 1949.
25. Minutes, Board, August 6, 1949; Minutes, Stockholders, August 6, 1949.
26. Robert W. Waddell Interview.
27. Minutes, Board, Delta Drilling Company, October 6, 1949.
28. Minutes, Stockholders, Delta Gulf Drilling Company, May 25, 1950.
29. Minutes, Board, Delta Drilling Company, December 29, 1952.
30. William D. Craig Interview.
31. Bevill Interview.
32. Andretta Interview.
33. Bevill Interview.

34. Samuel D. Myers, *The Permian Basin, Petroleum Empire of the Southwest: Era of Advancement, From the Depression to the Present* (El Paso: Permian Press. 1977), p. 371.
35. Leonard Phillips Interview.
36. John Justice Interview.
37. Alton Blow Interview.
38. Nixon Interview.
39. Ferguson Interview.
40. Nixon Interview.
41. Joe Zeppa Interview.
42. Nixon Interview.
43. J.M. Bevill and John D. Hall to J. Zeppa, September 9, 1950.
44. Minutes, Board, Delta Drilling Company, August 7, 1951; J. M. Gardner Interview.
45. Chris Zeppa Interview.
46. Minutes, Board, Delta Drilling Company, August 7, 1951.
47. Minutes, Board, Delta Drilling Company, December 29, 1952.
48. Nixon Interview.
49. Minutes, Board, Delta Drilling Company, December 12, 1956.
50. General Conveyance Assignment, Transfer, and Bill of Sale, Delta Gulf Drilling Company to Delta Drilling Company, December 29, 1956.
51. See Edward Kliewer, Jr., To Files, Inter-Office Correspondence, May 3, 1957; and Minutes, Board, Delta Drilling Company, March 19, 1957.

7. Domestic Outposts

1. Joe Bevill Interview.
2. Joe Zeppa Interview.
3. Minutes, Board, August 10, 1944.
4. Trent H. Zeppa Interview.
5. Chris Zeppa Interview.
6. William D. Craig Interview.
7. Keating V. Zeppa Interview.
8. Edward M. Gandy Interview.
9. Jack Elkins Interview.
10. Roger Choice Interview.
11. Homer Lee Terry Interview.
12. Craig Interview.
13. Thornton Tarvin Interview.
14. J. H. Gilleland Interview.
15. Gandy Interview.
16. W. Hugh McKenzie Interview.
17. Keating Zeppa Interview.
18. Warren Strahan Interview.
19. Guy B. White Interview.
20. H. J. Magner Interview.
21. Herman F. Murray Interview.
22. Gus E. Jones Interview.
23. Bert Gauntt, Gandy Interviews.
24. Magner Interview.
25. Minutes, Board, December 12, 1942.
26. *Ibid.* September 21, 1943.
27. R. C. Johnston to Board of Directors, Delta Gulf Company, August 1, 1944.
28. Trent Zeppa, Albert Sklar Interviews.
29. Magner, White, W. B. Kennedy Interviews.
30. Nick Andretta, J. M. Bevill Interviews.
31. Joe Zeppa to George T. Keating, August 21, 1943.
32. Jim Ewbank Interview.
33. Minutes, Board, December 19, 1961.
34. Tommy J. Blackwell Interview.
35. Minutes, Board, November 16, 1965; Minutes, Shareholders, March 15, 1966; *Dallas Morning News*, Sept. 14, 1967.
36. Gandy Interview.
37. Marvin Williams Interview.
38. J. M. Bevill Interview.
39. Dewey Strength Interview.
40. Minutes, Board, April 30, 1942.

41. Minutes, Board, August 24, 1942; Annual Board Meeting, January 9, 1945.
42. T. A. Everett, Lee Browning Interviews.
43. Robert W. Waddell Interview.
44. Everett Interview.
45. Waddell Interview.
46. Minutes, Board, December 23, 1946.
47. L.A. Bates, Melburn Miser, Guy B. White, Malcolm Eakins, Albert Sklar Interviews.
48. White Interview.
49. Auburn Bryant, Magner, White, Edward E. Kliewer, Jr., Interviews: Minutes, Board, December 1, 1949.
50. Magner Interview.
51. Minutes, Board, December 20, 1947.
52. *Ibid.* March 19, 1948; May 10, 1948; July 29, 1948.
53. Early F. Smith Interview; Minutes, Board, December 29, 1950; July 3, 1952; November 4, 1953.
54. William D. Craig Interview.
55. Mr. and Mrs. Frank Dykstra Interview.
56. Robert W. Waddell Interview; Minutes, Board, August 8, 1959; Garland Cox, Warren Strahan, Flo Bonham, J. W. Hardin Interviews; Flo Bonham, "The Orange Blossom Special," *The Delta Digger* (March-April, 1976), pp. 6-7
57. J. M. Bevill Interview.
58. J. Zeppa to George T. Keating, Jan. 14, 1943; Minutes, Board, August 10, 1944.
59. Ted T. Ferguson Interview.
60. Lynna Jo Edwards, Edward L. Doughty, J. H. Gilleland Interviews.
61. E. G. Durrett, Edwards Interviews.
62. John E. Faustlin, Doughty, T. C. Carlton Interviews.
63. Minutes, Board, December 21, 1954.
64. Ferguson Interview.
65. Edwards, Magner Interviews.
66. Durrett Interview.
67. Gilleland, Ferguson Interviews.
68. Minutes, Board, December 26, 1963.
69. Minutes, Shareholders, March 16, 1965.
70. Jody Conaway Interview.
71. Lonye Cain Interview.
72. Minutes, Board, June 23, 1950, February 7, 1952; H. J. Magner Interview; Minutes, Board, July 3, 1952.
73. Minutes, Board, November 12, 1953.
74. L. A. Bates Interview; Minutes, Board, November 4, 1953, May 9, 1955, March 17, 1970.
75. Don Carman Interview.
76. Auburn Bryant Interview.
77. Dallas S. Bryant Interview.
78. Melburn Miser Interview.
79. Magner Interview; "Air Drilling: Meeting the Challenge," *The Delta Digger* (July-Aug. 1979), pp. 10-11.
80. Miser, Malcolm Eakins, Auburn Bryant, Don Carman Interviews.
81. Magner, White Interviews.
82. Miser, Magner, Dallas Bryant, Eakins, Bates Interviews.
83. Waddell, Z. L. Spraggins, Flo Bonham Interviews; Minutes, Board, November 6, 1970.
84. Herman Gordon Interview.

8. Ups and Downs

1. Minutes, Board, December 23, 1955.
2. Minutes, Stockholders, March 20, 1956.
3. Minutes, Board, March 20, 1956.
4. Thornton Tarvin Interview.
5. W. H. Whitfield Interview.
6. Minutes, Stockholders, March 18, 1958.
7. Minutes, Board, December 23, 1959.
8. Minutes, Board, December 22, 1960.
9. Minutes, Stockholders, March 27, 1964.
10. Minutes, Board, December 15, 1966.
11. Minutes, Board, December 11, 1967, May 29, 1968, December 11, 1968; Stockholders, March 11, 1969; J. Zeppa to J. M. Gardner, October 16, 1969; Mark Gardner to Joe Zeppa, October 22, 1969; Minutes, Stockholders, March 17, 1970; J. Zeppa, President, Delta Machine Drilling Co. to First National City Bank, New York, June 11, 1970; J. Zeppa, President, Delta Drilling Co. to First National City Bank, New York, June 11,

50. Robert W. Waddell Interview.
51. McKenzie Interview.
52. Minutes, Board, December 23, 1959, December 22, 1960.
53. Minutes, Stockholders, March 21, 1961.
54. Waddell Interview.
55. Minutes, Board, December 19, 1961; Stockholders, March 20, 1962; Board, December 17, 1962; February 26, 1963.
56. Edward Kliewer, Jr., Interview.
57. Minutes, Board, December 26, 1963; Stockholders, March 27, 1964.
58. John D. Hall Interview.
59. Minutes, Shareholders, March 16, 1965.
60. *Ibid.* March 15, 1966, December 15, 1966; Minutes, Shareholders, March 21, 1967; Board, May 26, 1967.
61. Minutes, Board, December 11, 1968; Stockholders, March 18, 1969; Board, November 20, 1969.
62. Minutes, Board, December 7, 1971; March 21, 1972.
63. Bernie Wolford Interview.

10. South of the Border

1. Minutes, Board, July 21, 1959; J. M. Bevill to G. L. Downton, Washington Iron Works, June 14, 1963.
2. John Hall to J. Zeppa and Joe Bevill, undated (received August 14, 1959); *ibid.* (received August 17, 1959)
3. Keating V. Zeppa Interview.
4. Minutes, Board, December 23, 1959.
5. Minutes, Board, December 22, 1960.
6. Minutes, Stockholders, March 21, 1961; Minutes, Board, December 19, 1961.
7. Minutes, Stockholders, March 20, 1962; Minutes, Board July 9, 1962; *ibid.* Dec. 17, 1962.
8. Minutes, Board, December 26, 1963.
9. Minutes, Board, undated but preceding March, 1964 meeting (probably February, 1964); Minutes, Board, November 16, 1965.
10. Minutes, Board, May 26, 1967.
11. *Ibid.* December 11, 1967; Minutes, Stockholders, March 18, 1969.
12. Minutes, Board, November 20, 1969; *ibid.* December 7, 1971.
13. Minutes, Stockholders, March 21, 1972; *ibid.* March 20, 1973.
14. *Ibid.* March 18, 1975.
15. Minutes, Board, July 21, 1959.
16. J. M. Gardner Interview.
17. Edward Kliewer, Jr. Interview.
18. Minutes, Board, December 22, 1960.
19. Walter L. McElroy Interview.
20. Keating V. Zeppa Interview.
21. Minutes, Stockholders, March 21, 1961; Minutes, Board, August 18, 1961.
22. Minutes, Board, December 19, 1961.
23. Max W. Ball, Douglas Ball, Daniel S. Turner, *This Fascinating Oil Business* (Indianapolis, New York: Bobbs-Merrill Co., 1965; rev. ed.), p. 410.
24. Keating Zeppa Interview.
25. Gardner Interview.
26. Minutes, Stockholders, March 20, 1962; Board, July 9, 1962; Board, December 17, 1962.
27. Minutes, Stockholders, March 16, 1963; Board, December 26, 1963.
28. Keating Zeppa, John D. Hall, Jack D. Robinson Interviews.
29. Joe Zeppa to Jack Blalock and Clarence Lorman, December 27, 1961.
30. Charles C. Delton Interview.
31. Minutes, Board, December 23, 1964.
32. Minutes, Board, November 16, 1965.
33. Minutes, Shareholders, March 15, 1966.
34. Minutes, Board, December 15, 1966; Shareholders, March 21, 1967.
35. Minutes, Board, December 11, 1967.
36. *Ibid.* May 29, 1968, December 11, 1968; Minutes, Stockholders, March 18, 1969.
37. Minutes, Board, November 20, 1969.
38. Minutes, Stockholders, March 17, 1970, March 16, 1971.
39. Minutes, Board, July 21, 1971.
40. Minutes, Stockholders, March 21, 1972.
41. Minutes, Board, December 4, 1972.
42. Minutes, Stockholders, March 20, 1973.
43. Minutes, Board, October 31, 1974.
44. Joe Zeppa Interview.
45. Kliewer Interview.

46. Minutes, Stockholders, March 21, 1961.
47. Minutes, Board, August 17, 1963.
48. *Ibid,* March 27, 1964.
49. Charles C. Delton Interview.
50. Bernie Wolford Interview.
51. Nolan R. Jones, John D. Hall Interviews.
52. Jones Interview.
53. Delton Interview.·
54. Frank Dykstra Interview.
55. Delton Interview.
56. Dykstra Interview.
57. Delton Interview.
58. Dykstra Interview.
59. Wolford Interview.
60. Delton Interview.
61. Minutes, Shareholders, March 16, 1965.
62. Minutes, Board, November 16, 1965.
63. Minutes, Shareholders, March 15, 1966; Board, December 15, 1966, May 26, 1967.
64. Dykstra Interview.
65. Minutes, Directors, December 11, 1967; Board, May 29, 1968.
66. Minutes, Board, December 11, 1968.
67. *Ibid.* November 20, 1969.
68. Minutes, Stockholders, March 17, 1970.
69. *Ibid.* March 16, 1971.
70. Minutes, Board, July 21, 1971.
71. Minutes, Shareholders, March 20, 1973; Board, December 4, 1973.
72. Minutes, Board, October 31, 1974.
73. *Ibid.* December 17, 1974.
74. Minutes, Stockholders, March 18, 1975; Board, March 18, 1975.
75. Bernie Wolford Interview; Grace Baker to J. Presley, January 12, 1981.
76. Minutes, Board, December 23, 1957.
77. [Chairman] To the Board of Directors, Delta Drilling Company, June 1, 1973; Minutes, Board, June 5, 1973.
78. S. L. Florsheim, Jr., Sam Y. Dorfman, Jr. Interviews.
79. Joe Zeppa, Keating Zeppa Interviews.

11. Meanwhile, Out at the Farm

1. Roger Choice Interview.
2. Joe Milton Winston Interview.
3. *Ibid.*
4. Marguerite Marshall Interview.
5. Winston Interview.
6. John Murrell Interview.
7. Winston Interview.
8. Joe Zeppa Interview.
9. Winston Interview.
10. *Ibid.*
11. Murrell Interview.
12. Eric G. Schroeder, "Determination Paved Road from Italy to East Texas," *Dallas Morning News,* November 19, 1950.
13. John L. Merrill to Joe Zeppa, June 18, 1968; John L. Merrill to Joe Milton Winston, June 18, 1968.
14. Tyler *Morning Telegraph,* February 18, 1972.
15. Murrell Interview.
16. Roy Burchfield Interview.
17. John Robert Gunn Interview.
18. Avis Boren Interview.
19. Minutes, Board, December 17, 1962.
20. Minutes, Board, December 7, 1970; Shareholders, January 6, 1971.
21. Winston, Trent H. Zeppa, Keating V. Zeppa, Eric G. Schroeder Interviews.
22. Joe Zeppa to George T. Keating, March 10, 1944.
23. *Ibid.* January 18, 1951.
24. *Ibid.* February 14, 1952.
25. *Ibid.* May 8, 1952.
26. Joe Zeppa to Paola Zeppa, November 9, 1972.
27. Winston Interview.

12. **At The Center of an Empire**

1. Joe Zeppa Interview.
2. Keating Zeppa, H. J. Magner, Trent H. Zeppa Interview.
3. John H. Justice Interview.
4. Richard Gambino, *Blood of My Blood: The Dilemma of the Italian-Americans* (Garden City, N.Y.: Doubleday, 1974), pp. 80-81.
5. *Ibid*. p. 7.
6. Edward E. Kliewer, Jr., Interview.
7. Leonard Phillips Interview.
8. Trent Zeppa Interview.
9. Watson A. Wise, Trent Zeppa, C. L. Vickers, Homer Lee Terry, J. M. Bevill, Warren L. Baker, Tommy J. Blackwell Interviews.
10. J. M. Gardner Interview.
11. *Ibid*.
12. Stephen W. Schneider Interview.
13. Nicholas Andretta Interview.
14. Jim Ewbank Interview.
15. Keating Zeppa Interview.
16. John D. Hall Interview.
17. Bevill Interview.
18. Phillips Interview.
19. Schneider Interview.
20. Marguerite Marshall Interview.
21. Ewbank Interview.
22. Gardner Interview.
23. Keating Zeppa Interview.
24. Joe Milton Winston Interview.
25. Marshall Interview.
26. I am indebted to John F. Presley for this insight, December 14, 1980.
27. Joe Zeppa to Paola Zeppa, February 13, 1971.
28. Frederick Mayer Interview.
29. Schneider Interview.
30. Mayer Interview.
31. Joe Zeppa to George T. Keating, November 29, 1951.
32. Terry, Winston, Avis Boren, Robert W. Waddell, C. L. Vickers, Marshall Interviews.
33. E. G. Durrett Interview.
34. Chris Zeppa, Trent Zeppa, Terry, G. I. Nixon Interviews.
35. Joe Zeppa to Clarence Davisson, September 13, 1963.
36. Joe Zeppa to Edward M. Keating, March 27, 1952.
37. *Ibid*. January 5, 1951.
38. Marshall, Ralph Spence, Lynna Jo Edwards, Waddell Interviews.
39. Wise, Mayer, Henry Bell, Jr., Nixon, S. L. Florsheim, Jr., Terry, Marshall, Elizabeth Fischer, Winston Interviews.
40. Waddell, Burchfield, Terry Interviews.
41. Joe Zeppa Interview.
42. Keating Zeppa Interview.
43. H. J. Magner Interview.
44. Marshall Interview.
45. Winston Interview.
46. *Ibid*.
47. Keating Zeppa Interview.
48. Ewbank, Flo Bonham Interviews.
49. Terry Interview.
50. Marshall Interview.
51. Grace Baker Interview.
52. Boren Interview.
53. Hall, Waddell Interviews.
54. Gardner Interview.
55. *Ibid*.
56. *Ibid*.
57. Grace Baker Interview.
58. Minutes, Board, January 13, 1948, March 21, 1972.
59. Waddell Interview.
60. Keating Zeppa Interview.

61. *Ibid.*
62. Marshall Interview.
63. Winston Interview.
64. Keating Zeppa Interview.
65. Marshall Interview.
66. Gardner Interview.
67. Waddell Interview.
68. Warren L. Baker Interview.
69. Gardner Interview.
70. Marshall, Early F. Smith Interviews.
71. John Bainbridge, *The Super-Americans* (Garden City, N.Y.: Doubleday & Co., 1961), p. 65.
72. Trent Zeppa Interview.
73. Warren L. Baker Interview.
74. Joe Zeppa Interview.
75. Stephen Schneider Interview.
76. J. Zeppa to Robert B. Kamon, May 22, 1974.
77. Joe Zeppa to George T. Keating, May 28, 1953.
78. Joe Zeppa Interview.
79. Maxine Fant Interview.
80. Roy Burchfield Interview.
81. See O. Wayne Crisman to Joe Zeppa, July 10, 1970 (J. Z. Personal Files).
82. Magner Interview.
83. Letter in files, dated July 14, 1961.
84. Joe Zeppa Interview.
85. Trent Zeppa Interview.
86. J. M. Bevill Interview.
87. Keating Zeppa Interview.
88. Ralph Spence Interview.
89. Winston Interview.
90. Henry Bell, Jr., Interview.
91. Marshall Interview.
92. Warren Hirsch Interview.
93. Bell Interview.
94. Grace Baker Interview.
95. Suzette Dayson Shelmire, Joe Massie, J. R. Parten, Keating Zeppa, Trent Zeppa, J. M. Gardner, Frederick Mayer Interviews.
96. Grace Baker Interview.
97. Anthony Gibbon, Tiburon Adventure (manuscript).
98. Joe Zeppa to Anthony Gibbon, October 12, 1961; Anthony Gibbon to Guillermo Álvarez-Morphy, October 13, 1961.
99. Joe Zeppa to Anthony Gibbon, May 22, 1974.
100. Joe Zeppa to George T. Keating, December 10, 1969.
101. Marshall Interview.
102. George T. Keating, "Born a Collector," *The Cabellian: A Journal of the Second American Renaissance* (III, 2; Spring 1971), pp. 83-86.
103. *Ibid.* 85
104. *Ibid.* George T. Keating to Joe Zeppa, undated (probably 1964); February 22, 1964.
105. George T. Keating to Joseph Zeppa, March 3, 1964; *Ibid.* October 9, 1964; Joseph Zeppa to Robert M. Trent, April 23, 1965; George T. Keating to Joseph Zeppa, November 30, 1964.
106. George T. Keating to Joseph Zeppa, October 9, 1964; August 22, 1965.
107. "Mustang Report," *The Mustang* (18, 2; October, 1965), p. 7; *Announcing the Joseph Zeppa Collection of War, Diplomacy and Peace* (Dallas: Southern Methodist University, 1967); *Special Collections in The Libraries at Southern Methodist University* (Dallas: SMU, 1973).
108. Blackwell Interview.

13. Changing Times

1. Joseph Zeppa, "Present Drilling and Completion Methods and Their Effect on Costs of Exploratory and Development Wells," in the Southwestern Legal Foundation, *Proceedings of the 1962 National Institute for Petroleum Landmen* (Albany, N.Y.: Matthew Bender & Co., 1963), p. 245.
2. Flo Bonham, Bert Gauntt, L.A. "Gus" Mayton, Bill Goolsby Interviews.
3. Chris Zeppa Interview.
4. George T. Keating to Joe Zeppa, April 28, 1967.
5. Keating Zeppa Interview.
6. Robert W. Waddell Interview.

7. Edward Kliewer, Jr. Interview.
8. T. J. Blackwell Interview.
9. J. M. Bevill Interview.
10. "Participation Plan for Employees of Delta Drilling Company and Delta Marine Drilling Company," March 18, 1974.
11. Frank E. Schoof to Joe Zeppa, March 4, 1950.
12. J. M. Gardner Interview.
13. H. J. Magner Interview.
14. Waddell Interview.
15. J. W. Hardin Interview.
16. Fax C. Marshall Interview.
17. Gordon Burke Interview.
18. Charlie Lee Butler Interview.
19. W. Hugh McKenzie Interview.
20. Willis S. Howard Interview.
21. Doris Darbonne Interview.
22. Warren Hirsch Interview.
23. Willis and Mamie Howard Interview.
24. Edward M. Gandy Interview.
25. Donald G. Peters Interview.
26. *Safety Manual on the Handling of Sour Gases* (Tyler; Delta Drilling Company, 1973, rev. ed.).
27. Hirsch Interview.
28. Bill Goolsby Interview.
29. *Ibid.*
30. Esset Ates Interview.
31. Z. L. Spraggins Interview.
32. W. Hugh McKenzie Interview.
33. Finnis Cooper, Jr., Interview.
34. Charlie Butler Interview.
35. Keating Zeppa Interview.
36. Spraggins Interview.
37. T. A. Everett Interview.
38. Sarah Clark Interview.
39. Claudine Rogers, Trent Zeppa Interviews.
40. Stephen W. Schneider to All American Wildcatter, March 23, 1973.
41. "Introducing . . . Floorperson Jaculine [sic] Wilson LaCouture: Rig No. 31, " *The Delta Digger* (May-June, 1973), p. 3.
42. "half pint," *Ibid.* (May-June, 1974). p. 8.
43. Flo Bonham Interview.
44. H. J. Magner Interview.
45. Bernie Wolford Interview.
46. Ted T. Ferguson Interview.
47. Magner Interview.
48. J. M. Bevill Interview.
49. "Delta Checks Off Its 40th Year," *Drilling-DCW* (November, 1971), pp. 20-26.

14. The Death of a Patriarch

1. Joe Zeppa Interview.
2. Chris Zeppa Interview; Chris Zeppa to Paola Zeppa (telegram), May 20, 1963; Chris Zeppa to Paola Zeppa (telegram), November 11, 1963; Joe Zeppa to John Meditz, June 4, 1973.
3. Joe Zeppa to George T. Keating, July 8, 1964.
4. J. Zeppa to L. V. Portier, June 26, 1972.
5. *Shreveport Times,* July 31, 1961, p. 1.
6. *New York Times,* April 1, 1967; *Miriam Osborn Memorial Home: Sixtieth Anniversary, 1908-1968* (Rye, N.Y., 1968), pp. 9, 36.
7. Marvin Williams Interview.
8. Joe Zeppa to E. Constantin, Jr., May 15, 1965.
9. Keating Zeppa Interview.
10. Milton Winston Interview.
11. Keating Zeppa Interview.
12. Ray H. Skaggs, M.D., to Joseph Zeppa, April 10, 1967.
13. Marguerite Marshall, Grace Baker Interviews.
14. Milton Winston Interview.
15. Homer Lee Terry Interview.

16. Marguerite Marshall Interview.
17. Robert W. Waddell Interview.
18. Jim Ewbank Interview.
19. Marguerite Marshall Interview.
20. Joann Blalock Interview.
21. Joe Zeppa to Mr. and Mrs. Lawrence Barnard, November 9, 1971.
22. Joe Zeppa interview.
23. Joe Zeppa to Phil Russell, June 3, 1975.
24. Joe Zeppa Interview.
25. Minutes, Board, June 13, 1975.
26. Joe Zeppa to Henry M. Bell, June 17, 1975; Henry M. Bell, Jr., Interview; Aileen Wilkerson Interview.
27. T. J. Blackwell Interview.
28. Keating Zeppa Interview.
29. Suzette Dayson Shelmire Interview.
30. Milton Winston Interview.
31. Keating Zeppa Interview.
32. Marguerite Marshall Interview.
33. *Ibid.*
34. "Editorial: Joe Zeppa's Life . . . Story of Success," Tyler *Morning Telegram,* July 10, 1975.
35. Keating V. Zeppa to Calvin Clyde, Jr., July 10, 1975.
36. Keating Zeppa Interview.
37. Suzette Shelmire, Watson Wise Interviews.
38. Minutes, Board, July 9, 1975.
39. Keating Zeppa Interview.
40. *The Delta Digger* (July-August, 1975), p. 1.

15. A New Generation

1. Trent Zeppa Interview.
2. Minutes, Board, August 21, 1975.
3. Minutes, Shareholders, March 21, 1978
4. Minutes, Board, July 22, 1977.
5. Minutes Board, 1975-1980, *passim.*
6. Minutes, Board, May 23, 1980.
7. Keating Zeppa Interview.
8. Minutes, Board, March 21, 1978.
9. Sam Y. Dorfman, Jr., Interview.
10. Keating Zeppa Interview.
11. Dorfman Interview; *Prospectus, Delta Drilling Company* (New York: Morgan Stanley & Co., Rotan Mosle, Inc., 1981), p. 45.
12. Keating V. Zeppa, Transcript, Address to IADC-PESA, Conference, Dallas, 1980 (hereafter cited as K. Zeppa Transcript).
13. Flo Bonham Interview.
14. Keating Zeppa Interview.
15. Thornton Tarvin Interview.
16. Flo Bonham Interview.
17. Merrel Greenwell Interview.
18. Gloria Stackpole Interview.
19. T. A. Everett Interview.
20. Z. L. Spraggins Interview.
21. Flo Bonham Interview.
22. Grace Baker Interview.
23. Avis Boren Interview.
24. Edward Kliewer Interview.
25. Keating Zeppa Interview.
26. Keating Zeppa Transcript.
27. Glen Koonce Interview.
28. Bernie Wolford Interview.
29. Doris Darbonne Interview.
30. Flo Bonham Interview.
31. Z. L. Spraggins Interview.
32. Kenneth Fowler Interview.
33. Charles R. Martin Interview.
34. Jack D. Robinson Interview.
35. Edward Gandy Interview.

36. Kenneth Fowler Interview.
37. Flo Bonham Interview.
38. Thornton Tarvin Interview.
39. Fax C. Marshall Interview.
40. Keating Zeppa Interview.
41. Jack D. Robinson Interview.
42. Joe M. Baker, Bill Goolsby Interviews.
43. Claudine Rogers Interview.
44. John H. Justice Interview.
45. Keating Zeppa Interview.
46. Jack Robinson Interview.
47. Charles R. Martin Interview.
48. Minutes, Board, February 20, 1976.
49. *Ibid*. January 21, 1977.
50. *Ibid*. July 22, 1977.
51. Keating Zeppa Interview.
52. Minutes, Board, February 20, 1976.
53. H. J. Magner Interview.
54. Leonard Phillips Interview.
55. Jim Ewbank Interview.
56. Minutes, Board, January 27, 1978.
57. Ed Kliewer Interview.
58. T. C. Carlton Interview.
59. Merrel Greenwell Interview.
60. Minutes, Board, January 27, 1978.
61. Greenwell Interview.
62. Minutes, Board, March 21, 1978.
63. Richard J. McDonough Interview.
64. Kliewer Interview.
65. Frank Dykstra, Bill Goolsby Interviews.
66. Albert Sklar, John D. Hall, Leonard Phillips, Trent Zeppa, Milton Winston, Joann Blalock, Edward Kliewer Interviews.
67. Joann Blalock Interview.
68. Keating Zeppa Interview.
69. Gloria Stackpole Interview.
70. W. Hugh McKenzie Interview.
71. Keating V. Zeppa to John B. Wells, Jr., December 31, 1968.
72. Keating V. Zeppa to Allyn F. Taylor, July 18, 1977.
73. Leonard Phillips Interview.
74. Keating Zeppa Interview.
75. Minutes, Board, May 23, 1980.
76. Keating V. Zeppa, "Summary of Events Leading to Public Offering of Delta Drilling Company's Stock," March 26, 1981; Keating Zeppa interview; *Prospectus, passim;* Associated Press story, Texarkana *Gazette,* May 25, 1981; "View from the Top," *The Delta Digger* (March-April 1981), p. 2; Shawn Tully, "Sudden Wealth in Tyler, Texas," *Fortune* (June 1, 1981), pp. 40-43.
77. Edward Kliewer Interview
78. Leonard Phillips Interview

BIBLIOGRAPHY

This book is based primarily on two types of sources—corporate archival material and oral history—in an attempt to reconstruct the past. Obviously, each type has its limitations. Records do not always give the emotional tone to an event, frequently do not provide significant details, and, in many instances, fail completely to take note of some events. On the other hand, oral history has the weakness that human memory is rarely perfect, as the years color the reminiscences of long ago, sometimes editing, occasionally embellishing or ignoring. Furthermore, one person is unlikely to see everything that happens.

The approach has been to tie events to documents or correspondence whenever feasible but to rely heavily upon personal memories for reports of "how it was," particularly when eye-witness accounts were available. Occasionally, when it seemed appropriate and I deemed it reliable, I have admitted hearsay into the narrative. In certain sections I have resorted to secondary sources, particularly books about the Jewish and Italian migrations to this country.

Archival Material

I was permitted free access to Delta Drilling Company's central files in the corporate headquarters at Tyler. Company archives yielded, basically, three kinds of frequently consulted primary sources: corporate minutes, Joe Zeppa's personal files, and all other company files. I read all of the company minutes from 1931 on, as well as every carbon of Joe Zeppa's personal correspondence and other personal papers. I was considerably more selective on the "other" material, else I would still be researching the book.

Corporate Minutes

These records of board and stockholder meetings provide the foundation for the book; the minute books nail down dates and amounts, while at the same time charting any fluctuations in a particular operation. A weakness of the minute books, however, is that when a division or a subsidiary is doing well, it is likely to go unmentioned in the company minutes, while problems surface repeatedly. The minutes, however, are detailed and usually precise in describing matters which came before the board and at the annual shareholders meetings.

Joe Zeppa's Personal Files

These break down into:

Personal correspondence—Several file drawers are filled with carbons of Zeppa's personal letters from the last couple of decades of his life.

World War I journals—Zeppa's handwritten entries from shipboard, from the field during the hostilities, from Toulouse University after the Armistice, as well as when en route to the States, provide an insight into the young man's mind that few other businessmen have left behind.

Joe Zeppa's letters to Helen C. Adams—This large stack of letters, returned to Zeppa following Miss Adams's death, are filled with accounts of activities in which Zeppa participated, as well as of what he saw while in France. They also give us a window to his mind as he discusses his viewpoints to his old confidante.

Other Corporate Records

These include virtually everything in the central files that did not fit into the preceding categories. Subject matter might range from agreements on a specific well to merger proposals, from overseas operations to employee retirement plans. Because it was not physically possible to go through these long rows of filing cabinets systematically, I browsed through them at will, sometimes spot checking the contents of a file drawer, sometimes delving into stacks of documents. For the most part, however, I relied upon ever-efficient Grace Baker, curator of the central files, to find for me what I needed. She never failed me, and I marvel that one person can maintain so many facts in her head.

Oral History

I was able to talk with employees and retired employees, and had the cooperation of former employees whom I was able to locate. At the outset, Delta Drilling Company's public-relations department provided me with a list of retired employees and those employees whose connection with the company stretched far back or who had special vantage points in particular events. And as my conversations yielded new information, I was able to add more names to my interview list—some of whom proved to be among my most productive informants.

The interviews began, of course, with Joe Zeppa in 1975 for an earlier book; the remaining interviews began in 1980. Whenever possible, the employees were interviewed on company time. For those busy on a drilling rig, this was not possible, necessitating my seeing them on days off, often in their homes. The bulk of the interviews were conducted in Tyler, though I also traveled to the division headquarters in Lafayette, Louisiana, Odessa and Victoria, Texas, as well as to a number of cities such as Dallas, Houston, Shreveport, and Austin to see persons whose knowledge of the Delta's history was pertinent. The longest trip was to Port Townsend, Washington, to interview Trent H. Zeppa, several months before he died. I conducted a number of interviews by telephone.

The following interviews, practically all of which were tape-recorded, averaged close to two hours. In every location where the state is not identified, it is in Texas.

Andretta, Nicholas A. Tyler, February 21, 1980; (telephone), February 19, 1981.

Ates, Esset. Kilgore, March 31, 1980.

Bacle, Nathon. Tyler, March 4, 1980.

Baker, Grace. Tyler, May 13, 22, 1980.

Baker, Joe M. Tyler (telephone), May 1, 1981.

Baker, Warren L. Houston, May 8, 1980.

Bates, Leonard A. Indiana, Pennsylvania (telephone), August 12, 1980.

Beasley, Joe. Junction (telephone), February 14, 1981.

Bell, Henry M., Jr. Tyler, June 2, 1980.

Bevill, James M. "Joe." Tyler, February 21, 1980.

Bishkin, Rose Gold. Houston (telephone), November 11, 1980.

Blackwell, Tommy J. Tyler, May 21, 1980.

Blalock, Joann. Tyler, June 2, 1980.

Blow, Alton. Tyler, May 12, 1980.

Bonham, Flo. Lafayette, Louisiana, June 18, 1980; (telephone), May 1, 1981.

Boren, Avis. Tyler, May 20, 1980.

Browning, Elmer "Lee." Tyler, May 13, 1980.

Bryant, Auburn. Henderson, Kentucky (telephone), August 15, 1980.

Bryant, Dallas S. New Harmony, Indiana (telephone), August 13, 1980.

Burchfield, Roy W. Tyler, March 19, 1980.

Burke, Charles Gordon. Gueydan, Louisiana, June 18, 1980.

Butler, Charlie Lee. Jackson Oaks, near Tyler, June 13, 1980.

Cain, Lonye. Ozona, July 7, 1980.

Carlton, T. C. Tyler, July 1, 1980.

Carman, Donald E. Tyler, May 16, 1980.

Carter, Eula Brasher. El Dorado, Arkansas, August 25, 1980.

Choice, Roger L. Tyler, May 15, 1980.

Clark, Claude N. Rotan, July 11, 1980.

Clark, Sarah. Tyler, May 20, 1980.

Coats, Morris W. Longview, April 1, 1980.

Cole, Hiram S. Tyler, May 21, 1980.

Conaway, B. R. "Jody." Ozona, July 7, 1980.

Connolly, P. D. El Dorado, Arkansas (telephone) January 26, 27, 1981.

Cooper, Finnis, Jr. Lindale, Texas, November 17, 1980.

Cox, Garland. Pitkin, Louisiana, June 16, 1980.

Craddock, Joe. Winnsboro, March 21, 1980.

Craig, William D. Tyler, June 3, 1980.

Dailey, Willie N. "Bill." Sand Flat community north of Tyler, February 22, 1980.

Darbonne, Doris. Lafayette, Louisiana. June 18, 1980.

Delton, Charles C. Tyler, March 5, 1980.

Dorfman, Myron H. Austin, March 25, 1980.

Dorfman, Sam Y., Jr. Dallas, April 8, 1980; (telephone), February 13, 1981.

Doughty, Edward L. Odessa, July 8, 1980.

Durrett, E. G. "Eddie." Odessa, July 8, 1980.

Dykstra, Frank and Kathryn. Tyler, March 19, 1980.

Eakins, Malcolm. Henderson, Kentucky (telephone), August 14, 1980.

Edwards, Lynna Jo. Odessa, July 8, 1980.

Elkins, J. L. "Jack." Tyler, June 10, 1980.

Everett, T. A. Tyler, March 6, 1980.

Ewbank, Jim. Tyler, May 20, 1980.

Fant, Maxine. Longview, April 1, 1980.

Faustlin, John E. Odessa, July 10, 1980.

Ferguson, Ted T. Odessa, July 9, 1980.

Fischer, Elizabeth. Dallas, April 8, 1980.

Florsheim, Seymour L. "Bubba," Jr. Dallas, April 8, 1980.

Fowler, Burke "Pete." Lafayette, Louisiana, June 19, 1980.

Fowler, Kenneth. Lafayette, Louisiana, June 17, 1980.

Fullilove, Elizabeth Powell. Houston, November 12, 1980.

Gandy, Edward M. Tyler, May 14, 1980.

Gardner, J. M. "Mark." Houston, November 13, 1980.

Gauntt, Bertram S. Athens, February 11, 1980.

Gilleland, James H. "John." Odessa, July 9, 1980.

Goolsby, Bill M. Tyler, June 30, 1980.

Gordon, Herman. Victoria, May 7, 1980.

Greenwell, Merrel. Tyler, May 20, 1980.

Gunn, Mr. and Mrs. John Robert. Tyler, May 15, 1980.

Hall, John D. Tyler, February 12, 1980.

Hardin, J. W. "Turk." Murchison, June 9, 1980.

Hardin, Jeff. Murchison, June 9, 1980.

Hirsch, Warren A. Shreveport, Louisiana, April 15, 1980.

Hoppers, L. D. Tyler, March 20, 1980.

Howard, Willis S. Tenaha, June 20, 1980.

Hurt, L. D. Kilgore, June 27, 1980.

Johnson, J. C., Jr. Odessa, July 8, 1980.

Jones, Gus E. Marshall, March 7, 1980.

Jones, Marcus O. Kilgore, June 13, 1980; (telephone), January 26, 1981.

Jones, Nolan R. "Buddy" and Ethel Marie. Tyler, March 5, 1980.

Justice, John H. Tyler, May 12, 1980.

Keating, Edward M. Sunnyvale, California (telephone), February 2, 1981.

Kennedy, Mr. and Mrs. William B. "Jack." Tyler, March 17, 1980.

Kliewer, Edward, Jr. Dallas, April 10, 1980.

Koonce, Glen. Tyler, March 4, 1980.

LaCour, Betty. Shreveport, Louisiana (telephone), January 26, 1981.

Lewis, Raymond. Tyler, May 15, 1980.

Lovall, Leonard Weldon. Ozona (telephone), July 8, 1980.

Magner, H. J. "Red." Tyler, May 14, 1980.

Marshall, Fax C. Tyler, May 14, 1980.

Marshall, Marguerite. Tyler, May 13, 22, 1980.

Mayton, Louis A. "Gus." Gueydan, Lousiana, June 17, 1980.

Martin, Charles R. Odessa, July 8, 1980.

Massie, Joe. Dallas, August 22, 1980.

Mayer, Frederick M. Dallas, May 29, 1980.

McDonough, Richard J. Tyler, May 13, 1980.

McElroy, Walter L. Tyler, May 21, 1980.

McKenzie, W. Hugh. Alba, February 20, 1980.

McNeeley, R. L. Tyler, May 15, 1980.

Medlin, W. L. "Buster." Winnsboro, March 21, 1980.

Meyer, William O. "Shorty." Laird Hill (Kilgore), March 31, 1980.

Miears, Horace L. "Blackie." Ozona, July 7, 1980.

Miser, Melburn. Indiana, Pennsylvania (telephone), August 12, 1980.

Morris, W. S. "Buck." Kilgore, March 31, 1980.

Murray, Herman F. El Dorado, Arkansas, August 25, 1980.

Murrell, John. Dallas, May 28, 1980.

Nixon, G. I. "Red." Kilgore, June 5, 1980.

Parten, J. R. Madisonville, November 14, 1980.

Peters, Donald G. Tyler, March 6, 1980.

Pew, John G. "Jack." Dallas, May 27, 1980.

Phillips, Leonard W. Shreveport, Louisiana, April 14, 1980.

Portier, Ada Viola. Tyler, March 18, 1980.

Queen, Edward E. Longview, April 1, 1980.

Rappeport, Goldie. Austin, March 25, 1980.

Robinson, Jack D. Tyler, June 3, 1980.

Rogers, Claudine. Tyler, May 21, 1980.

Rolf, Jack. Dallas, April 9, 1980.

Rozeman, A. M. Shreveport, Louisiana (telephone), February 13, 1981.

Schneider, Stephen W. Dallas, May 27, 1980.

Schreiber, Mrs. Edward. Galveston (telephone), November 11, 1980.

Schroeder, Eric W. Dallas, May 28, 1980.

Sexton, Floyd. Texarkana, Arkansas, March 13, 1980.

Shelmire, Suzette Dayson. Dallas, August 22, 1980.

Sides, Leo. Palo Alto, California (telephone), February 3, 1981.

Sklar, Albert. Shreveport, Louisiana, April 14, 1980.

Smith, Early F. Tyler, March 5, 1980.

Smith, James Wylie. Longview, April 2, 1980.

Spence, Ralph. Tyler, June 2, 1980.

Spraggins, Z. L. "Nig." Tyler, May 20, 1980.

Stackpole, Gloria A. Tyler, July 17, 1980.

Stacy, Robert A., Jr. Shreveport, Louisiana, April 15, 1980; (telephone), March 6, 1981.

Strahan, Warren. Lafayette, Louisiana, June 17, 1980.

Strength, Dewey. Longview, April 1, 1980.

Tarvin, Thornton. Houston, May 8, 1980.

Terry, Homer Lee. Longview, April 1, 1980.

Vickers, C. L. Independence, May 9, 1980.

Waddell, Robert W. Tyler, February 22, 1980.

White, Guy B. Tyler, February 21, 1980.

White, Joe F. Tyler, May 16, 1980.

Whitfield, W. H. Palmetto, Louisiana, June 16, 1980.

Wilkerson, Aileen. Tyler, June 2, 1980.

Williams, Marvin R. Gladewater, June 11, 1980.

Wilson, Otis. Bullard, March 18, 1980.

Winston, Joe Milton. Winona, May 19, 1980.

Wise, Watson W. Tyler, May 22, 1980.

Wolford, Bernie. Jefferson, October 6, 1980.

Young, Mr. and Mrs. Elmer S. "Whitey." Kilgore, June 27, 1980.

Zeppa, Chris. Tyler, February 12, 1980; January 19, 1981; (telephone), February 2, 1981.

Zeppa, Joe. Tyler, March 25, 1975.

Zeppa, Keating V. Tyler, May 22, 26, 1980; April 27, 1981.

Zeppa, Trent H. Port Townsend, Washington, July 24, 1980.

Other Unpublished Material

Gibbon, Anthony. Tiburon Adventure. Undated manuscript. (Joe Zeppa's Personal Files).

Schroeder, Eric G. Notes on Joe Zeppa Interview; 1950. (Schroeder Personal Collection).

Zeppa, Keating V. "Summary of Events Leading to Public Offering of Delta Drilling Company's Stock." March 26, 1981. Manuscript.

Zeppa, Keating V. Transcript, Address to IADC-PESA Conference, Dallas, 1980.

Letters

Practically all of the letters cited are from Joe Zeppa's personal files or corporate records. However, these few below were helpful in clarifying certain points.

Baker, Grace, to James Presley, January 12, 1981.

Parten, J. R., to James Presley, January 29, 1981.

Pearlman, Samuel J., to James Presley, October 20, 1980, February 16, 1981.

Vollmer, Mrs. O. F., to James Presley, October 13, 1980.

Magazines

The Delta Digger, the company magazine (which received its title as a result of a contest which Joann Blalock won), was started while Doris Kivel was manager of public relations, and continues to the present. I have examined every issue, read practically every word, and, although I have cited its articles but infrequently in the notes, it was one of my first points of reference when I began this study. In many instances the *Digger* suggested to me additional interviewees who were not already on my list.

Books

Bainbridge, John. *The Super-Americans.* Garden City, N. Y.: Doubleday & Co., 1961.

Bell, Max W., Ball, Douglas, and Turner, Daniel S. *This Fascinating Oil Business.* Indianapolis, New York: Bobbs-Merrill Co., 1965; rev. ed.

Clark, James A. *The Chronological History of the Petroleum and Natural Gas Industries.* Houston: Clark Book Company, 1963.

Clark, James A., and Halbouty, Michel T. *The Last Boom.* New York: Random House, 1972.

Cordasco, Francesco, and Bucchioni, Eugene, eds. *The Italians: Social Backgrounds of an American Group.* Clifton, N. J.: Augustas McNelley, 1974.

Dinnerstein, Leonard, and Palsson, Mary Dale, eds. *Jews in the South.* Baton Rouge: Louisiana State University Press, 1973.

Gambino, Richard. *Blood of My Blood: The Dilemma of the Italian American.* Garden City, N.Y.: Doubleday, 1974.

Howe, Irving. *World of Our Fathers.* New York: Harcourt, Brace, Jovanovich, 1976.

Joseph, Samuel. *Jewish Immigration to the United States from 1881 to 1910.* New York: AMS Press, 1967; reprint.

Karp, Abraham J., ed. *Golden Door to America: The Jewish Immigrant Experience.* New York: Viking Press, 1976.

Kessner, Thomas. *The Golden Door: Italian and Jewish Immigrant Mobility in New York City 1880-1915.* New York: Oxford University Press, 1977.

LoGatto, Rev. Anthony F., compiler and editor. *The Italians in America, 1492-1972: A Chronology and Fact Book.* Dobbs Ferry, N.Y.: Oceana Publications, 1972.

Lopreato, Joseph. *Italian Americans.* New York: Random House, 1970.

Meltzer, Milton. *Taking Root: Jewish Immigrants in America.* New York: Farrar, Straus and Giroux, 1976.

Miriam Osborn Memorial Home: Sixtieth Anniversary, 1908-1968. Rye, N.Y.: [Osborn Memorial Home], 1968.

Myers, Samuel D. *The Permian Basin, Petroleum Empire of the Southwest: Era of Advancement, From the Depression to the Present.* El Paso: Permian Press, 1977.

Moquin, Wayne, ed., with Charles Van Doren. *A Documentary History of the Italian Americans.* New York and Washington: Praeger Publications, 1974.

Owen, Edgar Wesley. *Trek of the Oil Finders: A History of Exploration for Petroleum.* Tulsa, Oklahoma: The American Association of Petroleum Geologists, 1975.

Pares, Bernard. *Russia.* New York: New American Library, 1949.

Presley, James. *A Saga of Wealth: The Rise of the Texas Oilmen.* New York: G. P. Putnam's Sons, 1978.

Safety Manual on the Handling of Sour Gases. Tyler, Texas: Delta Drilling Company, 1973; rev. ed.

Staff, East Texas Salt Water Disposal Company. *Salt Water Disposal, East Texas Oil Field.* Austin: Petroleum Extension Service, The University of Texas, 1958; 2nd ed.

Articles

Beazley, Deanie. "El Dorado—Boomtown, 1921 and 1971." *Petroleum Independent* (May-June, 1971), 28-30 ff.

"Delta Checks Off Its 40th Year," *Drilling-DCW* (November, 1971), pp. 20-26.

[Joe Zeppa Profile]. *East Texas* (November, 1959), p. 13.

Keating, George T. "Born a Collector." *The Cabellian: A Journal of the Second American Renaissance* (III, 2; Spring 1971), pp. 83-86.

"Mustang Report," *The Mustang* (SMU) (18, 2; October, 1965), p. 7.

Schroeder, Eric G. "Determination Paved Road from Italy to East Texas." *Dallas Morning News,* November 19, 1950.

Tully, Shawn. "Sudden Wealth in Tyler, Texas." *Fortune* (June 1, 1981), pp. 40-43.

Zeppa, Joseph. "Present Drilling and Completion Methods and Their Effect on Costs of Exploratory and Development Wells." In *Proceedings of the 1962 National Institute for Petroleum Landmen.* (Albany, N.Y.: Matthew Bender & Co., 1963.)

Newspapers

Dallas Morning News, November 19, 1950; September 14, 1967.
New York Times, April 1, 1967.
Shreveport Times, July 31, 1961.
Tyler *Morning Telegraph,* February 18, 1972; July 10, 1975.

Miscellaneous Material

Announcing the Joseph Zeppa Collection of War, Diplomacy and Peace. Dallas: Southern Methodist University, 1967.

Special Collections in The Libraries at Southern Methodist University. Dallas: Southern Methodist University, 1973.

INDEX

salt-water problem in East Texas,
solving of, 128-132
selling Mexican oil in Italy, 60-61
at Shearman and Sterling, 41-43, 51,
59-60
in Shreveport, 63, 75
sibling of, 35
Stacey, renewing friendship with, 79
in stersage, 38
stroke of, 452
symbology in Delta's logo, 377-380
Trent, relationship with, 105, 233-235
tribute from Keating upon J. Z.'s
death, 463
Trice Production Company, rescuing,
395-396
Tyler, move to, 96, 106
understanding of Gold/Sklar/Dorfman
arrangement, 123
as U.S. citizen, 44, 58
at University of Toulouse, 51
visit to Fubine, 52, 345-346
visiting South Glastonbury, CT, 455
World War I, involvement in, 45-58
correspondence with Helen Adams,
46-49
Zeppa, Keating, 325, 343, 368, 403
in Argentina, 241, 307-312, 314
on bidding, 483-484
birth of, 106
on blacks, 438-439
in Brazil, 322
childhood of, 197-108, 236-238
on Chris Zeppa, 174, 177, 236-237
community activities, 493
on Delta, 494-496
on Delta Marine, 226-227

on Delta's symbology, 377, 380
as executive vice president, 177, 194,
412
as executor of parents' estates, 470
on foreign operations, 330-331
Fubine, visiting, 238
on J. Z., 347, 360, 363, 386-388,
395, 397, 398, 414
board meeting following his death,
461-463
at his death, 458
relationship with, 414-415
at his stroke, 452
tribute to, 463
on Keating Zeppa, 494-496
managerial style of, 469-470
in Mexico, 273-274, 303, 305
on modernization 445-447, 477-481
in Navy, 239-240
nickname of, 106
oil patch, introduced to, 106
organizational activity, 493
others' opinions of, 490-493
overriding J. Z., 457
on his parents, 389
at Pinehurst, 313
as president, 466 ff.
and Trent, 234
and Tyler Hotel Company, 230
visiting Fubine, 238
working for Blasingame,259
Zeppa, Pauline, 35, 53, 238, 345
Zeppa, Trent, 76, 105-106, 179, 228,
230, 233-235, 382 387, 392, 444
on J. Z., 349, 351-352, 354
on Keating Zeppa, 466, 491
Zeppa, Vincenzo, 35-38